Theory of Elementary Particles

edited by H. Dorn, D. Lüst, G. Weigt

WILEY-VCH

Theory of Elementary Particles

Proceedings of the 31st International Symposium
Ahrenshoop, September 2–6, 1997,
Buckow/Germany

edited by
H. Dorn, D. Lüst, G. Weigt

Berlin · Weinheim · New York · Chichester · Brisbane · Singapore · Toronto

Editors:

Dr. Harald Dorn
Institute of Physics
Humboldt University
Invalidenstraße 110
D-10115 Berlin
Germany

Prof. Dr. Dieter Lüst
Institute of Physics
Humboldt University
Invalidenstraße 110
D-10115 Berlin
Germany

Dr. Gerhard Weigt
DESY Zeuthen
Platanenallee 6
D-15738 Zeuthen
Germany

Die Deutsche Bibliothek – CIP-Einheitsaufnahme

Theory of elementary particles : proceedings of the 31st international symposium, Ahrenshoop, September 2–6, 1997, Buckow, Germany / ed. by H. Dorn ... –
Berlin ; Weinheim ; New York ; Chichester ; Brisbane ; Singapore ; Toronto :
WILEY-VCH, 1998
ISBN 3-527-40224-1

Printed on acid-free paper.
The paper used corresponds to both the U. S. standard ANSI Z.39.48 – 1984 and the European standard ISO TC 46.

Printing: GAM Media GmbH, Berlin
Bookbinding: Druckhaus „Thomas Müntzer“, Bad Langensalza

Printed in the Federal Republic of Germany

PREFACE

These proceedings contain the invited talks and short communications presented at the 31st International Symposium Ahrenshoop on the Theory of Elementary Particles which took place from September 2 to September 6, 1997 at Buckow near Berlin. The Symposium was jointly organized by DESY Zeuthen, the Institute of Physics of the Humboldt University Berlin, the Institute for Theoretical Physics of the University Hannover and the Section of Physics of the University Munich. At this place we would like to thank all the institutions that made this conference possible by financial support: the Stifterverband für die Deutsche Wissenschaft, the Walter and Eva Andrejewski Foundation, the Humboldt University Berlin and DESY Zeuthen. We especially thank Mrs Karin Pipke for her dedicated assistance in preparing this manuscript.

This year we were able to bring together a remarkable set of leading scientists in both string theory and lattice theory. The comparison of non-perturbative results obtained in (supersymmetric) field theories and superstring theory with results from lattice models was one of the central issues of this meeting. The conference was organized in plenary talks and two parallel sessions: Session A, devoted to strings, branes and M(atrix)-theory and Session B, devoted to lattice gauge theories and non-perturbative QCD.
The speakers presented a big part of the most active research in these extremely fast advancing fields. We are very grateful to the speakers for their talks and their efforts in preparing their interesting contributions to these proceedings. In this book the contributions are presented in the order Session A, Session B, plenary talks.

Our 1997 symposium followed an already long-standing tradition of conferences on elementary particle theory organized and held jointly by the groups of the Humboldt University in Berlin and the Institute of High Energy Physics in Zeuthen. Before the political changes in East Germany this conference was one of the very rare events where physicists from the East and the West could meet and discuss. In the course of the years these meetings held mostly in the little village Ahrenshoop at the baltic seashore became very efficient in providing scientific and personal contacts between the eastern and western particle physicists. After the year 1989 the Ahrenshoop Symposium went on - now away from the place of its naming - as an international meeting of high scientific quality and still with an emphasis on bringing together people from East and West in a stimulating environment.

Harald Dorn
Dieter Lüst
Gerhard Weigt

ORGANIZING COMMITTEE

Harald Dorn (Secretary)	Institute of Physics, Humboldt University of Berlin
Olaf Lechtenfeld	Institute for Theoretical Physics, University Hannover
Dieter Lüst	Institute of Physics, Humboldt University of Berlin
Michael Müller-Preussker	Institute of Physics, Humboldt University of Berlin
Gerhard Weigt	DESY Zeuthen
Julius Wess	Section of Physics of the University Munich

CONTENTS

Session B

Plenary Session

Multi-Matrix Models in the Double Scaling Limit

Smail Balaska [1] , Joachim Maeder [2] , Werner Rühl [3]

[1] Lab. de Physique Théorique, Université d'Oran ,
Oran - Algeria
[2], [3] Fachbereich Physik, Universität Kaiserslautern,
Postfach 3049, D-67653 Kaiserslautern,
E-mail: maeder@physik.uni-kl.de
E-mail: ruehl@physik.uni-kl.de

Abstract: A perturbative approach to solve f-matrix models, $f \geq 2$ is presented. The cases $f = 2$ and $f = 3$ are dealt with explicitly for polynomial potentials. Furthermore two-matrix models with non-polynomial potentials are examined.

We study matrix models whose action depends on hermitean $N \times N$ matrices as dynamical variables. They are coupled to a chain of r vertices and $r-1$ connecting links

$$S(M^{(1)}, M^{(2)}, M^{(3)} \ldots M^{(r)}) = \operatorname{Tr} \left\{ \sum_{\alpha=1}^{r} V_\alpha(M^{(\alpha)}) - \sum_{\alpha=1}^{r-1} c_\alpha M^{(\alpha)} M^{(\alpha+1)} \right\} \tag{1}$$

The two-matrix models ($r = 2$) exhibit the full richness of critical structures [1]. It turns out that they possess two series of critical points: the well-known "discrete series" for which the potentials V_α are polynomials, and a "continuous series" which was discovered in [2] and will be described below.

Statistical ensembles of matrices appeared first in connection with problems of nuclear physics [3]. As generalizations of vector sigma models they served as objects for the study of phase transitions and renormalization theory [4]. Recently they attracted interest as models for the coupling of conformal field matter with the gravitational field [5]. In this case they are analyzed in their critical domain defined by the "double scaling limit". We shall also apply this limiting procedure in this work.

If the critical potentials are polynomials, the matrix models can be solved perturbatively in the double scaling domain. This method has been applied to study all types of critical behaviour of the polynomial two-matrix models in [5, 6]. The final result can be described as follows. Let the polynomial degrees of the potentials be l_1 and l_2, $l_1 \geq l_2$. If l_2 does not divide l_1, the universality class of the maximal critical point of this model is

$$[p,q] = [l_1, l_2] \tag{2}$$

where p and q denote the degrees of differential operators of the generalized KdV hierarchy. If, however, l_2 divides l_1, but differs from two, then

$$[p,q] = [l_1 + 1, l_2] \tag{3}$$

The perturbative method can also be applied to polynomial actions with three and more independant matrices. An algorithm for their solution was presented in [7]. The main obstacle for such models with f multiplication operators $\{B_i\}_{i=1}^f$ is to understand why from f multiplication operators satisfying

$$[B_f, B_{f-1}] = \ldots = [B_3, B_2] = [B_2, B_1] = 1 \tag{4}$$

result only two independent differential operators P, Q (of degree p and q, $\frac{p}{q} \notin \mathbb{N}$, $\frac{q}{p} \notin \mathbb{N}$) as appearing in a generalized KdV hierarchy. Let the double-scaling asymptotic expansion of B_i be

$$B_i \to a^{-\lambda_i \gamma}(R_i + a^{-\gamma} R_i^{(1)} + a^{-2\gamma} R_i^{(2)} + \ldots) \tag{5}$$

($a^{-\gamma}$ is the expansion parameter, γ is the string susceptibility exponent, notations are as usual).

In the case of the two matrix model then

$$\lambda_1 = l_1, \quad \lambda_2 = l_2 \tag{6}$$

If l_2 divides l_1 then $R_2 = R_1^{l_1/l_2}$ and

$$[B_2, B_1] \to \left[R_2, R_1^{(1)}\right] + \left[R_2^{(1)}, R_1\right] = \text{const.}\ [P,Q] \tag{7}$$

$$Q = R_2, \quad P = (Q)_+^{(l_1+1)/l_2}. \tag{8}$$

This procedure generalizes in $f \geq 3$ matrix models. Three matrix models are solved by an ansatz $\sum_k (B_i)_{n,n+k}\, z^k \to r^{(i)}(z)$ at leading order

$$r^{(1)}(z) = (z-1)^{\lambda_1} z^{-(l_2-1)(l_3-1)} P_{(l_2-1)(l_3-1)+1-\lambda_1}(z) \tag{9}$$

$$r^{(2)}(z) = (z-1)^{\lambda_2} z^{-l_3+1} P_{(l_1+l_3-2)-\lambda_2}(z) \tag{10}$$

$$r^{(3)}(z) = (z-1)^{\lambda_3} z^{-1} P_{(l_1-1)(l_2-1)+1-\lambda_3}(z) \tag{11}$$

A maximal critical point is obtained, if all $\lambda_1, \lambda_2, \lambda_3$ are maximalized so that all parameters in the Schwinger-Dyson equations are fixed to critical values. A model with polynomial degrees (l_1, l_2, l_3) may have different maximal critical points $[\lambda_1, \lambda_2, \lambda_3]$, e.g. $(l_1, l_2, l_3) = (4, 3, 3)$ has the maximal critical points

$$[\lambda_1, \lambda_2, \lambda_3] \in \{[4,5,4],[4,4,5],[3,3,6]\}. \tag{12}$$

The maximal critical points appear in two types: type I with $\lambda_1 = \lambda_3$ and type II with $\lambda_1 \neq \lambda_3$. For type I where λ_1 does not divide λ_2 ($> \lambda_1$), we have a standard case.

$$R_1 = -R_3 = Q, \quad q = \lambda_1 \tag{13}$$

$$R_2 = P, \quad p = \lambda_2. \tag{14}$$

If λ_2 is a multiple of λ_1 we have to increase the perturbative order and thus $p = \lambda_2 + 1$.

Consider type II, $\lambda_1 < \lambda_3$. Then the minimal m for which

$$\sum_{i+j=m} \left[R_3^{(i)}, R_2^{(j)}\right] = 1 \tag{15}$$

determines the perturbative order n at which

$$[B_2, B_1] = 1 \tag{16}$$

namely

$$n = \lambda_2 + \lambda_3 + m. \tag{17}$$

The commutator

$$\sum_{i+j=m'} \left[R_2^{(i)}, R_1^{(j)}\right] \tag{18}$$

has to be evaluated at

$$m' = \lambda_3 - \lambda_1 + m > m \tag{19}$$

and vanishes for all smaller m' !

As an example we study the case $[3,3,6]$ in (12). The result is

$$R_1^{(i)} = R_2^{(i)}, \quad i \in \{0,1,2,3\}, \quad R_3 = -(R_2)^2 \tag{20}$$

$$Q = R_2, \quad q = 3 \tag{21}$$

$$P = R_2 R_2^{(1)} + R_2^{(1)} R_2 + R_3^{(1)}, \quad p = 7. \tag{22}$$

More examples, in fact complete lists, can be found in [7].

Three matrix models possess "rational" solutions which we obtain if in (10)

$$\lambda_2 = l_1 + l_3 - 2. \tag{23}$$

The critical coupling constants are rational numbers and only linear algebra is needed to solve the Schwinger-Dyson equations. All other maximal critical points are "algebraic".

In all polynomial models the string susceptibility is

$$\gamma = -\frac{2}{p+q-1}. \tag{24}$$

The continuous series of critical points necessitates nonpolynomial critical potentials that are holomorphic inside a circle of finite radius of convergence. Each depends on a parameter l_1 (respectively l_2). A third parameter, a natural number n, is connected with the perturbative order at which the equation

$$[B_2, B_1] = 1 \tag{25}$$

can be fulfilled and enters the string susceptibility exponent γ as

$$\gamma = \frac{-2}{n-1} \tag{26}$$

Whereas the discrete series is intimately connected with the theory of the Korteweg-de Vries equations [8] and positive integer powers of quasi-differential operators, we will use complex powers of such operators (as described in [9]) only marginally. The differential equations arising at the end are trivial and have polynomial solutions.

New in this work is that the multiplication operator $B_1(B_2)$ decomposes into "blocks" corresponding to a nondegenerate $\mathbb{N}^2$-lattice [1]

$$B_1 = \sum_{[n_1,n_2]\in\mathbb{N}^2} B_1^{[n_1,n_2]} \tag{27}$$

$$B_2 = \sum_{[n_1,n_2]\in\mathbb{N}^2} B_2^{[n_1,n_2]} \tag{28}$$

where each block possesses a double scaling expansion

$$B_1^{[n_1,n_2]} \cong \sum_{n=0}^{\infty} a^{-(l_2-\lambda+n)\gamma} Q_n^{[n_1,n_2]}(x;p) \tag{29}$$

$$B_2^{[n_1,n_2]} \cong \sum_{n=0}^{\infty} a^{-(l_1-\lambda+n)\gamma} P_n^{[n_1,n_2]}(x;p) \tag{30}$$

where $\lambda = n_1 l_1 + n_2 l_2$ and p is the differential symbol.

Obviously it is necessary that

$$Re\, l_{1,2} < 0 \tag{31}$$

in order to render the expansions (27)-(30) perturbative. Both Q_n and P_n are given as asymptotic expansions for $p \to +\infty$ (or $p \to -\infty$) and with p ordered to the right of x. They are quasidifferential operators involving complex powers of the

[1] In this paper $\mathbb{N}$ includes zero.

differential symbol p. For the block $[0,0]$ which is "basic" in some sense, we found simple expressions

$$Q_0^{[0,0]}(x;p) = (e^{i\frac{\pi}{2}}L(x;p))^{l_2} \tag{32}$$

$$P_0^{[0,0]}(x;p) = (e^{-i\frac{\pi}{2}}L(x;p))^{l_1} \tag{33}$$

where L has the form

$$L(x;p) = p + \sum_{n=1}^{\infty} u_n(x)p^{-n} \tag{34}$$

The meaning of the complex labels l_1 and l_2 can be fixed by (32), (33). The obvious commutativity of these operators can be extended to the whole perturbative series (29),(30)

$$[B_2^{[0,0]}, B_1^{[0,0]}] \cong 0 \tag{35}$$

The Schwinger-Dyson equations are solved by an ansatz

$$B_1 \to z(1-\frac{1}{z})^{l_2} + \sum_{m=1}^{\infty} a^{-(m+1)\gamma} U_m(x;z) \tag{36}$$

$$U_m(x;z) = \sum_{[n_1,n_2]\in\mathbb{N}^2} \sum_{r=0}^{\infty} U_{mr}^{[n_1,n_2]}(x) \times z\left(1-\frac{1}{z}\right)^{l_2-(n_1l_1+n_2l_2)-(m+1)+r} \tag{37}$$

$$B_2 \to \frac{1}{z}(1-z)^{l_1} + \sum_{m=1}^{\infty} a^{-(m+1)\gamma} V_m(x;z) \tag{38}$$

$$V_m(x;z) = \sum_{[n_1,n_2]\in\mathbb{N}^2} \sum_{r=0}^{\infty} V_{mr}^{[n_1,n_2]}(x) \times \frac{1}{z}(1-z)^{l_1-(n_1l_1+n_2l_2)-(m+1)+r}. \tag{39}$$

Since these expansions are given for $|z| > 1$, $|z| < 1$ respectively, the Schwinger-Dyson equations are defined on the common boundary $|z| = 1$. We evaluate (not solve) them by identification of contributions of singular terms $(1-z)^\alpha$, $\alpha \notin \mathbb{Z}$, on both sides. We construct a recursive procedure to achieve this. At the end we impose the commutator condition (16) for B_1, B_2 as additional constraint. The lowest perturbative order n at which $[B_2, B_1]$ is nonzero enters (26). For more details we refer to the original papers of the authors quoted.

References

[1] M.R. Douglas, *Phys. Lett* **B238** (1990) 176;
M.R. Douglas, *1990 Cargse workshop* on "Random surfaces and quantum gravity", Eds. O. Alvarez, E. Marinari, P. Windey. *NATO ASI Series B: Physics* **Vol. 262** (1992) 77.

[2] S. Balaska, J. Maeder, W. Rühl, The continuous series of critical points of the two-matrix model at $N \to \infty$ in the double scaling limit, hep-th/9703161.

[3] M.L. Metha, *Random matrices and the statistical theory of energy levels*, Acad. Press, New York 1967.

[4] C. Itzykson, J.-M. Drouffe, *Statistical field theory*, 2 Vols, Camb. Univ. Press, Cambridge 1989, Section 10.3

[5] E. Brézin, V.A. Kazakov, *Phys. Lett.* **B236** (1990) 144;
D.J. Gross, A.A. Migdal, *Phys. Rev. Lett.* **64** (1990) 127;
M.R. Douglas, S.H. Shenker, *Nucl. Phys.* **B335** (1990) 635;
Proceedings of *1990 Cargse workshop* on "Random surfaces and quantum gravity", Eds. O. Alvarez, E. Marinari, P. Windey. *NATO ASI Series B: Physics* **Vol. 262** (1990);
P. di Francesco, P. Ginsparg, J. Zinn-Justin, *Physics Reports* **254** (1995) 1.

[6] S. Balaska, J. Maeder, W. Rühl, *Nucl. Phys.* **B486** (1997) 673.

[7] S. Balaska, J. Maeder, W. Rühl, On the critical behaviour of hermitean f-matrix models in the double scaling limit with $f \geq 3$, hep-th/9707213.

[8] P. Ginsparg, M. Goulian, M.R. Plesser, J. Zinn-Justin, *Nucl. Phys.* **B342** (1990) 539.

[9] I.M. Gelfand, L.A. Dikii, *Funks. Anal. Prilozhen.* **10** (1976) No. 4, 13.

D-brane conformal field theory

Jürgen Fuchs [1], Christoph Schweigert [2]

[1] Max-Planck-Institut für Mathematik
Gottfried-Claren-Str. 26, D – 53225 Bonn
[2] CERN, CH – 1211 Genève 23

Abstract: We outline the structure of boundary conditions in conformal field theory. A boundary condition is specified by a consistent collection of reflection coefficients for bulk fields on the disk together with a choice of an automorphism ω of the fusion rules that preserves conformal weights. Non-trivial automorphisms ω correspond to D-brane configurations for arbitrary conformal field theories.

String theory and conformal field theory.

A complete understanding of string theory certainly requires many more ingredients than just conformal field theory, e.g. when it comes to finding a guiding principle that would tell what solitonic sectors (and with which multiplicities) must be included to arrive at a consistent theory. On the other hand, both at a conceptual and at a computational level, conformal field theory does lead very far indeed. While at the level of string perturbation theory this is more or less accepted knowledge in the case of closed strings, it is a prevailing prejudice that some of the more recently discovered structures that are tied to the presence of open strings with non-trivial boundary conditions are inaccessible to conformal field theory. This is of course a logical possibility, but before making a decision on this issue one should better inspect the tools that are summarized under the name 'conformal field theory' with sufficient care. It may well turn out that present day knowledge about these matters is as yet incomplete and that the uses of conformal field theory can be largely expanded by further efforts.

Indeed we claim that the basic new features of open as compared to closed strings, such as e.g. D-branes (possibly with field strength, or multiply wrapped) are well accessible to conformal field theory. Moreover, once a suitable framework for conformal field theory on closed orientable surfaces (*closed* conformal field theory, for short) is formulated [1], establishing the theory also on the open and / or non-orientable surfaces (*open* conformal field theory) that arise as world sheets of open strings does not pose any major conceptual problems any more, though there are several new ingredients which considerably complicate matters at a more technical level.

Building blocks.

Let us first recall a few facts about the world sheet picture of closed strings. The guiding principle for the construction of a string theory from some conformal field theory (consistently defined on all closed orientable Riemann surfaces C) is to discard all properties of the world sheet C while still keeping information about the field theory. This is achieved by eliminating first the (super-) Virasoro algebra via the relevant semi-infinite cohomology, then the choice of a conformal structure on C via integration over the moduli space of complex structures, and finally the choice of topology of C by a summation over topologies. The latter sum is weighted by the power $\gamma^{-\chi}$ of the string coupling constant γ, with $\chi = 2-2g$ the Euler number of C. In particular, string scattering amplitudes are obtained from the n-point correlation functions $\mathcal{F}_{g,n} \equiv \mathcal{F}_{g,n}(\vec{\lambda}; \vec{z}, \vec{\tau})$ of the field theory by integrating over the moduli $\vec{\tau}$ of the genus-g surface C and (modulo Möbius transformations) over the insertion points $\vec{z} \equiv (z_1, z_2, \dots, z_n)$, and afterwards multiplying with $\gamma^{-\chi}$ and summing over χ.

For a conformal field theory to be consistently defined on all surfaces C, the correlators $\mathcal{F}_{g,n}$ have to satisfy various locality and factorization constraints. The former require that the $\mathcal{F}_{g,n}$ are ordinary functions of the insertion points $\vec{z}$ and (up to the Weyl anomaly) of the moduli $\vec{\tau}$, while the latter implement compatibility with singular limits in the moduli spaces. These constraints are formulated in terms of the conformal field theory on C (which is orient*able*, but does *not* come naturally as an orient*ed* surface), to which we refer as the stage of *full* conformal field theory. This stage must be carefully distinguished from the stage of *chiral* conformal field theory, where in place of the correlation functions one is dealing with chiral blocks. Usually this stage is introduced by a somewhat heuristic recipe for 'splitting the theory into chiral halves'. A more appropriate, and for the present purposes more convenient, description of the chiral theory is as a conformal field theory on an orient*ed* covering surface $\hat{C}$ which has the structure of a complex curve and from which the surface C can be recovered by dividing out an anti-conformal involution [1].

For large classes of conformal field theories, in particular for WZW models, all correlation functions $\mathcal{F}_{g,n}$ can in principle be computed exactly (i.e., fully non-perturbatively in terms of the field theory on the world sheet). Moreover, in many interesting cases – including, but by no means exhausted by, free field theories – at least at string tree level this can also be achieved in actual practice. The reason is that the chiral blocks can be obtained as the solutions to the Ward identities of the theory. Let us note that even though conformal field theory is typically formulated in an operator picture, for establishing the Ward identities (and also for many other purposes) the existence of an operator formalism is not needed. Namely, they constitute identities for chiral blocks that can be formulated solely in terms of the representation theory of the relevant chiral algebra $\mathfrak{W}$, without making use of an operator formalism. Also, once the chiral blocks are known, the correlators are determined by the locality and factorization constraints which (are believed to) have a unique solution. Of course, in string theory one usually interprets the scattering amplitudes as expectation values for products of suitable vertex operators for the string modes. In conformal field theory terms this amounts to working with an operator formalism, in which the string modes are realized as (Virasoro-primary) chiral vertex operators in the chiral, respectively as corresponding fields in the full theory.

Via factorization, one can reduce many issues of interest to statements about only a small number of building blocks, namely the chiral 3-point blocks on $\mathbb{P}^1$, and these building blocks can be studied in terms of the representation theory of the chiral algebra $\mathfrak{W}$. For instance, the index set $\{\lambda\}$ (an n-tuple of which labels the $\mathcal{F}_{g,n}$) corresponds to a suitable set of irreducible modules of $\mathfrak{W}$, and in rational theories the numbers $\mathcal{N}_{\lambda_1\lambda_2}^{\ \ \lambda_3}$ of independent 3-point blocks of type $(\lambda_1, \lambda_2, \lambda_3)$ are related, via the Verlinde formula, to the modular behavior of the characters χ_λ of these modules.

For *open* strings, including *D-branes*, the situation is more complicated technically, but not conceptually. Some of the concepts are now realized in a somewhat different manner, but still they can be applied in much the same way as before. E.g.:

- The Euler characteristic χ still counts the order in the string perturbation theory. But now χ is given by $\chi = 2-2g-b-c$, where g, b and c are the numbers of handles, boundary components, and crosscaps of the surface C, respectively.
- One must still distinguish between the two stages of the chiral and the full conformal field theory. The full theory on C can again be expressed in terms of a chiral theory on some surface $\hat{C}$ by imposing locality and factorization constraints.
- Again $\hat{C}$ is an oriented cover of C from which one recovers C by modding out an anti-conformal involution I. But now $\hat{C}$ is connected, whereas in the closed case it consists of two connected components each of which is isomorphic to C as a real manifold [1]. Also, I may now possess fixed points, giving rise to boundaries of C.
- Again factorization allows to formulate the theory in terms of a few building blocks. But besides the 3-point blocks on $\mathbb{P}^1$, one now also needs the 1-point blocks on the disk $D = \mathbb{P}^1/_{z\mapsto 1/z^*}$ as well as the 1-point blocks on the crosscap $\mathbb{PR}^2 = \mathbb{P}^1/_{z\mapsto -1/z^*}$.

Boundary states and boundary conditions.

In contrast to the closed case, in open conformal field theory [2–4, 1] the locality and factorization constraints typically admit more than one solution, e.g. the 1-point correlators $\langle\phi_{\lambda,\tilde\lambda}\rangle_A$ of bulk fields $\phi_{\lambda,\tilde\lambda}$ on the disk depend on some additional label A. These correlators are simply proportional to the corresponding 1-point blocks; the constant of proportionality is the product of two factors N_0^{AA} and $R^A_{\lambda\tilde\lambda;0}$. The number N_0^{AA} is interpreted as the expectation value $\langle\Psi_0^{AA}\rangle$ of a 'boundary vacuum field' [2] Ψ_0^{AA}; roughly, the role of the boundary field is to make a geometric boundary component into a 'field theoretic boundary' that carries the boundary label A. Similarly, $R^A_{\lambda\tilde\lambda;0}$ is a *reflection coefficient*, defined via the expansion [3, 4]

$$\phi_{\lambda,\tilde\lambda}(r\mathrm{e}^{\mathrm{i}\sigma}) \sim \sum_\mu \sum_a (r^2-1)^{-\Delta_\lambda-\Delta_{\tilde\lambda}+\Delta_\mu}\, R^A_{\lambda\tilde\lambda;\mu}\, \Psi_\mu^{AA}(\mathrm{e}^{\mathrm{i}\sigma}) \qquad \text{for } r\to 1 \tag{1}$$

of $\phi_{\lambda,\tilde\lambda}$ in terms of boundary fields. Every consistent collection of 1-point correlators for all bulk fields, or equivalently, every consistent collection of reflection coefficients $R^A_{\lambda\tilde\lambda;0}$, is referred to [2] as a *boundary condition A*. For free fields these amount to boundary conditions in the ordinary geometric sense, but in the general case such an interpretation is not available. Roughly, one can interpret the relation (1) by imagining that to every bulk field there is associated a kind of mirror charge on $\mathbb{P}^1\backslash D$, which in turn corresponds to some charge distribution on the boundary.

In the literature it is common to denote the 1-point chiral blocks on the disk by $|\mathrm{B}_\lambda\rangle$ and to refer to them, as well as to their linear combinations

$$|\mathcal{B}^A\rangle := \sum_\lambda N_0^{AA}\, R^A_{\lambda\tilde\lambda;0}\, |\mathrm{B}_\lambda\rangle\,, \tag{2}$$

as *boundary states*. Such an object is, however, not a state in the usual sense. While formally it satisfies relations of the form

$$\big(W_n\otimes\mathbf{1} - (-1)^{\Delta(W)}\,\mathbf{1}\otimes W_{-n}\big)|\mathrm{B}_\lambda\rangle = 0\,, \tag{3}$$

and in concrete examples can be written [1] as an (infinite) sum of basis elements of the tensor product space $\mathcal{H}_\lambda\otimes\mathcal{H}_{\tilde\lambda}$ of the relevant $\mathfrak{W}$-modules, it is *not* an element of that space, nor even of the completion of the tensor product space with respect to its standard scalar product. Rather, the correct interpretation is indeed as a 1-point block on the disk. At a more technical level, this can be described as a so-called *co-invariant* of the space $\mathcal{H}_\lambda\otimes\mathcal{H}_{\tilde\lambda}$ with respect to the action $W_n\otimes\mathbf{1}-(-1)^{\Delta(W)}\,\mathbf{1}\otimes W_{-n}$ of the chiral algebra [1]. In place of these somewhat unfamiliar objects one may equivalently consider the singlets in the dual space $(\mathcal{H}_\lambda\otimes\mathcal{H}_{\tilde\lambda})^\star$; thus roughly, the boundary states may also be regarded as genuine vectors in this dual space.

In string theory, one often regards the boundary state $|\mathcal{B}^A\rangle$ as a synonym for the boundary condition A; its proper interpretation is that by saturating one leg of a multi-reggeon vertex with $|\mathcal{B}^A\rangle$ amounts to introducing a boundary of type A on the world sheet. The quantities $|\mathcal{B}^A\rangle$ also appear naturally in the vacuum amplitude for the annulus, which can be evaluated with the help of the formula

$$\langle\mathrm{B}_\lambda|\mathrm{e}^{2\pi\mathrm{i}\tau(L_0+\tilde L_0-c/12)}|\mathrm{B}_\lambda\rangle = \chi_\lambda(2\tau)\,, \tag{4}$$

where $\chi_\lambda(\tau)\equiv\chi_\lambda(\tau,0,0)$ is the Virasoro-specialized character of the $\mathfrak{W}$-module $\mathcal{H}_\lambda$.

Twisted actions of the chiral algebra.

A basic task in open conformal field theory is to determine all possible boundary conditions. The properties to be imposed depend on the application one has in mind. In the context of critical phenomena typically the boundary condition need to preserve just the Virasoro algebra; in special situations it may even be sufficient to respect only part of it. In string theory, one commonly requires to preserve the symmetry that is gauged, i.e. the Virasoro algebra respectively its relevant super extension; but boundary conditions for which the (super-)Virasoro algebra is preserved only up to BRST-exact terms seem to be perfectly admissible as well. Boundary conditions that violate part of the bulk symmetries can be roughly imagined as describing boundaries that carry some charge already in the absence of any fields.

[1] The formulæ in the literature actually describe the specific situation that the insertion point is at $z=0$ and that standard local coordinates on the covering surface $\hat C=\mathbb{P}^1$ of the disk are chosen.

The boundary blocks $|\mathrm{B}_\lambda\rangle$ introduced above do preserve the full chiral algebra $\mathfrak{W}$. Here the precise sense of 'preservation' is that $\mathfrak{W}$ acts on $\mathcal{H}_\lambda \otimes \mathcal{H}_{\tilde\lambda}$ as in (3), i.e. the action on $\mathcal{H}_{\tilde\lambda}$ is twisted by the automorphism σ_0: $W_n \mapsto (-1)^{\Delta(W)+1} W_{-n}$. It is then natural to look for other chiral blocks that constitute co-invariants for some differently twisted action of $\mathfrak{W}$. One way to achieve this is to replace σ_0 by $\sigma \circ \sigma_0$, with σ some other automorphism of $\mathfrak{W}$. One can check that (formal) solutions to

$$\big(W_n \otimes \mathbf{1} - (-1)^{\Delta(W)}\, \mathbf{1} \otimes \sigma(W_{-n})\big)|\mathrm{B}_\lambda\rangle_{(\sigma)} = 0 \tag{5}$$

(which replaces the condition (3)) are given by $|\mathrm{B}_\lambda\rangle_{(\sigma)} = (\mathbf{1} \otimes \theta_\sigma)|\mathrm{B}_\lambda\rangle$, where the map θ_σ which acts on $\mathcal{H}_{\tilde\lambda}$ is characterized by its 'σ-twining' property $\theta_\sigma \circ W_n = \sigma(W_n) \circ \theta_\sigma$.

Note that for non-trivial σ, such boundary conditions typically do *not* preserve the Virasoro algebra, and accordingly shouldn't play a role in applications to strings.

As a side remark, we mention that a large class of Virasoro non-preserving automorphisms σ, for which θ_σ still has reasonable properties, is provided by the automorphisms $\sigma = \sigma_\mathrm{J}$ that implement [5] the action of simple currents J of WZW models. When such a σ_J has order two, then e.g. analogues of the formula (4) are given by

$$\begin{aligned} {}_{(\sigma)}\langle \mathrm{B}_\lambda | \mathrm{e}^{2\pi i \tau (L_0 + \tilde L_0 - c/12)} | \mathrm{B}_\lambda\rangle_{(\sigma)} &= \mathcal{X}_\lambda(2\tau, -\bar\varpi_\mathrm{J}\tau, (\bar\varpi_\mathrm{J},\bar\varpi_\mathrm{J})\tau/2) \\ {}_{(\sigma)}\langle \mathrm{B}_\lambda | \mathrm{e}^{2\pi i \tau (L_0 + \tilde L_0 - c/12)} | \mathrm{B}_\lambda\rangle &= \begin{cases} 0 & \text{for } \ \mathrm{J}\star\lambda \neq \lambda\,, \\ \breve{\mathcal{X}}_\lambda(2\tau,0,0) & \text{for } \ \mathrm{J}\star\lambda = \lambda\,. \end{cases} \end{aligned} \tag{6}$$

Here $\bar\varpi_\mathrm{J}$ is the horizontal part of the fundamental weight of the relevant affine Lie algebra that characterizes the simple current J and $\breve{\mathcal{X}}_\lambda$ is a so-called twining character [5]. Similar formulæ hold when one twists in addition by an inner automorphism.

D-branes.

We now focus our attention on boundary conditions which are relevant to strings and D-branes. To this end we consider boundaries that respect the full chiral algebra. The natural structure underlying such boundary conditions is the one of *automorphisms ω of the fusion rules* that preserve conformal weights [1]. The origin of these automorphisms is the freedom that is present in relating the two labels λ and $\tilde\lambda$ of a bulk field $\phi_{\lambda,\tilde\lambda}$, and thus is quite similar to the origin of the appearence of fusion rule automorphisms in the classification of consistent torus partition functions. But in distinction to the case of closed conformal field theory, the factorization constraints do not require that this freedom is fixed in one and the same manner on all surfaces. Specifically, given a definite torus partition function, which (by taking $\mathfrak{W}$ sufficiently large) can be assumed to correspond to some fusion rule automorphism π, the pairing of λ and $\tilde\lambda$ is as prescribed by π on all closed orientable surfaces, but on the disk any other allowed fusion rule automorphism ω can appear as well. When $\omega = \pi$ one is dealing with an analogue of *Neumann* boundary conditions for free bosons, while the counterpart of *Dirichlet* boundary conditions of free bosons is given by $\omega = \pi \circ \omega_\mathrm{C}$, where ω_C: $\lambda \mapsto \lambda^+$ denotes charge conjugation.

Note that the choice of ω not only influences the values of the constants N_0^{AA} and $R^A_{\lambda\tilde\lambda;0}$ in (2), but also the explicit form of the 1-point block $|\mathrm{B}_\lambda\rangle$, which therefore should more precisely be denoted by $|\mathrm{B}_\lambda\rangle_\omega$. Adopting the terminology from the free boson case, one should refer to the $|\mathrm{B}_\lambda\rangle_\omega$ as *D-brane states*, or better as *D-brane blocks*. In the specific case of the theory of d uncompactified free bosons X^i with diagonal torus partition function and $\omega = \mathrm{diag}((+1)^{p+1},(-1)^{d-p-1}) \in \mathrm{O}(d)$ (acting on the X^i), $|\mathrm{B}_{\lambda=0}\rangle_\omega$ is indeed nothing but the usual Dirichlet p-brane with vanishing field strength on the p+1-dimensional world volume. The automorphisms ω form a group (which in some cases is a Lie group, e.g. $\mathrm{O}(d)$ for d free bosons). In a space-time interpretation, the choice of a connected component of that group looks like a topological information; thus ω encodes global topological features of the D-brane.

The choice of an automorphism ω does not refer to a boundary of C at all. Therefore this freedom is already present in the absence of boundaries, e.g. for $C = \mathbb{PR}^2$. In contrast, as soon as boundaries are present there is an additional freedom, namely the choice of a consistent collection of reflection coefficients $R^A_{\lambda\tilde\lambda;0}$. Thus a boundary condition A should be regarded as a *pair* $A \equiv (\omega,a)$, where ω is a fusion rule automorphism respecting conformal weights, while the label a is tied to the existence of the boundary. In a space-time interpretation, a characterizes local properties of the D-brane, such as its position or a field strength on it [1]. In [1], ω is called the *automorphism type* of the boundary condition, while a is referred to as the *Chan–Paton type* because in string theory one must attach a distinct Chan–Paton multiplicity N_a to each allowed a. (The numbers N_a are to be determined by string theoretic arguments, e.g. tadpole cancellation.) Note that the summation in (1) is over all possible Chan–Paton types a such that $A = (\omega,a)$ with fixed automorphism type ω.

So far we did not say much about the possible values of a. According to [2] in the Neumann case $\omega = \pi = \omega_C$ the allowed index set is equal to the set $\{\lambda\}$ and the associated reflection coefficients $R^A_{\lambda\tilde\lambda;0}$ furnish representations of the fusion algebra. In [1] evidence was collected for the fact that (for all rational theories, and similarly for certain non-rational ones), for fixed ω the number of labels a equals the dimension of some commutative associative algebra $\mathfrak{C}_\omega$ that generalizes the fusion algebra, and that the $R^A_{\lambda\tilde\lambda;0}$ furnish one-dimensional $\mathfrak{C}_\omega$-representations. The structure constants of $\mathfrak{C}_\omega$ are expected to satisfy some analogue of the Verlinde formula, related to structures similar to those uncovered in [5]. One class of examples for such classifying algebras had already been obtained before in [4] (for WZW models) and [6] (for arbitrary conformal field theories); several other examples are listed in [1].

References

[1] J. Fuchs and C. Schweigert, preprint hep-th/9712257

[2] J.L. Cardy, *Nucl. Phys. B* **324** (1989) 581

[3] D.C. Lewellen, *Nucl. Phys. B* **372** (1992) 654

[4] G. Pradisi, A. Sagnotti, and Ya.S. Stanev, *Phys. Lett. B* **381** (1996) 97

[5] J. Fuchs, B. Schellekens, and C. Schweigert, *Comm. Math. Phys.* **180** (1996) 39

[6] J. Fuchs and C. Schweigert, *Phys. Lett. B* **414** (1997) 251

Classifying algebras for boundary conditions and traces on spaces of conformal blocks

Christoph Schweigert [1] **, Jürgen Fuchs** [2]

[1] CERN, CH – 1211 Genève 23
[2] Max-Planck-Institut für Mathematik
Gottfried-Claren-Str. 26, D – 53225 Bonn

Abstract: The boundary conditions of a non-trivial string background are classified. To this end we need traces on various spaces of conformal blocks, for which generalizations of the Verlinde formula are presented.

Introduction.
It has been known for a long time that the low energy effective action of superstring theories has solitonic solutions. Recently, it has become apparent that string perturbation theory in such a background can be formulated in terms of world sheets with boundaries, where one imposes certain non-trivial boundary conditions. Therefore, theories of open strings and conformal field theories on two-dimensional surfaces with boundaries have received renewed interest. A central problem in these theories is the classification of all consistent boundary conditions. So far, however, most investigations have been limited either to models based on free world sheet theories or on orbifolds of such theories or to BPS-sectors of models with extended supersymmetry. In this contribution we discuss the structure of boundary conditions in an arbitrary rational conformal field theory with a specific type of non-diagonal modular invariant. For a general discussion of boundary conditions in two-dimensional conformal field theory we refer to [1].

Modular invariants.
A chiral conformal field theory typically admits several consistent torus partition functions. Any non-trivial modular invariant of a rational conformal field theory can be obtained [2] by first extending the chiral algebra and then superposing an automorphism of the fusion rules. The extension of chiral algebras is by now fairly well understood, at least in the case of extensions by so-called simple currents [3, 4, 5], and can be described entirely in terms of a chiral half of the theory. As

a consequence, such extensions do not raise any problem in the construction of open string theories that was not already encountered for closed strings so that we can assume that the modular invariant in question is of pure automorphism type. Hence the modular invariant describes the pairing between left-moving and right-moving fields, or, more precisely [1], the pairing between the two chiral conformal field theories on the oriented cover of the world sheet.

For simplicity, we also make a few more assumptions on the boundary conditions we consider: first, we assume that the pairing for bulk fields is the same as in the case of closed orientable surfaces. In the terminology of Ref. [1], we choose a trivial automorphism type, i.e. generalized Neumann boundary conditions. Next, we assume that the boundary preserves all symmetries of the bulk. Finally, we assume that we are dealing with the same chiral conformal field theory on every type of two-dimensional surface. (This implies [1] that D-brane configurations with multiple wrapping are excluded.)

Under these assumptions, the possible boundary conditions have been classified for a theory with the charge conjugation modular invariant [6]. A first investigation in the case of non-trivial modular invariants has been undertaken in [7] for WZW–models based on SU(2).

The type of modular invariant we focus on generalizes the modular invariants which in the *A-D-E* classification of SU(2) modular invariants are of D_{odd}-type. This modular invariant exists for level $k = 4l + 2$ with l integer:

$$Z(\tau) = \sum_{l=0}^{k/2} |\chi_{2l}|^2 + \sum_{l=0}^{k/2} \chi_{2l+1} \chi^*_{k-2l-1} \,. \tag{1}$$

The full conformal field theory described by this modular invariant can be regarded as a $\mathbb{Z}_2$-orbifold of the WZW–theory on SU(2) with the diagonal modular invariant. There are three types of primary fields:

- Primary fields with integral isospin form the untwisted sector of the orbifold. In the full theory, they are paired with themselves.
- The twisted sector consists of the primary fields with half-integral isospin. These fields Φ_l are paired with some other primary field Φ_{k-l} which is obtained by taking the fusion product with the primary field Φ_k of highest possible isospin, $\Phi_{k-l} = \Phi_k \star \Phi_l$. The primary field Φ_k is a *simple current*: its fusion product with any other primary field contains just one primary field with multiplicity one.
- The twisted sector contains in particular the *fixed point* $\Phi_{k/2}$ which is mapped by the simple current to itself, $\Phi_k \star \Phi_{k/2} = \Phi_{k/2}$. In the twisted sector, we distinguish between fixed points and non-fixed points.

In this note we consider conformal field theories with a similar $\mathbb{Z}_2$ symmetry: we assume that the theory contains a simple current J, i.e. a primary field such that its fusion product has the form $J \star \Phi_\Lambda = \Phi_{J\Lambda}$, for any primary field Φ_Λ. Moreover, we assume that J squares to the vacuum primary field, $J^2 = \Phi_0$, and that it has conformal weight $\Delta_J \in \mathbb{Z} + 1/2$. Given such a simple current J, we associate to every primary field Φ_Λ its *monodromy charge*

$$Q_J(\Lambda) := \Delta_\Lambda + \Delta_J - \Delta_{J\Lambda} \mod \mathbb{Z} \tag{2}$$

which generalizes the conjugacy class and is conserved in operator products. One can show that in this situation the following expression gives a modular invariant partition function.

$$Z = \sum_{Q(\Lambda)=0} \chi_\Lambda \chi^*_{\Lambda^+} + \sum_{Q(\Lambda)=1/2} \chi_\Lambda \chi^*_{J\Lambda^+} \tag{3}$$

(This partition function looks like a $\mathbb{Z}_2$ orbifold of the original theory. It has a similar structure as the one of the type IIA superstring in light cone gauge.) Again we have three types of primary fields: N_0 primary fields in the untwisted sector with monodromy charge $Q(\Lambda)=0$. They all form orbits of length 2 under the action of J. In the twisted sector, all fields have $Q(\Lambda)=1/2$. We have N_1 fields on full orbits and N_f fixed points.

Construction of the classifying algebra.

It has been shown in [1] that a consistent boundary condition in a conformal field theory can be described by an automorphism of the fusion rules which preserves conformal weights, the *automorphism type* of the boundary condition, and a further degeneracy label, the *Chan-Paton type*. Moreover, it has been shown [1] that each Chan-Paton type corresponds to an irreducible representation of a so-called *classifying algebra*. This classifying algebra $\tilde{\mathcal{A}}$ has been computed in [7] in the special case of WZW–theories based on SU(2), using the explicit form of fusing matrices and operator product coefficients. It was observed that the possible Chan-Paton types are in one-to-one correspondence to the orbits of the simple current J, where the fixed points of J are counted with multiplicity two.

Traces on the space of conformal blocks.

The construction of the correct classifying algebra in the general case (3) requires the knowledge of some traces on the spaces of conformal blocks. In general, we consider the finite-dimensional vector space $B_{\vec{\Lambda}}$ of conformal blocks, where $\vec{\Lambda} = (\Lambda_1, \ldots, \Lambda_n)$ stands for a finite sequence of primary fields. In the case of three-point blocks $B_{\lambda\mu\nu}$ its dimension is given by the Verlinde formula

$$\mathcal{N}_{\lambda\mu\nu} = \dim B_{\lambda\mu\nu} = \sum_\rho S_{\lambda\rho} S_{\mu\rho} S_{\nu\rho} \,/\, S_{0\rho} \,. \tag{4}$$

We now consider a collection $\vec{J} = (J_1, \ldots, J_n)$ of simple currents which fulfils $J_1 \star J_2 \star \cdots \star J_n = 1$. In this situation one can define a natural isomorphism between the spaces of conformal blocks

$$\Theta_{\vec{J}} : \quad B_{\vec{\Lambda}} \to B_{\vec{J}\vec{\Lambda}} \,, \tag{5}$$

where we introduced the short hand $\vec{J}\vec{\Lambda} = (J_1\Lambda_1, \ldots, J_n\Lambda_n)$. In the case of a simple current of order two, there are isomorphisms:

$$\Theta_J : \quad B_{\lambda\mu\nu} \to B_{J\lambda J\mu\nu} \,. \tag{6}$$

If $\lambda = f$ and $\mu = g$ are fixed points of J, then Θ_J is an *endomorphism* and we can consider the trace

$$\breve{\mathcal{N}}_{fg\nu} := \mathrm{Tr}_{B_{fg\nu}} \Theta_J \,. \tag{7}$$

The trace $\breve{\mathcal{N}}_{fg\nu}$ is an integer (surprisingly enough this is also true for simple currents of any arbitrary order), and can be used to compute the dimension of the eigenspaces of Θ_J to the eigenvalues ± 1:

$$\mathrm{Tr}_{B_{fg\mu}}(\tfrac{1}{2}(1 \pm \Theta_J)) = \tfrac{1}{2}(\mathcal{N}_{fg\mu} \pm \breve{\mathcal{N}}_{fg\mu}), \tag{8}$$

which are manifestly non-negative integers.

These traces have already played an important role in chiral conformal field theory, in the analysis of the twisted sector of extensions [3]. There is a *fixed point theory* with modular matrix $\breve{S}$ whose primary fields are in one-to-one correspondence with the fixed points. The traces are then given by a generalization of the Verlinde formula:

$$\breve{\mathcal{N}}_{fg\nu} = \sum_{h\ \mathrm{fix}} \breve{S}_{fh} \breve{S}_{gh} S_{\nu h} \,/\, S_{0h} \,. \tag{9}$$

For WZW–theories (and also for coset conformal field theories) the fixed point theories are obtained by folding Dynkin diagrams [8]. In general, it is conjectured that they describe the modular properties of the one-point blocks on the torus with insertion the simple current J. The structure of fixed point theories is found in many places: it occurs in the twisted sector of extension modular invariants [3], in the very definition of coset conformal field theories [9], in the Verlinde formula for non-simply connected Lie groups [3] and in the topologically non-trivial components of the moduli spaces of holomorphic principal bundles with non-simply connected structure groups over an elliptic curve [10].

As a side remark we mention some other traces on the space of conformal blocks which can be computed explicitly: on a four-point block on the sphere with two identical primary fields one can consider the trace $Y_{i,i,j,k}$ of the permutation acting on the two identical insertions. Such traces appear in the theory of permutation orbifolds [11] as well as in the description of amplitudes on the Möbius strip [12]. Again, one has a generalization of the Verlinde formula:

$$Y_{i,i,j,k} = \sum_n P_{jn} P_{kn} S_{in} \,/\, S_{0n} \,, \tag{10}$$

where we have introduced the matrix $P = T^{1/2} S T^2 S T^{1/2}$.

The classifying algebra and its representation theory.

We are now in a position to display the classifying algebra $\tilde{\mathcal{A}}$ [13] for the conformal field theory with torus partition function (3). The dimension of $\tilde{\mathcal{A}}$ equals the number of bulk fields, $\dim \tilde{\mathcal{A}} = N_0 + N_f$. Moreover $\tilde{\mathcal{A}}$ is $\mathbb{Z}_2$-graded and contains the fusion algebra of fields in the untwisted sector as a subalgebra, since the operator products in the untwisted sector of an orbifold theory are the same as in the original theory. This leads to the following structure constants for $\tilde{\mathcal{A}}$:

$$\tilde{\mathcal{N}}_{\lambda\mu}^{\nu} = \begin{cases} \mathcal{N}_{\lambda\mu}^{\nu} & \text{if } Q(\lambda) = Q(\mu) = Q(\nu) = 0 \\ \breve{\mathcal{N}}_{\lambda\mu}^{\nu} & \text{if there is precisely one full orbit} \\ 0 & \text{else} \end{cases} \tag{11}$$

One easily checks that this classifying algebra $\tilde{\mathcal{A}}$ is commutative and associative, that Φ_0 is a unit element and that the evaluation on the identity gives a conjugation on $\tilde{\mathcal{A}}$: $\tilde{\mathcal{N}}_{fg}^{0} = \delta_{f^+,g}$. As a consequence, $\tilde{\mathcal{A}}$ is still semi-simple, but in contrast to fusion algebras, its structure constants can also be negative. Notice that $\tilde{\mathcal{A}}$ is *not* a subalgebra of the fusion algebra $\mathcal{A}$.

The representation theory of $\tilde{\mathcal{A}}$ has the following structure: there are $N_0 + N_f = \frac{1}{2}(N_0 + N_1) + 2N_f$ irreducible representations, which are all one-dimensional. They are in correspondence to orbits of the simple current J: each of the orbits α of length two gives rise to a single irreducible representation $\mathcal{R}_{(\alpha)}$:

$$\mathcal{R}_{(\alpha)}(\Phi_\mu) = \begin{cases} S_{\alpha\mu} \,/\, S_{0\alpha} & \text{for} \quad Q(\mu) = 0\,, \\ 0 & \text{for} \quad J\mu = \mu\,. \end{cases} \tag{12}$$

Each of the orbits f of length one, the fixed points, gives rise to *two* different irreducible representations $\mathcal{R}_{(f+)}$ and $\mathcal{R}_{(f-)}$:

$$\mathcal{R}_{(f\pm)}(\Phi_\mu) = \begin{cases} S_{f\mu} \,/\, S_{0f} & \text{for} \quad Q(\mu) = 0\,, \\ \pm\breve{S}_{f\mu} \,/\, S_{0f} & \text{for} \quad J\mu = \mu\,. \end{cases} \tag{13}$$

Notice that fixed points are fields in the twisted sector, and accordingly the modular matrix $\breve{S}$ of the fixed point theory appears.

The annulus amplitude, consistency checks.
To be able to perform several consistency checks, we compute the amplitude $A_{ab}(t) = \sum_\mu A_{ab}^{\mu} \chi_\mu(\frac{it}{2})$ for an annulus, where we impose boundary condition a respectively b on the two boundaries. We obtain the following result for the tensor A_{ab}^{μ}:

$$\begin{aligned} A_{\alpha\beta}^{\mu} &= \mathcal{N}_{\beta\mu}^{\alpha} + \mathcal{N}_{\beta\mu}^{J\alpha}\,, & A_{\alpha(f\pm)}^{\mu} &= \mathcal{N}_{f\mu}^{\alpha}\,, \\ A_{(f\pm)(g\pm)}^{\mu} &= \tfrac{1}{2}(\mathcal{N}_{f^+g\mu} + \breve{\mathcal{N}}_{f^+g\mu})\,, & A_{(f\pm)(g\mp)}^{\mu} &= \tfrac{1}{2}(\mathcal{N}_{f^+g\mu} - \breve{\mathcal{N}}_{f^+g\mu})\,. \end{aligned} \tag{14}$$

The tensor A_{ab}^{μ} allows to perform the following checks. We first remark that all A_{ab}^{μ} are non-negative integers as it befits for an expansion of an open string partition function. This result is particularly non-trivial for fixed points, where it follows from (8). Next we observe that the fact that the conjugation on the classifying algebra that is the evaluation on the identity implies that the multiplicities of the vacuum in the open string partition functions are either 0 or 1. This was a consistency requirement in [6]. Finally we can check the consistency of several factorizations. In the case of a sphere with four boundary circles with boundary conditions a, b, c and d we have $\sum_\mu A_{ab}^{\mu} A_{cd}^{\mu^+} = \sum_\mu A_{ac^+}^{\mu} A_{b^+d}^{\mu^+}$ also, $A^\mu A^\nu = \sum_\lambda \mathcal{N}_{\mu\nu}^{\lambda} A^\lambda$, which gives the correct factorization of the annulus. We also remark that the heuristic argument used in [6] to derive the classifying algebra for the charge conjugation modular invariant can be generalized to the case of our interest; for details we refer to [13].

Conclusions.
The structure of the classifying algebra $\tilde{\mathcal{A}}$ is actually closely related to the fusion

algebra of another type of modular invariants, namely those of 'D_{even}-type' (which are also known as integer spin simple current extensions). In particular, it looks as if the boundary theory were extended by the *half*-integer spin simple current J.

This is indeed most remarkable, because in the case of extensions, this structure is a consequence of the powerful consistency requirements of modular invariance. But for the crosscap as well as for the annulus and the Möbius strip, there is no analogue of a modular group. In string theory it is usually argued that tadpole cancellation provides a substitute for such consistency conditions. Note, however, that for our investigations we did not have to assume that the conformal field theory is part of a string compactification (e.g., the central charge is not restricted), so that the conditions of tadpole cancellation cannot even be formulated. Still, it seems that already on a pure conformal field theory level there are similar powerful constraints; to unravel the underlying structure seems to be a promising task.

Finally we mention that the construction of a classifying algebra for modular invariants of automorphism type that are not simple current automorphisms is still an open problem. A particularly interesting case is the one of generalized Neumann boundary conditions for the true diagonal modular invariant, where the relevant automorphism is just charge conjugation.

References

[1] J. Fuchs and C. Schweigert, preprint hep-th/9712257

[2] G. Moore and N. Seiberg, *Nucl. Phys. B* **313** (1989) 16

[3] J. Fuchs, A.N. Schellekens, and C. Schweigert, *Nucl. Phys. B* **473** (1996) 323

[4] C. Dong, H. Li, and G. Mason, *Commun. Math. Phys.* **180** (1996) 671

[5] P. Bantay, preprint hep-th/9611124

[6] J.L. Cardy, *Nucl. Phys. B* **324** (1989) 581

[7] G. Pradisi, A. Sagnotti, and Ya.S. Stanev, *Phys. Lett. B* **381** (1996) 97

[8] J. Fuchs, A.N. Schellekens, and C. Schweigert, *Commun. Math. Phys.* **180** (1996) 39

[9] J. Fuchs, A.N. Schellekens, and C. Schweigert, *Nucl. Phys. B* **461** (1996) 371

[10] C. Schweigert, *Nucl. Phys. B* **492** (1997) 743

[11] L. Borisov, M.B. Halpern, and C. Schweigert, preprint hep-th/9701061

[12] G. Pradisi, A. Sagnotti, and Ya.S. Stanev, *Phys. Lett. B* **354** (1995) 279

[13] J. Fuchs and C. Schweigert, *Phys. Lett. B* **414** (1997) 251

Towards an exact off-shell solution of the O(3) nonlinear sigma-model

J. Balog [1], M. Niedermaier [2]

[1] Research Institute for Particle and Nuclear Physics
H-1525 Budapest Pf. 49, Hungary
[2] Max-Planck-Institut für Gravitationsphysik
(Albert-Einstein-Institut)
Schlaatzweg 1, D-14473 Potsdam, Germany

Abstract: A scaling hypothesis for the n-particle spectral densities of the O(3) nonlinear sigma-model is described. It states that for large particle numbers the n-particle spectral densities are "self-similar" in being rescaled copies of a universal shape function. Promoted to a working hypothesis, it allows one to compute the two point functions at *all* energy or length scales. Further the values of two non-perturbative constants (needed for a parameter-free matching of the perturbative and the non-perturbative regime) are determined exactly. In non-integrable perturbations of the model the hypothesis implies scaling laws for multi-particle production processes analogous to the KNO scaling in QCD.

1. Introduction: An efficient way to describe the two-point function of some local operator $\mathcal{O}$ in a relativistic QFT is in terms of a Källen-Lehmann spectral representation. The spectral density $\rho(\mu)$ of $\mathcal{O}$ can be viewed as a measure for the number of degrees of freedom coupling to $\mathcal{O}$ at energy μ. It decomposes into a sum of n-particle contributions

$$\rho(\mu) = \sum_n \rho^{(n)}(\mu) , \tag{1}$$

where, depending on the local operator under consideration, some of the n-particle contributions may vanish on the grounds of internal quantum numbers. In a theory with a single mass scale m one has $\rho^{(1)}(\mu) \sim \delta(\mu - m)$ and $\rho^{(n)}(\mu)$, $n \geq 2$ has support only above nm, i.e. above the n-particle threshold. Once $\rho(\mu)$ is known, the various (Minkowski space or Euclidean) two-point functions of $\mathcal{O}$ can be computed as convolution integrals with an appropriate kernel carrying only kinematical information. The dynamical problem consists in computing the n-particle spectral densities $\rho^{(n)}(\mu)$ of $\mathcal{O}$.

Here we shall be concerned with massive 1+1 dimensional QFTs and in particular with the O(3) nonlinear sigma (NLS) model. The four most interesting local operators in this model are: The spin field, the Noether current, the energy momentum (EM) tensor and the topological charge (TC) density. Their spectral densities can be grouped into two families $\rho_l^{(n)}(\mu)$, $n \geq 1$, $l = 0, 1$ according to their isospin. For n even/odd the $\rho_0^{(n)}$ are the spectral densities of the EM-tensor/TC-density, respectively; similarly $\rho_1^{(n)}$ for n even/odd are the spectral densities of the Current/Spin, respectively. The following pieces of information are available for these spectral densities: (i) For small particle numbers n the functions $\rho_l^{(n)}(\mu)$ can be computed *exactly* by means of the form factor approach. In [1] this has been done up to 6 particles. (ii) For all particle numbers n the $\mu \to \infty$ asymptotics of the n-particle spectral densities is known, and is given by

$$\rho_l^{(n)}(\mu) \sim \frac{A_l^{(n)}}{\mu(\ln \mu)^{4-2l}}\,, \qquad \mu \to \infty\,, \tag{2}$$

where the constants $A_l^{(n)}$ are computable from the integrals of the lower particle spectral densities [1]. The constants $A_l^{(n)}$ are rapidly increasing with n. This implies that the $\mu \to \infty$ asymptotics of the full spectral densities (1) cannot be computed by naively summing up the asymptotic expressions (2), which in fact would be divergent. (iii) The large μ asymptotics of the full spectral densities can however be computed in renormalized perturbation theory (PT).[1] One finds for the leading behavior

$$\begin{aligned} \text{EM \& top: } \rho(\mu) &\sim \frac{A^{\mathcal{O}}}{\mu}\left[\frac{1}{(\ln \mu)^2} + O\left(\frac{\ln\ln \mu}{(\ln \mu)^3}\right)\right], \\ \text{spin \& curr: } \rho(\mu) &\sim \frac{A^{\mathcal{O}}}{\mu}\left[1 + O\left(\frac{1}{\ln \mu}\right)\right]. \end{aligned} \tag{3}$$

Subleading terms can also be computed, but not all of the overall constants are accessible to PT. In particular $\lambda_1 := A^{\text{spin}}$ is an unkown non-perturbative constant. In the case of the TC density A^{top} is fixed by PT but its relation to the non-perturbatively defined spectral sum (1) is not. Equivalently the matrix elements of the TC density between the vacuum and some multi-particle state are defined only up to an unknown non-perturbative constant λ_0.

Missing pieces of information about the spectral densities are: (iv) One would like to be able to compute the full spectral densities for all $\mu \geq 0$, not only their large μ asymptotics. This would allow one to compute the two-point functions at *all* energy/length scales. In terms of the spectral resolution (1) this amounts to

[1]The correctness of PT in this model has been challenged in [4]. To simplify the exposition we shall assume the validity of PT for the UV asymptotics throughout this note.

knowing all the n-particle contributions, not only those with $n \leq n_0$ for which the computation can be done explicitly. (v) One would like to know the (exact) values of the non-perturbative constants λ_0 and λ_1. Knowledge of these constants would allow one to match non-perturbative and perturbative information unambiguously. In this respect their role is similar to that of the m/Λ ratio [2].

The aim in following is to bridge the gap between the perturbative and the non-perturbative regime and to provide the missing pieces of information (iv) and (v). It is based on a remarkable self-similarity property of the n-particle spectral densities. For large n they appear to be basically rescaled copies of a "universal shape function" $Y_l(z)$. Explicitly

$$\rho_l^{(n)}(\mu) \approx \frac{M_l^{(n)}}{\mu} Y_l\left(\frac{\ln(\mu/m)}{\xi_l^{(n)}}\right) , \quad l = 0, 1 , \tag{4}$$

where $M_l^{(n)}$ and $\xi_l^{(n)}$ are certain scaling parameters to be specified later. In the following we shall first give a precise formulation of the scaling law (4) and recall some of the evidence presented for it in [1]. Then we shall promote it to a working hypothesis and show that it has the following consequences: The UV behavior is consistent with PT; in particular those coefficients $A^{\mathcal{O}}$ in (3) accessible to PT are reproduced. The non-perturbative constants λ_0 and λ_1 are determined exactly and in the normalization [1] are given by

$$\lambda_0 = \frac{1}{4} , \quad \lambda_1 = \frac{4}{3\pi^2} . \tag{5}$$

Finally candidate results for the two-point functions at *all* energy or length scales can be obtained by performing the sum in (1) using (4).

2. Formulation of the Hypothesis: With hindsight to the asymptotics (2) let us introduce

$$R_l^{(n)}(x) := me^x \rho_l^{(n)}(me^x) , \quad l = 0, 1 . \tag{6}$$

Here $l = 0, 1$ as before correspond to the EM tensor & TC density and Spin & Current series, respectively. The graphs of these functions are roughly 'bell-shaped': Starting from zero at $x = \ln n$ they are strictly increasing, reach a single maximum at some $x = \xi_l^{(n)} > \ln n$ and then decrease monotonically for all $x > \xi_l^{(n)}$. The position $\xi_l^{(n)}$ of the maximum and its value $M_l^{(n)} = R_l^{(n)}(\xi_l^{(n)})$ are two important characteristics of the function, and hence of the spectral density. Defining

$$Y_l^{(n)}(z) := \frac{1}{M_l^{(n)}} R_l^{(n)}(\xi_l^{(n)} z) , \quad l = 0, 1 , \tag{7}$$

both the value and the position of the maximum are normalized to unity. Initially $Y_l^{(n)}(z)$ is defined for $(\ln n)/\xi_l^{(n)} \leq z < \infty$; in order to have a common domain of definition we set $Y_l^{(n)}(z) = 0$ for $0 \leq z \leq (\ln n)/\xi_l^{(n)}$. The proposed behavior of the spectral densities is as follows:

Scaling Hypothesis:

(a) *(Self-similarity) The functions* $Y_l^{(n)}(z)$, $n \geq 2$ *converge pointwise to a bounded function* $Y_l(z)$. *The sequence of* k*-th moments converges to the* k*-th moments of* $Y_l(z)$ *for* $k + l = 0, 1$, *i.e.*

$$\lim_{n\to\infty} Y_l^{(n)}(z) = Y_l(z)\ , \quad z \geq 0\ ,$$

$$\lim_{n\to\infty} \int_0^\infty dz\, z^k Y_l^{(n)}(z) = \int_0^\infty dz\, z^k Y_l(z)\ .$$

(b) *(Asymptotic scaling) The parameters* $\xi_l^{(n)}$ *and* $M_l^{(n)}$ *scale asymptotically according to powers of* n, *i.e.*

$$\xi_l^{(n)} \sim \xi_l\, n^{1+\alpha_l}\ , \qquad M_l^{(n)} \sim M_l\, n^{-\gamma_l}\ .$$

Feature (a) in particular means that for sufficiently large n the graphs of two subsequent members $Y_l^{(n-1)}(z)$ and $Y_l^{(n)}(z)$ should become practically indistinguishable. This appears to be satisfied remarkably well even for small $n = 4, 5, 6$, as is illustrated in Figure 1 for the $l = 1$ series.

The analysis of part (b) of the scaling hypothesis is more involved. It turns out that all but one of the exponents in (b) are fixed by self-consistency, and only this one has to be determined by fitting against the $n \leq 6$ particle data. The result is [1]

$$\begin{aligned} \gamma_1 &= 1\ , \qquad \alpha_0 = \alpha_1 =: \alpha\ , \\ \gamma_0 &= 3 + 2\alpha\ , \qquad \alpha \approx 0.273\ . \end{aligned} \tag{8}$$

3. Consequences of the Hypothesis: Let n_0 be the maximal particle number for which the spectral densities have been computed explicitly (at present $n_0 = 6$). Then (4) gives candidate expressions for all $n > n_0$ particle spectral densities so that one can evaluate their sum. Inserted into a Källen-Lehmann spectral representation candidate results for the exact two-point functions can be computed, some of which have been described elsewhere. Concerning the UV behavior of the sum (1) notice that a finite number of terms in the sum, each decaying according to (2), can never produce the different UV behavior in (3). However the infinite sum does. What is happening is that the partial sums $(\ln \mu)^{2-2l} \sum_{n_0+1}^{N} \rho_l^{(n)}(\mu)$ develop a plateau, i.e. are practically constant in a large interval $\mu_{min} \lesssim \mu \lesssim \mu_{max}(N)$. When N is increased the plateau is prolonged and eventually reaches out to infinity, i.e. $\mu_{max}(N) \to \infty$ for $N \to \infty$ (while μ_{min} is basically N-independent). The value of the plateau determines the asymptotic constants in (3). In those cases where the coefficients are accessible to PT the perturbative value is reproduced with an accuracy better than 1%. In addition one obtains the following two *exact* relations among the four constants

$$A^{\text{curr}} = \frac{\pi}{4} A^{\text{spin}}\ , \quad A^{\text{EM}} = 4 A^{\text{top}}\ . \tag{9}$$

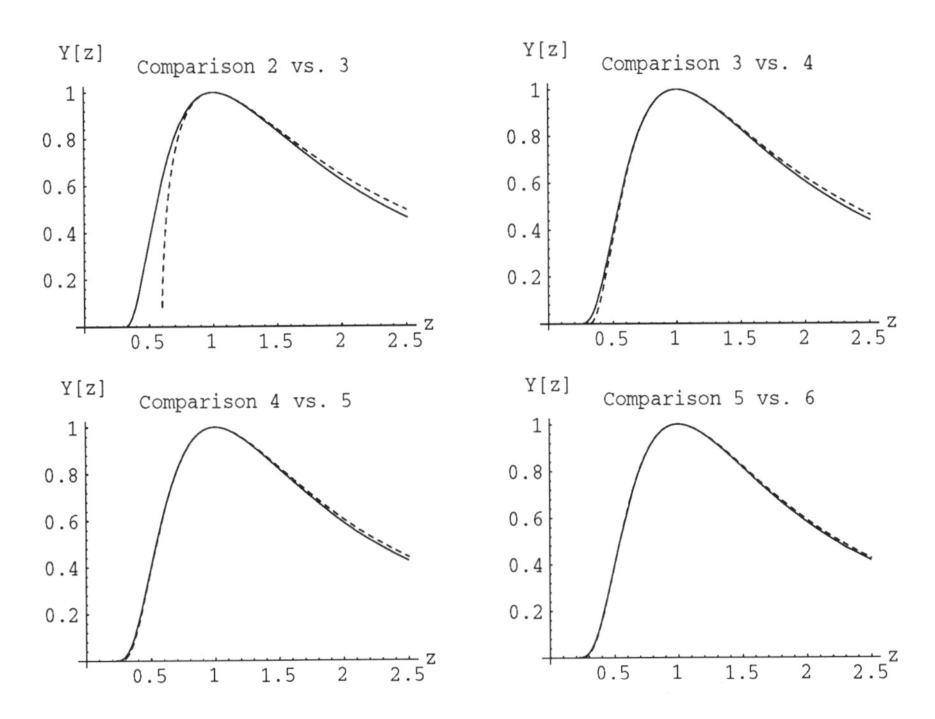

Figure 1: Illustration of the self-similarity property of the rescaled $l = 1$ spectral densities. The plots show $Y_1^{(n)}(z)$ (dashed) compared with $Y_1^{(n+1)}(z)$ (solid) for $n = 2, 3, 4, 5$.

These relations reflect the linking between the even and the odd particle members of an isospin family, which results from the clustering relations obeyed by the exact form factors[1]. Since $(A^{\rm curr})_{PT} = 1/3\pi$ the first eq. in (9) gives $A^{\rm spin} = \lambda_1$ as in (5), while the second eq. fixes the physical normalization of the TC matrix elements in terms of that of the EM tensor, which in our conventions amounts to $\lambda_0 = 1/4$.

The λ_0 value can be tested independently by means of MC simulations. In Figure 2 the results for the two-point functions of the TC density is shown – once computed via the form factor approach, with the absolute normalization fixed according to (5) and once via MC simulations. The simulations were done using the cluster algorithm of [3] with the standard action and the geometrical definition of the TC density. The data in Figure 2 correspond to a 460 square lattice at inverse coupling $\beta = 1.80$ (correlation length 65.05). The statistical errors are smaller than the size of the dots. The nice agreement confirms $\lambda_0 = 1/4$ and hence provides further support for the scaling hypothesis.

4. Relation to KNO Scaling: A bonus of the above scaling hypothesis is that it implies KNO-like scaling laws for multi-particle production processes in non-integrable perturbations of the model (but not vice versa). To allow for particle production we enlarge our model-world by a "weak" sector so that a general state will be of

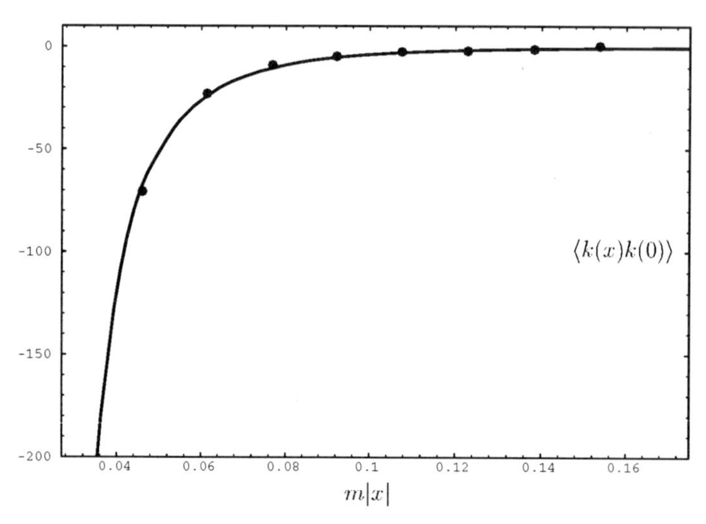

Figure 2: Two-point function of the topological charge density. Comparison MC data and form factor result. Test of $\lambda_0 = 1/4$.

the form $|s; w\rangle$, where s is the "strong" (sigma-model) part of the state and w is the "weak" part. Adding a current-current interaction term, production processes $|s; w\rangle \to |s'; w'\rangle$ (where the states s and s' have different particle numbers) become possible. The corresponding transition amplitude, to lowest order in the new interaction, reads

$$\mathcal{A} = \int d^2x\, l^{a\mu}(x)\, \langle s'|j^a_\mu(x)|s\rangle\,, \tag{10}$$

where the Fourier transform of $l^a_\mu(x)$ has support at the momentum Q, the "weak" momentum transfer.

The simplest production process is the two-dimensional analogue of the e^+e^- annihilation. We model this process by choosing $Q^2 > 0$ and $s = |0\rangle$. Summing over all (discrete and continuous) quantum numbers of the n-particle state s', the probability distribution for producing n particles at energy $\mu = \sqrt{Q^2}$ becomes independent of the "weak" part of the process and is proportional to the current spectral density. The proportionality factor involves the coupling constant of the external current-current interaction, which drops out when considering the conditional probability P_{2n} for having exactly $2n$ particles produced, once the process has taken place at all. One has

$$P_{2n} = \frac{\mu}{\kappa}\rho_1^{(2n)}(\mu)\,, \qquad \sum_{n=1}^{\infty} P_{2n} = 1\,, \tag{11}$$

where the second eq. fixes the normalization constant κ. Similarly production processes upon perturbation with the EM-tensor can be studied, in which case $\rho_0^{(2n)}(\mu)$ appears in (11). Using our scaling hypothesis the energy dependence of the probability distribution can in both cases be expressed, for large μ (where one can approximate the sums by integrals), in terms of $\bar{n} = \sum_{n=1}^{\infty} 2n\, P_{2n} \sim (\ln \mu)^{1/(1+\alpha)}$, the average number of particles produced. The asymptotic distribution takes the KNO-scaling form [5]

$$\bar{n}\, P_{2n} = f\left(\frac{n}{\bar{n}}\right) . \tag{12}$$

The scaling function $f(q)$ is given in terms of the universal shape function as

$$f(q) = \frac{2(1+\alpha)\tilde{h}_l}{h_l^2} \left(\frac{h_l}{2\tilde{h}_l q}\right)^{\gamma_l} Y_l\left(\left(\frac{h_l}{2\tilde{h}_l\, q}\right)^{1+\alpha}\right) , \tag{13}$$

where the parameters h_l and $\tilde{h}_l$ are the $(\frac{\gamma_l-1}{1+\alpha} - 1)^{\text{th}}$ and $(\frac{\gamma_l-2}{1+\alpha} - 1)^{\text{th}}$ moments of the universal shape function $Y_l(z)$, respectively, and the exponents are given in (8). The case $l = 1$ corresponds to the current perturbation and $l = 0$ to the perturbation by the EM-tensor. These KNO-type scaling laws, however, are only valid for simultaneously large particle numbers and large energies in the perturbed theory. In contrast, the scaling hypothesis (a), (b) for the spectral densities is valid already in the unperturbed theory for *all* energies, and in particular is non-perturbative in nature.

`Acknowledgements:` We are indebted to F. Niedermayer for making available to us his MC program. This research was supported by the Reimar Lüst fellowship of the Max-Planck-Society (M.N.) and in part by the Hungarian National Science Fund (OTKA) under T 016233 and T 019917.

References

[1] J. Balog and M. Niedermaier, Nucl. Phys. **B500** (1997) 421

[2] P. Hasenfratz, M. Maggiore and F. Niedermayer, Phys. Lett. **B245** (1990) 522.

[3] M. Blatter, R. Burkhalter, P. Hasenfratz and F. Niedermayer, Phys. Rev. **D53** (1996) 923.

[4] A. Patrascioiu and E. Seiler, Phys. Rev. Lett **74** (1995) 1924.

[5] Z. Koba, H.B. Nielsen and P. Olesen, Nucl. Phys. **B40** (1972) 317.
A.M. Polyakov, Sov. Phys. JETP **32** (1971) 296; **33** (1971) 850.

On N=2 SUSY gauge theories and integrable systems

A.Mironov [1]

[1] Theory Dept., Lebedev Physical Institute, Moscow, Russia
and ITEP, Moscow, Russia

Abstract: This note gives a brief review of the integrable structures presented in the Seiberg-Witten approach to the N=2 SUSY gauge theories with emphasize on the case of the gauge theories with matter hypermultiplets included (described by spin chains). The web of different N=2 SUSY theories is discussed.

General remarks. Since the paper by N.Seiberg and E.Witten [1], there have been a lot of attempts to get better understanding of the structures arising in the low-energy sector of N=2 SUSY gauge theories. In a sense, the paper [1] pointed out the importance of objects completely different from those typically dealt with in quantum field theory. In particular, one of the main quantities in the Seiberg-Witten (SW) approach is the prepotential giving the low-energy effective action of the theory.

One of the constituent parts of arena where the low-energy effective theory lives is, in accordance with [1], a Riemann surface, while a subspace of the moduli space of Riemann surfaces gives the moduli space of vacua of the physical theory. The whole world-sheet of the low-energy theory can be described in terms of 5-brane in M-theory [2, 3, 4, 5].

To these counterparts of the field theory objects, one should also add the analog of the symmetry principle which arises within the SW framework. Namely, it turns out that the symmetry properties of theory in the low-energy limit are encoded in the integrable system that underlines the low-energy dynamics.

The very fact of existence of an integrable system behind SW solution has been first realized in the paper [6] that dealt with the pure gauge theory. Since then, many more examples of the N=2 SUSY theories has been investigated, and corresponding integrable systems have been revealed [7]-[11].

The goal of the present short review is to sketch the general scheme of connection between the SW solutions and integrable systems. We also describe the correspondence (SUSY theory $\longleftrightarrow$ integrable system) in concrete examples discussing what deformations of integrable systems correspond to deformations of physical theories.

SW anzatz, prepotential and integrable system. For the $\mathcal{N}=2$ SUSY gauge theory the SW anzatz can be *formulated* in the following way. One starts with two *bare* spectral curves. One of them, with a holomorphic 1-form $d\omega$, is elliptic curve (torus)

$$E_1(\tau): \quad y^2 = \prod_{a=1}^{3}(x - e_a(\tau)), \quad \sum_{a=1}^{3} e_a(\tau) = 0, \qquad d\omega = \frac{dx}{y} \equiv d\xi, \tag{1}$$

when the YM theory contains the adjoint matter hypermultiplet, or its degeneration $\tau \to i\infty$ – the double-punctured sphere ("annulus"):

$$x \to w \pm \frac{1}{w}, \quad y \to w \mp \frac{1}{w}, \qquad d\omega = \frac{dw}{w} \tag{2}$$

otherwise. In particular, this latter possibility is the case for the theory with the fundamental matter hypermultiplets.

The second bare spectral curve is also elliptic curve $E_2(\tau')$ or its degenerations depending on the dimension of the space-time.

In the integrable framework, the two bare spectral curves are related by the full spectral curve that is just the Riemann surface emerging within the Seiberg-Witten construction. There are two different types of integrable system with the corresponding associated full spectral curve.

Integrable systems of the first type which could be naturally called Hitchin type systems are described as follows. First, one introduces the Lax operator $\mathcal{L}(x,y)$ that is defined as a $N \times N$ matrix-valued function (1-differential) on the first *bare* spectral curves. Then, the *full* spectral curve $\mathcal{C}$ is given by the Lax-eigenvalue equation: $\det(\mathcal{L}(x,y) - \lambda) = 0$, where λ is given on the second bare curve. As a result, $\mathcal{C}$ arises as a ramified covering over the *bare* spectral curves $E_1(\tau)$:

$$\mathcal{C}: \quad \mathcal{P}(\lambda; x, y) = 0 \tag{3}$$

The typical system of this type is the Calogero-Moser system. The Lax operators in the systems of the first type satisfy linear Poisson brackets with generally speaking *dynamical* elliptic r-matrix [12].

On contrary, integrable systems of the second type are characterized by the quadratic Poisson brackets with the *numerical* r-matrix (certainly, quadratic Poisson relations can be easily rewritten as the linear ones with dynamical r-matrix [13]). The typical systems of this type are lattice systems and spin chains [14]. They are described by 2×2 matrix-valued transfer-matrices $T_N(\xi)$, and the full spectral curve is given by the equation $\det(T_N(\xi) - w) = 0$, where w is given on the second bare curve. In fact, only the systems when at least one of the bare curves is degenerated are investigated in detail. Therefore, either w is the coordinate on the cylinder, or λ is the coordinate on the sphere or cylinder.

The function $\mathcal{P}$ in (3) depends also on parameters (moduli) s_I, parametrizing the moduli space $\mathcal{M}$. From the point of view of integrable system, the Hamiltonians (integrals of motion) are some specific co-ordinates on the moduli space. From the point of view of gauge theory, the co-ordinates s_I include s_i – (the Schur polynomials of) the adjoint-scalar expectation values $h_k = \frac{1}{k}\langle \mathrm{Tr}\phi^k \rangle$ of the vector $\mathcal{N} = 2$ supermultiplet, as well as $s_\iota = m_\iota$ – the masses of the hypermultiplets. One associates with the handle of $\mathcal{C}$ the gauge moduli and with punctures – massive hypermultiplets, masses being residues in the punctures.

The generating 1-form $dS \cong \xi d\omega$ is meromorphic on $\mathcal{C}$ (the equality modulo total derivatives is denoted by "$\cong$"), where $\xi \in E_2$ is associated with one of the bare curve and $d\omega$ – with another one E_1 (they are related via the spectral curve equation). In integrable system terms, this form is just the shorten action "pdq" along the non-contractible contours on the Hamiltonian tori, i.e. related to the symplectic form "$d\xi \wedge d\omega$".

The prepotential is defined in terms of the cohomological class of dS:

$$a_I = \oint_{A_I} dS, \qquad \frac{\partial F}{\partial a_I} = \int_{B_I} dS \qquad (4)$$
$$A_I \circ B_J = \delta_{IJ}.$$

The cycles A_I include the A_i's wrapping around the handles of $\mathcal{C}$ and A_ι's, going around the singularities of dS. The conjugate contours B_I include the cycles B_i and the *non-closed* contours B_ι, ending at the singularities of dS (see [15] for more details). The integrals $\int_{B_\iota} dS$ are actually divergent, but the coefficient of divergent part is equal to residue of dS at particular singularity, i.e. to a_ι. Thus, the divergent contribution to the prepotential is quadratic in a_ι, while the prepotential is normally defined *modulo* quadratic combination of its arguments (which just fixes the bare coupling constant). In particular models $\oint_{A,B} dS$ for some conjugate pairs of contours are identically zero on entire $\mathcal{M}$: such pairs are not included into our set of indices $\{I\}$.

Note that the data the period matrix of $\mathcal{C}$ $T_{ij}(a_i) = \frac{\partial^2 F}{\partial a_i \partial a_j}$ as a function of the action variables a_i gives the set of coupling constants in the effective theory.

The most important property of the differential $dS \cong \xi \frac{dw}{w}$ is that its derivatives w.r.t. moduli gives holomorphic differentials on $\mathcal{C}$ (see [15]). The prepotential in the context of integrable systems was also discussed in [16].

Spin chains: gauge theories with fundamental matter. The crucial difference between integrable systems of the two types is in interpretation of the spectral curve determinant equation. It is the general corollary of existence of the linear Poisson bracket (even with dynamical r-matrix) that the spectral determinant equation generates the conserved quantities [17]. Therefore, the coefficients of the spectral curve polynomial are the integrals of motion (and give some coordinates on the moduli space). However, in integrable system of the second type, there is some more direct meaning of the spectral determinant equation.

Namely, it can be described as periodicity condition that is imposed onto the transfer-matrix. In fact, the existence of the transfer-matrix describing the evolution into the discrete direction (see [14]) is the main peculiarity of this kind integrable systems.

The simplest example of the integrable system of the second type is the periodic Toda chain. This system is, at the same time, of the first (Hitchin) type. This surprising fact looks accidental and is due to the possibility of two different descriptions of the Toda chain [7]. The Lax operators in the first description satisfy linear Poisson brackets with the trigonometric *numerical* r-matrix, while in the second one – those with the rational numerical r-matrix. In this case, E_1 degenerates into the cylinder with coordinate w, while E_2 – into the sphere with coordinate λ.[1]

Let us now switch to the SUSY gauge theories that are underlined by the above described integrable systems. The periodic Toda chain is associated with the pure gauge $\mathcal{N} = 2$ theory [6, 8]. Possible deformations of this physical theory is to add matter hypermultiplets. In fact, one can add either one matter hypermultiplet in adjoint representation which gives rise to the UV-finite theory, or several fundamental matter hypermultiplets so that the theory is still asymptotically free, or UV-finite[2]. This corresponds to the two natural deformations of the periodic Toda chain. The first deformation (by adding the adjoint matter) is associated with the elliptic Calogero-Moser system [9] and is the deformation within the Hitchin-like approach to the Toda system. The other possible way to deform the Toda chain is to consider more general system admitting the transfer-matrix description. This system is the (inhomogeneous) periodic XXX spin chain and describes exactly the theory with fundamental matter [7].

In fact, there are more general spin chains that can be also associated with some physical theories. We return to this question in the next paragraph. Now let us just note that all periodic inhomogeneous chains admit the general description so that the chain of length n is given by the Lax matrices $L_i(\xi + \xi_i)$, ξ_i being the chain inhomogeneities, and periodic boundary conditions. Thus, integrable systems of the second type differ from each other only by different concrete Lax operators $L_i(\xi)$ [5, 18].

The linear problem in the spin chain has the following form

$$L_i(\xi)\Psi_i(\xi) = \Psi_{i+1}(\xi) \tag{5}$$

where $\Psi_i(\xi)$ is the two-component Baker-Akhiezer function. The periodic boundary

---

[1] This latter degeneration of the torus $E(x, y)$ is described as $x \to 0$, $y \to \lambda$, or $x \to \lambda$, $y \to 0$ and $d\omega \to d\lambda$.

[2] For the $SU(N_c)$ gauge group, there should be at most $2N_c$ fundamental hypermultiplets.

conditions are easily formulated in terms of the Baker-Akhiezer function and read as

$$\Psi_{i+n}(\xi) = -w\Psi_i(\xi) \tag{6}$$

where w is a free parameter (diagonal matrix). One can introduce the transfer matrix shifting i to $i+n$

$$T(\xi) \equiv L_n(\xi)\ldots L_1(\xi) \tag{7}$$

which provides the spectral curve equation

$$\det(T(\xi) + w \cdot \mathbf{1}) = 0 \tag{8}$$

and generates a complete set of integrals of motion. Note that in this approach the parameter w of the bare spectral curve E_1 just describes the periodicity conditions.

Integrability of the spin chain follows from *quadratic* r-matrix relations (see, e.g. [14])

$$\{L_i(\xi) \overset{\otimes}{,} L_j(\xi')\} = \delta_{ij}\left[r(\xi - \xi'),\ L_i(\xi) \otimes L_i(\xi')\right] \tag{9}$$

The crucial property of this relation is that it is multiplicative and any product like (7) satisfies the same relation

$$\{T(\xi) \overset{\otimes}{,} T(\xi')\} = \left[r(\xi - \xi'),\ T(\xi) \otimes T(\xi')\right] \tag{10}$$

Zoo of $\mathcal{N} = 2$ SUSY theories Thus far, we mentioned two possible generalizations of the periodic Toda chain: to the elliptic Calogero-Moser system, and to the inhomogeneous XXX spin chain. In fact, these integrable systems admit further deformations. Indeed, the elliptic Calogero-Moser system is the system with coordinate variables living on the torus, but momentum variables – on the sphere. One can naturally deform this system to involve momenta living on the cylinder (elliptic Ruijsenaars model [19]) or even on another torus (the second elliptic bare curve, double elliptic system [20]).

At the same time, XXX spin chains described by the rational r-matrix can be deformed to the either XXZ or XYZ spin chains, described correspondingly by the trigonometric or elliptic r-matrices [14]. In these theories the second bare curve is the cylinder or torus respectively, and this is the manifold where momenta of the system lives. It implies that the momentum variables of integrable system get restricted values. One might think of this as of a sort of Kaluza-Klein mechanism. This interpretation, indeed, turns out to be correct so that the XXZ spin chain describes the 5 dimensional SUSY gauge theories with fundamental matter[3] and one of the 5 dimensions compactified onto the circle [18, 22]. At the same time, the XYZ chain describes the 6 dimensional theory with fundamental matter and with 2 dimensions compactified onto the bare torus [11, 18, 22].

[3]The pure gauge theory in 5d is described by degeneration of the XXZ chain, namely, by the relativistic Toda chain [21, 15, 18].

Note that 6 dimensions exhaust the room for consistent theories. It perfectly matches the fact that the XYZ chain seems not to admit further deformations. Still there is yet another possible deformation of the spin chain. Namely, one can consider, instead of $sl(2)$, (inhomogeneous) $sl(p)$ spin chains that are described by the $p \times p$ matrix-valued Lax operators [14, 5, 18]. Such systems are associated [5] with the SUSY theory with the gauge group being the product of simple factors and with bi-fundamental matter hypermultiplets [2].

After these identifications made, one can build the whole picture of the correspondence (integrable systems $\longleftrightarrow$ SUSY gauge theories). As the starting point, one describes the pure gauge theory by the simple periodic Toda chain. Then, adding matter hypermultiplets leads to the spin chain and adding adjoint matter leads to the elliptic Calogero-Moser (Hitchin type) system. This latter procedure implies that the first spectral curve (target manifold of the coordinate variables) is elliptic. At the same time, increasing the dimension of the space-time leads to the cylindric (5d) or elliptic (6d) second bare curve (target manifold of the momentum variables). In 5 dimensions the corresponding pure gauge system is the relativistic Toda chain [21, 15, 18][4] and the theory with adjoint matter is the elliptic Ruijsenaars model [21]. Analogous systems in 6 dimensions are described less manifestly (see, however, [20]), excluding the case of XYZ spin chain associated with the fundamental matter [11, 18, 22]. At last, considering theories with the gauge group that is the product of simple factors, one should enlarge either the matrix dimensions of the Lax operator at the single site (within the spin chain framework) or the number of marked points (in the Hitchin-like approach) [5, 18].

I am grateful to A.Gorsky, A.Marshakov and A.Morozov for useful discussions. I also acknowledge the organizers of the 31st International Simposium Ahrenshoop on the Theory of Elementary Particles. The work is partially supported by the grant INTAS-RFBR-95-0690.

References

[1] N.Seiberg, E.Witten, Nucl.Phys. **B426** (1994) 19-52; hep-th/9407087; Nucl.Phys. **B431** (1994) 484-550; hep-th/9408099

[2] E.Witten, hep-th/9703166

[3] A.Gorsky, Phys.Lett. **B410** (1997) 22; hep-th/9612238

[4] A.Marshakov, M.Martellini, A.Morozov, hep-th/9706050

[5] A.Gorsky, S.Gukov, A.Mironov, hep-th/9707120

[4] Note that this system also admits two representations – as a spin chain and within the Hitchin-like approach [23].

[6] A.Gorsky, I.Krichever, A.Marshakov, A.Mironov, A.Morozov, Phys.Lett. **B355** (1995) 466-474; hep-th/9505035

[7] A.Gorsky, A.Marshakov, A.Mironov, A.Morozov, Phys.Lett. **B380** (1996) 75; hep-th/9603140

[8] E.Martinec, N.Warner, Nucl.Phys. **B459** (1996) 97-112; hep-th/9509161

[9] R.Donagi, E.Witten, Nucl.Phys. **B460** (1996) 299; hep-th/9510101;
E.Martinec, Phys.Lett. **B367** (1996) 91-96; hep-th/9510204;
A.Gorsky, A.Marshakov, Phys.Lett. **B375** (1996) 127, hep-th/9510224;
E.Martinec, N.Warner, hep-th/9511052;
H.Itoyama, A.Morozov, Nucl.Phys. **B477** (1996) 855-877; hep-th/9511126;
Nucl.Phys. **B491** (1997) 529-573; hep-th/9512161

[10] I.Krichever, D.Phong, J.Diff.Geom. **45** (1997) 349-389; hep-th/9604199; hep-th/9708170

[11] A.Gorsky, A.Marshakov, A.Mironov, A.Morozov, hep-th/9604078

[12] E.Sklyanin, hepth/9308060, Alg.Anal. **6** (1994) 227
H.W.Braden, T.Suzuki, hepth/9309033, Lett.Math.Phys. **30** (1994) 147
G.Arutyunov, P.Medvedev, hepth/9511070

[13] E.Sklyanin, solv-int/9504001

[14] L.Faddeev, L.Takhtadjan, *Hamiltonian Approach to the Theory of Solitons*, 1986

[15] A.Marshakov, A.Mironov, A.Morozov, hep-th/9701123

[16] T.Nakatsu, K.Takasaki, Mod.Phys.Lett. **A11** (1996) 157-168; hep-th/9509162;
Int.J.Mod.Phys. **A11** (1996) 5505-5518; hep-th/9603069;
C.Ahn, S.Nam, Phys.Lett. **B387** (1996) 304-309; hep-th/9603028

[17] O.Babelon, C.-M.Viallet, Phys.Lett. **B237** (1990) 411

[18] A.Gorsky, S.Gukov, A.Mironov, hep-th/9710239

[19] S.N.M.Ruijsenaars, Comm.Math.Phys. **110** (1987) 191

[20] V.Fock, A.Gorsky, N.Nekrasov, V.Rubtsov, Preprint ITEP/TH-36/96

[21] N.Nekrasov, hep-th/9609219

[22] A.Marshakov, A.Mironov, hep-th/9711156

[23] S.Kharchev, A.Mironov, A.Zhedanov, Int.J.Mod.Phys. **A12** (1997) 2675-2724; hep-th/9606144

Background Field Method and Structure of Effective Action in $N = 2$ Super Yang-Mills Theories

I.L.Buchbinder [1] , B.A.Ovrut [2]

[1] Department of Theoretical Physics, Tomsk State Pedagogical University, Tomsk 634041, Russia
[2] Department of Physics, University of Pennsylvania, Philadelphia, PA 19104 - 6396, USA
and
Institut für Physik, Humboldt Universität zu Berlin, Invalidenstraße 110, D-10115 Berlin, Germany
and
School of Natural Sciences, Institute for Advanced Study, Olden Lane, Princeton, NJ 08540, USA

Abstract: This paper is a brief review of background field method and some of its applications in $N = 2$ super Yang-Mills theories with a matter within harmonic superspace approach. A general structure of effective action is discussed, an absence of two-loop quantum corrections to first non-leading term in effective action is proved and $N = 2$ non-renormalization theorem in this approach is considered.

$N = 2$ supersymmetric field theories have attracted much attention due to significant progress in understanding their quantum aspects. Modern interest to such theories was inspired by seminal papers by Seiberg and Witten [1] where exact instanton contribution to low-energy effective action has been found. This result has demonstrated once more the wonderful features of the above theories and led to forming a research directions associated with study a general structure of effective action in $N = 2$ super Yang-Mills theories.

An adequate description of quantum $N = 2$ supersymmetric field theories should be based on formulating these theories in terms of unconstrained $N = 2$ superfields defined on an appropriate $N = 2$ superspace. Such a description is achieved within harmonic superspace approach [2].

The background field method is a powerful and highly efficient tool for study structure of quantum gauge theories (see f.e. [3]). The attractive features of the background field method is that it allows to preserve the manifest classical gauge invariance in quantum theory. Due to this circumstance the background field method is very convenient both for investigation of general properties of effective action in gauge theories and for carrying out the calculations in concrete field models with gauge symmetries.

This paper is a brief review of background field method for $N = 2$ super Yang-Mills theories in harmonic superspace and some of its applications [4, 5].

The harmonic superspace is defined as a supermanifold parametrized by the coordinates x_A^m, $\theta_\alpha^\pm$, $\bar{\theta}_{\dot{\alpha}}^\pm$, $u_i^\pm$ where x_A^m and $u_i^\pm$ are the bosonic coordinates and $\theta_\alpha^\pm$, $\bar{\theta}_{\dot{\alpha}}^\pm$ are the fermionic ones. The details of denotions are given in ref.[2]. The remarkable property of the harmonic superspace approach is that the set of coordinates x_A^m, θ_α^+, $\bar{\theta}_{\dot{\alpha}}^+$, $u_i^\pm$ transforms through each other under $N = 2$ supersymmetry transformations. It allows to treat the set of these coordinates as an independent superspace which is called an analytic subspace [2]. The analytic subspace is just that appropriate manifold for formulating the $N = 2$ supersymmetric field theories.

The pure $N = 2$ super Yang-Mills models are described in harmonic superspace approach by the superfield $V^{++} = V^{++a}T^a$ where V^{++a} is analytic superfield (that is it defined on analytic subspace), T^a are the internal symmetry generators and the denotion $++$ means that this superfield V^{++} has $U(1)$-charge $+2$. The action for the superfield V^{++} is given as follows [2, 6]

$$S_{SYM}[V^{++}] = \frac{1}{g^2}\int d^{12}z \sum_{n=0}^{\infty} \frac{(-i)^n}{n} \int du_1 \ldots du_n \frac{\mathrm{tr} V^{++}(z, u_1) \ldots V^{++}(z, u_n)}{(u_1^+, u_2^+) \ldots (u_n^+, u_1^+)} \tag{1}$$

Here $z \equiv (x^m, \theta_\alpha^i, \bar{\theta}_{\dot{\alpha}}^i)$; $i = 1, 2$; $(u_1^+, u_2^+) = u_1^{+i} u_{2i}^+$ and g is a coupling. This action is invariant under the gauge transformations [2]

$$\delta V^{++} = -D^{++}\Lambda - i[V^{++}, \Lambda] \tag{2}$$

where Λ is analytic superfield parameter and the operator D^{++} was defined in ref.[2].

$N = 2$ matter hypermultiplets are described by the analytic superfields $q^+(x_A, \theta^+, \bar{\theta}^+, u^\pm)$ or $\omega(x_A, \theta^+, \bar{\theta}^+, u^\pm)$. The corresponding actions have the forms

$$S_q[\breve{q}^+, q^+] = \int d\zeta^{(-4)} du \; \breve{q}^+ \; \nabla^{++} q^+ \tag{3}$$

and

$$S_\omega[\omega] = \int d\zeta^{(-4)} du (\nabla^{++}\omega)(\nabla^{++}\omega) \tag{4}$$

with $\nabla^{++} = D^{++} + iV^{++}$ and $d\zeta^{(-4)}$ be analytic measure [2]. Action $S_{SYM} + S_q + S_\omega$ describes interacting system of super Yang-Mills fields and q^+ and ω hypermultiplets.

To construct effective action $\Gamma[V^{++}]$ depending on V^{++} we split the superfield V^{++} into background V^{++} and quantum v^{++} superfields, $V^{++} \to V^{++} + gv^{++}$. The gauge transformations (2) can be realized as background gauge transformations $\delta V^{++} = -\nabla^{++}\Lambda$, $\delta v^{++} = +i[\Lambda, v^{++}]$ and as quantum gauge transformations

$$\begin{aligned} \delta V^{++} &= 0 \\ \delta v^{++} &= -\frac{1}{g}\nabla^{++}\Lambda - i[v^{++}, \Lambda] \end{aligned} \tag{5}$$

where $\nabla^{++}\Lambda = D^{++}\Lambda + i[V^{++}, \Lambda]$. It is worth to point out here that the form of background - quantum splitting and corresponding background and quantum gauge transformations are absolutely analogous to the conventional Yang-Mills theory but not to $N=1$ super Yang-Mills theory (see f.e.[7])

To quantize a gauge theory within background field method one should fix only quantum gauge transformations (5). We introduce the gauge fixing functions in the form

$$\mathcal{F}^{(4)} = \nabla^{++}v^{++} \tag{6}$$

and apply Faddeev-Popov procedure. As a result we obtain effective action $\Gamma[V^{++}]$ in the form (see the details in ref.[4]).

$$e^{i\Gamma[V^{++}]} = e^{iS_{SYM}[V^{++}]}\int \mathcal{D}v^{++}\mathcal{D}b\mathcal{D}c\mathcal{D}\phi\mathcal{D}\breve{q}^{+}\mathcal{D}q^{+}\mathcal{D}\omega \mathrm{Det}^{1/2}(\widehat{\Box})e^{iS_{total}} \tag{7}$$

where

$$\begin{aligned} S_{total}[v^{++},b,c,\phi,\breve{q}^{+},q^{+},\omega,V^{++}] &= S_2[v^{++},b,c,\phi,\breve{q}^{+},q^{+},\omega,V^{++}]+\\ &+ S_{int}[v^{++},b,c,\breve{q}^{+},q^{+},\omega,V^{++}] \end{aligned} \tag{8}$$

Here S_2 plays a role of action of free theory

$$\begin{aligned} S_2[v^{++},b,c,\phi,\breve{q}^{+},q^{+},\omega,V^{++}] &= -\frac{1}{2}\int d\zeta^{(-4)}du\,\mathrm{tr}\,v^{++}\,\widehat{\Box}\,v^{++} - \\ -\int d\zeta^{(-4)}du\,\mathrm{tr}\,(\nabla^{++}b)(\nabla^{++}c) &- \frac{1}{2}\int d\zeta^{(-4)}du\,\mathrm{tr}\,(\nabla^{++}\phi)(\nabla^{++}\phi) + \\ +\int d\zeta^{(-4)}du\,\breve{q}^{+}\,\nabla^{++}q^{+} &+ \int d\zeta^{(-4)}du\,(\nabla^{++}\omega)(\nabla^{++}\omega) \end{aligned} \tag{9}$$

The action S_{int} describes the interactions

$$\begin{aligned} S_{int}[v^{++},b,c,\breve{q}^{+},q^{+},\omega,V^{++}] &= -\int d^{12}z\sum_{n=2}^{\infty}\frac{(-ig)^{n-2}}{n}\int du_1\ldots du_n\times \\ \times\frac{\mathrm{tr}\,v_{\tau}^{++}(z,u_1)\ldots v_{\tau}^{++}(z,u_n)}{(u_1^+,u_2^+)\ldots(u_n^+u_1^+)} &+ \int d\zeta^{(-4)}du\,\breve{q}^{+}\,V^{++}q^{+}+ \\ +\int d\zeta^{(-4)}du(\nabla^{++}\omega v^{++}\omega &+ v^{++}\omega\nabla^{++}\omega + (v^{++}\omega)(v^{++}\omega)) \\ v_{\tau}^{++} &= e^{-i\Omega}v^{++}e^{i\Omega} \end{aligned} \tag{10}$$

Here Ω is a background bridge superfield [2]. The operator $\widehat{\Box}=\Box+$ terms depending on V^{++} is defined in ref.[4]. The analytic superfields b and c are Faddeev-Popov ghosts, the real analytic superfield ϕ is third (or Nilsen-Kallosh) ghost.

The path integral (9) for effective action $\Gamma[V^{++}]$ has the form standard for quantum field theory. The free action S_2 (9) defines the propagators of pure super Yang-Mills field, matter fields and ghosts fields. The interaction S_{int} (11) defines the vertices. Eqs. (7-11) completely determine the structure of perturbation expansion for calculating the effective action $\Gamma[V^{++}]$ in a manifestly $N=2$ supersymmetric and gauge invariant form.

As in conventional field theory one can suggest that the effective action $\Gamma[V^{++}]$ is described in terms of effective Lagrangians

$$\Gamma[V^{++}] = \int d^4xd^4\theta d^4\bar{\theta}\mathcal{L}_{eff} + (\int d^4xd^4\theta\mathcal{L}_{eff}^{(c)} + c.c.) \tag{11}$$

where the $\mathcal{L}_{eff}$ and $\mathcal{L}^{(c)}_{eff}$ can be called a general effective Lagrangian and chiral effective Lagrangian respectively.

If the theory under consideration is quantized with background field method the effective action $\Gamma[V^{++}]$ will be gauge invariant under initial classical gauge transformations (background gauge transformations). In this case this effective action should be constructed only from strengths W and $\bar{W}$ and their covariant derivatives. Therefore the effective Lagrangians must have the following general structure

$$\begin{aligned} \mathcal{L}_{eff} &= \mathcal{H}(W,\bar{W}) + \text{terms depending on covariant derivatives of } W \text{ and } \bar{W} \\ \mathcal{L}^{(c)}_{eff} &= \mathcal{F}(W) + \text{terms depending on covariant derivatives of strengths} \\ &\quad \text{and preserving chirality} \end{aligned} \tag{12}$$

The term $\mathcal{F}(\mathcal{W})$ in chiral effective Lagrangian depending only on W is called a holomorphic effective Lagrangian. This term is leading in low-energy limit and describes vacuum structure theory. Namely holomorphic effective Lagrangian was a main object of Seiberg-Witten theory [1]. The term

$$\int d^4x d^4\theta d^4\bar{\theta}\mathcal{H}(W,\bar{W}) \tag{13}$$

defines a first non-leading correction to low-energy effective action and describes an effective low-energy dynamics.

The structure of effective action (11-13) is turned out to be analogous to structure of effective action depending on chiral and antichiral superfields in $N = 1$ case. To be more precise, the chiral effective potential [8, 5] in $N = 1$ case is analogous to holomorphic effective Lagrangian $\mathcal{F}(W)$. The first non-leading correction $\mathcal{H}(W,\bar{W})$ in $N = 2$ case is analogous to Kählerian effective potential in $N = 1$ case [9, 5].

The explicit calculations of $\mathcal{F}(W)$ and $\mathcal{H}(W,\bar{W})$ in one-loop approximation for hypermultiplets coupled to abelian gauge superfield have been given within harmonic superspace formulation in [10]. It has been shown that $\mathcal{F}(W)$ is obtained in the form analogous to Seiberg one for pure $N = 2$, $SU(2)$ super Yang-Mills model [11]. The $\mathcal{H}(W,\bar{W})$ was given in a form of a series in a power of $W\bar{W}$ where a first term proportional to $(W\bar{W})^2$ is $N = 2$ generalization of known Heisenberg-Euler effective Lagrangian.

A simple consequence of the background field formulation is that there are no quantum corrections to $\mathcal{H}(W,\bar{W})$ at two loops in the pure $N = 2$ super Yang-Mills theory. All two-loop supergraphs contributing to the effective action within background field method are given in Fig.1

Fig.1

Here the wavy line corresponds to the super Yang-Mills propagator and the dotted line to the ghost propagator. These propagators are defined by the action S_2 (9) and have the form

$$
\begin{aligned}
\langle v_\tau^{++}(1)v_\tau^{++}(2)\rangle &= -\frac{i}{\widehat{\Box}}\,(\overrightarrow{\mathcal{D}_1^+})^4\,\{\delta^{12}(z_1-z_2)\delta^{(-2,2)}(u_1,u_2)\} \\
\langle c_\tau(1)b_\tau(2)\rangle &= -\frac{i}{\widehat{\Box}}\,(\overrightarrow{\mathcal{D}_1^+})^4\left\{\delta^{12}(z_1-z_2)\frac{(u_1^-u_2^-)}{(u_1^+u_2^+)^3}\right\}(\overleftarrow{\mathcal{D}_2^+})^4
\end{aligned}
\tag{14}
$$

Here v_τ^{++}, c_τ, b_τ and the derivatives $\mathcal{D}^+$ are given in so called τ-frame [2, 5] and the distributions $\delta^{(-2,2)}(u_1,u_2)$, $(u_1^+u_2^+)^{-3}$ were introduced in refs.[12].

As we have noted in ref.[5], in order to get a non-zero result in two-loop supergraphs we should use twice the identity $\delta^8(\theta_1-\theta_2)(\mathcal{D}_1^+)^4(\mathcal{D}_2^+)^4\delta^8(\theta_1-\theta_2) = (u_1^+u_2^+)\delta^8(\theta_1-\theta_2)$ [12]. This implies that we should have 16 spinor covariant derivatives to reduce the θ-integrals over the full $N=2$ superspace to a single one. All these spinor derivatives come or from $(\overrightarrow{\mathcal{D}^+})^4$ in the propagators (14) or from expansion the operator $\widehat{\Box}^{-1}$ in a power series of the W and $\bar{W}$. After we use one $(\mathcal{D}^+)^4$-factor from the ghost propagator to restore the full superspace measure, we see the propagators of both gauge and ghost superfields have at most a single factor $(\mathcal{D}^+)^4$. It is evident that the number of these $(\mathcal{D}^+)$-factors is not sufficient to form al 16 $(\mathcal{D}^+)$-factors we need in two-loop supergraphs. As to a possible way to get extra $(\mathcal{D}^+)$-factors from $\widehat{\Box}^{-1}$ we observe that the spinor covariant derivatives enter the $\widehat{\Box}$ always multiplied by the derivatives of W and $\bar{W}$ (see explicit form in of $\widehat{\Box}$ in ref.[4]). If we omit these derivatives the operator $\widehat{\Box}$ takes the form $\widehat{\Box}= \mathcal{D}^m\mathcal{D}_m + \frac{1}{2}\{\bar{W},W\}$ and does not contain the spinor covariant derivatives. Therefore, the two-loop supergraphs given in Fig.1 do not contribute to the function $\mathcal{H}(W,\bar{W})$ in effective Lagrangian (13). It is worth to point out that this result is a simple consequence of the $N=2$ background field method and does not demand any direct calculations of the supergraphs. Moreover, above result will be true even if we take into account the two-lop matter contributions to $\Gamma[V^{++}]$. This is almost obvious since, after restoring the full superspace measure, the matter superfield propagator following from action S_2 (9) have effectively the same structure as the gauge and ghost superfield propagators.

The $N=2$ background field method leads to a simple and clear proof of the $N=2$ non-renormalization theorem. See for comparison a consideration of problem of divergences in conventional $N=2$ superspace in ref.[13]. First of all, acting the same way as in the case of $N=1$ non-renormalization theorem (see f.e.[7]) we can use the $(\mathcal{D}^+)$-factors in the propagators (14) and in the matter superfield propagators and restore the full superspace measure $d^4xd^4\theta d^4\bar{\theta}$ in all vertices of all supergraphs. Then, using the identity $\delta^8(\theta_1-\theta_2)(\mathcal{D}_1^+)^4(\mathcal{D}_2^+)^4\delta^8(\theta_1-\theta_2) = (u_1^+u_2^+)\delta^8(\theta_1-\theta_2)$, and making integration by part we can transform any supergraph contributing to the effective action to the form containing only a single integral over $d^8\theta$.

Let us estimate a superficial degree of divergence for the theory under consideration. We consider an arbitrary L-loop supergraph G with P propagators, N_{MAT} external matter legs and an any number of gauge superfield external legs. We denote by N_D the number of spinor covariant derivatives acting on the external legs as a result of integration by parts in the process of transformating the contributions to a single integral over $d^8\theta$. Taking into account the dimensions of the factors $\widehat{\Box}$, $\mathcal{D}^+$ and the loop integrals over momenta we immediately obtain

$$
\omega(G) = 4L - 2P + (2P - N_{MAT} - 4L) - \frac{1}{2}N_D = -N_{MAT} - \frac{1}{2}N_D \tag{15}
$$

See the details of deriving eq.(15) in ref.[5]. The eq.(15) shows that all supergraphs with external matter legs are automatically finite. As to supergraphs with pure gauge superfield legs, they will be finite only if some non-zero number of spinor covariant derivatives acts on the external legs. We will show that this is always the case beyond one loop.

Let us consider the supergraph contributions after restoring the full superspace measure at all vertices. Then we transform these contributions to τ-frame [2, 5]. The propagators of gauge superfield, ghost superfields and matter superfields contain the background field V^{++} only via the $\widehat{\Box}$ and $\mathcal{D}^{+}$-factors, that is, only via u-independent connections A_M [2]. But al connections A_M contain at least one spinor covariant derivative acting on background superfield V^{++} [2]. Therefore, each external leg must contain at least one spinor covariant derivative. Thus, the number N_D in eq.(15) must be greater than or equal to one. It means that $\omega(G) < 0$ and, hence, all supergraphs are ultravioletly finite beyond the one-loop level. As to one-loop contributions to effective action they are given in terms of functional determinants [4, 5] and demand a special and independent investigation.

Acknowledgements We are grateful to our co-authors E.I.Buchbinder and S.M. Kuzenko for collaboration and valuable discussions. We would like to thank E.A. Ivanov for critical remarks and discussions. The work of ILB was partially supported by the grants of RFBR, project 96-02-16017, by grant of RFBR-DFG, project 96-02-00180 and by grant of INTAS, INTAS 96-0308. BAO acknowledges the POE Contract OE-AC02-76-ER-03072 and the Alexander fon Humboldt Foundation for partial support.

References

[1] N.Seiberg, E.Witten, *Nucl. Phys.* **B426** (1994) 19; *Nucl. Phys.* **B430** (1994) 485

[2] A.Galperin, E.Ivanov, S.Kalitzin, V.Ogievetsky, E.Sokatchev, *Class. Quant. Grav.* **1** (1984) 469

[3] I.L. Buchbinder, S.D.Odintsov, I.L.Shapiro, *Effective Action in Quantum Gravity*, IOP Publ., Bristol and Philadelphia (1992)

[4] I.L. Buchbinder, E.I.Buchbinder, S.M.Kuzenko, B.A.Ovrut, *Phys. Lett.* **B417** (1998) 61

[5] I.L. Buchbinder, S.M.Kuzenko, B.A.Ovrut, hep-th/9710

[6] B.Zupnik, *Teor. Mat. Fis. (Theoretical and Mathematical Physics, in Russian)* **69** (1986) 207; *Phys. Lett.* **B183** (1987) 175

[7] I.L. Buchbinder, S.M.Kuzenko, *Ideas and Methods of Supersymmetry and Supergravity*, IOP Publ., Bristol and Philadelphia (1995)

[8] P.West, *Phys. Lett.* **B258** (1991) 375
I.Jack, D.R.J.Jones, P.West, *Phys. Lett.* **B258** (1991) 382
I.L. Buchbinder, S.M.Kuzenko, A.Yu.Petrov, *Phys. Lett.* **B321** (1994) 372

[9] I.L. Buchbinder, S.M.Kuzenko, J.V.Yarevskaya, *Nucl. Phys.* **B411** (1994) 665

[10] I.L. Buchbinder, E.I. Buchbinder, E.A.Ivanov, S.M.Kuzenko, B.A.Ovrut, *Phys. Lett.* **B412** (1997) 309

[11] N.Seiberg, *Phys. Lett.* **B206** (1988) 75

[12] A.Galperin, E.Ivanov, V.Ogievetsky, E.Sokatchev, *Class. Quant. Grav.* **2** (1985) 601; 617

[13] P.West, *Supersymmetry and Finiteness*, in Proceedings of the 1983 Shelter Island II Conference on Quantum Field Theory and Fundamental Problems in Physics, edited by R.Jackiw, N.Kuri, S.Weinberg and E.Witten (M.I.T. Press)
P.Howe, K.Stelle, P.West, *Phys. Lett.* **B124** (1983) 55
P.S.Howe, K.S.Stelle, P.K.Townsend, *Nucl. Phys.* **B236** (1984) 125

Massive Branes and Creation of Branes

Eric Bergshoeff

Institute for Theoretical Physics
University of Groningen, Nijenborgh 4
9747 AG Groningen
The Netherlands

Abstract: We investigate the effective worldvolume theories of p-branes and D-branes in a massive background given by (the bosonic sector of) 10-dimensional massive IIA supergravity ("massive branes"). As an application we discuss the consequences of our results for the anomalous creation of branes.

1 Introduction

In order to get a better understanding of the dynamics of branes in string theory, it is important to know precisely what the effective worldvolume theory is that describes the dynamics of these objects. This worldvolume theory contains a great deal of information. For instance, fields propagating on a brane's worldvolume describe the dynamics of the intersections with other branes ending on it [1, 2]. In particular, the Born-Infeld (BI) vector field present in the worldvolume of all D-p-branes describes the $U(1)$ field whose sources are open string endpoints. The anti-selfdual 2-form potential living on the M-5-brane worldvolume describes the dynamics of an M-5-brane intersecting with an M-2-brane over a 1-brane. The dynamics of these worldvolume fields has recently received some attention [3, 4, 5, 6].

At present the structure of the worldvolume theory is quite well understood for the case in which the brane propagates in a supergravity background without a cosmological constant, the so-called "massless supergravity theories". As is well known, in the case of IIA supergravity, an extension is possible with a non-zero cosmological constant proportional to m^2 with m a mass parameter [7]. Such backgrounds are essential for the existence of D-8-branes whose charge is proportional to m [8, 9]. In this talk I will discuss the worldvolume theory of p-branes and D-branes that propagate in such a massive background. I will call such branes "massive branes"[1]. The

[1] I reserve the name "massive branes" for branes that propagate in a massive IIA supergravity background as opposed to branes that propagate in a background with zero mass parameter. Of course, all branes are massive in the sense that their physical mass is nonzero.

results presented here are based upon work done in collaboration with Tomas Ortín and Yolanda Lozano [10]. In that paper one can also find a discussion of massive M-branes and the M-theoretic origin of the massive p-branes and D-branes discussed here.

2 Massive p-Branes and D-Branes

The basic branes of string theory are the fundamental string (F1), the solitonic 5-brane (NS5-brane), the Brinkmann wave (W), the Kaluza-Klein monopole (KK) and the D-p-branes ($0 \leq p \leq 9$). The first two objects are called p-branes and their charge is carried by a Neveu-Schwarz/Neveu-Schwarz (NS/NS) field. The Brinkmann wave and the Kaluza-Klein monopole are purely gravitational solutions[2]. Finally, the D-p-branes carry a Ramond-Ramond (R-R) charge.

For objects that carry a NS/NS charge there are three possibilities: the brane may move in a Heterotic, IIA or IIB supergravity background. Each of these three possibilities is described by a different worldvolume theory. We distinguish between these cases using an obvious notation, e.g. for the different NS5-branes

$$\text{NS5-brane} \rightarrow \begin{cases} \text{Heterotic NS5-brane} \\ \text{NS5A-brane} \\ \text{NS5B-brane} \end{cases}$$

On the other hand the D-p-branes carry a R-R charge and therefore always propagate in a IIA (p even) or IIB (p odd) background.

Before discussing the massive p-branes and D-branes, it is instructive to first remind the *massless* case. The (0+1)-dimensional worldvolume theory corresponding to the Brinkmann wave is given by a kinetic term corresponding to a massless particle. The difference between the worldvolume theories of the Heterotic wave, IIA-wave and IIB-wave is in the fermionic terms and in the coupling to the dilaton.

In the case of the fundamental string F1 there is no difference between the worldvolume theories of the Heterotic F1, the F1A and the F1B strings at the bosonic level. These actions are given by a Nambu-Goto kinetic term and a WZ term that describes the coupling of the brane to the NS/NS 2-form B. The dilaton does not appear in any of these two terms, since the tension of fundamental objects is independent of the string coupling constant g.

In the case of the NS5-brane the kinetic term is again of the Nambu-Goto form but now there is an extra dilaton factor $e^{-2\phi}$ in front of it showing that the NS5-brane is a solitonic object whose mass is proportional to $1/g^2$. Concerning the WZ term there is a distinction between the three different cases indicated in the equation above already at the bosonic level. The reason for this is that the NS5-brane couples to

[2]The eleven-dimensional Brinkmann wave (Kaluza-Klein monopole) becomes, upon dimensional reduction, a D-0-brane (D-6-brane).

the 6-form dual of the NS/NS 2-form B. The definition of this dual 6-form depends on the background involved because both the field strengths and the couplings of B in the action are different. There are therefore three different dual 6-forms denoted by $\tilde{B}_{\rm H}, \tilde{B}_{\rm IIA}$ and $\tilde{B}_{\rm IIB}$.

The kinetic term of the (5+1)-dimensional worldvolume theory corresponding to the KK-monopole has only been recently constructed [11]. It is given by a gauged sigma model involving an auxiliary worldvolume 1-form. The kinetic term carries a factor $e^{-2\phi}k^2$ with k^2 the length of the Killing vector squared. Therefore, the effective tension of this object is proportional to $1/g^2$ (and so it is solitonic) and to R^2, R being the radius of the dimension to be compactified. This behavior is characteristic of KK monopoles. There is also a difference in the bosonic worldvolume content of the Heterotic KK, KK10A and KK10B monopoles [12]. The WZ terms for these monopoles have not been constructed explicitly so far (see, however, [13]).

Finally, the worldvolume theory of the D-p-branes is given by a Dirac-Born-Infeld (DBI) kinetic term with a $e^{-\phi}$ dilaton coupling in front of it and a WZ term whose leading term is the R-R (p+1)-form field. The dilaton coupling shows that the mass of these objects is proportional to $1/g$.

We now discuss the massive extension of the above branes. In the case of the D-p-branes, the result has already been given in the literature [14, 15]. The basic change in the worldvolume theory is that an extra Chern-Simons term has to be added to the WZ-term for p even. For instance, for p=0 this extra term is given by

$$S_{\rm massive\ D0-brane} \sim \int d\xi\ m\ b_\xi\,, \tag{1}$$

where b_ξ describes the tension of a fundamental string F1. Next, it turns out that the worldvolume theory of the massive F1A-brane is identical to that of the massless F1A-brane. In the case of the NS5A-brane, however, there are striking differences. It turns out that the massive NS5A-brane has, among other things, an additional coupling to a worldvolume 6-form $c^{(6)}$, describing the tension of a D-6-brane, with the strength of the coupling proportional to m:

$$S_{\rm massive\ NS5-brane} \sim \int d^6\xi\ m\ \epsilon^{i_1\cdots i_6} c^{(6)}{}_{i_1\cdots i_6}\,. \tag{2}$$

The reason for the presence of the worldvolume 6-form $c^{(6)}$ in eq. (2) is that in order to build the WZ term one has to dualize the *massive* NS/NS target space 2-form field B along the lines recently discussed in [16]. An interesting feature is that, whereas in the usual formulation of IIA supergravity the R-R 1-form $C^{(1)}$ is a Stueckelberg field that gets "eaten up" by the NS/NS 2-form B which becomes massive:

$$\begin{cases} C^{(1)} & \rightarrow \quad \text{Stueckelberg field}\,, \\ \\ B & \rightarrow \quad \text{massive field}\,, \end{cases}$$

in the dual formulation the situation is reversed: the dual NS/NS 6-form $\tilde{B}_{\rm IIA}$ becomes a Stueckelberg field giving mass to the dual R-R 7-form $C^{(7)}$:

$$\begin{cases} C^{(7)} & \to \quad \text{massive field}\,, \\ \tilde{B}_{\text{IIA}} & \to \quad \text{Stueckelberg field}\,. \end{cases}$$

An immediate consequence of the above observation is that, since $\tilde{B}_{\text{IIA}}$ occurs as the leading term of the WZ term of the NS5-brane, we must introduce an independent auxiliary 6-form worldvolume field $c^{(6)}$, as given in eq. (2), in order to cancel the Stueckelberg transformations of $\tilde{B}_{\text{IIA}}$. The situation is similar to that of the massive D-0-brane, see eq. (1). In that case the Stueckelberg variation of the leading term $C^{(1)}$ in the WZ term is canceled by the auxiliary BI 1-form b_ξ that couples to the D-0-brane with a strength proportional to the mass parameter m.

3 Anomalous Creation of Branes

One interesting outcome of our work is that the massive D-0-brane (NS5-brane) has an additional coupling to a worldvolume 1-form b_ξ (6-form $c^{(6)}$) describing the tension of a F1 string (D-6-brane) with the strength of the coupling proportional to m (see eqs. (1) and (2), respectively). In [17, 18] it was pointed out that, for the massive D-0-brane, this new coupling has implications for the anomalous creation of branes [19, 20] (for a more recent discussion see [21]). To be precise, if a D-0-particle crosses a D-8-brane a stretched fundamental string is created in the single direction transverse to the D-8-brane. This process is, via duality, related to the creation of a stretched D3-brane if a D-5-brane crosses a NS5-brane [19]. In the latter case the intersecting configuration is given by

$$\begin{array}{ll|l} \text{D5}: & \times & \times\times---\times\times\times- \\ \text{NS5}: & \times & \times\times\times\times\times---- \\ \text{D3}: & \times & \times\times-------\times \end{array} \qquad (3)$$

where we have used the notation of [22] [3]. The intersecting configuration of [17, 18] is obtained by first applying T-duality in the directions 1 and 2, next applying an S-duality and, finally, applying a T-duality in the directions 6,7 and 8 [17, 18]:

$$\begin{array}{ll|l} \text{D0}: & \times & --------- \\ \text{D8}: & \times & \times\times\times\times\times\times\times\times- \\ \text{F1}: & \times & --------\times \end{array} \qquad (4)$$

One would expect that the new coupling we find for the massive NS5-brane has similar implications[4]. To be precise, one would expect that if a NS5-brane crosses a D-8-brane a D-6-brane stretched between them is created. This process would be the dual of the one considered in [17, 18]. Indeed, this is exactly the process which

[3] Each horizontal line indicates the 10 directions $0, 1, \cdots 9$ in spacetime. A $\times(-)$ means that the corresponding direction is in the worldvolume of (transverse to) the brane.

[4] We thank C. Bachas for an illuminating discussion, during the conference, on this point.

has been considered in [23] where it was used to construct $N = 1$ supersymmetric gauge theories in four dimensions with chiral matter[5].

The intersecting configuration corresponding to this process is obtained by applying T-duality in the directions 3,4 and 5 on the configuration (3) [23]:

$$\begin{array}{rc|cccccccccc} \mathrm{D8}: & \times & \times & \times & \times & \times & \times & \times & \times & \times & - \\ \mathrm{NS5}: & \times & \times & \times & \times & \times & \times & - & - & - & - \\ \mathrm{D6}: & \times & \times & \times & \times & \times & \times & - & - & - & \times \end{array} \qquad (5)$$

A special feature of this configuration is that when the NS5-brane passes through the D-8-brane it is completely embedded within the D-8-brane. After passing through the D-8-brane a D-6-brane is created that has five of its directions on the worldvolume of the D-8-brane and in the single remaining direction is stretched transverse to the D-8-brane.

Acknowledgements

This talk is based upon the work of [10] where more details can be found. I thank T. Ortín and Y. Lozano for the many insightful discussions I shared with them. This work is supported by the European Commission TMR program ERBFMRX-CT96-0045, in which I am associated to the university of Utrecht.

References

[1] P.K. Townsend, *Brane Surgery*, to be published in the Proceedings of the European Research Conference on Advanced Quantum Field Theory, La Londe-les-Maures, France, September 1996, *Nucl. Phys. Proc. Suppl.* **58** (1997) 163.

[2] A. Strominger, *Open p-Branes*, *Phys. Lett.* **B383** (1996) 44.

[3] C.G. Callan Jr. and J.M. Maldacena, *Brane Dynamics from the Born-Infeld Action*, Report PUPT-1718 and `hep-th/9708147`.

[4] P.S. Howe, N.D. Lambert and P.C. West, *The Selfdual String Soliton*, Report KCL-TH-97-51 and `hep-th/9709014`.
ibid., *The Threebrane Soliton of the M-Fivebrane*, Report KCL-TH-97-54 and `hep-th/9710033`.

[5] G.W. Gibbons, *Born-Infeld Particles and Dirichlet p-Branes*, `hep-th/9709027`.

[6] E. Bergshoeff, G. Papadopoulos and J.P. van der Schaar, *Domain Walls on the Brane*, `hep-th/9801158`.

[7] L.J. Romans, *Massive N=2a Supergravity in Ten Dimensions*, *Phys. Lett.* **169B** (1986) 374.

[5] We thank P. Townsend for pointing out reference [23] to us.

[8] J. Polchinski and E. Witten, *Evidence for Heterotic - Type I String Duality*, *Nucl. Phys.* **B460** (1996) 525.

[9] E. Bergshoeff, M. de Roo, M.B. Green, G. Papadopoulos and P.K. Townsend, *Duality of Type II 7-Branes and 8-Branes*, *Nucl. Phys.* **B470** (1996) 113.

[10] E. Bergshoeff, T. Ortín and Y. Lozano, *Massive Branes*, `hepth/9712115`, to appear in Nucl. Phys. B.

[11] E. Bergshoeff, B. Janssen and T. Ortín, *Kaluza-Klein Monopoles and Gauged Sigma-Models*, Report CERN-TH/97-125, UG-5/97 and *Phys. Lett.* **B410** (1997) 132.

[12] C.M. Hull, *Gravitational Duality, Branes and Charges*, Report QMW-97-19, NI97028-NQF and `hep-th/9705162`.

[13] E. Bergshoeff, E. Eyras, B. Janssen, Y. Lozano and T. Ortín, in preparation.

[14] E. Bergshoeff and M. de Roo, *D-branes and T-duality*, *Phys. Lett.* **B380** (1996) 265.

[15] M.B. Green, C.M. Hull and P.K. Townsend, *D-p-brane Wess-Zumino Actions, T-Duality and the Cosmological Constant*, *Phys. Lett.* **B382** (1996) 65.

[16] F. Quevedo and C.A. Trugenberger, Condensation of p-Branes and Generalized Higgs/Confinement Duality, *Int. J. Mod. Phys.* **A12** (1997) 1227-1236.
F. Quevedo, *Duality and Global Symmetries*, lectures given at the 33rd Karpacz Winter School of Theoretical Physics: *Duality - Strings and Fields*, Karpacz, Poland, 13-22 Feb 1997, Report IFTUNAM FT97-07 and `hep-th/9706210`.

[17] I.R. Klebanov, *D-branes and Creation of Strings*, `hep-th/9709160`.

[18] U. Danielsson and G. Ferretti, *Creation of Strings in D-particle Quantum mechanics*, `hep-th/9709171`.

[19] A. Hanany and E. Witten, *Type IIB Superstrings, BPS Monopoles and three-dimensional Gauge Dynamics*, *Nucl. Phys.* **B492** (1997) 152.

[20] C.P. Bachas, M.R. Douglas and M.B. Green, *Anomalous Creation of Branes*, `hep-th/9705074`.

[21] C.P. Bachas and M.B. Green, *(8,0) Quantum Mechanics and Symmetry Enhancement in Type I' Superstrings*, `hep-th/9712086`.

[22] E. Bergshoeff, B. Janssen, M. de Roo and J.P. van der Schaar, *Multiple intersections of D-branes and M-branes*, *Nucl. Phys.* **B494** (1997) 119.

[23] A. Hanany and A. Zaffaroni, *Chiral Symmetry from Type IIA Branes*, `hep-th/9706047`.

Solitonic Fivebrane and its Orientifold

Stefan Förste[1], Debashis Ghoshal[1] and Sudhakar Panda[2]

[1] Sektion Physik, Universität München
Theresientraße 37, 80333 München, Germany
[2] Mehta Research Institute of Mathematics & Mathematical Physics
Chhatnag Road, Jhusi, Allahabad 221506, India

Abstract: We construct an orientifold of the non-trivial superconformal field theory of a solitonic (Neveu-Schwarz) fivebrane.

The remarkable duality of supersymmetric gauge theories[1] for classical groups has been reformulated as a set of simple maneuvre of branes in type IIA string theory. The prototypical set-up has solitonic (Neveu-Schwarz) fivebranes, D-6-branes and (finite) D-4-branes suspended between the fivebranes[2]. The duality for the orthogonal or symplectic gauge groups requires also the introduction of an orientifold (four or six) plane[3]. This motivates us to study the problem of an explicit orientifold construction in the background of an NS fivebrane soliton. We will give a brief report of this construction based on Ref.[4], where further details can be found.

The transverse space to the fivebrane in ten dimensional space-time is four dimensional. The four dimensional metric corresponding to the solution to the low energy effective field theory equations of motion describing the fivebrane is asymptotically flat, and close to the 'position' of the fivebrane develops a 'throat'. In addition, there is a linearly varying dilaton field[5]:

$$\begin{aligned} ds^2 &= \eta_{\mu\nu}dx^\mu dx^\nu + e^{-2(\varphi-\varphi_0)}\left(dr^2 + r^2 d\Omega_3^2\right) \\ e^{-2(\varphi-\varphi_0)} &= 1 + \frac{k}{r^2} \\ H &= -2k d\Omega_3. \end{aligned} \tag{1}$$

In the throat region, the geometry is that of a cylinder with an S^3 base whose asymptotic size is proportional to the (integer) charge k of the fivebrane (see figure 1). There is an exact CFT description of this region[5] in terms of a $(4,4)$ supersymmetric $SU(2)_k$ WZNW model times a scalar field with a background charge (Feigin-Fuchs model)[6]. The background charge is tuned to k such that the total central charge is that of a four dimensional flat space.

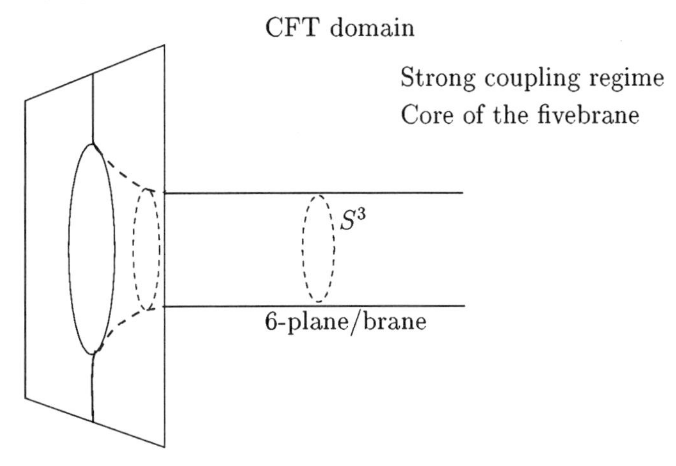

Figure 1: The geometry of the NS-5-brane orientifold.

The bosonic part of the $SU(2)_k$ WZNW model action is

$$S = \frac{k}{8\pi}\int_{\partial\mathcal{B}} \mathrm{Tr}(\partial g^{-1}\bar{\partial} g) - \frac{ik}{12\pi}\int_{\mathcal{B}} \mathrm{Tr}(g^{-1}dg)^3 \tag{2}$$

where $\mathcal{B}$ is a three manifold with the string worldsheet as its boundary. Apart from the continuous Kac-Moody symmetries generated by the currents $J(z) = (\partial g)g^{-1}$ and $\bar{J}(\bar{z}) = g^{-1}\bar{\partial} g$, this action has a discrete symmetry whose action is the worldsheet parity transformation Ω combined with a 'reflection' R taking the group element to its inverse:

$$\Omega R : \begin{array}{l} g(z,\bar{z}) \rightarrow g^{-1}(\bar{z}, z) \\ X(z,\bar{z}) \rightarrow X(\bar{z}, z) \end{array} \tag{3}$$

where $X(z,\bar{z})$ stands for the other bosons. The first term in the WZW action (2) is invariant separately under the action of Ω and R whereas the Wess-Zumino term, which describes the coupling to the NS-NS B-field, is invariant only under the combined action. Under this symmetry the Kac-Moody currents are interchanged $J(z) \leftrightarrow -\bar{J}(\bar{z})$.

The fixed points of the transformation (3) are the two polar points $g = \pm 1$ on S^3. The orientifold planes are therefore extended along the fivebrane times the Feigin-Fuchs direction. The O-6-planes are magnetically charged under the RR vector potential and may lead to the presence of D-6-branes. Clearly we are dealing with a type IIA theory. Taking some of the longitudinal directions of the fivebrane compact and performing T-duality on them, one can go back and forth from IIA to IIB theory.

The WZNW action admits another discrete symmetry[7][8], namely

$$\Omega R' : g(z,\bar{z}) \rightarrow -g^{-1}(\bar{z}, z). \tag{4}$$

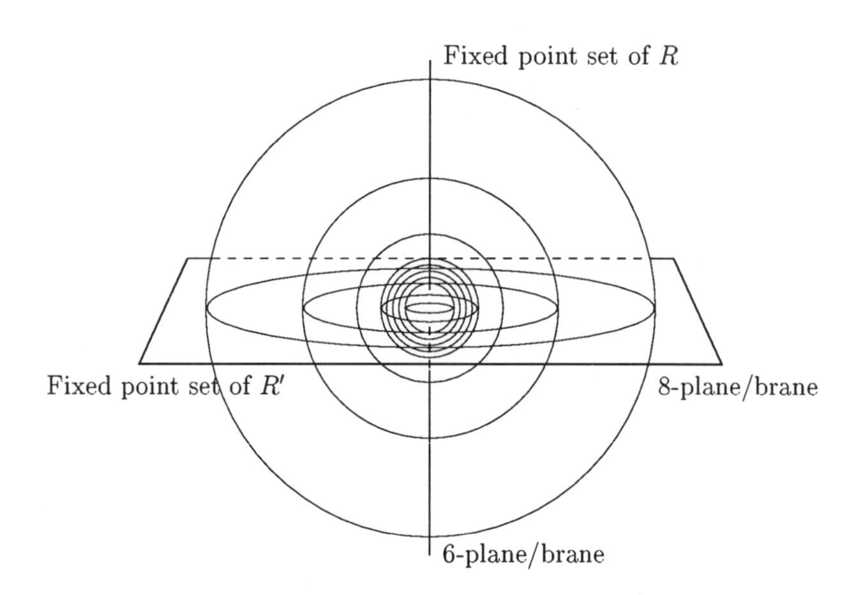

Figure 2: Fixed point sets of R and R'.

which results in a fixed S^2 surface at the equator of S^3. This would lead to orientifold 8-planes and D-8-branes, but we will not consider the details of this case. (Figure 2 gives a schematic picture of the fixed point sets of the two discrete symmetries.)

The orientifold construction we carry out in detail is the quotient of the SCFT of the fivebrane by the symmetry (3). Thus this is a generalization of the prescription of Ref.[9] to a non-trivial conformal field theory. (For earlier work on orientifolds, see [10].)

The projection by ΩR requires a truncation to states invariant under it, but does not give rise to any twisted sector. However one needs to include open strings which may have NN, ND, DN or DD boundary conditions. For open strings the mode expansion of a free boson is of the form (here we omit the zero mode part)

$$i\partial_\tau X = \sum_m \alpha_m \left(e^{im\sigma} \pm e^{-im\sigma}\right), \tag{5}$$

where + (respectively −) refers to NN (DD) boundary conditions for integer moded oscillators and to ND (DN) for half-integrally moded ones. For the WZNW part of the SCFT one cannot write down an action for worldsheets with boundaries. However the role of the $U(1)$ current (5) is played by the Kac-Moody currents. We will therefore take the intuitive approach of defining the mode expansion of the Kac-Moody current in the open string sector in analogy to (5) as

$$J(\sigma, 0) = \sum_m J_m \left(e^{im\sigma} \pm e^{-im\sigma}\right). \tag{6}$$

The fermions are free in all the sectors of SCFT. Their mode expansion depends on the boundary condition in the usual way[9].

Following [9], we now calculate the one-loop amplitude which gets contributions from the torus, Klein bottle, Möbius strip and cylinder diagrams. The last three diagrams are potentially divergent, while modular invariance of the torus eliminates all IR divergence. The one-loop amplitude is $\int_0^\infty \frac{dt}{t}\mathrm{Tr}\left(\Omega R e^{-2\pi t\mathcal{H}}\right)$, where t refers to the real modulus of the Klein bottle, Möbius strip or the cylinder. In flat (toroidal) background the trace takes schematically the following form

$$\begin{pmatrix}\text{integral over}\\ \text{momenta in}\\ \text{noncompact dim}\end{pmatrix}\times\begin{pmatrix}\text{weighted sum}\\ \text{over oscillator}\\ \text{states}\end{pmatrix}\times\begin{pmatrix}\text{sum over momenta}\\ \text{and windings in}\\ \text{compact dimensions}\end{pmatrix}.$$

In the background of the fivebrane (in the CFT regime) the above modifies to

$$\begin{pmatrix}\text{integral over momenta}\\ \text{in longitudinal}\\ \text{\& FF directions}\end{pmatrix}\times\begin{pmatrix}\text{weighted sum}\\ \text{over longitudinal}\\ \text{\& FF oscillators}\end{pmatrix}\times\sum_{\text{rep}}\begin{pmatrix}\text{function related to}\\ \text{the character of}\\ \text{WZNW model}\end{pmatrix}.$$

The character of the spin j representation of $SU(2)_k$ is

$$\chi_j(q) = \frac{q^{(2j+1)^2/4(k+2)}}{\eta^3(q)} \sum_{n=-\infty}^{\infty} (2j+1+2n(k+2))\, q^{n(2j+1+(k+2)n)} \tag{7}$$

with j taking values $0, \frac{1}{2}, \ldots, \frac{k}{2}$, and $\eta(q)$ is the Dedekind function. The functions related to the characters that appear in the one-loop amplitude depend on the specific diagram[4].

The leading IR behaviour of the amplitude turns out to be (here we restrict to the $k = 1$ case):

$$\int_{l\to\infty} dl \left(64 - 8\sum_{I=1,2}\mathrm{Tr}(\gamma_{I,\Omega R}^{-1}\gamma_{I,\Omega R}^{T}) + \frac{1}{2}\sum_{I=1,2}(\mathrm{Tr}\gamma_{I,\mathbf{1}})^2\right) e^{-\pi l/3}, \tag{8}$$

where the matrices γ are representations of the orientifold group elements on the Chan-Paton labels. Surprisingly, unlike the toroidal background there is no divergence!

The absence of any divergence can be traced to the constant curvature of S^3 which provides an IR regulator. The curvature induces a linear dilaton which causes mass-shift in the spectrum. This fact is familiar in the context of non-critical string theories[11]. The wave function of a physical state is the string vertex operator times the inverse string coupling. For a linear dilaton $\varphi = Q\cdot X$ the wave function thus looks like

$$\Psi = \zeta_{\mu_1\ldots\nu_n}\partial X^{\mu_1}\ldots X^{\mu_n}\bar{\partial}X^{\nu_1}\ldots\bar{\partial}X^{\nu_n}e^{i(k+iQ)\cdot X}. \tag{9}$$

Requiring the vertex operator to be of conformal dimension one provides the mass-shell condition, (after GSO projection $n \geq 1$),

$$m^2 = -(k+iQ)^2 = 2n - 2 + Q^2, \tag{10}$$

implying that for real non-vanishing Q there is no massless state in the spectrum. In the asymptotically flat region of the fivebrane background the dilaton assumes a constant value and there is no shift in the mass.

In the light of the above discussion, assuming that our calculation can be extrapolated to infinity by $e^{-\pi l/3} \to 1$ in (8) there is an IR divergence in the amplitude. This 'divergence' can be cancelled by demanding that at each of the two fixed planes there are four physical D-6-branes and their images leading to an $SO(8) \times SO(8)$ gauge symmetry. These D-branes are not allowed to move away from the fixed planes. Off the fixed points the branes are not only unable to cancel the 'divergence' but also break supersymmetry.

In the above we have provided an interpretation of the tadpole condition in orientifolding a non-trivial conformal field theory. However, this point of view is open to criticism. One may argue that since the transverse space is non-compact, no tadpole cancellation condition arises. But one should be careful since in the throat region the space *is* compact. It is not clear that a D-brane at infinity (not entering the throat) is able to cancel charges of branes or planes extending along the throat. A related point is the fact that translation invariance on the worldvolume of the branes is broken. Clearly, it will be desirable to construct orientifolds of other conformal field theories to shed more light on this. On this subject we should also mention the approach via the construction of boundary and crosscap states in conformal field theory (see [12] and the contributions of J. Fuchs and C. Schweigert to this proceedings).

In summary we have constructed an orientifold in the CFT regime of the fivebrane. Unlike in flat (toroidal) background, the one loop amplitude does not have any IR divergence. We trace this to the nontrivial dilaton which causes a mass-shift in the spectrum. We then argue that extrapolation to the asymptotically flat region leads to a divergence, which may be cancelled by the introduction of non-dynamical D-branes. Are these the extra D-branes at the core of the fivebrane[3] needed for duality in gauge theories? To answer this question, our analysis has to be extended to the strong coupling region. What is called for is an M-theoretic description of our configuration of branes and planes. Individually strong coupling limits of D-6-branes and planes in toroidal background[13], and D-6-branes in fivebrane background[14] are known. While it is not clear at present how this can be implemented, a combination of these two ingredients will perhaps give the desired configuration in M-theory.

Acknowledgement: The work presented here was supported by the German-Israeli Foundation for Scientific Research, the Alexander von Humboldt Foundation and by TMR program ERBFMX-CT96-0045. S.F. acknowledges discussions with Jürgen Fuchs and Christoph Schweigert and thanks the organizers for providing a stimulating atmosphere. We are grateful to Ashoke Sen and Stefan Theisen for helpful discussions.

References

[1] K. Intriligator and N. Seiberg, *Nucl. Phys. Proc. Suppl.* **45BC** (1996) 1, hep-th/9509066.

[2] A. Giveon, S. Elitzur and D. Kutasov, *Phys. Lett.* **B400** (1997) 269, hep-th/9602014.

[3] N. Evans, C. Johnson and A. Shapere, *Nucl. Phys.* **B505** (1997) 251, hep-th/9703210.

[4] S. Förste, D. Ghoshal and S. Panda, *Phys. Lett.* **B411** (1997) 46, hep-th/9706057.

[5] C. Callan, J. Harvey and A. Strominger, *Nucl. Phys.* **B359** (1991) 611.

[6] A. Sevrin, W. Troost and A. Van Proeyen, *Phys. Lett.* **B208** (1988) 447.

[7] P. Hořava, *Phys. Lett.* **B289** (1991) 293, hep-th/9203031.

[8] C. Johnson, *Phys. Rev.* **D56** (1997) 5160, hep-th/9705148.

[9] E. Gimon and J. Polchinski, *Phys. Rev.* **D54** (1996) 1667, hep-th/9601038.

[10] A. Sagnotti, in Proceedings of the 1987 Cargèse Summer Institute, Pergammon Press (1988);
G. Pradisi and A. Sagnotti, *Phys. Lett.* **B216** (1989) 59;
J. Govaerts, *Phys. Lett.* **B220** (1989) 77;
P. Hořava, *Nucl. Phys.* **B327** (1989) 461.

[11] P. Di Francesco and D. Kutasov, *Nucl. Phys.* **B375** (1992) 119, hep-th/9109005.

[12] J. Fuchs and C. Schweigert, hep-th/9712257, and references therein.

[13] A. Sen, *JHEP* **09** (1997) 001, hep-th/9707123.

[14] A. Tseytlin, *Nucl. Phys.* **B475** (1996) 149, hep-th/9604035;
M. Costa, *Nucl. Phys.* **B490** (1997) 202, hep-th/9609181.

A superspace approach to branes and supergravity

Bengt E.W. Nilsson

Institute of Theoretical Physics
Göteborg University and Chalmers University of Technology
S-412 96 Göteborg, Sweden

Abstract: Recent developments in string and M theory rely heavily on supersymmetry suggesting that a revival of superspace techniques in ten and eleven dimensions may be advantageous. Here we discuss three topics of current interest where superspace is already playing an important role and where an improved understanding of superspace might provide additional insight into the issues involved.

1 Introduction

It is well-known that besides the string (the $p = 1$ brane) itself also branes of other dimensionality ($d = p + 1$) play a central role in the non-perturbative structure of the theory. Using Dp-branes and T5-branes (having respectively vector and second rank antisymmetric tensor potentials on the world surface) non-perturbative relations between all string theories can be established as well as a connection with 11d supergravity. This hints at the existence of an underlying more profound formulation of the whole theory. Some aspects of this so called M theory are captured by M(atrix) theory which has its origin in D-particle (D0-brane) physics, but have many features in common with the 11d membrane (see e.g. the talk by B. de Wit).

The three topics discussed below are related to the role superspace is playing in this context. We start by discussing one of the direct implications of formulating the fundamental theory in terms of M(atrix) theory, namely the presence of higher order corrections (e.g. R^n) in 11d supergravity (for a recent review see [1]). Here we will have reason to recall certain facts in 10d supergravity established in the 1980's [2, 3, 4]. Then we review some recent results concerning the realization of κ-symmetry for various branes focusing on the 11d membrane ($p = 2$), the D3-brane in 10d type IIB supergravity, and the T5-brane in 11d superspace. In the latter case we present a new action [5] which avoids some of the problematic features of the lagrangian constructed previously in [6]. As a final topic we consider the generalization of this setup, based on bosonic world sheets embedded into target superspaces, to a situation where both target and the embedded world surface are superspaces [7, 8].

In [8] the relation between non-linearities of the tensor dynamics on the world sheet and non-linearly realized (super)symmetries is analyzed. This paper also contains a detailed account of general embeddings plus some new results for T5-branes in 7d.

2 Superspace and higher order corrections in 11d

One property of M theory that has been highlighted by the recent developments in M(atrix) theory is the higher order corrections in terms of e.g. the curvature tensor that must occur in the low energy supergravity lagrangian in 11d. Since 11d supersymmetry is tremendously restrictive (there is only one multiplet whose spin content does not exceed spin two, a cosmological term is not possible [9] , etc) but also very messy to deal with in terms of component fields, it might be worth the effort to develop further the superspace techniques that were introduced in the 1970's. The heart of the matter is the question of how to realize supersymmetry off-shell. This is fairly well understood in the case of $N = 1$ supersymmetry in 10d [2, 3, 4], a subject that we will therefore have reason to come back to.

The field content of 11d supergravity is $e_m{}^a$, ψ_m^α, c_{mnp} where $m, n, ..$ are 11d vector indices and $\alpha = 1, .., 32$ is a Majorana spinor index. These fields describe 128 bosonic and 128 fermionic degrees of freedom. The field equations are obtained from the corresponding superfields and their super-Bianchi identities. To this end we combine $e_m{}^a$ and ψ_m^α into the supervielbein $E_M{}^A$ where the superworld index $M = (m, \mu)$ and the supertangent space index $A = (a, \alpha)$. Normally one introduces also a superfield C_{MNP} with c_{mnp} as its first component. However, it was recently clarified [10] that this is not necessary, and we will refrain from doing so. Then the field strengths are just the supercurvature $R_A{}^B$ and the supertorsion $T^A = DE^A = dE^A - E^B \omega_B{}^A$. Their Bianchi identities read $DT^A = E^B R_B{}^A$ and $DR_A{}^B = 0$ but in fact it is only necessary to consider the first identity since the second one is automatically satisfied. The meaning of "solving the Bianchi identities" is as follows. By subjecting the torsion components ($T_{\alpha\beta}{}^c$, etc) to certain constraints the Bianchi identities cease to be identities and become equivalent to the field equations. One should remember however that a given set of gauge fields can often be given a variety of different kinds of dynamics, related either to different sets of constraints or to differences in the Bianchi identities. As shown recently by Howe [10] (on-shell) 11d supergravity is obtained from the *single* constraint (Γ^c is an 11d Dirac matrix)

$$T_{\alpha\beta}{}^c = 2i(\Gamma^c)_{\alpha\beta}$$

which is invariant under super-Weyl rescalings. Furthermore, no off-shell formulation of this theory is known.

This situation should be compared to what is known for $N = 1$ supergravity in 10d. When the Bianchi identities are solved on-shell [11] one finds that all physical fields appear at different θ levels in a scalar superfield denoted $\Phi(Z)$, where $Z = (x, \theta)$. Note that 10d supergravity contains a scalar and a spinor, apart from

$e_m{}^a$, ψ_m^α, B_{mn}. In contrast to 11d, one must in this case constrain several torsion components. This theory can be coupled to superYang-Mills using the same torsion constraints [12] and the Bianchi identity $DH = trF^2$ for the three-form $H = dB$. In fact, as proven in [13] even the R^2 term needed for anomaly cancellation can be dealt with without altering the constraints.

However, also R^4 and higher terms are present in the field theory of the low energy 10d superstring. In general such terms cannot be incorporated into the superspace Bianchi identities if the on-shell constraints are used. Fortunately, in this case it is known how to proceed since the off-shell field content is known. It consists of a superconformal gravity multiplet and an unconstrained scalar auxiliary superfield w [2]. To account for this new superfield the constraints must be modified to [3]

$$T_{\alpha\beta}{}^c = 2i(\Gamma^c)_{\alpha\beta} + 2i(\Gamma^{c_1\ldots c_5})_{\alpha\beta}X^c{}_{c_1\ldots c_5}$$

where X is in the representation 1050^+ of $so(1,9)$ appearing at level θ^4 in w.

All higher order corrections, like R^4, that can occur must be compatible with supersymmetry and fit somewhere in the solution of the Bianchi identities that follow from the off-shell constraints above. E.g. the R^4 term related by supersymmetry (see [14]) to the anomaly term BX_8 can be added as follows [4]:

$$S^{D=10,N=1} = -\frac{1}{(\kappa_{10})^2}\int d^{10}x d^{16}\theta E\Phi(w+c)$$

where c is a constant proportional to $\zeta(3)$, and where the w term is the kinetic one and the c term is the supersymmetrization of R^4. E is the superspace measure.

Turning to 11d the situation changes dramatically since it is not known how to solve the Bianchi identities off-shell or how to write down an off-shell action in components. This makes it very hard to address questions in 11d supergravity concerning the higher order corrections. We will here introduce the equivalent of w in 10d into the 11d supertorsion by means of the relaxed constraint

$$T_{\alpha\beta}{}^c = 2i\Gamma^c{}_{\alpha\beta} + 2i\Gamma^{d_1d_2}{}_{\alpha\beta}X^c{}_{d_1d_2} + 2i\Gamma^{d_1\ldots d_5}{}_{\alpha\beta}X^c{}_{d_1\ldots d_5}$$

where the tensors in the last two terms are in the representations 429 and 4290 which appear at level θ^4 in an unconstrained 11d scalar superfield. A preliminary analyzes of the Bianchi identities indicates an "off-shell situation" where new terms appear in the torsion which could account for the higher order corrections. In particular terms generated by anomalies and the presence of branes should be investigated. Note that the T5-brane produces in the supersymmetry algebra a five-form central charge, a fact that should be compared to the extra terms in the torsion.

Further studies will hopefully tell if these techniques can be utilized in the endeavour to extract an 11d supergravity theory from M(atrix) theory or perhaps directly from the 11d branes.

3 κ symmetric branes as bosonic surfaces

The relevant branes in 11d are the membrane and the T5-brane, while in 10d type II theories there are also Dp-branes intermediate between p-branes and T-branes. As we will see below, for branes with vector or tensor fields propagating on the world surface κ-symmetry is technically more complicated than for ordinary p-branes.

Let us as an example of an ordinary p-brane consider the 11d membrane [15]. After the elimination (see [16]) of the independent world sheet metric by means of its algebraic field equation the action reads:

$$S_3 = \int d^3\xi[-\sqrt{-detg} - \varepsilon^{ijk}B_{ijk}]$$

where g_{ij} is the pull-back of the target space metric, i.e. $g_{ij} = \Pi_i^a\Pi_j^b\eta_{ab}$, and the ξ^i's are three bosonic coordinates on the world sheet. The background superfields $E_M{}^A$ and B_{MNP} of 11d supergravity enter via the pull-backs $\Pi_i^A = \partial_i Z^M E_M{}^A$ and $B_{ijk} = \Pi_i^A\Pi_j^B\Pi_k^C B_{CBA}$. In order for this action to be supersymmetric the number of bosonic (here 11-3=8) and fermionic ($32\times\frac{1}{2}$) world sheet on-shell degrees of freedom must match. This requires the presence of the local fermionic κ-symmetry giving another factor of $\frac{1}{2}$ in the fermionic count. Its existence relies on the possibility to construct a projection operator $\frac{1}{2}(1+\Gamma)$ with $\Gamma^2 = 1$. Here $\Gamma = \frac{1}{6\sqrt{-g}}\epsilon^{ijk}\Gamma_{ijk}$ where Γ_{ijk} is the pull-back of $\Gamma_{abc} = \Gamma_{[a}\Gamma_b\Gamma_{c]}$.

This structure can be found also in the case of Dp-branes [17], but is now more involved due to the presence of the field strength F_{ij}. E.g. the D3-brane in the 10d type IIB theory has an action that reads

$$S_{D3} = -\int d^4\xi\sqrt{-det(g + e^{-\frac{\phi}{2}}\mathcal{F})} + \int e^{\mathcal{F}}C$$

In this case there are, apart from the dilaton ϕ, two kinds of background potentials B and C, coming from the NS-NS and the R-R sector, respectively. In S_{D3}, $\mathcal{F}_{ij} = F_{ij} - B_{ij}$ where B_{ij} is the pullback of B_{MN} in the 10d IIB target space theory. The last term in the action is constructed as a formal sum of forms of different rank and the integral is supposed to pick up only the four-form in this case. That the action is κ-symmetric can then be shown using the Γ matrix ($\Gamma^2 = 1$) [17]

$$\Gamma = \frac{\epsilon^{ijkl}}{\sqrt{-det(g+\mathcal{F})}}(\frac{1}{24}\Gamma_{ijkl}I - \frac{1}{4}\mathcal{F}_{ij}\Gamma_{kl}J + \frac{1}{8}\mathcal{F}_{ij}\mathcal{F}_{kl})$$

The $SL(2;Z)$ symmetry of the IIB theory mixes the B and C potentials and indeed a more symmetric version of the action exists [18]. In that version all background potentials have associated world sheet field strengths to which they couple as $F - B$.

The third case to be discussed here is the T5-brane in 11d. This brane has an additional complication in that the three-form field strength is self-dual. Finding a covariant action for such a field has been a long-standing problem. A first attempt at a solution was given recently by Bandos et al in [6] and involves an auxiliary scalar field a entering the lagrangian through a factor $\frac{1}{(\partial a)^2}$. Objections against using such an action in quantum calculations have been formulated (see also [19]). Another action that does not make use of such a scalar was subsequently presented in [5]. Although the problems associated with the scalar are gone, the other objections in

[19] probably remain. Nevertheless, since this action exhibits some new features it might be of some interest. The action is ($\mathcal{F}$ is an independent six-form field strength)

$$S_{T5} = \int d^6\xi\sqrt{-g}\lambda(1 + \frac{1}{12}F_{ijk}F^{ijk} - \frac{1}{24}k_{ij}k^{ij} + \frac{1}{72}(trk)^2 - (*\mathcal{F})^2)$$

where $k_{ij} = \frac{1}{2}F_i{}^{jk}F_{jkl}$. Note that, as explained in [5], the self-duality relation

$$-(*\mathcal{F}) * F_{ijk} = F_{ijk} - \frac{1}{2}k_{[i}{}^{l}F_{jk]l} + \frac{1}{6}(trk)F_{ijk}$$

does not arise as a field equation but as a result of demanding κ-symmetry, and must not be inserted into the action.

4 Superworld sheets in target superspace

It is possible to reformulate the brane dynamics, following from actions of the kinds described in the previous section, in terms of world sheet superfields. This can be done by embedding superworld sheets into target superspace. Besides making the world sheet supersymmetry manifest this has the further advantages of explaining the origin of κ-symmetry and the projection matrix [20] as well as of providing the connection between superembeddings on one hand, and Goldstone fermions and non-linearly realized supersymmetries on the supersheets on the other hand [21, 8].

In [21] Bagger and Galperin showed how the 4d Born-Infeld action for an abelian gauge field can be obtained by embedding a (4d,N=1) superspace into a (4d,N=2) one. The broken fermionic translations turn into non-linear supersymmetry transformations on the Goldstone fermions arising from half of the fermionic coordinates that are turned into dependent variables. The non-linearities of the Born-Infeld action are then seen to be a consequence of demanding consistency with the extra non-linear supersymmetries. In [8] this programme was taken over to the T5-brane in 7d superspace. This led to an equation for one of the supertorsion components whose rather complicated solution indicates that this way of analyzing this system is not the most efficient one. Fortunately once it is realized that this form of the torsion equation can also be obtained in the much more general formalism known as the embedding formalism developed in [7] these problems can be circumvented.

The central equation is the torsion pullback equation [7, 8]

$$\mathcal{D}_A\mathcal{E}_B{}^{\underline{C}} - (-1)^{AB}\mathcal{D}_B\mathcal{E}_A{}^{\underline{C}} + T_{AB}{}^{C}\mathcal{E}_C{}^{\underline{C}} = (-1)^{A(A+\underline{B})}\mathcal{E}_B{}^{\underline{B}}\mathcal{E}_A{}^{\underline{A}}T_{\underline{AB}}{}^{\underline{C}}$$

where underlined indices refer to target superspace while the other indices are connected with the superworld sheet. As shown in [20, 8] inserting constraints on the torsion components turns this equation into the equations of motion for the world sheet fields. In particular, the highly non-linear dynamics of the T5-branes in 11d [20] and in 7d [8] can be obtained this way.

Acknowledgement: I wish to thank my coauthors on refs. [4, 5, 8, 12, 16, 17] for very nice and fruitful collaborations.

References

[1] M.B. Green, hep-th 9712195

[2] P. Howe, H. Nicolai and A. Van Proeyen, *Phys. Lett.* **112B** (1982) 446

[3] B.E.W. Nilsson, *Phys. Lett.* **175B** (1986) 319
P.S. Howe and A. Umerski, *Phys. Lett.* **177B** (1986) 163

[4] B.E.W. Nilsson and A. Tollstén, *Phys. Lett.* **181B** (1986) 63

[5] M. Cederwall, B.E.W. Nilsson and P. Sundell, hep-th 9712059

[6] I. Bandos, K. Lechner, A. Nurmagambetov, P. Pasti, D. Sorokin and M. Tonin, *Phys. Rev. Lett.* **78** (1997) 4332, hep-th 9701037

[7] P.S. Howe and E. Sezgin, *Phys. Lett.* **B390** (1997) 133, hep-th 9607227
P.S. Howe and E. Sezgin, *Phys. Lett.* **B394** (1997) 62, hep-th 9611008

[8] T. Adawi, M. Cederwall, U. Gran, M. Holm and B.E.W. Nilsson, hep-th 9711203

[9] K. Bautier, S. Deser, M. Henneaux and D. Seminara, *Phys. Lett.* **B406** (1997) 49, hep-th 9704131

[10] P.S. Howe, hep-th 9707184

[11] B.E.W. Nilsson, *Nucl. Phys.* **B188** (1981) 176

[12] B.E.W. Nilsson and A. Tollstén, *Phys. Lett.* **171B** (1986) 212

[13] L. Bonora, P. Pasti and M. Tonin, *Phys. Lett.* **B188** (1987) 335

[14] M. de Roo, H. Suelmann and A. Wiedemann, *Phys. Lett.* **B280** (1992) 39

[15] E. Bergshoeff, E. Sezgin and P.K. Townsend, *Ann. of Phys.* **185** (1987) 330

[16] M. Cederwall, A. von Gussich, A. Mikovic, B.E.W. Nilsson and A. Westerberg, *Phys. Lett.* **330B** (1997) 148, hep-th 9606173

[17] M. Cederwall, A. von Gussich, B.E.W. Nilsson and A. Westerberg, *Nucl. Phys.* **B490** (1997) 163, hep-th 9610148
M. Cederwall, A. von Gussich, B.E.W. Nilsson, P. Sundell and A. Westerberg, *Nucl. Phys.* **B490** (1997) 179, hep-th 9611159

[18] M. Cederwall and P.K. Townsend, hep-th 9709002

[19] E. Witten, hep-th 9610234

[20] P.S. Howe, E. Sezgin and P. West, *Phys. Lett.* **B399** (1997) 49, hep-th 9702008

[21] J. Bagger and A. Galperin, *Phys. Rev.* **D55** (1997) 1091, hep-th 9608177

On the self-dual geometry of N=2 strings

Chandrashekar Devchand [1] , **Olaf Lechtenfeld** [2]

[1] Max-Planck-Institut für Mathematik in den Naturwissenschaften
Inselstraße 22-26, D-04103 Leipzig
[2] Institut für Theoretische Physik, Universität Hannover
Appelstraße 2, D-30167 Hannover

Abstract: We discuss the precise relation of the open N=2 string to a self-dual Yang-Mills (SDYM) system in 2+2 dimensions. In particular, we review the description of the string target space action in terms of SDYM in a "picture hyperspace" parametrised by the standard vectorial $\mathbb{R}^{2,2}$ coordinate together with a commuting spinor of $SO(2,2)$. The component form contains an infinite tower of prepotentials coupled to the one representing the SDYM degree of freedom. The truncation to five fields yields a novel one-loop exact lagrangean field theory.

1. Introduction The relation to self-dual Yang-Mills (SDYM) of the critical open N=2 string has recently been elaborated by us [1] in view of the particular picture degeneracy and global $SO(2,2)$ properties of the physical spectrum of string states. There has been much discussion in the literature of this relationship since Ooguri and Vafa [2] first mooted the idea that the self-duality equations,

$$F_{\mu\nu} - \tfrac{1}{2}\epsilon_{\mu\nu}{}^{\rho\lambda}F_{\rho\lambda} \;=\; 0 \;, \tag{1}$$

with the field strengths taking values in the Chan-Paton Lie algebra, describe what at that stage appeared to be the single dynamical degree of freedom of the open N=2 string. (We shall not give all relevant references here, referring to [1] for further references). The comparison has been based on determinations of tree-level amplitudes for the two theories, so light-cone gauge action principles for SDYM have played a central role in the discussion. In two-spinor notation, using the splitting of the $\mathbb{R}^{2,2}$ "Lorentz algebra",

$$so(2,2) \;\cong\; sl(2,\mathbb{R})\oplus sl(2,\mathbb{R})' \quad\Longleftrightarrow\quad x^\mu\sigma_\mu^{\alpha\dot\beta} \;=\; x^{\alpha\dot\beta} \;=\; \begin{pmatrix} x^0{+}x^3 & x^1{+}x^2 \\ x^1{-}x^2 & x^0{-}x^3 \end{pmatrix}, \tag{2}$$

the three (real) SDYM equations take the form

$$F_{\alpha\beta} \;\equiv\; \tfrac{1}{2}\left(\partial_{(\alpha}{}^{\dot\gamma}A_{\beta)\dot\gamma} \;+\; [A_\alpha{}^{\dot\gamma}, A_{\beta\dot\gamma}]\right) \;=\; 0 \tag{3}$$

In components, with the spinor indices α, β taking values $+$ and $-$, we have

$$
\begin{aligned}
F^{++} &\equiv \partial^{+\dot{\gamma}} A^{+}_{\dot{\gamma}} + \tfrac{1}{2}[A^{+\dot{\gamma}}, A^{+}_{\dot{\gamma}}] = 0 \\
F^{+-} &\equiv \tfrac{1}{2}\left(\partial^{+\dot{\gamma}} A^{-}_{\dot{\gamma}} + \partial^{-\dot{\gamma}} A^{+}_{\dot{\gamma}} + [A^{+\dot{\gamma}}, A^{-}_{\dot{\gamma}}]\right) = 0 \\
F^{--} &\equiv \partial^{-\dot{\gamma}} A^{-}_{\dot{\gamma}} + \tfrac{1}{2}[A^{-\dot{\gamma}}, A^{-}_{\dot{\gamma}}] = 0 \quad .
\end{aligned} \tag{4}
$$

Clearly, the $(++)$ equation affords the generalised light-cone gauge $A^{+}_{\dot{\gamma}}=0$ in which F^{+-} becomes homogeneous. Two strategies now suggest themselves. First, resolving the (inhomogeneous) $(--)$ equation in the Yang fashion,

$$
A^{-}_{\dot{\alpha}} = e^{-\phi}\, \partial^{-}_{\dot{\alpha}}\, e^{+\phi} \,, \tag{5}
$$

the $(+-)$ equation describes the ϕ-dynamics in the form of the (non-polynomial) Yang equation

$$
\partial^{+\dot{\alpha}}\, (e^{-\phi}\, \partial^{-}_{\dot{\alpha}}\, e^{+\phi}) = 0 \,. \tag{6}
$$

Second, the (homogeneous) $(+-)$ equation is instead fulfilled in terms of a Leznov prepotential, writing

$$
A^{-}_{\dot{\alpha}} = \partial^{+}_{\dot{\alpha}}\, \varphi^{--} \,, \tag{7}
$$

which then must satisfy $F^{--}=0$, tantamount to the (quadratic) Leznov equation,

$$
\Box \varphi^{--} - \tfrac{1}{2}[\partial^{+\dot{\alpha}}\varphi^{--}, \partial^{+}_{\dot{\alpha}}\varphi^{--}] = 0 \,. \tag{8}
$$

The light-cone gauge explicitly breaks the global $SO(2,2)$ covariance of eq. (3) to $GL(1,\mathbb{R}) \otimes SL(2,\mathbb{R})'$. In a Cartan-Weyl basis for $sl(2,\mathbb{R})$ consisting of a diagonal hyperbolic generator L_{+-} and two parabolic generators $L_{\pm\pm}$, the unbroken $gl(1,\mathbb{R})$ generator is L_{+-} in the Yang but L_{++} in the Leznov case.

Non-covariant action principles for (6) or (8) yield themselves using merely the prepotentials,

$$
S_{\text{Yang}} = \mu^2 \int d^4x \; \text{Tr}\Big\{-\tfrac{1}{2}\, \phi \Box \phi + \tfrac{1}{3}\, \phi\, \partial^{(+\dot{\alpha}}\phi\, \partial^{-)}_{\dot{\alpha}}\phi + \mathcal{O}(\phi^4)\Big\} \tag{9}
$$

$$
S_{\text{Leznov}} = \mu^2 \int d^4x \; \text{Tr}\Big\{-\tfrac{1}{2}\, \varphi^{--} \Box \varphi^{--} + \tfrac{1}{6}\, \varphi^{--}\big[\partial^{+\dot{\alpha}}\varphi^{--}, \partial^{+}_{\dot{\alpha}}\varphi^{--}\big]\Big\} \tag{10}
$$

with some mass scale μ. Alternatively, Lagrange multipliers facilitate the construction of dimensionless actions, for example,

$$
S_{\text{CS}} = \int d^4x \; \text{Tr}\Big\{-\varphi^{++} \Box \varphi^{--} + \tfrac{1}{2}\varphi^{++}\big[\partial^{+\dot{\alpha}}\varphi^{--}, \partial^{+}_{\dot{\alpha}}\varphi^{--}\big]\Big\} \tag{11}
$$

which was shown to be even one-loop exact by Chalmers and Siegel [3].

The tree-level amplitudes following from these actions are extremely simple. Since we are dealing with massless fields in 2+2 dimensions, the on-shell momenta factorise,

$$
k^{\alpha\dot{\beta}}\, k_{\alpha\dot{\beta}} = 0 \qquad \Longleftrightarrow \qquad k^{\alpha\dot{\beta}} = \kappa^{\alpha}\, \kappa^{\dot{\alpha}} \quad . \tag{12}
$$

The on-shell three-point functions $A_3(k_1, k_2, k_3)$ can be read off as

$$A_3^{\text{Yang}} = f^{abc}\, \kappa_1^{(+} \kappa_2^{-)}\, \kappa_1^{\dot\alpha} \kappa_{2\dot\alpha} \quad , \quad A_3^{\text{Leznov}} = f^{abc}\, \kappa_1^{+} \kappa_2^{+}\, \kappa_1^{\dot\alpha} \kappa_{2\dot\alpha} \tag{13}$$

where $\sum_i k_i = 0$ and f^{abc} are the structure constants of the gauge group. Surprisingly, the four-point Feynman diagrams sum to zero, in virtue of a quartic contact interaction in the Yang case. It is believed that all higher tree amplitudes vanish on-shell. The version of Chalmers and Siegel leads to the same tree-level amplitudes as the Leznov action, although in the former case one of the legs needs to be the multiplier field. Interestingly, this two-field theory does not allow diagrams beyond one loop. As we shall describe below, both (10) and (11) are related to the target space effective action for the open N=2 string.

2. N=2 Open Strings The spectrum of world-sheet fields in the NSR formulation of N=2 strings consists of the 2d N=2 supergravity multiplet, whose conformal gauge fixing produces the standard set of N=2 superconformal ghosts, plus N=2 matter fields $(X^{\alpha\dot\beta}, \Psi^{\alpha\dot\beta})$. The computation of open string amplitudes requires the evaluation of correlation functions for appropriate choices of physical external states on Riemann surfaces with handles, boundaries, punctures, and a harmonic $U(1)$ gauge field background with instantons. The result is to be integrated over the moduli of the Riemann surface and the $U(1)$ gauge field, and finally one is to sum over the topologies labelled by the Euler and $U(1)$ instanton numbers. The relative cohomology of the BRST operator determines the string external states, which are annihilated by the commuting N=2 Virasoro and the anticommuting N=2 antighost zero modes. The resulting spectrum has the following quantum numbers:

- total ghost number $u \in \mathbb{Z}$
- target space momentum $k^{\alpha\dot\beta}$
- total picture $\pi \in \mathbb{Z}$
- picture twist $\Delta \in \mathbb{R}$
- $gl(1,\mathbb{R}) \oplus sl(2,\mathbb{R})'$ quantum numbers m, (j', m')

These quantum numbers are however redundant, since they have interrelationships: $k{\cdot}k = 0$ (i.e. $k^{\alpha\dot\beta} = \kappa^\alpha \kappa^{\dot\beta}$), $u - \pi = 1$, $j' = m' = 0$, and m runs in integral steps from $-j$ to $+j$, where $j := \frac{\pi}{2}+1$. The physical spectrum consists of just one $SL(2,\mathbb{R})'$ singlet for each value of π, Δ, and $(\kappa, \dot\kappa)$. There is still a certain redundancy, since the pictures $(\pi \geq \pi_0, \Delta)$ can be reached from $(\pi_0, 0)$ by applying spectral flow $\mathcal{S}$ and picture raising $\mathcal{P}^\alpha$, which commute with the BRST operator and effect the mappings

$$(\pi, \Delta) \xrightarrow{\mathcal{S}(\rho)} (\pi, \Delta{+}2\rho) \quad , \quad (\pi, \Delta) \xrightarrow{\mathcal{P}^\alpha} (\pi{+}1, \Delta) \tag{14}$$

with $\rho \in \mathbb{R}$. Because the string path integral integrates over the twists of the $U(1)$ gauge bundle, it averages over the spectral flow orbits. The $\mathcal{S}$-equivalent states therefore ought to be identified and we may choose the Δ=0 representative. Picture lowering can also be constructed, except on zero-momentum states. In essence, all physical states (with $k \neq 0$) can be generated starting from the canonical picture $\pi = -2$ (i.e. $j=0$), and the result is symmetric under the "Poincaré duality" $\pi \stackrel{*}{\to} -4-2\pi$ (i.e. $j \stackrel{*}{\to} -j$):

$$\begin{array}{c|cccccccccc} \pi & \cdots & -5 & -4 & -3 & -2 & -1 & 0 & +1 & \cdots \\ j & \cdots & -\frac{3}{2} & -1 & -\frac{1}{2} & 0 & +\frac{1}{2} & +1 & +\frac{3}{2} & \cdots \\ \hline \text{states} & \cdots & |\alpha\beta\gamma\rangle^* & |\alpha\beta\rangle^* & |\alpha\rangle^* & |\,\rangle & |\alpha\rangle & |\alpha\beta\rangle & |\alpha\beta\gamma\rangle & \cdots \end{array} \quad . \tag{15}$$

The states form $SL(2,\mathbb{R})$ tensors of rank $2|j|$ (spin $|j|$), because the picture-raising operator $\mathcal{P}^\alpha$ carries a spinor index. There is no contradiction with the above statement of unit multiplicity, since all states in a given $SL(2,\mathbb{R})$ multiplet are related to each other, albeit in a non-local fashion. The open spinor indices are just carried by normalisation factors multilinear in κ^α, with $\mathcal{P}^\alpha$ increasing the spin by $\frac{1}{2}$:

$$\mathcal{P}^{\alpha_1}\mathcal{P}^{\alpha_2}\ldots\mathcal{P}^{\alpha_{2j}}\,|(0);k\rangle \;=\; |\alpha_1\alpha_2\ldots\alpha_{2j};k\rangle \;\propto\; \kappa^{\alpha_1}\kappa^{\alpha_2}\ldots\kappa^{\alpha_{2j}}\;|(j);k\rangle \quad . \tag{16}$$

The NSR formulation of $N{=}2$ strings introduces a complex structure in the target space, which explicitly breaks $SO(2,2) \to GL(1,\mathbb{R}) \otimes SL(2,\mathbb{R})'$. Individual pieces of an n-point amplitude are only $SL(2,\mathbb{R})'$ invariant, and contributions from the M-instanton $U(1)$ background carry a $gl(1,\mathbb{R})$ weight equal to M. Surprisingly, the path integral measure constrains the instanton sum to $|M| \leq J \equiv n{-}2$ at tree level. Moreover, the weight factors built from the string coupling e and the instanton angle θ conspire to restore $SO(2,2)$ invariance of the instanton sum if $\sqrt{e}(\cos\frac{\theta}{2}, \sin\frac{\theta}{2})$ is assumed to transform as an $SL(2,\mathbb{R})$ spinor! This spinor simply parametrises the choices of complex structure, and it may be Lorentz-rotated to $(1,0)$. Henceforth we shall remain in such a frame where $e{=}1$ and $\theta{=}0$. It has the virtue that only the highest $SL(2,\mathbb{R})$ weights $m_i{=}j_i$, $i=1,\ldots,n$, occur and only the maximal instanton number sector, $M=J$, contributes.

The tree-level open string on-shell amplitudes may then be found to be

$$A_3^{\text{string}} \;=\; f^{abc}\;\kappa_1^+\kappa_2^+\;\kappa_1^{\dot\alpha}\kappa_{2\dot\alpha} \;=\; A_3^{\text{Leznov}} \tag{17}$$

$$A_4^{\text{string}} \;\propto\; \kappa_1^+\kappa_2^+\kappa_3^+\kappa_4^+\,(\dot\kappa_1{\cdot}\dot\kappa_2\;\dot\kappa_3{\cdot}\dot\kappa_4\;t + \dot\kappa_2{\cdot}\dot\kappa_3\;\dot\kappa_4{\cdot}\dot\kappa_1\;s) \;=\; 0 \tag{18}$$

$$A_{n>4}^{\text{string}} \;=\; 0 \quad . \tag{19}$$

They are independent of the external $SL(2,\mathbb{R})$ spins j_i, as long as $\sum_{i=1}^n j_i = J \equiv n{-}2$. Clearly, the Leznov version (13) of SDYM is reproduced. However, a covariant description needs to take the entire tower of states in (15) into account.

3. Target Space Actions Physical string states correspond to target space (background) fields, whose on-shell dynamics is determined by the string scattering amplitudes. In particular, the string three-point functions directly yield cubic terms in the effective target space action. In the present case, the correspondence reads

$$|{+}{+}\ldots{+}\rangle \;\longleftrightarrow\; \varphi^{--\ldots-}\quad (j{\geq}0) \quad , \qquad |{-}{-}\ldots{-}\rangle^* \;\longleftrightarrow\; \varphi^{++\ldots+}\quad (j{<}0) \tag{20}$$

and we denote the fields by φ_j. Then, the target space effective action for the infinite tower $\{\varphi_j\}$ is

$$
\begin{aligned}
S_\infty &= \int d^4x \,\mathrm{Tr}\Big\{-\tfrac{1}{2}\sum_{j\in\mathbb{Z}/2}\varphi_{(-j)}\Box\varphi_{(+j)} + \tfrac{1}{3}\sum_{j_1+j_2+j_3=1}\varphi_{(j_1)}\left[\partial^{+\dot\alpha}\varphi_{(j_2)}\,,\,\partial^{+}_{\dot\alpha}\varphi_{(j_3)}\right]\Big\} \\
&= \int d^4x\,\mathrm{Tr}\Big\{-\tfrac{1}{2}\,\Phi^{--}\Box\Phi^{--} + \tfrac{1}{6}\,\Phi^{--}\left[\partial^{+\dot\alpha}\Phi^{--},\partial^{+}_{\dot\alpha}\Phi^{--}\right]\Big\}_{\eta^4}
\end{aligned}
\qquad (21)
$$

where we have introduced a "picture hyperfield",

$$
\Phi^{--}(x,\eta^-) = \sum_j (\eta^-)^{2j}\,\varphi_{1-j}(x) \quad , \qquad (22)
$$

depending on an extra commuting coordinate η^-, and we project the Lagrangean onto the part quartic in η. It is remarkable that the action (21) has the Leznov form in terms of the hyperfield. It not only reproduces all (tree-level) string three-point functions (17) but also yields vanishing four- and probably higher-point functions for the same reason that the Leznov action (10) does. Picture raising induces a dual action on the component fields,

$$
Q^+ : \quad \varphi_j \longrightarrow (3-2j)\,\varphi_{j-\frac{1}{2}} \quad , \qquad (23)
$$

which is nothing but the η^- derivative on the hyperfield!

Three successive truncations to a finite number of fields are possible. First, keeping only $\{\varphi_{-1},\varphi_{-\frac{1}{2}},\varphi_0,\varphi_{+\frac{1}{2}},\varphi_{+1}\}$, a consistent five-field model ensues, viz.,

$$
\begin{aligned}
S_5 = \int d^4x\,\mathrm{Tr}\Big\{ & \tfrac{1}{2}\partial^{+\dot\alpha}\varphi\,\partial^-_{\dot\alpha}\varphi + \partial^{+\dot\alpha}\varphi^+\partial^-_{\dot\alpha}\varphi^- + \partial^{+\dot\alpha}\varphi^{++}\partial^-_{\dot\alpha}\varphi^{--} \\
& + \tfrac{1}{2}\varphi\,[\partial^{+\dot\alpha}\varphi^-,\partial^+_{\dot\alpha}\varphi^-] + \tfrac{1}{2}\varphi^{--}[\partial^{+\dot\alpha}\varphi\,,\partial^+_{\dot\alpha}\varphi] \\
& + \varphi^{--}[\partial^{+\dot\alpha}\varphi^+,\partial^+_{\dot\alpha}\varphi^-] + \tfrac{1}{2}\varphi^{++}[\partial^{+\dot\alpha}\varphi^{--},\partial^+_{\dot\alpha}\varphi^{--}]\Big\} \ .
\end{aligned}
\qquad (24)
$$

Second, eliminating also the fermions leaves us with three fields. Third, we may in addition kill φ_0 as well, resulting in the two-field model of Chalmers and Siegel [3]! All truncations share the one-loop exactness mentioned before.

4. Self-Duality in Hyperspace The infinite tower of higher-spin fields which arise from the picture degeneracy parametrise simply SDYM in a hyperspace with coordinates $\{x^{\alpha\dot\alpha},\eta^\alpha,\bar\eta^{\dot\alpha}\}$, with η and $\bar\eta$ *commuting* spinors. This commutative variant of superspace exhibits a $\mathbb{Z}_2$-graded Lie-algebra variant of the super-Poincare algebra (i.e. with all anti-commutators replaced by commutators). So the covariant target space symmetry is effectively the extension of the $\mathbb{R}^{2,2}$ Poincaré algebra by two *Grassmann-even* spinorial generators squaring to a translation, i.e., $[Q_{\dot\alpha},Q_\alpha]=P_{\alpha\dot\alpha}$ (see [1] for details). Hyperspace self-duality allows compact expression in a chiral subspace independent of the $\bar\eta$ coordinates. In terms of chiral subspace gauge-covariant derivatives $\mathcal{D}_\alpha = \partial_\alpha + A_\alpha(x,\eta)$ and $\mathcal{D}_{\alpha\dot{\alpha}} = \partial_{\alpha\dot{\alpha}} + A_{\alpha\dot{\alpha}}(x,\eta)$, the self-duality conditions take the simple form

$$[\mathcal{D}_\alpha, \mathcal{D}_\beta] = \epsilon_{\alpha\beta} F \quad , \quad [\mathcal{D}_\alpha, \mathcal{D}_{\beta\dot{\beta}}] = \epsilon_{\alpha\beta} F_{\dot{\beta}} \quad , \quad [\mathcal{D}_{\alpha\dot{\alpha}}, \mathcal{D}_{\beta\dot{\beta}}] = \epsilon_{\alpha\beta} F_{\dot{\alpha}\dot{\beta}} \quad . \tag{25}$$

Jacobi identities yield the equations

$$\mathcal{D}_\alpha{}^{\dot{\alpha}} F_{\dot{\alpha}\dot{\beta}} = 0 \quad , \quad \mathcal{D}_\alpha{}^{\dot{\alpha}} F_{\dot{\alpha}} = 0 \quad , \quad \mathcal{D}_{\alpha\dot{\alpha}} F = \mathcal{D}_\alpha F_{\dot{\alpha}} \quad . \tag{26}$$

The first two are respectively the Yang-Mills and Dirac equations for a SDYM multiplet, and the third implies the scalar field equation $\mathcal{D}^2 F = [F^{\dot{\alpha}}, F_{\dot{\alpha}}]$. All chiral hyperfields have η expansions, e.g.

$$A_\alpha(x,\eta) \;=\; A_\alpha(x) \;+\; \eta^\beta A_{\alpha\beta}(x) \;+\; \eta^\beta\eta^\gamma A_{\alpha\beta\gamma}(x) \;+\; \dots \quad . \tag{27}$$

Choosing the light-cone gauge, $A^+ = 0 = A^+_{\dot{\alpha}}$, we note that all fields are defined in terms of a generalised Leznov prepotential,

$$A^- \;=\; \partial^+\, \Phi^{--} \quad , \quad A^-_{\dot{\alpha}} \;=\; \partial^+_{\dot{\alpha}}\, \Phi^{--} \quad , \tag{28}$$

$$F \;=\; \partial^+\partial^+\, \Phi^{--} \quad , \quad F_{\dot{\alpha}} \;=\; \partial^+\partial^+_{\dot{\alpha}}\, \Phi^{--} \quad , \quad F_{\dot{\alpha}\dot{\beta}} \;=\; \partial^+_{\dot{\alpha}}\partial^+_{\dot{\beta}}\, \Phi^{--} \quad . \tag{29}$$

Since ∂^- does not occur in the above, all fields are determined by the chiral (η^+–independent) part of Φ^{--}. The dynamics is determined by the remaining constraints

$$[\mathcal{D}^-_{\dot{\alpha}}, \mathcal{D}^-_{\dot{\beta}}] \;=\; 0 \qquad \text{and} \qquad [\mathcal{D}^-, \mathcal{D}^-_{\dot{\beta}}] \;=\; 0 \;, \tag{30}$$

where the former equation is precisely the Leznov equation for Φ^{--}. Choosing this to be chiral, $\Phi^{--} = \Phi^{--}(x,\eta^-)$, allows identification with (22), with action given by (21). The second equation above then merely determines the η^- dependence of Φ^{--}.

The restricted system of five fields (24) has the $SO(2,2)$-invariant action

$$S_5^{\rm inv} \;=\; \int d^4x\; {\rm Tr}\Big\{ \tfrac{1}{4} g^{\alpha\beta} F_{\alpha\beta} \;+\; \tfrac{1}{3}\chi^\alpha \mathcal{D}_{\alpha\dot{\alpha}} F^{\dot{\alpha}} \;+\; \tfrac{1}{8}\mathcal{D}^{\alpha\dot{\alpha}} F\, \mathcal{D}_{\alpha\dot{\alpha}} F \;+\; \tfrac{1}{2} F\, [F^{\dot{\alpha}}, F_{\dot{\alpha}}] \Big\} \tag{31}$$

where $g^{\alpha\beta}$ and χ^α are (propagating) multiplier fields for $A_{\alpha\dot{\alpha}}$ and $F_{\dot{\alpha}}$, respectively. The similarity with N=4 supersymmetric SDYM [4] is evident, however with commuting single-multiplicity fermions replacing multiplicity 4 anticommuting ones.

To conclude, we note that theories of N=2 closed as well as N=(2, 1) heterotic strings are also intimately related to self-dual geometry and our covariant hyperspace description generalises to both these cases.

References

[1] C. Devchand and O. Lechtenfeld,
Extended self-dual Yang-Mills from the N=2 string, hep-th/9712043.

[2] H. Ooguri and C. Vafa, *Mod. Phys. Lett.* **A5** (1990) 1389.

[3] G. Chalmers and W. Siegel, *Phys. Rev.* **D54** (1996) 7628, hep-th/9606061.

[4] W. Siegel, *Phys. Rev.* **D46** (1992) 3235.

Modifying N=2 supersymmetry via partial breaking

Evgeny Ivanov [1] **, Boris Zupnik** [1]

[1] Bogoliubov Laboratory of Theoretical Physics, JINR,
141980 Dubna, Moscow Region, Russia

Abstract: We study realization of $N = 2$ SUSY in $N = 2$ abelian gauge theory with electric and magnetic FI terms within a manifestly supersymmetric formulation. We find that after dualization of even one FI term $N = 2$ SUSY is realized in a partial breaking mode off shell. In the case of two FI terms, this regime is preserved on shell. The $N = 2$ SUSY algebra is shown to be modified on gauge-variant objects.

1. A celebrated mechanism of spontaneous breakdown of rigid $N = 2$ SUSY consists in adding a Fayet-Iliopoulos (FI) term to the action of $N = 2$ gauge theory. Recently, Antoniadis, Partouche and Taylor (APT) [1] have found that the dual formulation of $N = 2$ abelian gauge theory [2] provides a more general framework for such a spontaneous breaking due to the possibility to define two kinds of the FI terms (see also [3]). One of them ('electric') is standard, while another ('magnetic') is related to a dual $U(1)$ gauge supermultiplet. APT show that a partial spontaneous breakdown of $N = 2$ SUSY to $N = 1$ becomes possible, if one starts with an effective $N = 2$ Maxwell action and simultaneously includes two such FI terms.

Here we report on the results of studying the invariance properties of $N = 2$ Maxwell action with the two types of FI-terms in a manifestly supersymmetric $N = 2$ superfield approach. Our basic observation is that after duality transformation of a system even with one sort of the FI term, off-shell $N = 2$ SUSY starts to be realized in a *partial* spontaneous breaking mode. When *both* FI terms are included, this partial breaking is retained on shell by the APT mechanism. We study how these modified $N = 2$ SUSY transformations act on the gauge potentials and find that the $N = 2$ SUSY algebra also undergoes a modification.

2. Let us start from the following representation of the superfield action of $N = 2$ Maxwell theory with the FI term [4]

$$
\begin{aligned}
S(W, L) &= \frac{i}{4}\int d^4x d^4\theta\, [\mathcal{F}(W) - WW_L + \frac{i}{2}E^{ik}(\theta_i\theta_k)W] + \text{c.c.} \\
&\equiv S(W) + S_L + S_e\,, \qquad W_L = (\bar{D})^4 D_{ik} L^{ik}\,. \qquad (1)
\end{aligned}
$$

Here, $\mathcal{F}(W)$ is an arbitrary holomorphic function, W is a chiral $N=2$ superfield, L^{ik} is a real unconstrained $N=2$ superfield Lagrange multiplier, E^{ik} is a real $SU(2)$ triplet of constants and $D^{ik}=D^{i\alpha}D^k_\alpha$. Varying L^{ik} yields the constraint [5]

$$D^{ik}W-\bar{D}^{ik}\overline{W}=0 \tag{2}$$

which can be solved in terms of Mezincescu prepotential V^{ik} [6]

$$W\equiv W_V=(\bar{D})^4 D_{ik}V^{ik}\ . \tag{3}$$

Upon substituting this solution back into (1), the latter becomes

$$S(W,L)\Rightarrow S(V)=S(W_V)+S_e\ ,\quad S_e=\int d^4x d^4\theta d^4\bar{\theta}\ E_{ik}V^{ik} \tag{4}$$

that is the standard 'electric' form of the action of $N=2$ Maxwell theory with FI term. On the other hand, one can first vary (1) with respect to W, which yields

$$\frac{\partial\mathcal{F}}{\partial W}=W_L-(i/2)(\theta_k\theta_l)E^{kl}\equiv\hat{W}_L\ , \tag{5}$$

$$S(W,L)\Rightarrow\frac{i}{4}\int d^4x d^4\theta\ \hat{\mathcal{F}}(\hat{W}_L)+\text{c.c.}\ ,\quad \hat{\mathcal{F}}(\hat{W}_L)\equiv\mathcal{F}(W(\hat{W}_L))-\hat{W}_L\cdot W(\hat{W}_L)\ . \tag{6}$$

In this dual, 'magnetic' representation the FI term-modified $N=2$ Maxwell action is expressed through the dual ('magnetic') superfield strength and prepotential W_L and L^{ik}. Thus, (1) is a sort of 'master' action from which both the electric and magnetic forms of the $N=2$ Maxwell action can be obtained (see [2, 8] for a similar discussion of the standard $E^{ik}=0$ case).

Let us discuss peculiarities of realization of $N=2$ SUSY in the magnetic representation. The electric action (4) is invariant under the standard $N=2$ SUSY. The same is true for the dual actions (1), (6) in the absence of the FI term. However, if $E^{ik}\neq 0$, the conventional $N=2$ SUSY gets broken in (1), (6). The only way to restore it is to modify the transformation law of W_L

$$\delta_\epsilon W_L=i(\epsilon_k\theta_l)E^{kl}+i(\epsilon Q+\bar{\epsilon}\bar{Q})W_L\ , \tag{7}$$

where $Q^i_\alpha,\bar{Q}^i_{\dot{\alpha}}$ are standard generators of $N=2$ SUSY. It is easy to find the appropriate modified transformation of L^{ik}.

An inhomogeneous term in the $N=2$ SUSY transformation law means spontaneous breakdown of $N=2$ SUSY. To see which kind of breaking occurs in the case at hand, let us firstly discuss the standard electric action (4).

Spontaneous breaking of $N=2$ SUSY by the FI term is related to the possibility of a non-zero vacuum solution for the auxiliary component $X^{ik}=-\frac{1}{4}D^{ik}W|_0$

$$<X^{ik}>\equiv x^{ik}\sim\ E^{ik}\ . \tag{8}$$

Provided that such a solution of the equations of motion exists and corresponds to a stable classical vacuum, there appears an inhomogeneous term in the on-shell SUSY transformation law of the $N=2$ gaugini doublet $\lambda^{i\alpha}$

$$\delta\lambda^{i\alpha}\sim\epsilon^\alpha_k E^{ik}\ , \tag{9}$$

ϵ_k^α being the transformation parameter. Thus there are Goldstone fermions in the theory, which is a standard signal of spontaneous breaking of $N=2$ SUSY. Since the inhomogeneous terms in (9) appear as a result of solving equations of motion, it is natural to call $\lambda^{i\alpha}$ *on-shell* Goldstone fermions. As the matrix E^{ik} is non-degenerate, *both* $\lambda^{1\alpha}, \lambda^{2\alpha}$ are shifted by independent parameters, and so they both are Goldstone fermions. Thus, with the standard electric FI term, only *total* spontaneous breaking of $N=2$ SUSY can occur (actually, for one vector multiplet this is possible only in the free case, with quadratic function $\mathcal{F}$ [7]).

In the dual, magnetic representation of the same theory corresponding to the actions (1) or (6) the situation is radically different: the off-shell transformation law (7) contains an 'inborn' inhomogeneous piece. This leads us to interpret the magnetic gaugini as *off-shell* Goldstone fermions.

Both gaugini are shifted in (7), so at first sight we are facing the total off-shell spontaneous breaking of $N=2$ SUSY in this case. However, by a proper shift

$$W_L \rightarrow \widetilde{W}_L = W_L + \frac{1}{2}(\theta_i\theta_k)C^{ik}, \tag{10}$$

one can restore a homogeneous transformation law with respect to *one* of two $N=1$ supersymmetries contained in $N=2$ SUSY (it is easy to find the appropriate redefinition of L^{ik}). The object $\widetilde{W}_L$ transforms as follows

$$\delta_\epsilon \widetilde{W}_L = (\epsilon_k\theta_l)(C^{kl} + iE^{kl}) + i(\epsilon Q + \bar{\epsilon}\bar{Q})\widetilde{W}_L\,. \tag{11}$$

One can always choose C^{ik} so that

$$\det\,(C + iE) = 0\,. \tag{12}$$

This means that $C^{ik} + iE^{ik}$ is a degenerate symmetric 2×2 matrix, so it can be brought to the form with only one non-zero entry. As a result, $\widetilde{W}_L$ is actually shifted under the action of only *one* linear combination of the modified $N=2$ SUSY generators $\hat{Q}^{1,2}_\alpha$. The same is true for the physical fermionic components: only *one* their combination is the genuine off-shell Goldstone fermion.

Thus we arrive at the important conclusion: in the dual, magnetic representation of $N=2$ Maxwell theory with FI term $N=2$ SUSY is realized *off shell* in a *partial spontaneous breaking mode*, so that some $N=1$ SUSY remains unbroken.

It is straightforward to show that the action (6) leads to the same vacuum structure as the original electric action (4). Thus on shell in the magnetic representation we again encounter the total spontaneous breaking of $N=2$ SUSY.

3. Let us show that the phenomenon of partial breaking of $N=2$ SUSY becomes valid both off and on shell upon adding to the 'master' action (1) the new sort of FI term, the 'magnetic' one:

$$S(W,L) \Rightarrow S(W,L)' = S(W,L) + S_m\,, \quad S_m = \frac{1}{8}\int d^4x d^4\theta\ M^{kl}(\theta_k\theta_l)W_L + \text{c.c.}\,, \tag{13}$$

M^{ik} being another triplet of real constants. It is easy to show that S_m is invariant under the Goldstone-type transformation (7).

When one descends to the electric representation of (13) (by varying L^{ik}), the only effect of the magnetic FI term S_m is the modification of the constraint (2):

$$D^{ik}W - \bar{D}^{ik}\overline{W} = 4iM^{ik} . \tag{14}$$

It suggests the redefinition

$$W = W_V - \frac{i}{2}(\theta_i\theta_k)M^{ik} , \tag{15}$$

with W_V satisfying eq. (2) and hence given by eq. (3). This shift amounts to the appearance of the constant imaginary part $-(i/2)M^{ik}$ in the auxiliary field of W,

$$\hat{X}^{ik} \equiv -\frac{1}{4}D^{ik}W|_0 = X^{ik} - \frac{i}{2}M^{ik} .$$

The inclusion of magnetic FI term cannot change the standard transformation properties of W under $N = 2$ SUSY. Then the relation (15) requires to modify the transformation law of W_V, and, respectively, of V^{ik} on the pattern of eq. (7)

$$\delta_\epsilon W_V = i(\epsilon_k\theta_l)M^{kl} + i(\epsilon Q + \bar{\epsilon}\bar{Q})W_V \tag{16}$$

(it is easy to find the appropriate transformation law of V^{ik}). In other words, $N = 2$ SUSY is now realized in a Goldstone-type fashion in the electric representation as well, but with M^{ik} instead of E^{ik} as the 'structure' constants. So, when both FI terms are present, there is no way to restore the standard $N = 2$ SUSY off shell. The same arguments as in the previous Section show that $N = 2$ SUSY in both representations is realized off shell in *the partial breaking mode.*

A general electric effective action of the abelian gauge model with the (E, M)-mechanism of the spontaneous breaking can be obtained by substituting the expression for W, eq.(15), into the action (13):

$$S_{(E,M)} = \left[\frac{i}{4}\int d^4x d^4\theta \, \mathcal{F}(W) + \text{c.c.}\right] + \int d^4x d^4\theta d^4\bar{\theta} \, E_{ik}V^{ik} . \tag{17}$$

Taking the standard vacuum ansatz

$$< W_V >_0 = a + (\theta_i\theta_k) \, x^{ik} , \tag{18}$$

it is easy to show that the superfield equation of motion following from (17) implies the following equations for moduli a, x^{ik}

$$(i) \;\; x^{kl} = \frac{1}{2\tau_2(a)}\left(\tau_1(a)M^{kl} - E^{kl}\right) , \quad (ii) \;\; \tau' \, \hat{x}^{ik}\hat{x}_{ik} = 0 , \tag{19}$$

where

$$\tau = \mathcal{F}'' = \tau_1 + i\tau_2 , \qquad \hat{x}^{ik} =< \hat{X}^{ik} >= x^{ik} - (i/2)M^{ik} .$$

A crucial new point compared to the case of $M^{ik} = 0$ is that the vector $\hat{x}^{ik} = x^{ik} - (i/2)M^{ik}$ is *complex*, so the vanishing of its square does not imply it to vanish.

As a result, besides the trivial solution $\tau' = 0$ [7], eq.*(ii)* in (19) possesses the non-trivial one

$$\tau' \neq 0\ , \qquad \hat{x}^{ik}\hat{x}_{ik} = 0\ . \tag{20}$$

This solution amounts to the following relations

$$\tau_1(a) = \frac{\vec{E}\vec{M}}{\vec{M}^2}\ , \quad |\tau_2(a)| = \frac{\sqrt{\vec{E}^2\vec{M}^2 - (\vec{E}\vec{M})^2}}{\vec{M}^2} = \frac{|\,\vec{E}\times\vec{M}\,|}{\vec{M}^2}\ . \tag{21}$$

This is just the vacuum solution found in [1] based on the component version of the action (17). It triggers a *partial* spontaneous breaking of $N = 2$ SUSY down to $N = 1$. Only one combination of gaugini is the Goldstone fermion in this case.

Thus, the phenomenon of the off-shell partial breaking of $N = 2$ SUSY is preserved on shell, provided that both electric and magnetic FI terms are included.

4. The modified $N = 2$ SUSY, when realized on the superfield strengths (eqs. (7), (16)), still closes on space-time translations. One can wonder how it is realized on the gauge-variant objects: gauge potentials and prepotentials. The $N = 2$ SUSY algebra itself proves to be modified in this case. We will demonstrate this in the $N = 1$ superfield formalism, on the example of the model considered in Sect. 3.

We define

$$W = W_V + \hat{x}^{ik}(\theta_i\theta_k)\ , \quad < W_V >_0 = 0\ . \tag{22}$$

Decomposing W_V in powers of $\theta_2, \bar{\theta}^2$

$$W_V = \varphi(x,\theta_1) + i\theta_2^\alpha W_\alpha(x,\theta_1) + (\theta_2)^2(1/4)(\bar{D}_1)^2\bar{\varphi}\ , \tag{23}$$

it is easy to find how the $N = 1$ superfield components $\varphi(x,\theta_1)$, $W^\alpha(x,\theta_1) = (\bar{D}_1)^2 D^{\alpha 1}V$ behave under the modified $N = 2$ SUSY transformations (16). We will be interested in how the latter are realized on the $N = 1$ gauge prepotential $V(x,\theta_1,\bar{\theta}^1)$. Modulo gauge transformations, the $N = 2$ SUSY acts on V as (we choose the $SU(2)$ frame so that $M^{12} = 0, M^{11} = M^{22} = m$)

$$\delta V = m(\bar{\theta}^1)^2\theta_1^\alpha\epsilon_{\alpha 2} + (i/2)\theta_1^\alpha\epsilon_{\alpha 2}\bar{\varphi} + \text{c.c} + i(\epsilon_1 Q^1 + \bar{\epsilon}^1\bar{Q}_1)V\ . \tag{24}$$

Defining $Q_\alpha \equiv Q_\alpha^1$, $S_\alpha \equiv Q_\alpha^2$, one finds that the (anti)commutation relations of the $N = 2$ SUSY algebra are modified as follows:

$$\{Q_\alpha, S_\beta\} = \varepsilon_{\alpha\beta}G\ , \quad \{Q_\alpha, \bar{S}_{\dot{\beta}}\} = G_{\alpha\dot{\beta}}\ , \quad (\text{ and c.c. })\ . \tag{25}$$

The newly introduced generators possess the following action on V

$$GV = (i/2)m(\bar{\theta}^1)^2\ , \quad \bar{G}V = (i/2)m(\theta^1)^2\ , \quad G_{\alpha\dot{\beta}}V = im\theta_{1\alpha}\bar{\theta}^1_{\dot{\beta}}\ , \tag{26}$$

and are easily seen to be particular $N = 1$ gauge transformations. Thus, off-shell $N = 2$ SUSY algebra turns out to be non-trivially unified with the gauge group algebra. Such a unification does not contradict the famous Coleman-Mandula theorem.

Indeed, in gauge theories it is impossible to simultaneously satisfy two important conditions of this theorem: manifest Lorentz covariance and positive definiteness of the metric in the space of states. Note that the $N = 2$ transformation (24) is defined up to pure gauge terms which are capable to change the above algebra. However, in the closure of spinor generators one will always find some gauge generators in parallel with 4-translations. An interesting problem is to construct a non-linear realization of this modified $N = 2$ algebra along the lines of ref. [9] and to compare it with that constructed in [9] based on the standard $N = 2$ algebra. It is also tempting to elaborate on a possible stringy origin of the modified algebra.

Finally, we would like to point out that the difficulty with the self-consistent incorporation of charged matter hypermultiplets into the $N = 2$ Maxwell theory with the two sorts of FI terms recently discussed in [10] is directly related to the modification of $N = 2$ SUSY algebra in the presence of these terms. The standard minimal coupling of the q^+ hypermultiplets to the harmonic superspace $N = 2$ Maxwell potential V^{++} [11] is not invariant under the modified $N = 2$ SUSY.

Acknowledgement. We are grateful to J. Bagger, E. Buchbinder, A. Galperin, S. Krivonos, O. Lechtenfeld, D. Lüst, A. Pashnev and M. Vasiliev for useful discussions. This work was partially supported by the grants RFBR-96-02-17634, RFBR-DFG-96-02-00180, INTAS-93-127-ext and INTAS-96-308. B.Z. is grateful to Uzbek Foundation of Basic Research for the partial support (grant N 11/97).

References

[1] I. Antoniadis, H. Partouche and T.R. Taylor, Phys. Lett. B 372 (1996) 83

[2] N. Seiberg and E. Witten, Nucl. Phys. B 426 (1994) 19

[3] S. Ferrara, L. Girardello and M. Porrati, Phys. Lett. B 376 (1996) 275

[4] E.A. Ivanov and B.M. Zupnik, Preprint JINR E2-97-322, hep-th/9710236

[5] R. Grimm, M. Sohnius and J. Wess, Nucl. Phys. B 133 (1978) 275 ; M.Sohnius, Nucl. Phys. B 136 (1978) 461

[6] L. Mezincescu, "On the superfield formulation of $O(2)$-supersymmetry", Preprint JINR, P2-12572, Dubna, 1979 (In Russian)

[7] A. Galperin and V. Ogievetsky, Nucl. Phys. Proc. Suppl. 15 B (1990) 93

[8] M. Henningson, Nucl. Phys. B 458 (1996) 445

[9] J. Bagger and A. Galperin, Phys. Rev. D 55 (1997) 1091

[10] H. Partouche and B. Pioline, Nucl. Phys. Proc. Suppl. 56 B (1997) 322

[11] A. Galperin, E. Ivanov, S. Kalitzin, V. Ogievetsky and E. Sokatchev, Class. Quant. Grav. 1 (1984) 469

Integrating Flow in d=3 Higher-Spin Gauge Theories with N=2 SUSY

Sergey Prokushkin and Mikhail Vasiliev

I.E.Tamm Department of Theoretical Physics, Lebedev Physical Institute, Leninsky Prospect 53, 117924 Moscow

Abstract: We discuss properties of non-linear equations of motion which describe higher-spin gauge interactions for massive spin-0 and spin-1/2 matter fields in 2+1 dimensional anti-de Sitter space. An integrating flow is found which allows one to construct a non-local Bäcklund-Nicolai–type mapping which reduces the full non-linear system to the free field equations. The model is shown to describe higher-spin interactions of a d3 N=2 massive hypermultiplet.

1 Introduction and Preliminaries

In this talk, we report on some recent results in the study of the higher-spin (HS) interactions of massive matter fields in 2+1 dimensional space-time. The main new result consists of the constructive definition of a sort of a non-linear non-local mapping which allows one to reduce the problem to the free one. As this talk is a direct continuation of the report [1] at the XXX Ahrenshoop Symposium, we refer the reader to [1] for more detailed motivations and references. A most important argument in favor of the consideration of the theories of HS gauge fields is that they may serve as an alternative way towards a fundamental theory presently identified with M-theory. Although a route to M-theory passes through models in higher dimensions, particularly d=11 and d=12, it is useful to study a d=3 model which has much simpler dynamics because d3 HS gauge fields do not propagate [2]. As shown below, the simplicity of the model indeed allows one to study it in great details, thus leading to some general conclusions on a structure of more complicated HS models.

Now let us summarize the results reported in [1]. The full nonlinear system of equations is formulated in terms of the generating functions $W(z,y;\psi,k,\rho|x)$, $B(z,y;\psi,k,\rho|x)$ and $S_\alpha(z,y;\psi,k,\rho|x)$, which depend on the space-time coordinates x^ν ($\nu = 0 \div 2$), auxiliary commuting spinors z_α, y_α ($\alpha = 1,2$), $[y_\alpha, y_\beta] = [z_\alpha, z_\beta] = [z_\alpha, y_\beta] = 0$, a pair of Clifford elements $\{\psi_i, \psi_j\} = 2\delta_{ij}$ ($i = 1,2$), which commute to all other generating elements, and another pair of Clifford type elements k and ρ

which have the following properties [1]

$$k^2 = 1\,,\, \rho^2 = 1\,,\, k\rho + \rho k = 0\,,\, k y_\alpha = -y_\alpha k\,,\, k z_\alpha = -z_\alpha k\,,\, \rho y_\alpha = y_\alpha \rho\,,\, \rho z_\alpha = z_\alpha \rho\,. \quad (1)$$

The generating function W for the HS gauge fields is a space-time 1-form, $W = dx^\nu W_\nu(z, y; \psi, k, \rho|x)$,

$$W_\mu(z, y; \psi, k, \rho|x) = \sum_{A,B,C,D=0}^{1} \sum_{n=0}^{\infty} \frac{1}{m!n!} W^{ABCD}_{\mu,\,\alpha_1\ldots\alpha_m\beta_1\ldots\beta_n}(x) k^A \rho^B \psi_1^C \psi_2^D z^{\alpha_1} \ldots z^{\alpha_m} y^{\beta_1} \ldots y^{\beta_n}. \quad (2)$$

$B = B(z, y; \psi, k, \rho|x)$ is the generating function for the matter fields. The components of its expansion similar to (2) are identified with the d3 matter fields and all their on-mass-shell non-trivial derivatives. $S_\alpha(z, y; \psi, k, \rho|x)$ plays an auxiliary role. The generating functions are treated as elements of an associative algebra with the product law[2]

$$(f*g)(z, y; \psi, k, \rho) = \frac{1}{(2\pi)^2} \int d^2u d^2v \exp(i u_\alpha v^\alpha) f(z+u, y+u; \psi, k, \rho) g(z-v, y+v; \psi, k, \rho)\,, \quad (3)$$

where $v^\alpha = \epsilon^{\alpha\beta} v_\beta$, $\epsilon^{\alpha\beta} = -\epsilon^{\beta\alpha}$, $\epsilon^{12} = \epsilon_{12} = 1$, and the variables u and v are required to satisfy the commutation relations similar to those of y and z in (1).

The full system of equations can be written in the form

$$dW = W * W\,, \quad dB = W * B - B * W\,, \quad dS_\alpha = W * S_\alpha - S_\alpha * W\,, \quad (4)$$

$$S_\alpha * S_\beta - S_\beta * S_\alpha = -2i\epsilon_{\alpha\beta}(1 + B * K)\,, \quad S_\alpha * B = B * S_\alpha\,. \quad (5)$$

Here $d = dx^\nu \frac{\partial}{\partial x^\nu}$ and $K = k e^{i(zy)}$, $(zy) = z_\alpha y^\alpha$. With the aid of an obvious involutive automorphism $S_\alpha \to -S_\alpha$ one can consider a truncation of the system (4), (5) to the one with the fields W and B independent of ρ and S_α linear in ρ,

$$W(z, y; \psi, k, \rho|x) = W(z, y; \psi, k|x)\,, \quad B(z, y; \psi, k, \rho|x) = B(z, y; \psi, k|x)\,, \quad (6)$$

$$S_\alpha(z, y; \psi, k, \rho|x) = \rho s_\alpha(z, y; \psi, k|x)\,. \quad (7)$$

It is this reduced system which was considered in [1] and which is discussed below.

Eqs. (4), (5) are invariant under the infinitesimal HS gauge transformations

$$\delta W = d\varepsilon - W * \varepsilon + \varepsilon * W\,, \quad \delta B = \varepsilon * B - B * \varepsilon\,, \quad \delta S_\alpha = \varepsilon * S_\alpha - S_\alpha * \varepsilon\,, \quad (8)$$

where $\varepsilon = \varepsilon(z, y; \psi, k|x)$ is an arbitrary gauge parameter.

To elucidate the dynamical content of the system (4), (5), one first of all has to find an appropriate vacuum solution. Such a solution presented in [1] has a form

$$B_0 = \nu = const\,, \quad (9)$$

[1] Note that these commutation relations can be reduced to the more familiar ones by virtue of the substitution $y \to \rho y$ and $z \to \rho z$. The definition (1) is chosen for future convenience.

[2] This product law yields a particular realization of Heisenberg-Weyl algebra in the sector of the variables y and z, $[y_\alpha, y_\beta]_* = -[z_\alpha, z_\beta]_* = 2i\epsilon_{\alpha\beta}$, $[y_\alpha, z_\beta]_* = 0$, where $[a, b]_* = a * b - b * a$.

$$S_{0\alpha} = \rho \left(z_\alpha + \nu(z_\alpha + y_\alpha) \int_0^1 dt t e^{it(zy)} k \right) , \tag{10}$$

$$W_0(z, y; \psi, k|x) = W_0(\tilde{y}; \psi, k|x) , \tag{11}$$

where $\tilde{y}_\alpha$ are the elements with the defining property $[\tilde{y}_\alpha \,, S_{0\beta}]_* = 0$, which have a form

$$\tilde{y}_\alpha = y_\alpha + \nu(z_\alpha + y_\alpha) \int_0^1 dt(t-1) e^{it(zy)} k \,. \tag{12}$$

It is assumed that an arbitrary "function" $W_0(\tilde{y}; \psi, k|x)$ on the r.h.s. of (11) contains star-products of $\tilde{y}_\alpha$. It is interesting to note that $\tilde{y}_\alpha$ possess the deformed oscillator algebra commutation relations

$$[\tilde{y}_\alpha, \tilde{y}_\beta]_* = 2i\epsilon_{\alpha\beta}(1 + \nu k) \,, \quad \tilde{y}_\alpha k = -k \tilde{y}_\alpha \,. \tag{13}$$

Eqs. (9)-(11) solve all the equations (4), (5) except for the first one in (4),

$$dW_0 = W_0 \wedge W_0 \,, \tag{14}$$

which requires a further specification of W_0. An appropriate anzats is

$$W_0 = \omega_0 + \lambda h_0 \psi_1 \,, \quad \omega_0 = \frac{1}{8i} \omega_0^{\alpha\beta}(x) \{\tilde{y}_\alpha, \tilde{y}_\beta\}_* \,, \quad h_0 = \frac{1}{8i} h_0^{\alpha\beta}(x) \{\tilde{y}_\alpha, \tilde{y}_\beta\}_* \,, \tag{15}$$

where $\omega_0^{\alpha\beta}(x)$ and $h_0^{\alpha\beta}(x)$ are identified with Lorentz connection and dreibein of the background space and are required to solve (14). Here the properties of the deformed oscillators (13) play a crucial role, guaranteeing that the anticommutators $\{y_\alpha, y_\beta\}_* = y_\alpha * y_\beta + y_\beta * y_\alpha$ satisfy [3] the $sp(2)$ commutation relations for all ν. As a consequence, the gauge fields (15) take values in the d3 anti-de Sitter (AdS) algebra $so(2,2) \sim sp(2) \oplus sp(2)$ and, therefore, (15) leads to the zero-curvature and zero-torsion conditions for $\omega_0^{\alpha\beta}$ and $h_0^{\alpha\beta}$ respectively, thus describing AdS background.

Once a vacuum solution is known, one can study the system (4), (5) perturbatively expanding the fields as

$$B = B_0 + B_1 + \ldots \,, \qquad S_\alpha = S_{0\alpha} + S_{1\alpha} + \ldots \,, \qquad W = W_0 + W_1 + \ldots \,. \tag{16}$$

Substitution of these expansions into (4), (5) gives in the lowest order

$$D_0 W_1 = 0 \,, \tag{17}$$

$$D_0 C = 0 \,, \tag{18}$$

$$D_0 S_{1\alpha} = [W_1 \,, S_{0\alpha}]_* \,, \tag{19}$$

$$[S_{0\alpha} \,, S_{1\beta}]_* - [S_{0\beta} \,, S_{1\alpha}]_* = -2i\epsilon_{\alpha\beta} C * K \,, \tag{20}$$

$$[S_{0\alpha} \,, C]_* = 0 \,, \tag{21}$$

where we denote $C = B_1$ and D_0 is the background covariant derivative which acts on a r–form P as $D_0 P = dP - W_0 \wedge P + (-)^r P \wedge W_0$.

To analyze the system (17)-(21) one proceeds as follows. From (21), one concludes that C has a form similar to (11), i.e. $C = C(\tilde{y}; \psi, k|x)$. Expanding C as $C = C^{aux}(\tilde{y}; \psi_1, k|x) + C^{dyn}(\tilde{y}; \psi_1, k|x)\psi_2$, one identifies [1, 4] C^{aux} with some topological fields which carry no degrees of freedom, and C^{dyn} with the generating function for the spin 0 and spin 1/2 matter fields. Namely, in accordance with the normal spin-statistics, $\tilde{y}$-even (odd) part of C^{dyn} identifies with the generating function for spin 0 (1/2) matter fields, along with all their on-mass-shell non-trivial derivatives [4]. The equation (18) amounts to free field equations. Resolving the constraints (20), one reconstructs the auxiliary field $S_{1\alpha}$ as a linear functional of C, $S_{1\alpha} = S_{1\alpha}(C)$, up to a gauge ambiguity. Then, (19) allows one to express a part of degrees of freedom in W_1 via C, while the rest modes, which belong to the kernel of the mapping $[S_{0\alpha}, \ldots]_*$, remain free. These free modes are again arbitrary functions of $\tilde{y}_\alpha$, i.e.

$$W_1 = \omega(\tilde{y}; \psi, k|x) + \Delta W_1(C)\,, \tag{22}$$

where $\omega(\tilde{y}; \psi, k|x)$ corresponds to the HS gauge fields, and the dynamical equations for them are imposed by eq. (17) after (19) is solved. Eq. (17) describes the C-dependent first order corrections to the HS strengths for ω, which are argued below to vanish. In principle, one can proceed similarly in the highest orders. However, the computation complicates enormously in the second order, and one needs some efficient methods to proceed. The main goal of this talk is to develop a scheme which allows one to analyze the system systematically order by order.

2 Integrating Flow

A remarkable property of eqs. (4), (5) is that they admit a flow which allows one to express constructively solutions of the full system in terms of free fields. The main idea is based on the observation that, from the point of view of the system (4), (5), B behaves like a constant: it commutes to S_α and satisfies covariant constancy condition. On the other hand, a vacuum solution with $B = const$ is known explicitly which fact can be used to reconstruct the full dependence of B.

Since our perturbation expansion is just an expansion in powers of the physical fields which are identified with the deviation C of B from its vacuum value ν, let us introduce a formal perturbation expansion parameter η as follows

$$B(\eta) = \nu + \eta\mathcal{B}(\eta)\,. \tag{23}$$

Simultaneously, the rest of the fields acquire a formal dependence on η, $W = W(\eta)$ and $S = S(\eta)$. The system (4), (5) takes a form

$$dW = W * W\,, \quad d\mathcal{B} = W * \mathcal{B} - \mathcal{B} * W\,, \quad dS_\alpha = W * S_\alpha - S_\alpha * W\,, \tag{24}$$

$$S_\alpha * S^\alpha = -2i(1 + \nu K + \eta\mathcal{B} * K)\,, \quad S_\alpha * \mathcal{B} = \mathcal{B} * S_\alpha\,. \tag{25}$$

Now, one observes that for the limiting case $\eta = 0$ the system (24), (25) reduces to the free one. Indeed, setting

$$\nu = B_0\,, \quad \mathcal{B}(0) = B_1 \equiv C\,, \quad W(0) = W_0 \equiv \omega\,, \quad S_\alpha(0) = S_{0\alpha}\,,$$

we see that at the point $\eta = 0$, the system (24), (25) acquires the form of the vacuum system which is solved in terms of $S_\alpha = S_{0\alpha}$ with $W = \omega(\tilde{y};\psi,k|x)$ and $\mathcal{B} = C(\tilde{y};\psi,k|x)$ satisfying the free field equations $d\omega = \omega \wedge \omega$, $dC = \omega * C - C * \omega$. This situation is similar to the one with contractions of Lie algebras. For all values of $\eta \neq 0$, the systems of equations (24), (25) are pairwise equivalent since the field redefinition (23) is non-degenerate. On the other hand, although the field redefinition (23) degenerates at $\eta = 0$, eqs. (24), (25) still make sense for $\eta = 0$. This limiting system describes the free field dynamics.

Remarkably the two inequivalent systems are still related to each other. To show this let us define a flow with respect to η as follows:

$$\frac{\partial}{\partial\eta}W = \mathcal{B} * \frac{\partial}{\partial\nu}W\,, \quad \frac{\partial}{\partial\eta}\mathcal{B} = \mathcal{B} * \frac{\partial}{\partial\nu}\mathcal{B}\,, \quad \frac{\partial}{\partial\eta}S_\alpha = \mathcal{B} * \frac{\partial}{\partial\nu}S_\alpha\,. \tag{26}$$

By applying $\frac{\partial}{\partial\eta}$ to the both sides of eqs. (24), (25), one can easily check that eqs. (26) are compatible with (24), (25). Therefore, solving the system (26) with the initial data $\mathcal{B}(\eta=0) = C$, $W(\eta=0) = \omega$, $S_\alpha(\eta=0) = S_{0\alpha}$, we can express solutions of the full nonlinear system at $\eta = 1$ via solutions of the free system at $\eta = 0$. The fact of the existence of the integrating flow (26) is a formal realization of the original argument that B behaves as a constant: the meaning of (26) is that a derivative with respect to $\eta\mathcal{B}$ is the same as that with respect to[3] ν.

Such an approach is very efficient and allows one to derive order by order the relevant field redefinitions since the r.h.s.-s of (26) contain one extra power of $\mathcal{B}$. In particular, one can easily derive in the first order a field redefinition necessary to show that the HS gauge field strengths do not admit nontrivial sources linear in fields. (Let us note, that even at the linearized level, it is a complicated technical problem to find a form of such a transformation without using the flow (26).) In the first order this field redefinition can be shown to be essentially local. This result is expected since in the lowest order the non-trivial r.h.s.-s of the equations for HS gauge fields are nothing else as the HS currents which are expected to be bilinear in the matter fields.

Remarkably, the method works in all higher orders thus reducing the full non-linear problem to the free one. The point, however, is that beyond the first order one has to be careful in making statements on the locality of the mapping induced by the flow (26). Actually, although it does not contain explicitly space-time derivatives, it contains them implicitly via highest components $C_{\alpha_1\ldots\alpha_n}$ of the generating function $C(\tilde{y})$ which are identified with the highest derivatives of the matter fields according to the equations (18) [4]. For example, in the second order in C one gets $\frac{\partial}{\partial\eta}B_2(z,y) = C(\tilde{y}) * \frac{\partial}{\partial\nu}C(\tilde{y})$. As a result of using the $*$-product, for each fixed rank multispinorial component of the l.h.s. of this formula, its r.h.s. is an infinite series involving bilinear combinations of the components $C_{\alpha_1\ldots\alpha_n}$ with all n. Therefore, the r.h.s.-s of (26) effectively contain all orders of the space-time derivatives, i.e. the transformation

[3] All fields acquire a non-trivial dependence on ν via the vacuum solution as it follows e.g. from eqs. (22), (12).

laws resulting from (26) can describe some non-local transformations. This means that one cannot treat the system (4), (5) as locally equivalent to the free system.

To illustrate this issue it is instructive to consider an example of some matter field C interacting with the gravitational field fluctuating near the AdS vacuum solution. Schematically, the mechanism is as follows. Linearized Einstein equations have a form (with appropriate gauge fixings)

$$(L^C - \Lambda^2)h_{\mu\nu} = T_{\mu\nu}(C)\,, \tag{27}$$

where $h_{\mu\nu}$ is the fluctuational part of the metric tensor, L^C is the linear operator corresponding to the l.h.s. of the free field equations of the matter fields $L^C C = 0$, while $\Lambda = \alpha\lambda$ with some numerical coefficient $\alpha \neq 0$. It is important that when the cosmological constant is non-vanishing, the term with Λ^2 turns out to be non-vanishing too. This property allows one to solve formally (27) by a field redefinition

$$h_{\mu\nu} \to h'_{\mu\nu} = h_{\mu\nu} - (L^C - \Lambda^2)^{-1}T_{\mu\nu}(C) = h_{\mu\nu} + \Lambda^{-2}\sum_{n=0}^{\infty}(\Lambda^{-2}L^C)^n T_{\mu\nu}(C)\,. \tag{28}$$

It is also clear that a non-vanishing dimensionful constant, the cosmological constant, plays an important role in this analysis. In [5] we show that this field redefinition admits a natural realization in terms of the generating function C, which is similar to the one to be derived with the aid of the integrating flow (26).

3 N=2 Supersymmetry

The full system (4), (5) is explicitly invariant under the infinite-dimensional HS gauge transformations (8). Fixation of the vacuum solution (9), (10), (15) breaks this local symmetry down to some global symmetry, the symmetry of the vacuum. This global symmetry is generated by the parameter $\varepsilon_{gl}(x)$ obeying the conditions $d\varepsilon_{gl} = [W_0\,,\, \varepsilon_{gl}]_*$, $[\varepsilon_{gl}\,,\, S_{0\alpha}]_* = 0$, which follow from the requirement that $\delta W_0 = 0$, $\delta S_{0\alpha} = 0$ under the transformations (8) ($\delta B_0 = 0$ holds automatically). The condition $[\varepsilon_{gl}\,,\, S_{0\alpha}]_* = 0$ implies that $\varepsilon_{gl} = \varepsilon_{gl}(\tilde{y};\psi,k|x)$. The dependence of ε_{gl} on the space-time coordinates x_μ is fixed by the differential equation $d\varepsilon_{gl} = [W_0\,,\, \varepsilon_{gl}]_*$ with arbitrary initial data $\varepsilon_{gl}(x_0) = \varepsilon^0_{gl}$ at any space-time point x_0. This global symmetry is also the symmetry of the linearized system (17)-(21).

The full global symmetry algebra contains elements ε_{gl} linear in ψ_2. This part of the symmetry mixes the matter and topological modes in C and does not allow unitary realization on quantum states [5]. We therefore analyze the subalgebra A of the full global symmetry algebra with the ψ_2-independent parameters[4].

The algebra A is infinite-dimensional due to the dependence of $\varepsilon_{gl}(\tilde{y};\psi_1,k|x)$ on $\tilde{y}_\alpha$. A maximal finite-dimensional subalgebra of A, $osp(2,2)\oplus osp(2,2)$, is spanned by the generators $\Pi_\pm T^A$, where $\Pi_\pm = \frac{1}{2}(1\pm\psi_1)$ and $T^A = \{T_{\alpha\beta} = \frac{1}{4i}\{\tilde{y}_\alpha,\tilde{y}_\beta\}_*\,, Q^{(1)}_\alpha =$

[4] Also we factor out a trivial central element corresponding to a constant parameter $\varepsilon_{gl}(\tilde{y};\psi_1,k|x)=\varepsilon_{gl}$.

$\tilde{y}_\alpha, Q^{(2)}_\alpha = \tilde{y}_\alpha k, J = k + \nu\}$. The fact that the generators T^A close to $osp(2,2)$ was first shown in [6]. As shown in [4], the dynamical components $C^{dyn}(\tilde{y}; k, \psi_1|x)\, \psi_2$ decompose into two bosonic and two fermionic infinite-dimensional representations of the AdS algebra $o(2,2)$, each describing a single AdS particle. The doubling of fields is due to the presence of the operator k which allows one to project out invariant subspaces as $C_\pm = P_\pm C$, $P_\pm = \frac{1\pm k}{2}$. The values of mass are related to the parameter ν as follows: $M^2_\pm = \frac{\nu(\nu\mp 2)}{2}$ for bosons, and $M^2_\pm = \frac{\nu^2}{2}$ for fermions [4]. Here $\pm$ originates from $C_\pm$ and the parameter of mass is scaled in units of the inverse AdS radius λ which for simplicity was set equal to unity in the most of this talk. These free fields form altogether an irreducible d3 N=2 hypermultiplet. Thus the proposed field equations describe HS interactions of a N=2 massive hypermultiplet.

4 Conclusions

The main conclusion of this talk is that dynamical systems based on infinite sets of HS gauge fields admit deep structures which allow their constructive perturbative solvability. It is argued that the resulting solution is essentially non-local. A non-local character of the transformation manifests itself in the appearance of an infinite series in the inverse cosmological constant. Taking into account that HS gauge interactions are known [7] to require non-analyticity in the cosmological constant, we conjecture that HS gauge theories are indeed non-local in a certain sense. This conjecture agrees with the light-cone analysis of [8] and fits the ideology of the modern string theory as well as the property that HS models possess N=2 SUSY.

This research was supported in part by INTAS, Grant No.96-0538 and by the RFBR Grant No.96-01-01144. S. P. acknowledges a partial support from the Landau Scholarship Foundation.

References

[1] M. A. Vasiliev, *Nucl. Phys. B (Proc. Supplement)* **56B** (1997) 241-252

[2] M. P. Blencowe, *Class. Quantum Grav.* **6** (1989) 443

[3] M. A. Vasiliev, *JETP Lett.* **50** (1989), 374; *Int. J. Mod. Phys.* **A6** (1991) 1115

[4] A. V. Barabanschikov, S. F. Prokushkin, and M. A. Vasiliev, *Rus. Theor. Math. Phys.* **110** (1997) 295; hep-th/9609034

[5] S. F. Prokushkin and M. A. Vasiliev, in preparation.

[6] E. Bergshoeff, B. de Wit and M. A. Vasiliev, *Nucl. Phys.* **B366** (1991) 315

[7] E. S. Fradkin and M. A. Vasiliev, Phys. Lett. **B189** (1987) 89

[8] R. R. Metsaev, *Mod. Phys. Lett.* **A4** (1991) 359

D-instantons and black holes in N=2, D=4 supergravity

K. Behrndt [1] , I. Gaida [2] , D. Lüst [1] , S. Mahapatra [3] and T. Mohaupt [4]

[1] Institut für Physik, Humboldt-Universität, Invalidenstr. 110, D-10115 Berlin, Germany,
[2] DAMTP, University of Cambridge, Silver street, Cambridge CB3 9EW, UK,
[3] Physics Department, Utkal University, Bhubaneswar-751004, India,
[4] Martin-Luther Universität, Fachbereich Physik, D-06099 Halle, Germany.

Abstract: We discuss T-duality between the solutions of type II A versus type II B superstrings compactified on Calabi-Yau threefolds. Within the context of $N = 2$, $D = 4$ supergravity effective Lagrangian, the T- duality transformation is equivalently described by the c-map, which transforms the IIA black hole solutions into the IIB D-instanton solutions. We construct via this mapping a broad class of D-instanton solutions in four dimensions.

1 Introduction

Much of the recent progress in non-perturbative string and field theory is due to the insight that open strings with Dirichlet boundary conditions can be used to describe type I and type II string theory in non-perturbative, solitonic or instanton backgrounds [1]. Specifically, the (D)-p-branes, carrying the charges of the Ramond-Ramond gauge potentials, arise as solutions of type IIA supergravity in ten dimensions for p even, whereas the D-p-branes, with p odd, appear as solitons in the type IIB superstring.

An interesting class of non-perturbative solutions of type IIB supergravity breaking half of the supersymmetries, are given by the ten-dimensional D-instantons [2], *i.e.* (-1)-branes in Euclidean space. We shall investigate the T-duality between the supersymmetric solitonic solutions of the IIA and IIB theories, compactified on the same Calabi-Yau threefold. In particular, we will map the four-dimensional $N = 2$ IIA black holes to D-instanton solutions of the type IIB $N = 2$ effective actions. In fact, T-duality with respect to time cordinate among the four-dimensional $N = 2$ IIA and IIB superstrings, compactified on the same Calabi-Yau space, is completely equivalent to the $N = 2$ c-map between these two theories [3, 4], which (almost) exchanges the number of $N = 2$ vector and $N = 2$ hypermultiplets. So the T-duality transformations relate the special Kähler moduli space $K^{(A)}_{h_{1,1}}$ of complex dimension $h_{1,1}$ of the IIA vector multiplets to the quaternionic moduli space $Q^{(B)}_{h_{1,1}+1}$ of quaternionic dimensions $h_{1,1} + 1$ of the hypermultiplets (and vice versa). Applying the T-duality or c-map on known type IIA black hole solutions, we find a full new class of type IIB D-instanton type of solutions. Just as the IIA black hole solutions, the IIB D-instanton solutions are then determined by a set of harmonic functions plus the topological data of the underlying Calabi-Yau space.

2 D-branes and D-instantons in ten dimensions

The notion of a D-brane arises when one considers open strings with Dirichlet boundary conditions along $9-p$ directions. D-branes can exist for any values $-1 \leq p \leq 9$ of p. Within the low energy effective action, D-branes can be identified with BPS saturated, R-R charged p-brane solutions. Such p-brane solutions can be characterized in terms of a harmonic function given by

$$ds_p^2 = \frac{1}{\sqrt{H}}(-dt^2 + (dx^1)^2 + \ldots + (dx^p)^2) + \sqrt{H}((ds^{p+1})^2 + \ldots + (dx^9)^2 \quad (1)$$

and the dilaton is

$$e^{-2\Phi} = H^{\frac{p-3}{2}} \quad (2)$$

where, $H = H(r) = c_1 + \frac{c_2}{r^{7-p}}$ and the transversal radius r is given by, $r = \sqrt{x_t.x_t}$. The gauge field can also be expressed in terms of the harmonic function. A p-brane is T-dual to $(p \pm 1)$-brane by dualizing over a transversal direction or a world volume direction respectively. The -1-brane is interpreted as a D-instanton [2]. The associated gauge field is a 0-form having an axion like shift symmetry. The dual theory is described in terms of a 7-brane which couples to a 8-form gauge field.

The ten-dimensional action in terms of the R-R 0-form A is given by,

$$S = \int d^{10}x\sqrt{G}\{R - \frac{1}{2}(\partial\Phi)^2 - \frac{1}{2}e^{2\Phi}(\partial A)^2\} \quad (3)$$

This action can also be written in terms of the dual 9-form $F^{(9)} = e^{2\Phi} \star dA$,

$$S' = \int d^{10}x\sqrt{G}\{R - \frac{1}{2}(\partial\Phi)^2 - \frac{1}{2\cdot 9!}e^{-2\Phi}(F^{(9)})^2\} \quad (4)$$

These actions are written in Einstein frame with signature $(-+\cdots+)$. The Euclidean action is obtained by Wick rotation and is given by,

$$S_{Euc} = \int d^{10}x\sqrt{G}\{-R + \frac{1}{2}(\partial\Phi)^2 - \frac{1}{2}e^{2\Phi}(\partial A)^2\} + \oint e^{2\Phi}A \wedge \star dA \quad (5)$$

The boundary term is crucial in finding the correct instanton action. The D-instanton solution can be derived either from the above action or the dual Euclidean 9-form action. The instanton ansatz is given by [2],

$$dA = \pm e^{-\Phi} d\Phi \tag{6}$$

With this ansatz, the equation of motion are solved by taking the Einstein metric to be flat and e^{Φ} to be harmonic. Thus a single D-instanton is described by,

$$e^{\Phi} = e^{\Phi_{\infty}} + \frac{c}{r^8} = H(r) \tag{7}$$

Here, e^{Φ} is singular at $r = 0$ and the metric in the string frame is asymptotically flat for $r \to 0$ and describes a finite neck wormhole connecting two asymptotically flat regions at $r \to 0, \infty$. The instanton action is computed to be

$$S_{Inst} = \frac{|Q^{(-1)}|}{g} \tag{8}$$

where, $Q^{(-1)}$ is the electric charge of the solution with respect to A. The bulk action vanishes and the instanton action comes purely from the boundary term.

By T-duality, the D-instanton solution of IIB can be related to a (Euclideanized) 0-brane solution of IIA, *i.e.* to a black hole. In order to get non-degenerate black holes and instantons in four dimensions, one has to compactify the ten dimensional 0-brane and (-1)-brane solutions as well as to add further 4-branes and 3-branes with a particular intersection pattern. A four dimensional Reissner-Nordstrom like black hole results from a $(0, 4, 4, 4)$ pattern with 4-branes pairwise intersecting along planes and triple intersection in a point. In terms of the four dimensional dilaton φ, $e^{2\varphi} = \sqrt{H_0 H^1 H^2 H^3}$ the metric is $ds_4^2 = -e^{-2\varphi}dt^2 + e^{2\varphi}d\mathbf{x}^2$. The T-dual configuration has a $(-1, 3, 3, 3)$ pattern and its four dimensional metric $ds^2 = e^{2\varphi}(dt^2 + d\mathbf{x})^2$ which is a finite neck wormhole. In each case the harmonic functions H_0, H^1, H^2, H^3, which describe the component D-branes of the solution, depend only on the respective overall transversal radius in order to describe a localized black hole or instanton.

The simplest solution is obtained by taking all four harmonic functions to be equal and the solution is given by,

$$e^{\varphi} = e^{\varphi(\infty)} + \frac{q_0}{r^2}; \quad A = \mp \frac{1}{e^{\phi_{\infty}} + \frac{q_0}{r^2}} \tag{9}$$

This represents the direct four dimensional analogue of the ten dimensional finite neck wormhole.

The black hole solutions we have obtained here by a toroidal compactification from ten to four dimensions, correspond to solutions of $N = 8$ supergravity in ten dimensions. Since we are interested in $N = 2$ supersymmetry corresponding to compactification on a Calabi-Yau threefold, it is appropriate to start with the general $D = 4$ action and solve the corresponding equation of motion.

3 N=2, D=4 type II string theory on Calabi-Yau manifolds

The action of the effective $N = 2$, $D = 4$ supergravity theory contains a gravity multiplet $(e_\mu^m, \mathcal{A}_\mu, \psi_\mu^i), m, \mu = 0, \cdots, 3,\ i = 1, 2$ containing graviton, the graviphoton and two gravitini. Second there are N_V vector multiplets $(\mathcal{A}_\mu^A, z^A, \lambda_i^A), A = 1, \cdots, N_V, i = 1, 2$ consisting of a vector, a complex scalar and two gauginos (weyl spinors). Third there are N_H hypermultiplets $(q^u, \xi_\alpha), u = 1, \cdots, 4N_H, a = 1, \ldots, 2N_H$ consisting of 4 real scalars and two weyl spinors. The structure of $N = 2$ supergravity coupled to vector and hypermultiplets is governed by special geometry [5, 6] and the corresponding action is completely specified in terms of the special Kähler manifold K_{N_V} parametrized by the vector multiplet scalars z^A with metric $g_{A\bar{B}}$, the quaternionic manifold Q_{N_H} parametrized by the hypermultiplet scalars q^u with metric $g_{u\bar{v}}$ and the gauge group.

For ungauged supergravity, the gauge group is $\mathcal{G} = U(1)^{N_V+1}$, where the additional $U(1)$ is due to the graviphoton. The moduli space is then a product space

$$\mathcal{M} = K_{N_V} \otimes Q_{N_H} \tag{10}$$

and the special Kähler manifold has the properties that the Kähler potential and all other physical quantities can be obtained from a holomorphic section $(X^I(z), F_I(z))$. In the hypermultiplet moduli space, the holonomy group must be contained in $SU(2) \otimes Sp(2N_H)$ and the $SU(2)$ part of the curvature must be non-trivial [7]. For these manifolds, the curvature scalar must be constant, negative and is fixed by the number of hypermultiplets.

The bosonic part of the action of ungauged $N = 2$ supergravity with N_V vector and N_H hypermultiplets is given by [6],

$$S = \int d^4x\sqrt{G}\{R - 2g_{A\bar{B}}\partial_\mu z^A \partial^\mu \bar{z}^B - \frac{1}{4}\left(\Im\mathcal{N}_{IJ}F_{\mu\nu}^I F^{J|\mu\nu} + \Re\mathcal{N}_{IJ}F_{\mu\nu}^I \star F^{J|\mu\nu}\right) \tag{11}$$

$$-\tilde{g}_{uv}\partial_\mu q^u \partial^\mu q^v\}$$

Here, $G_{\mu\nu}$ is the space-time metric in the Einstein frame and R is the curvature scalar. $F_{\mu\nu}^I$ are the field strength of the $N_V + 1$ vectors in a symplectic basis and $\star F_{\mu\nu}^I$ are the (Hodge-)dual field strength. $\mathcal{N}_{IJ}(z, \bar{z})$ is the gauge kinetic matrix. The Kähler metric is derived from the Kähler potential by $g_{A\bar{B}} = \partial_A\partial_B K$ with $K = -\log(i(\bar{X}^I(\bar{z})F_I(z) - X^I(z)\bar{F}_I(\bar{z})))$, whereas the gauge kinetic matrix is computed by,

$$\mathcal{N}_{IJ} = \bar{F}_{IJ} + 2i\frac{\Im F_{IK}\Im F_{JL}X^K X^L}{\Im F_{KL}X^K X^L} \tag{12}$$

where we have assumed that a prepotential F exists and $F_{IJ} = \frac{\partial^2 F}{\partial X^I \partial X^J}$.

The equations of motion following from the above action for the case of vanishing fermions can now be written down.

Consider now the type IIA theory compactified on a Calabi-Yau threefold with Hodge numbers $h_{1,1}$ and $h_{2,1}$. The resulting spectrum is given by $N = 2$ gravity multiplet, $N_V^{(A)} = h_{1,1}$ vector multiplets and $N_H^A = h_{2,1} + 1$ hypermultiplets. Now considering IIB supergravity on the same Calabi-Yau threefold, the resulting spectrum comes to be the gravity multiplet, $N_V^{(B)} = h_{2,1}$ vector multiplets and $N_H^{(B)} = h_{1,1} + 1$ hypermultiplets. Comparing the two spectra one realizes that the number of vector and hypermultiplets are (almost) interchanged:

$$N_V^{(A)} = h_{1,1} = N_H^{(B)} - 1, \quad N_H^{(A)} = h_{2,1} = N_V^{(B)} \tag{13}$$

This is explained by the c-map which relates the two moduli spaces:

$$c : K_{h_{1,1}}^{(A)} \times Q_{h_{2,1}+1}^{(A)} \leftrightarrow Q_{h_{1,1}+1}^{(B)} \times K_{h_{2,1}}^{(B)} \tag{14}$$

We call the above action (involving the gravity and vector multiplet sector and setting the fermions to zero) as the type IIA theory action and corresponding dual action as the type IIB action. In order to obtain the dual action, one compactifies the D=4 type IIA theory to D=3 along an isometry direction and decompactifies over the inverse radius [4]. At the end, we have $4(N_V + 1)$ real scalars, which are identified with the scalars coming from the compactification of the hypermultiplet section in the dual type IIB theory. After decompactification, the hypermultiplet part of the dual action is given by [4],

$$\begin{aligned} S_{IIB}^H = \int d^4x\sqrt{G^{(IIB)}}\{ & -2g_{A\bar{B}}\partial_\mu z^A\partial^\mu\bar{z}^{\bar{B}} - \frac{1}{2\phi^2}(\partial\phi)^2 - \frac{1}{2\phi^2}(\partial\tilde{\phi} + \zeta^I\partial\tilde{\zeta}_I - \partial\zeta^I\tilde{\zeta}_I)^2 \\ & -\frac{1}{\phi}\partial\zeta^I\Im\mathcal{N}_{IJ}\partial\zeta^J - \frac{1}{\phi}(\partial\tilde{\zeta}_I + \Re\mathcal{N}_{IK}\partial\zeta^K)\Im\mathcal{N}^{IJ}(\partial\tilde{\zeta}_J + \Re\mathcal{N}_{JL}\zeta^L)\} \end{aligned} \tag{15}$$

where $\mathcal{N}^{IJ}$ is the inverse of $\mathcal{N}_{IJ}$. One can combine the $2N_V^{(A)} + 4$ real scalars ϕ, $\tilde{\phi}$, ζ^I, $\tilde{\zeta}_I$ into $N_V^{(A)} + 2$ complex scalars

$$\begin{aligned} S' &= \phi - i\zeta^I\mathcal{N}_{IJ}\zeta^J + i\tilde{\phi} - i\zeta^I\tilde{\zeta}_I, \\ C_I &= -\Im\mathcal{N}_{IJ}\zeta^J + i(\tilde{\zeta}_I + \Re\mathcal{N}_{IJ}\zeta^J) \end{aligned} \tag{16}$$

and write down the above action in terms of S', C_I, z^A and show that the target space is quaternionic by computing the corresponding curvature and curvature scalar.

4 Reissner-Nordstrom black holes in pure supergravity and D-instantons in D=4

We now discuss the construction of explicit solutions of the equations of motion. During the last year, various interesting static and stationary solutions to the gravity and vector multiplet equations have been obtained. Using the T-duality or c-map, we can relate any such solution to configuration with nontrivial hypermultiplet scalars.

As an example, here we relate the Reissner-Nordstrom black hole solutions in pure supergravity (IIA) to the D-instanton solution in IIB by c-map, which in this case exchanges the gravity multiplet containing the graviphoton with the universal hypermultiplet. The Lagrangian for pure supergravity reduces to,

$$e^{-1}\mathcal{L} = R - \frac{1}{4}F_{\mu\nu}F^{\mu\nu} \tag{17}$$

The solution of the equations of motion is given by the Reissner- Nordstrom solution and the metric in Einstein frame is given by,

$$ds_E^2 = -e^{2U(r)}dt^2 + e^{-2U(r)}d\mathbf{x}^2 \tag{18}$$

Restricting ourselves to pure electric solution $F_{\mu\nu} \sim F_{0m} = -\partial_m A$, we find from the equation of motion that the pure electric type Reissner- Nordstrom black hole is specified by the harmonic function

$$e^{-U(r)} = e^{-U_\infty} + \frac{q_0}{r} \tag{19}$$

In the following we assume flat space at infinity ($e^{-U_\infty} = 1$). The entropy and the **ADM** mass are given by,

$$S_{BH} = \pi(r^2 g_{rr})_{r=0} = \pi q_0^2, \quad M_{\mathbf{ADM}} = q_0 \tag{20}$$

These expressions agree with the recent observations that in N=2, D=4 supergravity, the entropy and the mass of a BPS saturated black hole are completely determined by the corresponding prepotential [8], namely,

$$S_{BH} = \pi|Z|^2; \quad M_{BPS}^2 = |Z|^2 = e^{K(z,\bar{z})}|q_I X^I(z) - p^I F_I(z)|^2. \tag{21}$$

where Z is the central charge, q_I and p^I are the electric and magnetic charges respectively. Hence, the corresponding electric Reissner-Nordstrom black hole preserves one half of the $N = 2$ supersymmetry.

Now performing the c-map, the dual Lagrangian in terms of real fields for the type IIB theory is given by,

$$e^{-1}\mathcal{L} = R - 2(\partial\varphi)^2 - e^{2\varphi}(\partial\zeta^0)^2 \tag{22}$$

The Euclidean version of this action represents the D-instanton action in type IIB theory. By performing a Wick rotation $\zeta^0 \to i\zeta^0$, the Euclidean action is given by,

$$e^{-1}\mathcal{L}_E = R - 2(\partial\varphi)^2 + e^{2\varphi}(\partial\zeta^0)^2 + \text{boundary terms} \tag{23}$$

We consider the instanton ansatz $d\zeta^0 = \sqrt{2}e^{-\varphi}d\varphi$ and flat Euclidean metric $g_{\mu\nu} = \delta_{\mu\nu}$. The general spherically symmetric solution depending on all four coordinates ($r^2 = x^2 + y^2 + z^2 + t^2$) is given by,

$$e^{\varphi} = e^{\varphi_\infty} + \frac{q_0}{r^2} \tag{24}$$

This is the D-instanton solution in four dimensions. The dilaton is singular at the origin. The D-instanton solution in string frame is given by,

$$ds_S^2 = (e^{\varphi_\infty} + \frac{q_0}{r^2})^2 (dr^2 + r^2 d\Omega_3^2) \quad (25)$$

which is invariant under the transformation, $r \to \frac{q_0 e^{-\varphi_\infty}}{r}$. In the string frame, the instanton solution is a wormhole solution connecting two asymptotically flat Euclidean regions by a neck. This solution is expected to be sypersymmetric as the corresponding black hole solution in IIA theory is supersymmetric. The non zero instanton action comes only from the boundary term and is given by,

$$S_{Inst} = 8\pi \frac{|q_0|}{g} \quad (26)$$

The $\frac{1}{g}$ dependence shows its origin from the R-R sector.

To explore the quaternionic geometry associated with D- instanton, it is convenient to express the real scalar fields in terms of the complex scalar fields S' and C_0 and the complex fields parametrize the coset $\frac{SU(1,2)}{U(1)\times SU(2)}$. One can compute the coset metric as well as the curvatrure two form etc. by introducing the complex one forms and the quaternionic vierbeins. One then obtains the result that the scalar manifold of the N=2 universal hypermultiplet is quaternionic in agreement with [4].

We then consider the more general case of N_V vector multiplets. Because of the complexity of the computation, we restrict ourselves to the case when $z^A = const$, which corresponds to the case of double extreme black holes given by the configurations where the moduli take constant values all the way from the horizon upto spatial infinity [8]. In this case, the nontrivial scalars parametrize the Kähler manifold $\frac{SU(2+N_V,1)}{SU(2+N_V)\otimes U(1)}$. One can also compute the coset metric and associated connection in terms of the shifted scalar field S and C_I, where $S' \to S = S' - \frac{1}{2}C_I(\Im\mathcal{N})^{IJ}C_J$. The Kähler potential is given by,

$$\tilde{K} = -\log(S + \bar{S} - C_I(\Im\mathcal{N})^{IJ}\bar{C}_J) \quad (27)$$

Now consider dualizing an axion-free double extreme black hole solution corresponding to a static solution via c-map. The hypermultiplet part of the dual type IIB action is given by,

$$S_{IIB}^H = \int d^4x \{\sqrt{G^{(IIB)}}\{-2(\partial\varphi)^2 - e^{2\varphi}\partial\zeta^0\Im\mathcal{N}_{00}\partial\zeta^0 - 2\varphi\partial\tilde{\zeta}_A\Im\mathcal{N}^{AB}\partial\tilde{\zeta}_B\} \quad (28)$$

The Euclidean version of this action is the generalization of the D-instanton action. For a cubic prepotential of the form $F = d_{ABC}\frac{X^A X^B X^C}{X^0}$ and the axion-free case ($\Re z^A = 0$), the Bekenstein-Hawking entropy is given in terms of the minimized central charge

$$\frac{1}{\pi}S_{BH} = \frac{1}{4\pi}A = |Z_{min}|^2 = \sqrt{4q_0 d_{ABC}p^A p^B p^C} \quad (29)$$

In the type IIB side, the D-instanton solution is characerized by,

$$e^{2\varphi} = (1 + \frac{|Z_{min}|}{r^2})^2 \tag{30}$$

The solution has an inversion symmetry $r \to \frac{|Z_{min}|}{r}$ and the neck of the wormhole is localized at the self-dual radius $r = |Z_{min}|^{-1/2}$, which implies that the IIA central charge indeed characterizes the wormhole geometry.

We have also considered the examples of axionic and stationary solutions on IIA side [9] and the associated c-map. For details, we refer to our paper [10].

5 Conclusion

Here, we discussed the T-duality between the extremal black hole type solutions in compactified four-dimensional IIA strings and the D-instanton solution in IIB superstrings compactified on the same Calabi-Yau space. In the context of the effective four-dimensional N=2 supergravity, the T-duality precisely acts like the c-map which exchanges the special Kähler moduli space of IIA N=2 vector multiplets to the quaternionic moduli space of the N=2 hypermultiplets on the type IIB side. We explicitly worked out several interesting examples to show this correspondence. It is worth exploring the meaning of IIA black hole entropy in the context of IIB D-instantons as well as the relation of the self-dual radius to the macroscopic D-brane counting. Finally, these D-instanton solutions will be relevant for the computation of stringy instanton effects.

Acknowledgement

We thank the organizers for giving us the opportunity to report our work in the Symposium.

References

[1] J. Dai, R. G. Leigh and J. Polchinski, *Mod. Phys. Lett.* **A4** (1989) 2073
J. Polchinski, *Phys. Rev. Lett.* **75** (1995) 4724.

[2] G. W. Gibbons, M. B. Green and M. J. Perry, *Phys. Lett.* **B370** (1996) 37

[3] S. Cecotti, S. Ferrara and L. Girardello, *Int. Jour. Mod. Phys.* **A10** (1989) 2475

[4] S. Ferrara and S. Sabharwal, *Nucl. Phys.* **B322** (1990) 317.

[5] B. de Wit, P. G. Lauwers and A. van Proeyen, *Nucl. Phys.* **B255** (1985) 569
B. Craps, R. Roose, W. Troost and A. Van Proeyen, hep-th/9703082
B. de Wit, F. Vanderseypen and A. Van Proeyn, *Nucl. Phys.* **B400** (1993) 463.

[6] L.Andrianopoli, M. Bertolini, A. Ceresole, R. D'Auria, S. Ferrara, P. Fré and T. Magri, hep-th/9605032.

[7] J. Bagger and E. Witten, *Nucl. Phys.* **B222** (1983) 1.

[8] S. Ferrara, R. Kallosh, *Phys. Rev.* **D52** (1995) 5412; A. Strominger, *Phys. Lett.* **B383** (1996)39; S. Ferrara and R. Kallosh, *Phys. Rev.* **D54** (1996) 1514; K. Behrndt, G. Lopes Cardoso, B. de Wit, R. Kallosh, D. Lüst and T. Mohaupt, *Nucl. Phys.* **B488**(1997) 236.

[9] K. Behrndt, D. Lüst and W. Sabra, hep-th/9705169; K. Behrndt, *Phys. Lett.* **B396** (1997) 77.

[10] K. Behrndt, I. Gaida, D. Lüst, S. Mahapatra and T. Mohaupt, *Nucl. Phys.* **B508** (1997) 659.

The hypermultiplet low-energy effective action, N=2 supersymmetry breaking and confinement

Sergei V. Ketov

Institut für Theoretische Physik, Universität Hannover,
Appelstrasse 2, D–30167 Hannover

Abstract: Some exact solutions to the hypermultiplet low-energy effective action in $N = 2$ supersymmetric four-dimensional gauge field theories with massive 'quark' hypermultiplets are discussed. The need for a spontaneous $N = 2$ supersymmetry breaking is emphasized, because of its possible relevance in the search for an ultimate theoretical solution to the confinement problem.

1 Introduction

Despite of remarkable recent advances in string duality and brane technology, our knowledge about the non-perturbative string theory (= M-theory) is still very much dependent upon our understanding of non-perturbative quantum field theory like QCD. At high energies the QCD is well described by a perturbation theory because of its asymptotic freedom, in a good agreement with well-known experimental data about deep inelastic scattering and jet production. However, at low energies, the QCD vacuum is essentially non-perturbative, so that some of the most obvious experimental facts about strong interactions, e.g., the confinement of quarks inside hadrons, are still waiting for an ultimate theoretical solution. On the theoretical side, the quantum generating functional (or the effective action) of a non-abelian gauge field theory should be defined in practical terms, which would allow one to get a non-perturbative solution to the theory. Unfortunately, the corresponding path integral is usually defined in many ways beyond the perturbation theory (e.g., lattice field theory, instantons, duality), which makes getting an exact solution to be extremely difficult, if ever possible.

Therefore, it seems to be quite natural to take advantage of the existence of *exact* solutions to the low-energy effective action in certain $N = 2$ supersymmetric gauge field theories since the remarkable discovery of Seiberg and Witten [1], and apply them to the old problem of color confinement in QCD. In fact, it was one of the main motivations in the original work [1].

The most attractive mechanism for color confinement is known to be the dual Meissner effect or the dual (Type II) superconductivity [2]. It takes just three steps to connect an ordinary BCS superconductor to the simplest Seiberg-Witten model in quantum field theory: first, define a relativistic version of the superconductor, known as the (abelian) Higgs model in field theory, second, introduce a non-abelian version of the Higgs model, known as the Georgi-Glashow model, and, third, $N = 2$ supersymmetrize the Georgi-Glashow model in order to get the Seiberg-Witten model [1]. Since the t'Hooft-Polyakov monopole of the Georgi-Glashow model belongs to a (HP) hypermultiplet in its $N = 2$ supersymmetric (Seiberg-Witten) generalisation, it is quite natural to explain confinement as the result of a monopole condensation (dual Higgs effect), i.e. a non-vanishing vacuum expectation value for the magnetically charged (dual Higgs) scalars belonging to the HP hypermultiplet.

In fact, the exact solutions to the low-energy effective action in quantum gauge field theories are only available in $N = 2$ supersymmetry, and neither in $N = 1$ supersymmetry nor in the bosonic QCD. Hence, on the one side, it is the $N = 2$ supersymmetry that crucially simplifies an evaluation of the low-energy effective action. However, on the other side, it is the same $N = 2$ supersymmetry that is obviously incompatible with phenomenology e.g., because of equal masses of bosons and fermions inside $N = 2$ supermultiplets (it also applies to any $N \geq 1$ supersymmetry), and the non-chiral nature of $N = 2$ supersymmetry (e.g. 'quarks' then appear in real representations of the gauge group). Therefore, if we believe in the $N = 2$ supersymmetry, we should find a way of judicious $N = 2$ supersymmetry breaking.

The $N = 2$ supersymmetry can be broken either softly or spontaneously, if one wants to preserve the benefits of its presence (e.g. for the full control over the low-energy effective action) at high energies. As regards the *gauge* low-energy effective action, the information about it in the Seiberg-Witten approach is encoded in terms of holomorphic functions defined over the quantum moduli space whose modular group is identified with the duality group, while the functions themselves can be calculated exactly. In the $N = 2$ supersymmetric QCD, one has to add 'quark' hypermultiplets, which have some bare (BPS) masses, flavour and color, i.e. belong to the fundamental representation of the gauge group. In the full theory, one expects an appearance of additional (e.g. magnetically charged) degrees of freedom, to be described by some effective action via strong-weak coupling duality and depending upon the (Coulomb, Higgs or confinement) branch under consideration. The full low-energy effective action in the $N = 2$ super-QCD is given by a sum of the gauge and the hypermuliplet parts.

We would like to find a vacuum solution to the full $N = 2$ supersymmetric low-energy effective action, which would break supersymmetry due to the non-vanishing vacuum expectation value of a magnetically charged (Higgs) scalar, similarly to that in ref. [1]. The same dual Higgs mechanism may also be responsible for the chiral symmetry breaking and the appearance of the pion effective Lagrangian if the dual Higgs field has flavor charges also [1]. In fact, Seiberg and Witten used a mass term for the $N = 1$ chiral multiplet, which is a part of the $N = 2$ vector multiplet, in order to *softly* break $N = 2$ supersymmetry to $N = 1$ supersymmetry. As a result, they found a non-trivial vacuum solution with a monopole condensation and, hence, a confinement. The weak point of their approach is an *ad hoc* assumption about the existence of the mass gap, i.e. the mass term itself. It would be nice to derive the mass gap from the fundamental theory instead of postulating it. The $N = 2$ supersymmetry may be useful here since it severely constrains all possible ways of its soft (or spontaneous) breaking.

The *soft* breaking of $N = 2$ supersymmetry is a very practical approach to analyse the consequences of the Seiberg-Witten exact solution towards its possible phenomenological applications like a derivation of the pion lagrangian or the confinement problem in QCD. The general analysis of all possible soft $N = 2$ supersymmetry breaking patterns in the $N = 2$ supersymmetric QCD was recently given by Alvarez-Gaumé, Mariño and Zamora in ref. [3]. Though being quite pragmatic, the soft susy breaking has, however, a limited predictive power and too many free parameters. Hence, it makes sense to search for the patterns of *spontaneous* $N = 2$ supersymmetry breaking. In practice, this means finding a non-supersymmetric vacuum solution for the $N = 2$ supersymmetric scalar potential at the level of the low-energy effective action in $N = 2$ gauge theories. Since the $N = 2$ supersymmetry remains unbroken for any exact Seiberg-Witten solution in the gauge sector, we should consider the induced (i.e. quantum generated) scalar potentials in the hypermultiplet sector of an $N = 2$ gauge theory. Moreover, once we accepted $N = 2$ supersymmetry in field theory, we can also take into account those brane configurations of the underlying M-theory that are relevant for the four-dimensional $N = 2$ supersymmetric effective physics in the limit $M_{\text{Planck}} \to \infty$. The related brane-technology [4] can provide us with some additional insights into the non-perturbative field theory, as well as supply us with its geometrical interpretation. Since the relevant M-theory brane configurations with eight supercharges arise as the solitonic solutions to the effective equations of motion in the M-theory, their 'soft' deformation, which breaks some more of the supersymmetries but still remains to be a solution to the M-theory effective equations of motion, should be interpreted as a spontaneous supersymmetry breaking (see an example in sect. 3).

In sect. 2 we analyse the general problem of constructing the low-energy hypermultiplet effective action in $N = 2$ rigid (global) supersymmetry, by using the $N = 2$ harmonic superspace. In sect. 3 we give two simple (toy) examples of the non-trivial induced scalar potentials for a single matter hypermultiplet.

2 The hypermultiplet low-energy effective action

There are only two basic $N = 2$ supermultiplets (modulo classical duality transformations) in the rigid $N = 2$ supersymmetry (with the $SU(2)_A$ internal symmetry): an $N = 2$ vector multiplet and a hypermultiplet. The $N = 2$ vector multiplet components (in a WZ-like gauge) are $(A, \lambda^i_\alpha, V_\mu, D^{(ij)})$, where A is a complex Higgs scalar, λ^i is a chiral spinor ('gaugino') $SU(2)_A$ doublet, V_μ is a real gauge vector field, and $D^{(ij)}$ is an auxiliary $SU(2)_A$ scalar triplet $(i, j = 1, 2)$. Similarly, the on-shell physical components of the Fayet-Sohnius (FS)-type hypermultiplet are $(q^i, \psi_\alpha, \bar{\psi}_{\dot{\alpha}})$, where q^i is a complex scalar $SU(2)_A$ doublet, and ψ is a Dirac spinor. There exists another (dual) Howe-Stelle-Townsend (HST)-type hypermultiplet, whose on-shell physical components are $(\omega, \omega^{(ij)}, \chi^i_\alpha)$, where ω is a real scalar, $\omega^{(ij)}$ is a scalar $SU(2)_A$ triplet, and χ^i is a chiral spinor ('quark') $SU(2)_A$ doublet.

The universal (i.e. most general and off-shell) and manifestly $N = 2$ supersymmetric formulation of all $N = 2$ supersymmetric four-dimensional field theories is only possible in the $N = 2$ harmonic superspace (HSS) [5] (see e.g, ref. [6] for a recent introduction). The $N = 2$ HSS coordinates include extra bosonic variables (called harmonics $u^\pm_i$), which parametrize the sphere $S^2 \sim SU(2)/U(1)$, in addition to the standard $N = 2$ superspace coordinates. The harmonics play the role of twistors or spectral parameters known in the theory of integrable systems. In particular, an off-shell FS hypermultiplet in HSS is described by an analytic superfield q^+ of the $U(1)$-charge $(+1)$, whereas the HST hypermultiplet in HSS is described by a real analytic superfield ω of vanishing $U(1)$ charge. An $N = 2$ vector gauge multiplet is similarly described by an analytic HSS superfield V^{++} of the $U(1)$-charge $(+2)$, which is introduced as a connection to the basic HSS harmonic derivative D^{++} present in the kinetic terms of the hypermultiplet actions (see below).

The power of $N = 2$ superspace is clearly seen in the most general form of the $N = 2$ gauge low-energy effective action in the Coulomb branch,

$$\Gamma_V[W, \bar{W}] = \int_{\text{chiral}} \mathcal{F}(W) + \text{h.c.} + \int_{\text{full}} \mathcal{H}(W, \bar{W}) + \dots \,, \tag{1}$$

where the abelian field strength $W(V)$, which is a harmonic-independent, $N = 2$ chiral and gauge-invariant superfield, has been introduced. The leading term in eq. (1) is given by the chiral $N = 2$ superspace integral over a *holomorphic* function $\mathcal{F}$ of W, with the latter being valued in the Cartan subalgebra of the gauge group. The Seiberg-Witten approach provides a solution to the holomorphic function $\mathcal{F}$ in terms of the auxiliary Riemann surface Σ_{SW}. It appears to be a solution to the particular Riemann-Hilbert problem of fixing a holomorphic multi-valued function $\mathcal{F}$ by its given monodromy and singularities. The number (and nature) of the singularities is the physical input: they are identified with the appearance of massless non-perturbative BPS-like physical states (dyons) like the t'Hooft-Polyakov magnetic monopole. The monodromies are supplied by perturbative renormalization-group β-functions and S-duality. The next-to-leading-order term in eq. (1) is given by the full $N = 2$ superspace integral over a real function $\mathcal{H}$ of W and $\bar{W}$. Some partial results about this function are known [7]. The dots in eq. (1) stand for higher-order terms containing the derivatives of W and $\bar{W}$.

The most general form of the leading term in the hypermultiplet low-energy effective action can be written down in the $N = 2$ HSS as follows:

$$\Gamma_H[q^+, \overset{*}{q}{}^+; \omega] = \int_{\text{analytic}} \mathcal{K}^{(+4)}(q^+, \overset{*}{q}{}^+; \omega; u^\pm_i) + \dots \,, \tag{2}$$

where $\mathcal{K}^{(+4)}$ is a function of the FS analytic superfield q^+, its conjugate $\overset{*}{q}{}^+$, the HST analytic superfield ω and the harmonics $u^\pm_i$, with the overall $U(1)$-charge $(+4)$.

The action (2) is supposed to be added to the kinetic hypermultiplet action whose analytic Lagrangian is quadratic in q^+ or ω, and of $U(1)$-charge (+4). A free FS hypermultiplet action is given by

$$S[q] = -\int d\zeta^{(-4)} du \stackrel{*}{q}{}^{+} D^{++} q^{+} , \tag{3}$$

whereas its minimal coupling to an $N = 2$ gauge superfield reads

$$S[q, V] = -\int d\zeta^{(-4)} du \stackrel{*}{q}{}^{+} (D^{++} + iV^{++}) q^{+} . \tag{4}$$

Similarly, a free action of the HST hypermultiplet is given by

$$S[\omega] = -\tfrac{1}{2} \int d\zeta^{(-4)} du \, (D^{++}\omega)^2 , \tag{5}$$

and it is on-shell equivalent to the standard $N = 2$ tensor (or linear) multiplet action in the ordinary $N = 2$ superspace [6].

The function $\mathcal{K}$ is called the *hyper-Kähler potential.* In components, it automatically leads to the $N = 2$ supersymmetric non-linear sigma-model for the scalars with a *hyper-Kähler* metric, just because of the $N = 2$ supersymmetry by construction (see the examples in sect. 3). When being expanded in components, the first term in eq. (1) also leads to the certain Kähler non-linear sigma-model in the Higgs sector $(A, \bar{A})$. The corresponding Kähler potential $K_{\mathcal{F}}(A, \bar{A})$ is dictated by the holomorphic function $\mathcal{F}$ as $K_{\mathcal{F}} = \mathrm{Im}[\bar{A}\mathcal{F}'(A)]$, so that the function $\mathcal{F}$ plays the role of a potential for this *special* Kähler (but not hyper-Kähler) geometry $K_{\mathcal{F}}(A, \bar{A})$. As regards the hypermultiplet non-linear sigma-model of eqs. (2)–(5), a relation between the hyper-Kähler potential $\mathcal{K}$ and the corresponding Kähler potential $K_{\mathcal{K}}$ is much more involved. It is easy to see that the hyper-Kähler condition on a Kähler potential amounts to a non-linear (Monge-Ampere) partial differential equation. It is remarkable that the HSS approach allows one to get a formal 'solution' to any hyper-Kähler geometry in terms of an analytic scalar potential $\mathcal{K}$. However, the real problem is now translated into finding the relation between $\mathcal{K}$ and the corresponding Kähler potential (or metric) in components, whose determination amounts to solving infinitely many linear differential equations altogether, just in order to eliminate an infinite number of HSS auxiliary fields (sect. 3).

The gauge-invariant functions $\mathcal{F}(W)$ and $\mathcal{H}(W, \bar{W})$ receive both perturbative and non-perturbative contributions,

$$\mathcal{F} = \mathcal{F}_{\text{per.}} + \mathcal{F}_{\text{inst.}} , \qquad \mathcal{H} = \mathcal{H}_{\text{per.}} + \mathcal{H}_{\text{non-per.}} , \tag{6}$$

while the non-perturbative corrections to the holomorphic function $\mathcal{F}$ are entirely due to instantons. This is an important difference from the (bosonic) non-perturbative QCD whose low-energy effective action is dominated by instanton-antiinstanton contributions.

It is remarkable that the perturbative contributions to the leading and subleading terms in the $N = 2$ gauge effective action (1) come from the one loop only. As regards the leading holomorphic contribution, $N = 2$ supersymmetry puts the trace of the energy-momentum tensor $T_\mu{}^\mu$ and the axial or chiral anomaly $\partial_\mu j^\mu_R$ of the abelian R-symmetry into one $N = 2$ supermultiplet. The $T_\mu{}^\mu$ is essentially determined by the perturbative renormalization group β-function, $T_\mu{}^\mu \sim \beta(g)FF$, whereas the one-loop contribution to the chiral anomaly, $\partial \cdot j_R \sim C_{1-\text{loop}}F^*F$, is known to saturate the exact solution to the Wess-Zumino consistency condition for the same anomaly. Hence, $\beta_{\text{per.}}(g) = \beta_{1-\text{loop}}(g)$ by $N = 2$ supersymmetry also. Since the $\beta_{\text{per.}}(g)$ is effectively determined by the second derivative of $\mathcal{F}_{\text{per.}}$, one concludes that $\mathcal{F}_{\text{per.}} = \mathcal{F}_{1-\text{loop}}$ too. This simple component argument can be extended to a proof when using the $N = 2$ HSS approach [8]. It than becomes clear that the non-vanishing central charges of the $N = 2$ supersymmetry algebra are of crucial importance for the non-vanishing holomorphic contribution to the gauge effective action (1).

Similarly, the BPS mass of a hypermultiplet can only come from the central charges since, otherwise, the number of the massive hypermultiplet components has to be increased. The most natural way to introduce central charges $(Z, \bar{Z})$ is to identify them with spontaneously broken $U(1)$ generators of dimensional reduction from six dimensions via the Scherk-Schwarz mechanism [9]. It naturally leads to the additional 'connection' term in the four-dimensional harmonic derivative as

$$\mathcal{D}^{++} = D^{++} + v^{++} , \quad \text{where} \quad v^{++} = i(\theta^+\theta^+)\bar{Z} + i(\bar{\theta}^+\bar{\theta}^+)Z . \tag{7}$$

Therefore, the $N = 2$ central charges can be equally treated as a non-trivial $N = 2$ gauge background, with the covariantly constant $N = 2$ chiral superfield strength $\langle W \rangle = Z$.

3 Examples

We are still far from presenting a convincing pattern of spontaneous $N = 2$ supersymmetry breaking via the hypermultiplet low-energy effective action. Nevertheless, the examples that we already have, give some reasons for optimism. Our point here is quite simple: given non-trivial kinetic terms in the hypermultiplet low-energy effective action to be represented by the non-linear sigma-model, in a presence of non-vanishing central charges it leads to a non-trivial hypermultiplet scalar potential whose form is entirely determined by the hyper-Kähler metric of the kinetic terms and $N = 2$ supersymmetry.

The first example of this interesting connection was given in ref. [9]. Consider a single charged FS hypermultiplet q^+ in the Coulomb branch of the $N = 2$ gauge theory. As was shown in ref. [9], it has a unique non-trivial self-interaction whose form in the $N = 2$ HSS reads

$$LEEA[q^+]_{\text{Taub-NUT}} = \int_{\text{analytic}} \left[\overset{*}{q}{}^+ D^{++} q^+ + \frac{\lambda}{2} (q^+)^2 (\overset{*}{q}{}^+)^2 \right] , \tag{8}$$

where the induced coupling constant λ is given by

$$\lambda = \frac{g^4}{\pi^2} \left[\frac{1}{m^2} \ln\left(1 + \frac{m^2}{\Lambda^2}\right) - \frac{1}{\Lambda^2 + m^2} \right] , \tag{9}$$

in terms of the gauge coupling constant g, the hypermultiplet BPS mass $m^2 = |Z|^2$, and the IR-cutoff Λ. When using the parametrization

$$q^+\Big|_{\theta=0} = f^i(x)u_i^+ \exp\left[\lambda f^{(j}(x)\bar{f}^{k)}(x)u_j^+u_k^-\right] , \tag{10}$$

the bosonic terms take the form of the one non-linear sigma-model,

$$\begin{aligned} LEEA_{\text{bosonic}}[f] &= \\ \int d^4x \, \left\{g_{ij}(f)\partial_m f^i \partial^m f^j + \bar{g}^{ij}(f)\partial_m \bar{f}_i \partial^m \bar{f}_j + h^i{}_j(f)\partial_m f^j \partial^m \bar{f}_i - V(f)\right\} , \end{aligned} \tag{11}$$

whose metric turns out to be that of Taub-NUT or a KK-monopole (modulo field redefinitions), whereas the induced scalar potential is [9]

$$V(f) = |Z|^2 \frac{f\bar{f}}{1+\lambda f\bar{f}} . \tag{12}$$

A non-trivial hypermultiplet self-interaction for a single neutral HST-type ω-hypermultiplet can be non-perturbatively generated in the presence of non-vanishing constant $N=2$ Fayet-Iliopoulos (FI) term $\left\langle D^{(ij)}\right\rangle \equiv \xi^{(ij)} = \frac{1}{2}(\vec{\tau}\cdot\vec{\xi})^{ij}$, where $\vec{\tau}$ are Pauli matrices. The FI-term has a nice geometrical interpretation in the underlying ten-dimensional type-IIA superstring brane picture made out of two solitonic 5-branes located at particular values of $\vec{w} = (x^7, x^8, x^9)$ and some Dirichlet 4- and 6-branes, all having the four-dimensional spacetime (x^0, x^1, x^2, x^3) as the common macroscopic world-volume [4]. The values of $\vec{\xi}$ can then be identified with the *angles* at which the two 5-branes intersect, $\vec{\xi} = \vec{w}_1 - \vec{w}_2$, in the type-IIA picture [6]. The three hidden dimensions $(\vec{w})$ are identified by the requirements that they do not include the two hidden dimensions (x^4, x^5) already used to generate central charges in the effective four-dimensional field theory, and that they are to be orthogonal (in the effectively $N=2$ supersymmetric configuration) to the direction (x^6) in which the Dirichlet 4-branes are finite and terminate on 5-branes.

The unique low-energy effective action for the (dimensionless) ω-hypermultiplet in the presence of the FI-term reads [9]:

$$S_{EH}[\omega] = -\frac{1}{2\kappa^2}\int d\zeta^{(-4)}du \left\{\left(D^{++}\omega\right)^2 - \frac{(\xi^{++})^2}{\omega^2}\right\} , \tag{13}$$

where $\xi^{++} = u_i^+u_j^+\xi^{(ij)}$ is the FI-term, and κ is the coupling constant of dimension one (in units of length). After changing the variables to $q_a^+ = u_a^+\omega + u_a^- f^{++}$, and eliminating the Lagrange multiplier f^{++} via its algebraic equation of motion, one can rewrite eq. (13) to the equivalent gauge-invariant form

$$S_{EH}[q,V] = -\frac{1}{2\kappa^2}\int d\zeta^{(-4)}du \left\{q_A^{a+}D^{++}q_{aA}^+ + V^{++}\left(\tfrac{1}{2}\varepsilon^{AB}q_A^{a+}q_{Ba}^+ + \xi^{++}\right)\right\} , \tag{14}$$

in terms of *two* FS hypermultiplets q_{aA}^+ $(A=1,2)$ and the auxiliary real analytic $N=2$ vector superfield V^{++} [9], where we have introduced the pseudo-real notation

$q_a = (\overset{*}{\bar{q}}{}^{+}, q^{+})$ and $\varepsilon^{ab} q_b^{+} = q^{a+}$, $a = 1, 2$. It is now straightforward to calculate the bosonic terms in the HSS action (14), in terms of the scalar fields, $q_A^{+}\big| = f_A^i u_i^{+}$, and $f_A^i \equiv m_A^i \exp(i\varphi_A^i)$. One finds the constraint

$$\xi^{(ij)} = \bar{f}_1^{(i} f_2^{j)} - f_1^{(i} \bar{f}_2^{j)} , \tag{15}$$

leading to the Eguchi-Hanson metric for the kinetic terms, as well as the scalar potential [10]

$$V = \frac{Z\overline{Z}}{(f_1\bar{f}_1 + f_2\bar{f}_2)} \left[(f_1^i \bar{f}_{2i} - f_2^i \bar{f}_{1i})^2 + (f_1^i \bar{f}_{1i} + f_2^i \bar{f}_{2i})^2 \right] . \tag{16}$$

When choosing the direction $\xi^2 = \xi^3 = 0$ and $\xi^1 = 2i$, it is not difficult to solve the constraint (15) in terms of four independent fields $|f_2^1| \equiv m$, $|f_2^2| \equiv n$, $\varphi_1^1 \equiv \theta$, $\varphi_2^2 \equiv \phi$, where the local $U(1)$ invariance has been fixed by the gauge condition $\varphi_2^1 + \varphi_2^2 = \varphi_1^1 + \varphi_1^2$. One finds [10]

$$V = \frac{|Z|^2 \sin^2(\theta+\phi)}{m^2+n^2} \left[\frac{4(m^2-n^2)^2}{1+(m^2+n^2)^2 \sin^2(\theta+\phi)} + \frac{1+(m^2+n^2)^2 \sin^2(\theta+\phi)}{\sin^4(\theta+\phi)} \right] . \tag{17}$$

It is clear that the potential V is positively definite, and it is only non-vanishing due to the non-vanishing central charge $|Z|$. It signals the spontaneous breaking of $N = 2$ supersymmetry in our model.

References

[1] N. Seiberg and E. Witten, *Nucl. Phys.* **B426** (1994) 19; **B431** (1994) 484

[2] S. Mandelstam, *Phys. Rep.* **C23** (1976) 245
G. t'Hooft, *Phys. Scr.* **25** (1982) 133

[3] L. Alvarez-Gaumé, M. Marino and F. Zamora, *Softly broken N=2 QCD with massive quark hypermultiplets*, CERN preprints TH/97–37 and TH/97–144, hep-th 9703072 and 9707017

[4] A. Hanany and E. Witten, *Nucl. Phys.* **B492** (1997) 152
E. Witten, *Nucl. Phys.* **B500** (1997) 3

[5] A. Galperin, E. Ivanov, S. Kalitzin, V. Ogievetsky and E. Sokatchev, *Class. and Quantum Grav.* **1** (1984) 469

[6] S. Ketov, *On the exact solutions to quantum N=2 gauge theories*, DESY and Hannover preprint, DESY 97–199 and ITP-UH-26/97, hep-th 9710085

[7] M. Henningson, *Nucl. Phys.* **B458** (1996) 445
M. Matone, *Phys. Rev. Lett.* **78** (1997) 1412

S. Ketov, *On the next-to-leading-order correction to the effective action in N=2 gauge theories*, DESY and Hannover preprint, DESY 97–103 and ITP-UH-18/97, hep-th 9706079
J. de Boer, K. Hori, H. Ooguri and Y. Oz, *Kähler potential and higher derivative terms from M theory fivebrane*, Berkeley preprint, hep-th 9711143

[8] I. Buchbinder, S. Kuzenko and B. Ovrut, *On the D=4, N=2 non-renormalization theorem*, Philadelphia, Princeton and Hannover preprint, UPR-775T, IASSNS-97/109 and ITP-UH/25/97, hep-th 9710142

[9] E. Ivanov, S. Ketov and B. Zupnik, *Induced hypermultiplet self-interactions in N=2 gauge theories*, DESY, Hannover and Dubna preprint, DESY 97–094, ITP-UH-10/97 and JINR–97–164, hep-th 9706078

[10] S. Ketov and Ch. Unkmeir, *Induced scalar potentials for hypermultiplets*, DESY and Hannover preprint, DESY 97–206 and ITP-UH-28/97, hep-th 9710185

A detailed case study of the rigid limit in Special Kähler geometry using $K3$.

Marco Billó [1], **Frederik Denef** [1], **Pietro Frè** [2], **Igor Pesando** [2], **Walter Troost** [1], **Antoine Van Proeyen** [1] **and Daniela Zanon** [3]

[1] Instituut voor theoretische fysica,
Katholieke Universiteit Leuven, B-3001 Leuven, Belgium
[2] Dipartimento di Fisica Teorica dell' Università,
via P. Giuria 1, I-10125 Torino, Italy
[3] Dipartimento di Fisica dell'Università di Milano and
INFN, Sezione di Milano, via Celoria 16, I-20133 Milano, Italy.

Abstract: This is a résumé of an extensive investigation of some examples in which one obtains the rigid limit of $N = 2$ supergravity by means of an expansion around singular points in the moduli space of a Calabi-Yau 3-fold. We make extensive use of the $K3$ fibration of the Calabi-Yau manifolds which are considered. At the end the fibration parameter becomes the coordinate of the Riemann surface whose moduli space realises rigid $N = 2$ supersymmetry.

The vector multiplet of $N = 2$, $d = 4$ supersymmetry often occurs in the study of string dualities. This is related to the fact that $N = 2$ is the minimal supersymmetry to connect the scalars to the vectors, which in four dimensions have duality transformations between their electric and magnetic field strengths. These transformations are in a symplectic group, and therefore the structure of the manifold of the scalars also inherits this symplectic structure. The resulting geometry of these complex scalars is denoted as 'special Kähler geometry', and exists as well for rigid as for local supersymmetry. In both cases this geometry can be realised on complex structure moduli spaces: for rigid geometry on moduli spaces of a class of Riemann surfaces (RS), for local supersymmetry (supergravity) on moduli spaces of Calabi-Yau 3-folds (CY). The relevant objects which build the supersymmetric actions are 'periods', of 1-forms over 1-cycles in the case of RS, of 3-forms over 3-cycles in the case of CY. These forms depend on the moduli which are identified with the scalar fields of the supersymmetric theory.

Many relevant CY manifolds in string theory are $K3$-fibrations, and we will restrict to these. That means that the manifold can be described as a (complex 2-dimensional) $K3$ surface for which the moduli depend on the moduli of the CY but also on a complex variable, denoted as ζ, which can be viewed as the third complex dimension of the CY. Thus for any fixed value of ζ, the CY is a $K3$-manifold. As such e.g. the unique CY (3,0)-form will be represented as $d\zeta \wedge \Omega^{(2,0)}$, where the latter is the $(2,0)$-form of the $K3$. The 3-cycles of the CY manifolds on the other hand can be obtained in 2 different ways. One can consider a path between two (singular) points in the ζ-base space where the same $K3$ cycle vanishes. Combining the 2-cycle above these points leads to one type of CY 3-cycles. Another one can be constructed by considering in the base space a loop around such a singular point combined again with the $K3$ 2-cycle which vanishes at that point. We can often calculate the integral over the $K3$ 2-cycles. Then the CY period reduces to the integral of a 1-form over the 1-cycle in the ζ plane. The latter is not yet a Riemann surface however.

As already mentioned, the CY moduli space has singular points where cycles degenerate. We consider an expansion around certain singular points. The expansion is as well an expansion in moduli space as in the CY coordinates. The CY moduli z^α become in this way a function of the expansion parameter ϵ and variables u^i, which will become the moduli of a Riemann surface. In this way the local geometry is expanded so that a rigid special Kähler geometry remains. In this expansion the $K3$ manifold reduces to an ALE manifold. By performing the 2-dimensional integrals, the periods of the CY reduce to periods of an element of this class of Riemann surfaces. We have made an expansion from a supergravity model to a rigid supersymmetric one.

In supergravity a rigid limit is not defined a priori. In the present framework this is reflected in that different singular points may give rise to different rigid limits. In [1] a procedure was set up to reduce the CY to an ALE manifold, leading to such rigid limits. See [2] for further references. Rather than using this reduction, we computed [3] all the periods in the picture of the $K3$-fibration. This shows explicitly how the full supergravity model approaches its rigid limit. Some cycles which do not occur in the ALE manifolds lead to periods whose contribution give in the limit $\epsilon \to 0$ (infinite Planck mass) a diverging renormalisation of the rigid Kähler potential. Thus these renormalisation effects are included in our computation. For full references see [3].

1 Special Kähler geometry and CY moduli spaces

First we summarise the relevant geometric concepts, both for the rigid and for the local case. We consider symplectic vectors $V(u)$ (rigid), resp. $v(z)$ (local) which are holomorphic functions of r, (resp. n) complex scalars $\{u^i\}$ (resp. $\{z^\alpha\}$). These are $2r$-vectors for the rigid case (in correspondence with the electric and magnetic field strengths), and $2(n+1)$-vectors in the local case (because in that case there is also the graviphoton). A symplectic inner product is defined as

$$\langle V, W\rangle = V^T Q^{-1} W \; ; \qquad \langle v, w\rangle = v^T q^{-1} w \; , \tag{1.1}$$

where Q (and q) is a real, invertible, antisymmetric matrix (we wrote Q^{-1} in (1.1) in view of the meaning which Q will get in the moduli space realisations).

The Kähler potential is respectively for the rigid and local manifold

$$K(u,\bar{u}) = i\left\langle V(u), \bar{V}(\bar{u})\right\rangle \; ; \qquad \mathcal{K}(z,\bar{z}) = -\log(-i\left\langle v(z), \bar{v}(\bar{z})\right\rangle) \; . \qquad (1.2)$$

In the rigid case there is a rigid invariance $V \to e^{i\theta}V$, but in the local case there is even a symmetry with a holomorphic function: $v(z) \to e^{f(z)}v(z)$, because this gives a Kähler transformation $\mathcal{K}(z,\bar{z}) \to \mathcal{K}(z,\bar{z}) - f(z) - \bar{f}(\bar{z})$. Because of this local symmetry we have to introduce covariant derivatives $\mathcal{D}_{\bar{\alpha}}v = \partial_{\bar{\alpha}}v = 0$ and $\mathcal{D}_\alpha v = \partial_\alpha v + (\partial_\alpha \mathcal{K})\, v$ (There exists also a more symmetrical formulation). In any case we still need one more constraint (leading to the 'almost always' existence of a prepotential), which is for rigid, resp. local supersymmetry:

$$\langle \partial_i V, \partial_j V\rangle = 0 \; ; \qquad \langle \mathcal{D}_\alpha v, \mathcal{D}_\beta v\rangle = 0 \; . \qquad (1.3)$$

There are further global requirements; for an exact formulation we refer to [4].

Local special Kähler geometry is realised in moduli spaces of CY manifolds. Consider a CY manifold with $h^{21} = n$. It has n complex structure moduli to be identified with the complex scalars z^α. There are $2(n+1)$ 3-cycles c_Λ, whose intersection matrix will be identified with the symplectic metric $q_{\Lambda\Sigma} = c_\Lambda \cap c_\Sigma$. One identifies v with the 'period' vector formed by integration of the $(3,0)$ form over the $2(n+1)$ cycles:

$$v = \int_{c_\Lambda} \Omega^{(3,0)} \; ; \qquad \mathcal{D}_\alpha v = \int_{c_\Lambda} \Omega^{(2,1)}_{(\alpha)} \; . \qquad (1.4)$$

Rigid special Kähler geometry is realised in moduli spaces of RS. A RS of genus g has g holomorphic $(1,0)$ forms. Now in general we need a family of Riemann surfaces with r complex moduli u^i, such that one can isolate r (1,0)-forms which are the derivatives of a meromorphic 1-form λ up to a total derivative:

$$\gamma_i = \partial_i \lambda + d\eta_i \; ; \qquad \alpha = 1,\ldots,r \leq g \; . \qquad (1.5)$$

Then one should also identify $2r$ 1-cycles c_A forming a complete basis for the cycles over which the integrals of λ are non-zero. We identify $V = \int_{c_A} \lambda$, but it should be clear that all this is much less straightforward then in the CY moduli space.

2 Description of a Calabi-Yau moduli space

We present here the description of one of the examples which we use in [3], i.e. a CY space with $n = h^{(12)} = 3$. First one introduces a complex 4-dimensional *weighted projective space* in which points are equivalence classes $(x_1, x_2, x_3, x_4, x_5) \sim (\lambda x_1, \lambda x_2, \lambda^2 x_3, \lambda^8 x_4, \lambda^{12} x_5)$. The CY manifold $X_{24}[1,1,2,8,12]$ is a 3-dimensional submanifold of this projective space, determined by a *polynomial equation* $W = 0$, of degree 24. The manifold which we use, $X^*_{24}[1,1,2,8,12]$, has *global identifications*

$$\begin{aligned} &x_j \simeq \exp(n_j \frac{2\pi i}{24})\, x_j \\ &(n_1, n_2, n_3, n_4, n_5) = m_1(1,-1,0,0,0) + m_2(-1,-1,2,0,0) \; , \end{aligned} \qquad (2.1)$$

where $m_i \in \mathbb{Z}$. The most general polynomial of degree 24 which is invariant under these identifications depends on 11 parameters, i.e. moduli of the CY manifold. However, they are not independent: there are still *compatible redefinitions* of the x variables, i.e. compatible with (2.1) and the weights. This leaves at the end in this example 3 independent moduli. We learned the advantages of not restricting immediately to one gauge choice in this moduli space.

Now *the* $K3$ *fibration* is exhibited by performing the change of variables

$$x_0 = x_1 x_2 \; ; \qquad \zeta = (x_1/x_2)^{24} \; . \tag{2.2}$$

We take a partial gauge choice with one remaining scale invariance, corresponding to a rescaling of x_0. Then the polynomial looks like

$$W = \tfrac{1}{12} B' \, x_0^{12} + \tfrac{1}{12} x_3^{12} + \tfrac{1}{4} x_4^3 + \tfrac{1}{2} x_5^2 - \psi_0 \, (x_0 x_1 x_3 x_5) - \tfrac{1}{6} \psi_1 \, (x_0 x_3)^6 \; , \tag{2.3}$$

with

$$B' = \tfrac{1}{2} B \left(\zeta + \zeta^{-1} \right) - \psi_s \; . \tag{2.4}$$

We thus have a description as a $K3$ manifold $X^*_{12}[1,1,4,6]$, with a projective moduli space $\{B', \psi_0, \psi_1\}$. Here B' is a function of moduli B and ψ_s of the CY manifold and contains the dependence on the base of the $K3$ fibration ζ.

The manifolds are singular when simultaneously $W = 0$ and $dW = 0$. For the $K3$ this happens for

$$a) \; B' = (\psi_1 + \psi_0^6)^2 \; ; \qquad b) \; B' = \psi_1^2 \; ; \qquad c) \; B' = 0 \; . \tag{2.5}$$

The singularities then occur for a specific point on $K3$. The cases a) and b) are A_1-type singularities. They coincide if $\psi_0^6 = 0$, in which case the singularity becomes of type A_2, and if $\psi_0^6 = -2\psi_1$, in which case the singularity becomes of type $A_1 \times A_1$. We will concentrate here on the first possibility: the A_2 singularity.

For the CY to be singular, we should also have that the derivative of W with respect to ζ is zero, which leads to $\frac{\partial B'}{\partial \zeta} = 0$, satisfied for all ζ if $B = 0$. So we have a full $\mathbb{P}^1$ of singularities. In the CY moduli space, an appropriate expansion is obtained by taking

$$B = 12\epsilon \; ; \qquad \psi_1^2 = -\psi_s + \epsilon u_1 \; ; \qquad (\psi_1 + \psi_0^6)^2 = -\psi_s + \epsilon (u_1 + 2u_0^6) \; . \tag{2.6}$$

After a corresponding expansion of the variables of the CY space around its singular point, the polynomial reduces to one defining an ALE manifold of type A_2, with moduli u_1 and u_2.

3 Periods, monodromies, intersection matrix

The main work is to obtain the periods for the $K3$ fibre, after which the CY periods remain as integrals over 1-cycles. First we use the *Picard-Fuchs equations*, which are differential equations for the integrals of the $(2,0)$ form over the 2-cycles. It turns out that working in the enlarged moduli space (where gauges are not yet fixed) simplifies the derivation of these equations, using toric geometry in disguise without introducing all the formalism, and avoiding its higher order differential equations. The independent solutions give a basis for the periods. The periods are functions of the moduli appearing in the polynomial, which have branch points in singular points, and we have to choose the position of the cuts in this moduli space. We choose a basis of solutions in one sheet of this moduli space. Continuing the periods around such singular points, we cross the cuts, and arrive to the same values of the moduli. Reexpressing the analytically continued periods in the previously chosen basis gives rise to the *monodromy matrices.* A generic basis of solutions to the differential equations does, however, not correspond to integrals over integer cycles. Therefore we need also supplementary methods.

In some examples we can *integrate over one cycle and analytically continue.* We start by integrating the $(2,0)$ form over a cycle which is known in the neighbourhood of the 'large complex structure' singular point. Then this period is analytically continued to other regions. By following its analytic continuation we also obtain the other periods. Because we start here from an integral cycle, we obtain the monodromy matrices in an integral basis.

In the example described above, the strategy which we plan to use for CY, can already be used for the $K3$ periods themselves. Indeed in this case the $K3$ itself is a *torus fibration.* The forms and cycles can be decomposed in forms and cycles on the torus, fibred over a $\mathbb{P}^1$. It has the advantage that we start from the torus, where we know already a basis of cycles and its intersection matrix.

The result is that we find expressions corresponding to 4 $K3$-cycles of which one vanishes at singularity a), called v_α, one at b), called v_β, and two vanish at c), which we will call t_α and t_β. The points of singularity of the $K3$ manifold each occur at two points in the ζ-plane, one inside the circle $|\zeta| = 1$, and one outside. We will take the cuts from the former to $\zeta = 0$, and from the latter to ∞. We can then construct 4 CY cycles by taking the paths in ζ between the two points with the same vanishing $K3$-cycle and combining these with the corresponding 2-cycle in $K3$. These are called V_{v_α}, V_{v_β}, V_{t_α} and V_{t_β}. On the other hand we can combine the circle $|\zeta| = 1$ with the 4 $K3$ cycles, obtaining the CY-cycles T_{v_α}, T_{v_β}, T_{t_α} and T_{t_β}.

4 The rigid limit

Considering then again the expansion of the moduli as in (2.6), we see that (2.4) implies that the singularities are in leading order of ϵ at

$$\begin{aligned} v_\alpha &: \tfrac{1}{2}(\zeta+\zeta^{-1}) = \tfrac{1}{12}(u_1 + 2u_0^6)\ ; \qquad v_\beta \ : \ \tfrac{1}{2}(\zeta+\zeta^{-1}) = \tfrac{1}{12}u_1 \\ t_\alpha, t_\beta &: \tfrac{1}{2}(\zeta+\zeta^{-1}) = \frac{1}{12\epsilon}\psi_s\ . \end{aligned} \tag{4.1}$$

The former thus keep their position, while the latter move to $\zeta = 0$ and $\zeta = \infty$ when $\epsilon \to 0$. The cycles V_{t_α} and V_{t_β} thus become infinitely stretched, and the corresponding periods will get a $\log\epsilon$ dependence. The dependence on ϵ is obtained from studying the ϵ-monodromies.

By a complex basis transformation one can isolate the different types of small ϵ behaviour of the periods, and one rewrites the period vector v as

$$v = v_0(\epsilon) + \epsilon^{1/3} v_1(u) + v_2(\epsilon) \ . \tag{4.2}$$

The relevant term will be the v_1 term, which has only 4 non-zero components, corresponding to the cycles V_{v_α}, V_{v_β}, T_{v_α} and T_{v_β}, i.e. related to the singularities which remain at finite ζ in the limit. v_0 is independent of the moduli. It contains 2 non-zero components, one of which starts with a constant, and the other one has a logarithmic dependence alluded to above. Finally $v_2(\epsilon)$ has as lowest order terms with $\epsilon^{2/3}$ and $\epsilon^{2/3}\log\epsilon$. These appear in the two remaining components of v. The intersection matrix is in this basis (complex antihermitian, not an integral basis) block diagonal in the mentioned $4+2+2$ components. For the Kähler potential this leads to

$$\begin{aligned} \mathcal{K} &= -\log\left(-i < v, \bar{v} >\right) \\ &= -\log\Big(-i < v_0(\epsilon), \bar{v}_0(\epsilon) > -i|\epsilon|^{2/3} < v_1(u), \bar{v}_1(u) > + R(\epsilon, u, \bar{u})\Big) \\ &\approx -\log\left(-i < v_0(\epsilon), \bar{v}_0(\epsilon) >\right) - \frac{|\epsilon|^{2/3}}{< v_0(\epsilon), \bar{v}_0(\epsilon) >} < v_1(u), \bar{v}_1(u) > + \ldots \ , \end{aligned} \tag{4.3}$$

where $R(\epsilon, u, \bar{u})$ are higher order terms. The first term in the final expression is irrelevant, not depending on the moduli. The second one is with a diverging renormalisation (produced by the cycles which disappear in the ALE limit) of the form which it should have for a rigid special Kähler manifold: the first of (1.2). Moreover it coincides with the one the $SU(3)$ Seiberg–Witten Riemann surface.

Acknowledgments.
Work supported by the European Commission TMR programme ERBFMRX-CT96-0045, in which D.Z. is associated to U. Torino. F.D., W.T. and A.V.P. thank their employer, the FWO Belgium, for financial support.

References

[1] A. Klemm, W. Lerche, P. Mayr, C. Vafa, N. Warner, *Nucl. Phys.* **B477** (1996) 746, hep-th/9604034

[2] W. Lerche, hep-th/9611190; A. Klemm, hep-th/9705131.

[3] M. Billó, F. Denef, P. Frè, I. Pesando, W. Troost, A. Van Proeyen and Daniela Zanon, in preparation

[4] B. Craps, F. Roose, W. Troost and A. Van Proeyen, *Nucl. Phys.* **B503** (1997) 565, hep-th/9703082

Subleading terms in interactions between branes and supergravity – super Yang-Mills correspondence

A.A. Tseytlin

Theoretical Physics Group, Blackett Laboratory
Imperial College, London SW7 2BZ, U.K.

Abstract: We compare long-distance interaction potentials between branes and their bound states in supergravity and in super Yang-Mills descriptions. We first consider the supergravity-SYM correspondence at the level of the leading term in the interaction potential, and then describe some recent results concerning the subleading term and their implications for the structure of the 2-loop F^6 term in the SYM effective action.

1 Introduction

In certain cases of BPS bound states of branes (having non-trivial 0-brane content in IIA case or instanton content in IIB case) one may expect a correspondence between their description in terms of curved supersymmetric backgrounds of type II supergravity compactified on T^p and their description as SYM backgrounds on the dual torus $\tilde{T}^p$. On SYM side, it may seem that one may not actually need to assume that N is large in order to have this correspondence. As discussed in [1, 2], on supergravity side, this corresponds to viewing IIA configurations of branes as resulting from a $D = 11$ theory in which a null direction $x_- = x_{11} - t$ is compactified [3]. This is formally equivalent to the prescription of computing the interaction potentials between branes by taking the 0-brane harmonic function without its standard asymptotic value 1. A similar $H \to H - 1$ prescription applies in the IIB instanton case.

On the SYM side, the interaction potential between two different BPS configurations of branes is represented by the SYM effective action $\mathbf{\Gamma}$ computed in an appropriate SYM background [4, 5, 6]. In general, both the vectors A_a ($a = 0, ..., D-1$) and the scalars X_i ($i = D, ..., 9$) may have non-trivial background values. Consider, for example, a system of a 0-brane probe interacting with a BPS bound state of branes (wrapped over T^p), containing, in particular, N_0 0-branes. Under T-duality this becomes a system of a Dp-brane probe with charge n_0 and a Dp-brane source with charge N_0 bound to some other branes of lower dimensions (both probe and source being wrapped over $\tilde{T}^p$). If the probe and the source are separated by a distance r in the direction 8 and the probe has velocity v along the transverse direction 9, this configuration may be described by the following $u(N)$, $N = n_0 + N_0$, SYM background on $\tilde{T}^p$: $\bar{A}_a = \begin{pmatrix} 0_{n_0\times n_0} & 0 \\ 0 & A_a \end{pmatrix}$, $\bar{X}_i = \begin{pmatrix} 0_{n_0\times n_0} & 0 \\ 0 & X_i \end{pmatrix}$, $\bar{X}_8 = \begin{pmatrix} r\ I_{n_0\times n_0} & 0 \\ 0 & 0_{N_0\times N_0} \end{pmatrix}$, $\bar{X}_9 = \begin{pmatrix} vt\ I_{n_0\times n_0} & 0 \\ 0 & 0_{N_0\times N_0} \end{pmatrix}$, where A_a and X_i are $N_0 \times N_0$ matrices in the fundamental representation of $u(N_0)$ which describe the source bound state. The dependence on derivatives of the scalar fields X_i may be formally determined from the dependence of the effective action $\mathbf{\Gamma}$ on the gauge field in a higher-dimensional background representing T-dual ($X_i \to A_i$) configuration. The background value of X_8 plays the role of an IR cutoff $M \sim r$ (in the open string theory picture it is related to the mass of the open string states stretched between the probe and source branes). The long-distance interaction potential $\mathcal{V}$ will be given by the leading IR terms in the expansion of $\mathbf{\Gamma}$ in powers of $1/M$.

On the supergravity side, the interaction potential may be determined from the action of a brane probe moving in a curved background produced by a brane source. For example, the action for a 0-brane probe in a background produced by a marginal bound state of branes $1||0$, $4||0$, $4\perp 1||0$ or $4\perp 4\perp 4||0$ (which is essentially the same as the action for a $D = 11$ graviton scattering off the corresponding M-brane configuration 2+wave, 5+wave, $2\perp 5$+wave or $5\perp 5\perp 5$+wave) has the following general structure [7, 8, 9]

$$I_0 = -T_0 \int dt\ H_0^{-1}[\sqrt{1 - H_0 H_1 ... H_k v^2} - 1] \equiv \int dt\ [\tfrac{1}{2} T_0 v^2 - \mathcal{V}(v,r)] \ . \qquad (1)$$

Here H_0 and $H_{(1)}, ..., H_{(k)}$ ($k = 1$ for 1/4 supersymmetric bound states and $k = 2$ or 3 for 1/8 supersymmetric bound states) are the harmonic functions $H_i = 1 + Q_i/r^{7-p}$ representing the constituents of the bound state. Since for D-branes $Q_i \sim g_s N_i$ and $T_0 \sim n_0 g_s^{-1}$ (g_s is the string coupling constant), the long-distance expansion of the classical supergravity interaction potential $\mathcal{V}$ has the following form

$$\mathcal{V} = \sum_{L=1}^{\infty} \mathcal{V}^{(L)} = \frac{n_0}{g_s} \sum_{L=1}^{\infty} (\frac{g_s}{r^{7-p}})^L k_L(v, N_i) \ , \qquad (2)$$

so that the $(1/r^{7-p})^L$ term has the same g_s dependence as in the L-loop term in SYM theory with coupling $g_{\rm YM}^2 \sim g_s$.

To have a precise agreement between the supergravity and SYM expressions for the potential (already at the leading level) one needs to assume that in expanding (1) in powers of $1/r^{7-p}$ one should set $H_0 = Q_0/r^{7-p}$, $Q_0 \sim g_s N_0$. This prescription may be interpreted in two possible ways (which are equivalent for the present purpose of comparing interaction potentials). One may assume (as was done in [6, 10, 11]) that

N_0 is large for fixed r (larger than $N_1, ..., N_k$ and any other charge parameters that may be present in non-marginal brane configurations), so that $H_0 = 1 + Q_0/r^{7-p} \approx Q_0/r^{7-p}$. Alternatively, one may keep N_0 finite but consider the $D = 10$ brane system as resulting from an M-theory configuration with $x_- = x_{11} - t$ being compact [3]: as was pointed out in [1], the dimensional reduction of a $D = 11$ gravitational wave combined with M-brane configurations along x_- results in supergravity backgrounds with $H_0 = Q_0/r^{7-p}$.

The formal technical reason why the leading-order SYM and supergravity potentials happen to agree in certain simple cases is related to the fact that the combination of F^4 terms that appears in the 1-loop SYM effective action is also the same as the one in the expansion of the Born-Infeld action [12], but the latter is closely related to the action of a D-brane probe moving in a supergravity background. This becomes especially clear in the type IIB (instanton model) context [11], provided one takes into account that because of the T-duality involved, the relevant gauge field backgrounds which appear in the SYM and supergravity descriptions are related by $F_{ab} \to (F_{ab})^{-1}$ [11].

The known (weak-coupling string theory) explanation [4, 6] of the precise agreement between the leading-order supergravity and 1-loop SYM potentials in certain simple cases uses the observation that for configurations of branes with sufficient amount of underlying supersymmetry, the long-distance and short-distance limits of the string-theory potential (represented by the annulus diagram) are the same. That implies that the leading-order (long-distance) interaction potential determined by the classical supergravity limit of the closed string theory is the same as the (short-distance) one-loop potential produced by the massless (SYM) open string theory modes.

The results of [13, 1, 2] suggest that this supergravity-SYM correspondence should extend beyond the leading-order level. One may conjecture that, in general, the existence of the open string theory description of D-branes combined with enough supersymmetry implies again the agreement between long-distance and short-distance limits of higher open string loop terms in the interaction potential. Equivalently, that would mean that (i) the leading IR part of the L-loop term in the $SU(N)$ SYM effective action in $D = 1 + p$ dimensions has a universal $F^{2L+2}/M^{(7-p)L}$ structure, and (ii) computed for a SYM background representing a configuration of interacting branes, the $F^{2L+2}/M^{(7-p)L}$ term should reproduce the $1/r^{(7-p)L}$ term in the corresponding classical supergravity potential.

2 Leading-order interaction potentials

In general, the effective action of the $D = p + 1$ dimensional $U(N)$ SYM theory on $\tilde{T}^p$ ($\frac{1}{2g_{\rm YM}^2} \int d^{p+1}\tilde{x}\ {\rm tr}F^2$, $g_{\rm YM}^2 = (2\pi)^{-1/2} g_s \tilde{V}_p$) for a purely gauge field background and with an explicit IR cutoff M has the following structure

$$\Gamma = \sum_{L=1}^{\infty} (g_{\rm YM}^2 N)^{L-1} \int d^{p+1}\tilde{x} \sum_n \frac{c_{nL} F^n}{M^{2n-(p-3)L-4}} \ .$$

We will be interested only a special subset of terms in $\mathbf{\Gamma}$ (generalising the 'diagonal terms' in [1]) which have the same coupling g_s, 0-brane charge N_0 and distance $r = M$ dependence as the terms in the long-distance expansion (2) of the classical supergravity interaction potential $\mathcal{V}$ between a Dp-brane probe (with tension $\sim n_0/g_s$) and a Dp-brane source (with charge parameter $\sim g_s N_0$). We shall assuming that the SYM backgrounds describing individual branes are supersymmetric, so that the effective action vanishes when evaluated on each of them separately (i.e. its non-vanishing part will represent the interaction between branes).

One may conjecture that due to maximal underlying supersymmetry of the SYM theory, the terms $F^{2L+2}/M^{(7-p)L}$ represent, in fact, the *leading* IR (large M) contribution to $\mathbf{\Gamma}$ at L-th loop order. This is indeed true for $L = 1$ [12] and, in view of the results of [13] (for $p = 0$) and [14] (for $p = 3$) this should be true also for $L = 2$. The sum of such leading IR terms at each loop order will be denoted as Γ. Thus

$$\Gamma = \frac{1}{2g_{\rm YM}^2 N} \sum_{L=1}^{\infty} \int d^{p+1}\tilde{x} \; (\frac{a_p g_{\rm YM}^2 N}{M^{7-p}})^L \hat{C}_{2L+2}(F),$$

where $\hat{C}_{2L+2}(F) \sim F^{2L+2}$ and a_p are universal coefficients not depending on N or L.

At the 1-loop level, $\Gamma^{(1)} = \frac{a_p}{2M^{7-p}} \int d^{p+1}\tilde{x} \; \hat{C}_4(F)$ where $\hat{C}_4 = \mathrm{STr}\; C_4 = -\frac{1}{8}\mathrm{STr}[F^4 - \frac{1}{4}(F^2)^2]$, i.e.

$$\hat{C}_4 = \; -\tfrac{1}{12}\mathrm{Tr}(F_{ab}F_{bc}F_{cd}F_{da} + \tfrac{1}{2}F_{ab}F_{bc}F_{da}F_{cd} - \; \tfrac{1}{4}F_{ab}F_{ab}F_{cd}F_{cd} - \tfrac{1}{8}F_{ab}F_{cd}F_{ab}F_{cd}) \, .$$

Here $a_p \; = \; 2^{2-p}\pi^{-(p+1)/2}\Gamma(\frac{7-p}{2})$ and STr is the symmetrised trace in the adjoint representation (for $SU(N)$ $\mathrm{Tr}Y^4 = 2N\mathrm{tr}Y^4 + 6\mathrm{tr}Y^2\mathrm{tr}Y^2$, so one gets the expression containing terms with single and double traces in the fundamental representation [11]). The polynomial C_4 is the same one that appears in the expansion of the BI action, $\sqrt{-\det(\eta_{ab} + F_{ab})} = \sum_{n=0}^{\infty} C_{2n}(F)$,, i.e.

$$C_0 = 1 \, , \quad C_2 = -\frac{1}{4}F^2 \, , \quad C_4 = -\frac{1}{8}[F^4 - \frac{1}{4}(F^2)^2] \, ,$$
$$C_6 = -\frac{1}{12}[F^6 - \frac{3}{8}F^4F^2 + \frac{1}{32}(F^2)^3] \, , \quad .... \, .$$

One may consider several examples of different brane configurations which admit a SYM description and compute the leading-order potentials $\mathcal{V}^{(1)}$ by substituting the corresponding gauge field backgrounds into Γ. We shall assume that $\Gamma^{(1)} = \int dt \mathcal{V}^{(1)}$ in the cases involving 0-branes.[1]

A 0-brane (with velocity v) interacting with $p + ... + 0$ IIA bound state described by $F_{mn} = \mathrm{diag}(0_{n_0 \times n_0}, \; \mathrm{F}_{mn} I_{N_0 \times N_0})$ leads to [10, 11]

$$\hat{C}_4 = -\tfrac{1}{4}n_0 N_0[\mathrm{F}^4 - \tfrac{1}{4}(\mathrm{F}^2)^2 - v^2\mathrm{F}^2 + v^4]. \tag{3}$$

An example of a bound state with $1/4$ of supersymmetry is $4||0$ which may be described by a self-dual $SU(N)$ background on $\tilde{T}^4$: $F_{mn} = F^*_{mn}$, $\int d^4\tilde{x}\mathrm{tr}(F_{mn}F_{mn}) =$

[1] In what follows we shall set $2\pi\alpha' = 1$ and assume for simplicity that the volumes of the tori take self-dual values, $V_p = \tilde{V}_p = (2\pi)^{p/2}$.

$(4\pi)^2 N_4$, or, explicitly, by a constant background: $F_{12} = F_{34} = qJ_1$, $q^2 = \frac{N_4}{N_0}$, $J_1 \equiv \text{diag}(0_{n_0\times n_0}, I_{\frac{1}{2}N_0\times\frac{1}{2}N_0}, -I_{\frac{1}{2}N_0\times\frac{1}{2}N_0})$. In the case of the $(4||0)-(4||0)$ system of two parallel $4||0$ states with charges (n_4, n_0) and (N_4, N_0) we find [10]

$$\mathcal{V}^{(1)} = -\frac{n_0 N_0}{16r^3}[4(\frac{N_4}{N_0} + \frac{n_4}{n_0})v^2 + v^4]. \tag{4}$$

To determine the leading-order potentials for configurations involving 1/8 supersymmetric states one needs to find their SYM description and substitute the resulting backgrounds into Γ. The configuration of a 0-brane interacting with $4\perp1||0$ state wrapped over T^5 (corresponding to extremal $D=5$ black hole) may be described by a combination of an instanton and a momentum wave (carried, in general, by vectors *and* scalars), or, explicitly (after T-duality trading scalar backgrounds for the vector ones) [2]

$$F_{09} = vJ_0\ , \qquad F_{12} = F_{34} = qJ_1\ , \qquad F_{51} = F_{01} = hJ_0\ , \qquad F_{56} = F_{06} = wJ_0\ , \tag{5}$$

where the $su(n_0+N_0)$ matrix J_1 is $J_0 = \frac{1}{n_{-1}+N_{-1}}\text{diag}(N_{-1}I_{n_{-1}\times n_{-1}}, -n_{-1}I_{N_{-1}\times N_{-1}})$ and the periodic 'vector wave' and 'scalar wave' functions $h = h(\tilde{x}_5 + t)$ and $w = w(\tilde{x}_5+t)$ satisfy $< h^2 + w^2 > = \frac{1}{L_5}\int d\tilde{x}_5\ (h^2+w^2) = g_s\frac{N_1}{N_0}$. The $\hat{C}_4$-coefficient of the corresponding leading-order potential is found to be (equivalent results were obtained in [9, 15]) $\hat{C}_4 = -\frac{1}{4}n_0N_0[4v^2q^2 + 4v^2(h^2+w^2) + v^4]$. The same expression is found for the 0-brane interaction with $4\perp4\perp4||0$ bound state wrapped over T^6 (corresponding to extremal $D=4$ black hole), or for $6-(6||2\perp2\perp2)$ interaction on $\tilde{T}^6$. The $6||2\perp2\perp2$ configuration may be described by an 'overlap' of the three 4d instantons on $\tilde{T}^6$, i.e. by the following $su(N_0)$ constant gauge field strength background [2]

$$F_{14} = F_{23} = q_1\lambda_1\ , \qquad F_{45} = F_{36} = q_2\lambda_2, \qquad F_{15} = -F_{26} = q_3\lambda_3\ , \tag{6}$$

where $q_k^2 = \frac{N_{4(k)}}{N_0}$ and λ_k are some three independent $su(N_0)$ matrices normalised so that this gauge configuration produces the right 2-brane charges (and only them) on 6-brane. One possible choice of λ_k is the following *'commuting'* one: $\lambda_k = \mu_k \otimes I_{\frac{N_0}{4}\times\frac{N_0}{4}}$, where μ_k are the diagonal 4×4 matrices $\mu_1 = \text{diag}(1,1,-1,-1)$, $\mu_2 = \text{diag}(1,-1,-1,1)$, $\mu_3 = \text{diag}(1,-1,1,-1)$. A *'non-commuting'* choice is to set λ_k to be proportional to the Pauli matrices σ_k, i.e. $\lambda_k = \sigma_k \otimes I_{\frac{N_0}{2}\times\frac{N_0}{2}}$. Both choices represent 1/4 supersymmetric configurations in the $D = 6+1$ SYM theory [2]. The leading-order potential in this case is proportional to $\hat{C}_4 = -\frac{1}{4}n_0N_0[4v^2(q_1^2+q_2^2+q_3^2)+v^4]$. Comparing with various special cases discussed above we conclude that the *leading-order* SYM interaction potentials for *marginal* bound states are essentially the sums of pair-wise interactions between constituent branes. The same will, of course, be true on the supergravity side (cf. (1)), and the potentials will be in full agreement.

To find the supergravity potentials we shall use the probe method, i.e. consider the action of a D-brane probe

$$I_p = -T_p[\int d^{p+1}x e^{-\phi}\sqrt{\det(G_{ab} + G_{ij}\partial_a X^i \partial_b X^j + \mathcal{F}_{ab})} - \sum C e^{\mathcal{F}}]$$

in a curved background produced by a brane bound state as a source. The key example in the 1/2 supersymmetric brane case is the interaction of a D-instanton with a type IIB non-marginal bound state $p+(p-2)+...+(-1)$. The action for the latter considered as a probe may be found by switching on a constant $\mathcal{F}_{ab}$ background on the Dp-brane world volume. The fluxes produced by $\mathcal{F}_{ab}$ determine the numbers of branes [17] of each type in the bound state.

The $(p+...+0)$ probe action in the 0-brane background is

$$I_p = -T_p V_p \int dt\ H_0^{-1} \times [\sqrt{(1-H_0 v^2)\det(H_0^{1/2}\delta_{mn} + \mathcal{F}_{mn})} - \sqrt{\det \mathcal{F}_{mn}}],$$

where $\mathcal{F}_{mn}$ describes other brane charges, e.g., $n_0 = n_p (2\pi)^{-p/2} V_p \sqrt{\det\ \mathcal{F}_{mn}}$. This action may be rewritten as

$$I_p = -T_0 \int dt\ H_0^{-1} \times [\sqrt{1-H_0 v^2}\sqrt{\det(\delta_{mn} + H_0^{1/2}\mathrm{F}_{mn})} - 1], \tag{7}$$

where $T_0 = n_0 g_s^{-1}(2\pi)^{1/2}$ and $\mathrm{F}_{mn} \equiv (\mathcal{F}_{mn})^{-1}$. In this form it corresponds to a T-dual configuration, i.e. to the interaction of a p-brane source (with charge N_0) with parallel $(0+...+p)$-brane probe (with 0-brane charge n_0) moving in a relative transverse direction. Introducing the velocity component $\mathrm{F}_{09} = v$ we can put this action in the BI form $I_p = -T_0 \int dt H_0^{-1}[\sqrt{-\det(\eta_{ab} + H_0^{1/2}\mathrm{F}_{ab})} - 1]$ were again $H_0 = Q_0/r^{7-p}$ so that the agreement between the leading-order long-distance interaction potential and the SYM result (3) is manifest. Similar agreement is found in other cases. For example, using the explicit form of the $4\perp4\perp4||0$ background [16] one finds that the $0-(4\perp4\perp4||0)$ interaction is described by

$$I_0 = -T_0 \int dt H_0^{-1}(\sqrt{1-H_0 H_{4(1)} H_{4(2)} H_{4(3)} v^2} - 1)\ ,$$

where $H_{4(k)} = 1 + Q_{4(k)}/r$, so that $\mathcal{V}^{(1)} = -\frac{n_0}{16r}[4v^2(N_{4(1)} + N_{4(2)} + N_{4(3)}) + v^4 N_0]$, which is in agreement with the SYM expression.

To summarize, the SYM–supergravity correspondence observed on the above examples is formally due to (i) the BI-type structure of the actions of the non-marginal bound state branes, (ii) the 'product of harmonic functions' structure of the actions in the marginal bound state case (implying additive dependence of the leading-order potential on constituent charges), and (iii) a combination of these two features in more general cases of interactions with non-marginal 1/4 or 1/8 supersymmetric bound states.

3 Subleading term in interaction potentials

The result of [1] may be interpreted as implying that the subleading term $\mathcal{V}^{(2)} \sim n_0 N_0^2 g_s v^6/r^{14}$ in the 0-brane - 0-brane interaction potential (2) is reproduced by the leading 2-loop term in the $D=1$ SYM effective action Γ. This is easy to check by assuming that the 2-loop coefficient $\hat{C}_6$ in Γ has the same structure as the 1-loop one $\hat{C}_4$, i.e. is given by the (symmetrized) trace in the adjoint representation of the F^6 term appearing in the expansion of the BI action, $\hat{C}_6 = \text{STr}\ C_6(F)$. Indeed, interpreting the velocity as an electric field component in $D = 2$ SYM theory and substituting $F_{09} = vJ_0$, using that $\text{Tr}J^{2n} = 2n_0N_0$ and separating the $n_0N_0^2$-term as required [1] to match the supergravity result one finds the precise agreement with the supergravity potential.

In [2] we attempted to test the above ansatz by studying the subleading terms in the potentials in more complicated examples of 0-brane interacting with bound states of branes wrapped over tori. Since an explicit computation of the 2-loop term in Γ for arbitrary non-abelian gauge field in a higher-dimensional SYM looks as a complicated problem, we followed an indirect route: making a plausible ansatz for the 2-loop term in Γ and then trying to compare it with the supergravity expressions for $\mathcal{V}^{(2)}$ on different examples with varied amount of supersymmetry, *assuming* that the supergravity-SYM correspondence should continue to hold beyond the leading order as it does in the simplest 0-brane scattering case. The consistency of the resulting picture supports the basic assumption.

The first non-trivial example is the 0-brane interaction with 1/2 supersymmetric non-marginal bound state $(p+...+0)$. Since the action (7) has the BI structure, the subleading term in its potential part (the one which is quadratic in $H_0 = Q_0/r^{7-p}$) has the $C_6 \sim F^6$ form. Plugging the corresponding SYM background $F_{09} = vJ_0$, $F_{mn} = \text{F}_{mn}J_0$ into Γ we find the precise agreement between the SYM and the supergravity expressions for $\mathcal{V}^{(2)}$ for *arbitrary* F_{mn}.

A test then comes from the cases of 0-brane interaction with 1/4 supersymmetric 1||0 and 4||0 bound states. Though the supergravity action $I_0 = -T_0 \int dt H_0^{-1}(\sqrt{1 - H_0H_1v^2} - 1)$ and thus $\mathcal{V}^{(2)} = -\frac{1}{16r^{12}}Q_0(4Q_1v^4 + Q_0v^6)$ in the $0-(1||0)$ case have different structure than (7), $\mathcal{V}^{(2)}$ is still reproduced [2] by the 2-loop term in Γ after one substitutes into $\hat{C}_6$ the relevant $SU(n_0 + N_0)$ background $F_{09} = vJ_0$, $F_{12} = F_{02} = h(\tilde{x}_1 + x_0)J_0$.

In the $0 - (4||0)$ case the supergravity potential has the same form as in the $0 - (1||0)$ case, but the corresponding SYM background is now more complicated: it is parametrised by two independent commuting matrices J_0 and J_1. Substituting $F_{09} = vJ_0$, $F_{12} = F_{34} = qJ_1$ into Γ gives $\text{Tr}C_6 = -\frac{1}{8}n_0N_0(2v^4q^2 + v^6)$ instead of $\hat{C}_6 = -\frac{1}{8}n_0N_0(4v^4q^2 + v^6)$ which is needed for agreement with the supergravity potential.

It is natural to try to modify our ansatz in order to correct the factor of 2 discrepancy in the v^4 term, without changing, however, the result for $\hat{C}_6$ in all the previous cases (which were represented by more primitive gauge field backgrounds depending on a single $SU(n_0 + N_0)$ matrix J_0). Remarkably, it turns out [2] that there is a *unique* way of achieving that goal (up to terms involving commutators of F which are discussed below and which vanish on the backgrounds we considered so far): one is to keep the same Lorentz-index structure of the F^6 terms as in C_6 but should replace the internal index trace STr by a different combination of tr, $(\text{tr})^2$ and $(\text{tr})^3$ terms (all such fundamental trace structures may in general appear at 2 loops)

$$\widehat{\text{STr}}(Y_{s_1}...Y_{s_6}) \equiv 2N\text{tr}[Y_{(s_1}...Y_{s_6)}] + \ \alpha_1\text{tr}[Y_{(s_1}...Y_{s_4}]\text{tr}[Y_{s_5}Y_{s_6)}]$$

$$+ \alpha_2 \mathrm{tr}[Y_{(s_1}...Y_{s_3}]\mathrm{tr}[Y_{s_4}...Y_{s_6)}] + \alpha_3 N^{-1}\mathrm{tr}[Y_{(s_1}Y_{s_2}]\mathrm{tr}[Y_{s_3}Y_{s_4}]\mathrm{tr}[Y_{s_5}Y_{s_6)}].$$

Y_s are the $SU(N)$ generators and $\alpha_1 = 30,\ \alpha_2 = -20,\ \alpha_3 = 0$ in the case when $\widehat{\mathrm{STr}}Y^6 = \mathrm{STr}Y^6$, but we need to choose $\alpha_1 = 60,\ \alpha_2 = -50,\ \alpha_3 = -30$ in order for the modified ansatz $\hat{C}_6 = \widehat{\mathrm{STr}}\ C_6(F)$ to reproduce the supergravity expressions in all of the above examples, including the $0-(4||0)$ one (which is the only case among them where $\widehat{\mathrm{STr}}C_6(F) \neq \mathrm{STr}C_6(F)$). Indeed, one finds that for the gauge field background representing the $0-(4||0)$ configuration the relevant $n_0 N_0^2$ term in the 2-loop coefficient $(n_0+N_0)\hat{C}_6$ in Γ is now in agreement with the supergravity expression for the subleading potential $\mathcal{V}^{(2)}$.

A consistency test is provided by further examples of 0-brane interactions with 1/8 supersymmetric bound states. In the $0-(4\perp 1||0)$ case the subleading term in the supergravity potential (1),(2) is

$$\mathcal{V}^{(2)} = -\frac{T_0}{16r^4}[8v^2 Q_4 Q_1 + 4v^4(Q_4+Q_1)Q_0 + v^6 Q_0^2],$$

where $Q_1 = g_s Q_0 \frac{N_1}{N_0}$. The same expression should be reproduced by the 2-loop SYM effective action Γ with the coefficient $\hat{C}_6$ computed for the background (5). One finds that

$$\hat{C}_6 = -\frac{1}{8(n_0+N_0)} n_0 N_0^2 [16v^2 w^2 q^2 + 4v^4(q^2+h^2+w^2) + v^6] + O(n_0^2),$$

so that the v^4 and v^6 terms are indeed the same as in $\mathcal{V}^{(2)}$ for any distribution of the total momentum between the vector and scalar oscillations (representing the momentum wave along the D-string bound to D5-brane in the T-dual configuration 5||1+wave). However, the coefficient of the v^2 term is *non-vanishing* (cf. [9]) only if $w \neq 0$, i.e. only if the scalar background is excited. The v^2 term has the required coefficient provided we assume that the momentum is distributed equally between the scalar and the vector waves, i.e. if $< h^2 >=< w^2 >= \frac{1}{2} g_s \frac{N_1}{N_0}$.

Finally, in the $0-(4\perp 4\perp 4||0)$ case the v^4 and v^6 terms in the corresponding supergravity potential

$$\mathcal{V}^{(2)} = -\frac{n_0 g_s}{64(2\pi)^{1/2} r^2}[\tfrac{1}{2}v^2(N_{4(1)}N_{4(2)} + N_{4(1)}N_{4(3)}) + N_{4(2)}N_{4(3)}) + 4v^4(N_{4(1)} + N_{4(2)} + N_{4(3)}) + v^6 N_0^2],$$

are correctly reproduced by the $\hat{C}_6$ evaluated on the background (6) supplemented by the velocity component $F_{09} = vJ_0$. However, the coefficient of the v^2 term in the resulting expression for $\Gamma^{(2)}$ is vanishing for both commuting and non-commuting choice of the matrices λ_k in (6). One should note, however, that up to this point we were ignoring the terms with commutators of F_{ab} which may, of course, be present in the 2-loop effective action, but which were vanishing on the commuting backgrounds we were discussing above. It is easy to see that such commutator terms $C_6 \sim \mathrm{Tr}(F_{ab}F_{ab}[F_{cd},F_{ef}]F_{cd}F_{ef})$ may indeed produce the required v^2 term.

As we have discussed above, the supergravity-SYM (matrix theory) correspondence is manifest for the leading term in the long-distance interaction potential between appropriate configurations of branes in $D = 10$ (having large 0-brane number N_0 or finite N_0 but obtained from $D = 11$ using 'null' reduction).

We have suggested that this correspondence holds also for the subleading terms in the long-distance potential between extended branes, i.e. not only for the $D = 1$ SYM (0-brane scattering) case considered in [1]. It would be important to perform a string-theory computation of the subleading (2-loop) terms in the interaction potential, checking that the $r \to 0$ and $r \to \infty$ limits of the string result continue to agree (for relevant configurations of branes) beyond the leading 1-loop level considered in [18, 4, 6]. This would provide an explanation for the supergravity-SYM correspondence at the subleading level observed in [1, 2].

I would like also to thank the organizers for the invitation to give a talk. This work was supported in part by PPARC and the European Commission TMR programme grant ERBFMRX-CT96-0045.

References

[1] K. Becker, M. Becker, J. Polchinski, and A. Tseytlin, *Higher order graviton scattering in M(atrix) theory*, Phys. Rev. D56 (1997) 3174, hep-th/9706072.

[2] I. Chepelev and A.A. Tseytlin, *Long-distance interactions of branes: correspondence between supergravity and super Yang-Mills descriptions*, hep-th/9709087.

[3] L. Susskind, *Another Conjecture about M(atrix) Theory*, hep-th/9704080.

[4] M.R. Douglas, D. Kabat, P. Pouliot and S.H. Shenker, *D-branes and short distances in string theory*, Nucl. Phys. B485 (1997) 85, hep-th/9608024.

[5] T. Banks, W. Fischler, S.H. Shenker and L. Susskind, *M Theory as a matrix model: a conjecture*, Phys. Rev. D55 (1997) 5112, hep-th/9610043.

[6] G. Lifschytz and S.D. Mathur, *Supersymmetry and Membrane Interactions in M(atrix) Theory*, hep-th/9612087.

[7] A.A. Tseytlin, *No-force condition and BPS combinations of p-branes in 11 and 10 dimensions*, Nucl. Phys. B487 (1997) 141, hep-th/9609212.

[8] A.A. Tseytlin, *Composite BPS configurations of p-branes in 10 and 11 dimensions*, Class. Quant. Grav. 14 (1997) 2085, hep-th/9702163.

[9] M. Douglas, J. Polchinski and A. Strominger, *Probing Five-Dimensional Black Holes with D-Branes*, hep-th/9703031.

[10] I. Chepelev and A.A. Tseytlin, *Long-distance interactions of D-brane bound states and longitudinal 5-brane in M(atrix) theory* , Phys. Rev. D56 (1997) 3672, hep-th/9704127.

[11] I. Chepelev and A.A. Tseytlin, *Interactions of type IIB D-branes from D-instanton matrix model*, hep-th/9705120.

[12] R.R. Metsaev and A.A. Tseytlin, *On loop corrections to string theory effective actions*, Nucl. Phys. B298 (1988) 109.

[13] K. Becker and M. Becker, *A two loop test of M(atrix) theory*, hep-th/9705091.

[14] M. Dine and N. Seiberg, *Comments on Higher Derivative Operators in Some SUSY Field Theories*, hep-th/9705057.

[15] J. Maldacena, *Probing near extremal black holes with D-branes*, hep-th/9705053.

[16] I.R. Klebanov and A.A. Tseytlin, *Intersecting M-branes as four-dimensional black holes*, Nucl. Phys. B475 (1996) 179, hep-th/9604166.

[17] M.R. Douglas, *Branes within Branes*, hep-th/9512077.

[18] C. Bachas, *D-Brane Dynamics*, Phys. Lett. B374 (1996) 37, hep-th/9511043.

Branes in $N = 2$, $D = 4$ supergravity and the conformal field theory limit

Klaus Behrndt

Institut für Physik, Humboldt-Universität,
Invalidenstr. 110, D-10115 Berlin

Abstract: In this article we summarise the brane solutions (instantons, black holes and strings) in 4 dimensions when embedded in $N = 2$ supergravity. Like in 10 dimensions these solutions are related by duality transformations (T-duality, c-map). For the case that the graviphoton has no magnetic charge, we discuss the conformal field theory description of the near-horizon geometry. The decoupling of the massless modes in the matrix limit of the M-theory compactified on a Calabi-Yau threefold is discussed.

In the last years we made substantial progress in the understanding of N=2 black holes in 4 and 5 dimensions [1, 2, 3, 4, 5]. In order to obtain these black holes one has to compactify the 10-dimensional supergravity on an internal space with unknown metric, $K3$ for heterotic and Calabi-Yau threefold (CY_3) on type II side. As consequence, one cannot obtain them simply by dimensional reduction and hence one has to solve either the 4-d equations of motion or the Killing spinor equations.

Considering only the ungauged bosonic part the equations of motion follow from the N=2 supergravity action given by

$$S = \int d^4x \sqrt{G} \left\{ R - \frac{1}{4} F^I_{\mu\nu} {}^\star G_I^{\ \mu\nu} - 2 g_{A\bar{B}} \partial_\mu z^A \partial^\mu \bar{z}^B - \tilde{g}_{uv} \partial_\mu q^u \partial^\mu q^v \right\} . \tag{1}$$

The n_v complex scalars z^A coming from vector multiplets parameterize a special Kähler manifold K_{n_v} with metric $g_{A\bar{B}}(z, \bar{z})$, whereas the $4n_h$ real scalars q^u a quaternionic manifold Q_{n_h} with metric $\tilde{g}_{uv}(q)$. The gauge field strengths $(F^I_{\mu\nu}, G_{I\,\mu\nu})$ form a symplectic vector ($I = 0...n_v$) and therefore are not independent

$$G_{I\,\mu\nu} = \mathrm{Re}\mathcal{N}_{IJ} F^J_{\mu\nu} - \mathrm{Im}\mathcal{N}_{IJ} {}^\star F^J_{\mu\nu} \, . \tag{2}$$

Like in 10 dimensions we can classify the 4-d supergravity theory in terms brane-solutions (related to elementary and solitonic states); for the 6-d case an analog discussion can be found in [6]. These are the membranes, strings, black holes (0-branes) and instantons (-1-branes). The membranes introduce a mass term and are the analog of the 8-brane solution of massive IIA supergravity. We will ignore this solution furthermore and focus mainly on the other three branes. The string solution is naturally supported by antisymmetric tensors, that couples to their worldvolume. In the action (1) the antisymmetric tensors have been dualized to scalars which are part of the hyper scalars. Black holes couple naturally to vector fields and therefore are solutions with non-trivial vector multiplets. Finally the instantons are dual to the strings and are related to non-trivial hyper scalars. In addition to these counterparts of the 10-d D-branes, we have also purely gravitational solutions, the G-branes as discussed in [7], the wave and Taub-NUT soliton. The branes can also be seen as a classification of the solutions in the number of translational isometries.

These are the "basic" branes of 4-d supergravity. In addition, (threshold) bound states should exist, but so far no solution could be constructed which includes non-trivial vector as well as hyper multiplet scalars. But we can adopt the 10-d duality rules to transform all "basic" solutions into one another.

A good starting point is the black hole or 0-brane. In this case we can neglect the quaternionic part by taking $q^u = constant$. Let us summarise here only the main points, for details we refer to [3]. The gauge field equations and Bianchi identities are given by $dF^I = dG_I = 0$ and have the solution

$$F^I_{mn} = \frac{1}{2}\epsilon_{mnp}\partial_p \tilde{H}^I \quad , \quad G_{I\,mn} = \frac{1}{2}\epsilon_{mnp}\partial_p H_I \tag{3}$$

where $m, n, p = 1, 2, 3$ and $(\tilde{H}^I, H_I)$ are harmonic functions. Note, that the timelike components F^I_{0m} and $G_{I\,0m}$ are fixed in terms of the spatial ones using (2). As next step we need the scalars and the metric, which can be obtained by solving the Killing spinor equations [3]

$$ds^2 = -e^{2U}(dt + \omega_m dx^m)^2 + e^{-2U} dx^m dx^m \quad , \quad z^A = \frac{X^A}{X^0} \tag{4}$$

with

$$\begin{aligned} e^{-2U} &= e^{-K} \equiv i(\bar{X}^I F_I - X^I \bar{F}_I) \, , \\ \tfrac{1}{2} e^{2U} \epsilon_{mnp}\partial_n \omega_p &= Q_m \equiv \tfrac{1}{2} e^{2U}(H_I \partial_m \tilde{H}^I - \tilde{H}^I \partial_m H_I) \end{aligned} \tag{5}$$

and the symplectic section is constraint by

$$-i(X^I - \bar{X}^I) = \tilde{H}^I(x^\mu) \quad , \quad -i(F_I - \bar{F}_I) = H_I(x^\mu) \tag{6}$$

where the harmonic functions $(\tilde{H}^I, H_I)$ are introduced in (3) and $F_I = \frac{\partial}{\partial X^I} F(X)$. This solution is expressed completely in terms of duality invariant quantities, the Kähler potential K and the Kähler connection Q_m. But it is *not* written in a Kähler invariant way, instead we have fixed the Kähler invariance in a proper way. If $\omega_m = 0$ we get the expected static black hole and for special choices for the harmonic functions we obtain a rotating black hole or generalisations of Taub-NUT and Eguchi-Hanson spaces.

By T-dualizing different directions of the 4-d space time, one obtains the other brane solutions. In order to be concrete let us consider one simple example, which is axion-free, i.e. $\mathrm{Re} z^A = 0$. This can be done by setting $H^0 = H_A = 0$, i.e. the graviphoton is electric and all vector multiplets are magnetic. As prepotential we consider

$$F(X) = \frac{d_{ABC} X^A X^B X^C}{X^0} \tag{7}$$

and the solution is given by [4]

$$e^{-2U} = \sqrt{H_0 \, d_{ABC} H^A H^B H^C} \quad , \qquad z^A = i \, H_0 H^A e^{2U} \tag{8}$$

with $H_0 = h_0 + q_0/r$, $H^A = h^A + p^A/r$. This black hole is a compactification of the $5 \times 5 \times 5 + mom.$ intersection, where each of the 5-brane wraps a 4-cycle of the CY_3 and H_0 corresponds to the momentum modes travelling along the common string. A detailed microscopic analysis has been given recently in [2].

In order to obtain the instanton solution we have first to T-dualize the Euclidean time, which is also known as c-map has been used in [5] to map the general solution (5). Note, because the black hole solution considered here, is a type IIA compactification, after this mapping we obtain a type IIB compactification, where the wrapped 5-branes become (instantonic) wrapped 3-branes and the momentum modes are converted into a IIB (-1)-brane, i.e. the intersection is $(-1) \times 3 \times 3 \times 3$. The resulting solution reads [5]

$$\begin{array}{lll} ds^2_{inst.} = e^{-2U} dx^\mu dx^\mu \quad , & e^{2\phi} = e^{-2U} \quad , & \zeta^0 = \frac{1}{\sqrt{2}} A^0 = \frac{1}{\sqrt{2} H_0} \\ \partial_m \zeta_A = -\frac{1}{2\sqrt{2}} e^{2U} \mathrm{Im} \mathcal{N}_{AB} \partial_m H^A \quad , & & z^A \to q^A \end{array} \tag{9}$$

where the first line contains the string-metric and the universal hypermultiplet, i.e. the dilaton ϕ and the former graviphoton gauge potential. In the second line are the scalars coming from the dualization of the magnetic vector fields and also the original vector multiplet scalars. All these scalars group together in $n_v + 1$ hypermultiplets each containing 4 scalars. After having done this dualization one can allow the fields to depend on all 4 coordinates in a symmetric way, i.e. the new harmonic functions are $H_0 = h_0 + q_0/r^2$, $H^A = h^A + p^A/r^2$. Note, the Einstein metric is flat.

As next step we consider the string solution, obtained by T-dualizing one of the spatial direction. This is again a type IIB solution corresponding intersection is $1 \times 5 \times 5 \times 5$

$$\begin{array}{c} ds^2_{string} = e^{2U} \left(-dt^2 + dx^2 \right) + e^{-2U} dw d\bar{w} \\ e^{2\phi} = e^{2U} \ , \ \partial_m a \sim \epsilon_{mnp} \partial_p H_0 \ , \ \mathcal{H}_A \sim e^{2U} \mathrm{Im} \mathcal{N}_{AB} \, {}^\star dH^B \ , \ \tilde{\mathcal{H}}_A \sim -i \, e^{2U} g_{AB} \, {}^\star dz^B \end{array} \tag{10}$$

where we choose a parameterization in terms of one complex scalar containing the axion a and the dilaton ϕ. In addition, the compactified D-5-branes and the Hodge dualization of the scalars give $2n_v$ antisymmetric tensors $(\mathcal{H}_A, \tilde{\mathcal{H}}_A)$. Equivalently, one could dualize them into (hyper) scalars. In doing this T-duality we have to assume a further isometry, i.e. the harmonic functions depend on one coordinate less. Thus the new harmonic functions are general holomorphic functions $H = f(w) + f(\bar{w})$, i.e. they are undetermined from the equations of motion, but restricted due to the proper behaviour at infinity and finite energy, see [8]. Notice however, because of the Kähler structure it is not obvious that the dilaton factorizes in harmonic factors. We will leave a detailed analysis for the future.

One has to keep in mind, that for all these solutions the internal space is fixed (branes wrapped around 4-cycles). Our T-duality transformation acts in the 4-d space time. Duality with respect to the internal manifold would of course not change the 4-d solution, it is given by the duality invariant Kähler potential.

Having these "basic" brane solutions we have to ask for singularities. Let us focus on the instantons and black holes. The harmonic function that we consider here have a pole of $1/r$. As consequence of the constraint for the symplectic section (6) also (X^I, F_I) have a $1/r$ pole. Thus the Kähler potential behaves like $1/r^2$ and the metric is non-singular on the horizon. This property is independent of the number of charges, as long as we keep a non-trivial graviphoton. So, *any generic CY black hole has a non-singular horizon.* This is an important difference to $N = 4, 8$ black holes that need at least 4 charges to get a non-singular horizon. Here, in the CY case, the selfintersection regularize the horizon, which has a throat geometry. Of complete different nature is the instanton solution. Again using the scaling behaviour of the Kähler potential one can show, that for the $r \to 0$ one reaches a second asymptotic flat region, where the charges and moduli are interchanged. Both regions are connected by wormhole at the minimum of re^{-U} and one of them is strongly coupled ($\phi \to \infty$), which is very similar analog to the situation in 10 dimensions [9].

Let us end this article with the conformal field theory (CFT) description of the near horizon region of the black hole. Basically it is not necessary to approach the horizon, instead one can scale the physical parameter approprietely (conformal limit). The result is the same – one neglects the constant parts in the harmonic functions in this limit. All non-singular brane configurations are expected to have a super conformal field theory description [10]. As example we consider the solution (8). This is a special case, but if we convince ourselfes that it is described by an exact CFT, due to duality also the other charge configurations should be exact. This solution is a compactified 5-d magnetic string [4]. Near the horizon, the spherical part becomes constant and a reduction over this spherical part yields a (negative) cosmological constant giving an AdS_3 geometry. By proper boundary conditions a conformal algebra is realized as diffeomorphism group at spatial infinity [11] (this is also nicely summarized in [12]) and the corresponding solution is the BTZ black hole [13] (for a discussion of the entropy and duality see especially [14, 12, 15]).

In the following we will sketch this procedure for our solution, a more detailed analysis will be given elsewhere. Near the horizon the magnetic string becomes [4]

$$ds^2 = \frac{r}{D^{1/3}}\left[-dt^2 + dx^2 + \frac{q_0}{r}(dx - dt)^2\right] + D^{2/3}\left(\frac{dr}{r}\right)^2 + D^{2/3} d\Omega_2 \ , \qquad (11)$$

where $D = d_{ABC} p^A p^B p^C$. As next step one compactifies the S^2 part. The 5-d sugra action contains besides the Einstein-Hilbert term, a gauge field part and a scalar part. Near the horizon the scalar part drops out and the gauge field part contributes to the cosmological constant $G_{IJ}F^I \cdot F^J = D^{-2/3}$ with $G_{IJ}(X)$ as the 5-d gauge coupling matrix. After this reduction the 3-d Einstein metric reads

$$ds_E^2 = D\,r\left[-dt^2 + dx^2 + \frac{q_0}{r}(dx-dt)^2\right] + D^2\left(\frac{dr}{r}\right)^2 \tag{12}$$

or in other coordinates ($\rho^2 = D(r+q_0)$, $\tau = 2Dt$ and $x = \phi$)

$$ds_E^2 = -\frac{(\rho^2 - q_0 D)^2}{4\rho^2 D^2}d\tau^2 + \rho^2\left(d\phi - \frac{q_0}{2\rho^2}d\tau\right)^2 + \frac{4D^2\rho^2}{(\rho^2 - q_0 D)^2}d\rho^2\,. \tag{13}$$

This is the extreme 3-d BTZ black hole [13], with the angular momentum $J = q_0$, the mass $M = \frac{q_0}{2D}$ and the cosmological constant $\Lambda^{-1} \sim l = 2D$. Thus excited momentum modes along the 5-d string correspond to angular momentum for this BTZ black hole. As it has been shown in [16], 3-d AdS gravity can be written as a Chern-Simons theory and furthermore, due to the cylindric symmetry ($\phi \simeq \phi + 2\pi$) it reduces to a sum of two SL(2,R) WZW theories for the left and right moving modes (in the BPS limit only one sector survive). Finally, implementing the boundary condition at spatial infinity we end up with a Liouville theory living on the boundary [17]. So, summarizing all steps we have the sequence for the actions

$$S_{sugra_5} \to S_{AdS_3} \sim k\,S_{CS} \sim k\,S_{WZW} \sim k\,S_L \tag{14}$$

where up to numerical factors, the central charge of the CFT is given by $k \sim \frac{D}{G_N}$, where we introduced the 3-d Newton constant G_N to get a dimensionless quantity.

Let us conclude with implications of this CFT for the Matrix description. It has been suggested recently that the matrix limit of M-theory compactified on CY_3 is given by the conifold point, which is mirror dual to the vanishing 6-cycle [18]. An important point in the matrix limit is to ensure the decoupling of massless states (decoupling from the bulk). For a T^6 compactification this is in general not the case (although as discussed in [19] there is may be a loophole by a gauged world volume theory). In order to discuss this question here we consider only 5-branes which wrap vanishing cycles. Notice, in our setup we approaching the conifold point from the other side, i.e. we consider vanishing 4-cycles. As consequence, the corresponding 4-d black hole besomes massless [20] or, for the magnetic sting, $D \to 0$ in this limit (note, if L counts the vanishing cycles $X^L \sim p^L \to 0$ at this point). As consequence the central charge of the CFT vanish, which coincides with the expectation that at the transition point the worldvolume theory is scale invariant. But more important, in this limit the boundary degrees of freedom (encoded in the Liouville action S_L) decouple, like the Liouville mode in critical string theory. This supports the idea that the matrix description for M-theory compactified on CY_3 [18] yields an consistent picture.

Acknowledgements

I am grateful I. Brunner, G. Cardoso, S. Kachru, A. Karch and R. Schimmrigk for helpful discussions. This work is support by the DFG.

References

[1] S. Ferrara, R. Kallosh and A. Strominger, hep-th/9508072; K. Behrndt, G. Lopes-Cardoso, B. de Wit, R. Kallosh, D. Lüst and T. Mohaupt, hep-th/9610105.

[2] J. Maldecena, A. Strominger and E. Witten, hep-th/9711053; C. Vafa, hep-th/9711067.

[3] K. Behrndt, D. Lüst and W.A. Sabra, hep-th/9705169.

[4] K. Behrndt, hep-th/9610232.

[5] K. Behrndt, I. Gaida, D. Lüst, S. Mahapatra and T. Mohaupt, hep-th/9706096.

[6] K. Behrndt, E. Bergshoeff and B. Janssen, hep-th/9604168.

[7] C. Hull, hep-th/9705162.

[8] B.R. Greene, A. Shapere, C. Vafa and S.-T. Yau, *Nucl. Phys.* **B337** (1990) 1, M. Bourdeau and G. Lopes Cardoso, hep-th/9709174.

[9] G.W. Gibbons, M.B. Green and M.J. Perry, hep-th/9511080.

[10] P. Claus, R. Kallosh and A. Van Proeyen, hep-th/9711161; J. Maldecena, hep-th/9711200; R. Kallosh, J. Kumar and A. Rajaraman, hep-th/9712073; S. Ferrara and C. Fronsdal, hep-th/9712239.

[11] J.D. Brown and M. Henneaux, *Comm. Math. Phys.* **104** (1986) 207.

[12] A. Strominger, hep-th/9712251.

[13] M. Banados, C. Teitelboim and J. Zanelli, hep-th/9204099; M. Banados, M. Henneaux, C. Teitelboim and J. Zanelli, gr-qc/9302012.

[14] S. Carlip, gr-qc/9409052.

[15] S. Hyun, hep-th/9704005; K. Sfetson and K. Skenderis, heo-th/9711138; H.J. Boonstra, B. Peeters and K. Skenderis, hep-th/9706192.

[16] A. Achucarro and P.K. Townsend, *Phys. Lett.* **B180** (1986) 89; E. Witten, *Nucl. Phys.* **B311** (1988) 46.

[17] O. Coussaert, M. Henneaux and P.van Driel, gr-qc/9506019

[18] S. Kachru, A. Lawrence and E. Silverstein, hep-th/9712223.

[19] K. Behrndt, hep-th/9710228.

[20] A. Strominger, hep-th/9504090; K. Behrndt, D. Lüst and W.A. Sabra, hep-th/9708065

The SL(2,ℝ)/U(1) WZNW Field Theory

Uwe Müller [1], Gerhard Weigt [2]

[1] Institut für Physik, Johannes-Gutenberg-Universität Mainz,
Staudingerweg 7, D-55099 Mainz, Germany
[2] Deutsches Elektronen-Synchrotron DESY-Zeuthen
Platanenallee 6, D–15738 Zeuthen, Germany

Abstract: We discuss the general classical solution of the gauged SL(2,ℝ)/U(1) Wess-Zumino-Novikov-Witten (WZNW) model in terms of canonical free fields. Although this transformation is non-local, the energy-momentum tensor assumes the canonical free-field form. The Poisson bracket structure of the theory is completely determined by a differential equation of the Gelfand-Dikii type.

1 Introduction

The talk mainly follows ref. [1] and we add some unpublished results. Our intention is to find a canonical transformation of the non-linear physical fields of the SL(2,ℝ)/U(1) WZNW model onto free fields to prepare canonical quantization.

Witten has shown [2] that the classical action of this model [3] describes string propagation in the background of a two-dimensional space-time black hole. This conformal field theory was intensively discussed (see e.g. [4, 5, 6, 7]), however, there is left considerable uncertainty in the calculation of an effective action. Usually the U(1) gauge field is eliminated by path integration [8, 2, 5]. But such integrations are incomplete because functional determinants remain uncalculated [8]. On the other hand, the theory has classically a traceless energy-momentum tensor, but the β-function is non-zero [9, 2, 4, 5] as the gauged WZNW action corresponds to a σ-model with curvature. Therefore, a dilaton appears perturbatively [2], and additional quantum mechanical corrections of the dilaton and the curvature arise [4, 5].

We eliminate the gauge field, as in ref. [3], purely classically by its (algebraic) equation of motion, and obtain an integrable classical action. We expect that the non-local structure of the theory will finally settle the question about the appearance of a dilaton. But this requires, obviously, a complete analytical solution of the classical theory, and its exact quantization.

In this talk we shall give the complete analytical solution of the classical theory.

2 Integration of the SL(2,ℝ)/U(1) Theory

In light-cone coordinates $z = \tau + \sigma$, $\bar{z} = \tau - \sigma$, $\partial_z = (\partial_\tau + \partial_\sigma)/2$, $\partial_{\bar{z}} = (\partial_\tau - \partial_\sigma)/2$ the gauged SL(2,ℝ)/U(1) WZNW action [3] is given by

$$S[r,t] = \frac{1}{\gamma^2} \int\limits_M \left(\partial_z r \partial_{\bar{z}} r + \tanh^2 r\, \partial_z t \partial_{\bar{z}} t\right) \mathrm{d}z\mathrm{d}\bar{z}. \tag{1}$$

This action has the euclidean signature black hole [2] target-space metric

$$\mathrm{d}s^2 = \mathrm{d}r^2 + \tanh^2 r\, \mathrm{d}t^2 \tag{2}$$

and a traceless energy-momentum tensor showing its conformal invariance.

The equations of motion of (1) are

$$\begin{aligned} \partial_z \partial_{\bar{z}} r &= \frac{\sinh r}{\cosh^3 r} \partial_z t\, \partial_{\bar{z}} t, \\ \partial_z \partial_{\bar{z}} t &= -\frac{1}{\sinh r\, \cosh r} \left(\partial_z r\, \partial_{\bar{z}} t + \partial_z t\, \partial_{\bar{z}} r\right). \end{aligned} \tag{3}$$

They are integrable because there exists a Lax pair representation

$$\left[\partial_{\bar{z}} - \bar{C}, \partial_z - C\right] = \partial_z \bar{C} - \partial_{\bar{z}} C - \left[C, \bar{C}\right] = 0. \tag{4}$$

C and $\bar{C}$ take values in the Lie algebra of the group SL(2,ℝ)

$$C = C_a T^a, \quad \bar{C} = \bar{C}_a T^a, \quad (a = 1,2,3). \tag{5}$$

Here T^1 is the matrix

$$T^1 = \begin{pmatrix} 0 & 1 \\ -1 & 0 \end{pmatrix}, \tag{6}$$

T^2 the Pauli matrix σ_3, and $T^3 = T^1 T^2$. We found that

$$\begin{aligned} &C_1 = -\frac{1}{2} \tanh^2 r\, \partial_z t, \quad C_2 = C_3 = 0, \quad \bar{C}_2 = \frac{1}{2} \tanh^2 r\, \partial_{\bar{z}} t, \\ &\bar{C}_2 = -\frac{1}{\cosh r} \partial_{\bar{z}} \left(\sinh r \cos t\right), \quad \bar{C}_3 = \frac{1}{\cosh r} \partial_{\bar{z}} \left(\sinh r \sin t\right), \end{aligned} \tag{7}$$

makes the flatness condition (4) equivalent to the equations of motion (3).

Instead of starting from the Lax pair (4), we found in ref. [10] the analytical solution of (3) by asymptotically relating it to the solution of the non-abelian Toda theory of ref. [6, 7]. It can easily be seen that

$$\begin{aligned} \sinh^2 r &= X\bar{X}, \\ t &= \mathrm{i}(B - \bar{B}) + \frac{\mathrm{i}}{2} \ln \frac{X}{\bar{X}} \end{aligned} \tag{8}$$

solve the equations of motion (3) if

$$\begin{aligned} X &= A + \frac{\bar{B}'}{\bar{A}'}(1 + A\bar{A}), \\ \bar{X} &= \bar{A} + \frac{B'}{A'}(1 + A\bar{A}). \end{aligned} \tag{9}$$

$A(z)$, $B(z)$, $A'(z)$, $B'(z)$ ($\bar{A}(\bar{z})$, $\bar{B}(\bar{z})$, $\bar{A}'(\bar{z})$, $\bar{B}'(\bar{z})$) are arbitrary chiral (anti-chiral) functions and their derivatives, respectively. The solution (8, 9) is invariant under $\mathrm{GL}(2, \mathbb{C})$ transformations of these functions. Therefore, these (anti-)chiral functions are determined by the physical fields only up to four complex constants.

3 Conservation Laws and Gelfand-Dikii Equation

The equations of motion guarantee the conservation of the chiral component of the energy-momentum tensor (whenever possible, anti-chiral parts will not be indicated)

$$T = \frac{1}{2}\left(T_{\tau\tau} + T_{\tau\sigma}\right) = \frac{1}{\gamma^2}\left((\partial_z r)^2 + \tanh^2 r\,(\partial_z t)^2\right). \tag{10}$$

But there are two further conserved chiral quantities on shell, the Wilson line modified Kac-Moody currents of the ungauged WZNW model [3]

$$V_\pm = \frac{1}{\gamma^2}\mathrm{e}^{\pm\mathrm{i}\nu}\left(\partial_z r \pm \mathrm{i}\tanh r\;\partial_z t\right). \tag{11}$$

Here the (non-local) ν is sufficiently defined by the derivatives

$$\partial_z \nu = (1 + \tanh^2 r\,)\partial_z t, \quad \partial_{\bar{z}}\nu = \cosh^{-2} r\;\partial_{\bar{z}} t. \tag{12}$$

Since the integrability condition of (12) is just the second equation of motion (3), the general solution (8, 9) provides the integrated expression for ν

$$\nu - t = \mathrm{i}(B + \bar{B}) - \frac{\mathrm{i}}{2}\ln\frac{1 + X\bar{X}}{(1 + A\bar{A})^2}, \qquad \nu + \bar{\nu} = 2t. \tag{13}$$

Surprisingly, the chiral component of the energy-momentum tensor factorizes in terms of the $V_\pm(z)$ and looks like a Sugawara construction

$$T = \gamma^2 V_+ V_-, \tag{14}$$

although the conformal spin-one quantities $V_\pm$ are no usual Kac-Moody currents.

From the conserved quantities we can derive a differential equation of the Gelfand-Dikii type with $W^{(1)} = (\partial_z V_-/V_-)$

$$y'' - (\partial_z V_-/V_-)y' - \gamma^2 T y = 0. \tag{15}$$

It has two fundamental solutions

$$\begin{aligned} y_1(z) &= e^{B(z)}, \\ y_2(z) &= y_1(z)\int^z \frac{W(z')}{y_1^2(z')}\mathrm{d}z'. \end{aligned} \tag{16}$$

The Wronski determinant

$$W = y_1 y_2' - y_2 y_1' \tag{17}$$

is related by means of eq. (15) to V_-

$$W = \gamma^2 V_- = A' e^{2B}, \tag{18}$$

and we find $A(z)$ and $B(z)$ explicitly in terms of the fundamental solutions of (15)

$$A(z) = \frac{y_2}{y_1}, \qquad B(z) = \ln y_1. \tag{19}$$

Therefore, the solutions y_1 and y_2 parametrize the general solution (8,9) as well.

4 Poisson Brackets and Free-Field Transformations

We assume canonical Poisson brackets for the physical fields r, t and their canonical conjugated momenta π_r, π_t. The differential equation (15) determines the dependence of the y_i on these fields. In order to derive the Poisson brackets for the y_i, we need the variations δy_i in terms of the variations δr, δt, $\delta\pi_r$, and $\delta\pi_t$, which enter the definition of the Poisson bracket. Varying eq. (15) we obtain

$$\delta y'' - (\partial V_-/V_-)\delta y' - \gamma^2 T\delta y = \delta(\partial V_-/V_-)y' + \gamma^2\delta T y. \tag{20}$$

The homogeneous part of this differential equation is identical to eq. (15). To find a unique solution of the inhomogeneous equation we need subsidiary conditions. The simplest case uses the field-theoretic initial conditions $\delta y_i(-\infty) = 0$, $\delta y_i'(-\infty) = 0$. In this case, there are no zero modes and we find the solution of the inhomogeneous equation in terms of the two solutions y_1 and y_2 of the homogeneous equation

$$\begin{aligned} \delta y_i(\sigma) &= \int_{-\infty}^{\sigma} \mathrm{d}\sigma'\Omega(\sigma,\sigma')\left(\delta(\partial V_-/V_-)(\sigma')y_i'(\sigma') + \gamma^2\delta T(\sigma')y_i(\sigma')\right), \\ \Omega(\sigma,\sigma') &\equiv \frac{y_1(\sigma')y_2(\sigma) - y_2(\sigma')y_1(\sigma)}{y_1(\sigma')y_2'(\sigma') - y_2(\sigma')y_1'(\sigma')}. \end{aligned} \tag{21}$$

This leads directly to the Poisson brackets of the y_i

$$\{y_i(\sigma), y_j(\tilde\sigma)\} = \frac{\gamma^2}{2}\left(y_i(\sigma)y_j(\tilde\sigma) - y_j(\sigma)y_i(\tilde\sigma)\right)\epsilon(\sigma-\tilde\sigma). \tag{22}$$

$\epsilon(\sigma)$ denotes here the sign function.

The Poisson brackets for the $A(z)$, $B(z)$ are found by taking into consideration eqs. (19), and we read off from them, finally, the following free-field realizations

$$\begin{aligned} A'(z) &= \gamma\,(\partial_z\phi_1 - \mathrm{i}\partial_z\phi_2)\exp\left(-2\gamma\phi_1(z)\right), \\ B(z) &= \gamma\,(\phi_1 - \mathrm{i}\phi_2)\,. \end{aligned} \tag{23}$$

The conserved quantities $V_\pm(z)$ then become

$$V_\pm(z) = \frac{1}{\gamma}\,(\partial_z\phi_1 \pm \mathrm{i}\partial_z\phi_2)\exp\left(\pm 2\mathrm{i}\gamma\phi_2(z)\right), \tag{24}$$

and, as expected, the energy-momentum tensor T assumes the canonical free field form

$$T = (\partial_z\phi_1)^2 + (\partial_z\phi_2)^2\,. \tag{25}$$

With these representations, the Gelfand-Dikii equations are identically fulfilled in terms of the free fields. We fixed the GL$(2,\mathbb{C})$ invariance by choosing

$$\begin{aligned} &y_k|_{\sigma=-\infty} = y_k(-\infty) = C_k, \quad \bar{y}_k|_{\sigma=-\infty} = \bar{y}_k(\infty) = \bar{C}_k, \\ &C_k, \bar{C}_k \in \mathbb{C}, \quad k \in \{1,2\}. \end{aligned} \tag{26}$$

It is worth mentioning here the $V_\pm(z)$ satisfy non-linear Poisson brackets

$$\begin{aligned} \{V_\pm(\sigma), V_\pm(\sigma')\} &= \gamma^2 V_\pm(\sigma)V_\pm(\sigma')\epsilon(\sigma-\sigma'), \\ \{V_\pm(\sigma), V_\mp(\sigma')\} &= -\gamma^2 V_\pm(\sigma)V_\mp(\sigma')\epsilon(\sigma-\sigma') + \frac{1}{\gamma^2}\delta'(\sigma-\sigma') \end{aligned} \tag{27}$$

which are characteristic for parafermions [11, 3]. This non-linear algebra also provides the Virasoro algebra, and implies conformal weight one for the $V_\pm$

$$\{T(\sigma), V_\pm(\sigma')\} = -\left(\partial_{\sigma'}V_\pm(\sigma')\delta(\sigma-\sigma') - V_\pm(\sigma')\delta'(\sigma-\sigma')\right). \tag{28}$$

Most of the calculations are very similar in the periodic case, after including the corresponding zero modes, the periodic sawtooth function

$$h(\sigma) = \epsilon(\sigma) - \frac{1}{\pi}\sigma, \tag{29}$$

instead of the non-periodic sign function, and the periodic δ-function

$$\delta_{2\pi}(\sigma) \equiv \frac{1}{2}\partial_\sigma\epsilon(\sigma) = \sum_n \delta(\sigma - 2\pi n). \tag{30}$$

$\epsilon(\sigma)$ denotes now the stair-step function

$$\epsilon(\sigma) = 2n+1 \quad \text{for} \quad 2n\pi < \sigma < 2(n+1)\pi, \tag{31}$$

which coincides with sign(σ) for $-2\pi < \sigma < 2\pi$.

Instead of (26), we now require $A(z)$ (and the other functions) to be periodic up to a factor

$$A(z+2\pi) = \exp(-\gamma p_1)\, A(z), \tag{32}$$

in order to fix the GL(2, $\mathbb{C}$) invariance. The integration of $A'(z)$ then yields

$$A(z) = -\frac{\gamma}{2}\sinh^{-1}\left(\frac{\gamma p_1}{2}\right) \times \int_0^{2\pi} \mathrm{d}z' \exp\left(-\frac{\gamma p_1}{2}\epsilon(z-z')\right)(\partial_{z'}\phi_1 - \mathrm{i}\partial_{z'}\phi_2)\exp(-2\gamma\phi_1(z')). \tag{33}$$

This shows the non-locality of the free-field transformation we were looking for. The free-field transformation is only complete after including the anti-chiral parts.

5 Conclusion

Although the classical solution of the SL(2,$\mathbb{R}$)/U(1) gauged WZNW model (1) bears strong resemblance to Liouville or Toda theories its quantum structure might be different because its energy-momentum tensor does not have a 'central charge' term classically.

We would like to see whether the black hole singularity survives quantization, a dilaton appears, and whether the Hawking radiation might be discussed in some manner based on this analytical solution.

References

[1] U. Müller and G.Weigt, *Phys. Lett.* **B400** (1997) 21.

[2] E. Witten, *Phys. Rev.* **D44** (1991) 314.

[3] K. Bardakci, M. Crescimanno, E. Rabinovici, *Nucl. Phys.* **344** (1990) 344.

[4] R. Dijkgraaf, E. Verlinde, H. Verlinde, *Nucl. Phys.* **B371** (1992) 269.

[5] A. A. Tseytlin, *Nucl. Phys.* **B399** (1993) 601.

[6] J.-L. Gervais, M. Saveliev, *Phys. Lett.* **B286** (1992) 271.

[7] A. Bilal, *Nucl. Phys.* **B422** (1994) 258.

[8] T.H. Buscher, *Phys.Lett.* **B201** (1988) 466.

[9] C.G. Callan, D. Friedan, E.J. Martinec, and M.J. Perry, *Nucl. Phys.* **B262** (1985) 593.

[10] U. Müller and G.Weigt, in Proceedings of the 30th International Symposium Ahrenshoop, 1996, Buckow, Germany, p. 328.

[11] V.A. Fateev and A.B. Zamolodchikov, *Zh.Eksp.Teor.Fiz.* **89** (1985)380.

String Amplitudes and Prepotential in Heterotic $K3 \times T^2$ Compactifications

K. Förger [1] **, S. Stieberger** [2]

[1] Centre de Physique Théorique, Ecole Polytechnique,
F-91128 Palaiseau Cedex, France
[2] CERN, Theory Division
CH-1211 Geneva 23, Switzerland

Abstract: We calculate one loop string amplitudes of gauge couplings which appear after torus compactifications of $d = 6$ dimensional $N = 1$ heterotic string vacua. The results in $d = 4$ dimensions are compared with the corresponding expressions of $N = 2$ special geometry. Moreover we find relations between three-point amplitudes, which are related to Yukawa couplings in the type II dual picture and derivatives of the prepotential and point out general features that appear in any T^2 compactification.

1 Introduction

Recent developments have shown that theories with extended supersymmetries have a rich dynamical structure which allow for an exact analysis. The special property of $N = 2$ supersymmetric string vacua in $d = 4$ dimensions is that the sector of vector multiplets of the low energy effective action is completely determined by a holomorphic prepotential up to second order derivatives. In particular the Kähler metric and the gauge coupling are expressed in terms of derivatives of the prepotential. String vacua with these properties can arise from heterotic $K3 \times T^2$ compactifications, type II compactifications on Calabi-Yau threefolds or type I models which are orientifold reductions of type IIB compactifications on $K3 \times T^2$. In the following we will focus on heterotic models.

The heterotic gauge group G consists of a non Abelian factor G', which depends on the specific instanton embeddings (e.g. $E_7 \times E_8$ for the standard embedding of the spin connection into the gauge group), and Abelian factors $\left[U(1)^2\right]_{\text{left}} \times \left[U(1)^2\right]_{\text{right}}$ which arise from the two left-moving and two right-moving $U(1)$ gauge groups after

T^2 compactification. The moduli T and U of the torus are the scalar components of the $U(1)$ $N = 2$ vector multiplets. At special points in the (T, U) moduli space additional vector multiplets become massless where the prepotential exhibits logarithmic singularities and the gauge group is enhanced.

The loop corrections to the prepotential are constrained by target space duality symmetry and by the Peccei-Quinn symmetry of the dilaton, which prohibits higher than one-loop corrections. The aim is to calculate one loop gauge threshold corrections associated to the $U(1)^2$ by string amplitudes and to relate the resulting τ–integrals, where τ is the modulus of the worldsheet torus, to the underlying prepotential and its derivatives. This calculation admits not only to verify $N = 2$ special geometry emerging from heterotic $K3 \times T^2$ compactification but also to get some insight into the general structure of string amplitudes and its relation to the prepotential for T^2 compactifications in any dimension.

2 Review of $N = 2$ Supergravity

The moduli space of $N = 2$ heterotic vacua in $d = 4$ dimensions is locally described by a direct product of the moduli space of vector multiplets, which is a special Kähler manifold [1], and the moduli space of hypermultiplets which is quaternionic [2]. In the following we focus on the vector multiplet moduli space. The dilaton S and the moduli T and U are scalar fields of the $N = 2$ vector multiplets [3, 4]. Since the dilaton, which governs string perturbation is part of a vector multiplet, the prepotential F receives perturbative as well as non-perturbative corrections from space–time instantons. However, thanks to $N = 2$ non-renormalization theorems, there are no higher than one loop corrections to the prepotential $F(X)$, where X^I with $I = 0, \ldots, 3$ denote the complex scalar components of the vector fields. Neglecting non-perturbative corrections the tree-level part and the one-loop correction to the prepotential reads [4, 5]

$$F(X) = (X^0)^2 STU + (X^0)^2 f(T, U) \tag{1}$$

where the special coordinates are $S = \frac{X^1}{X^0}$, $T = \frac{X^2}{X^0}$ and $U = \frac{X^3}{X^0}$ and X^0 accounts for the vector multiplet of the graviphoton. The one loop-corrections to the prepotential $f(T, U)$ is constrained by the duality group $SL(2, \mathbf{Z})_T \times SL(2, \mathbf{Z})_U \times \mathbf{Z}_2^{T \leftrightarrow U}$.

The gauge couplings are expressed in terms of a field dependent tensor $\mathcal{N}_{IJ}$ as

$$g_{IJ}^{-2} = \mathcal{N}_{IJ} - \bar{\mathcal{N}}_{IJ} \tag{2}$$

where

$$\mathcal{N}_{IJ} = \bar{F}_{IJ} + 2i \frac{\mathrm{Im} F_{IK} \mathrm{Im} F_{JL} X^K X^L}{\mathrm{Im} F_{MN} X^M X^N} \tag{3}$$

and $F_{IJ} = \frac{\partial^2 F}{\partial X^I \partial X^J}$. Due to the second term of $\mathcal{N}_{IJ}$ the gauge coupling are generically non-harmonic functions of the moduli. Expanding (3) for $S - \bar{S} \to \infty$ the gauge coupling e.g. for $I = J = 2$ turns out to be:

$$\begin{aligned} g_{22}^{-2} &= \frac{(S-\bar{S})(U-\bar{U})}{(T-\bar{T})} - \frac{1}{4}\left[\partial_T D_T f - \partial_{\bar{T}} D_{\bar{T}} \bar{f}\right] - \frac{1}{4}\frac{(U-\bar{U})^2}{(T-\bar{T})^2}\left[\partial_U D_U f - \partial_{\bar{U}} D_{\bar{U}} \bar{f}\right] \\ &+ \frac{1}{2}\frac{(U-\bar{U})}{(T-\bar{T})}\left[\partial_T \partial_U f - \partial_{\bar{T}} \partial_{\bar{U}} \bar{f}\right] + \mathcal{O}\left(\frac{f^2, (\partial f)^2}{S-\bar{S}}\right) \end{aligned} \tag{4}$$

where we introduced the covariant derivatives $D_T = \partial_T - \frac{2}{T-\bar{T}}$. Notify, that the one loop correction to g_{22}^{-2} is dilaton independent although the expansion of $\mathcal{N}_{IJ}$ allows for higher powers in f as well as terms of the order $\mathcal{O}\left(\frac{1}{S-\bar{S}}\right)$. At this point one might wonder about the non-renormalization theorem. This puzzle is solved by going to a new basis via symplectic transformation [4] in which the loop counting works, i.e. the perturbative corrections stop after one loop.

3 String Amplitudes and Prepotential

There are various approaches in the literature to determine derivatives of the prepotential F which are related to physical quantities such as gauge couplings and the Kähler metric. In [4] the derivatives of the one loop correction f to the prepotential have been found from the field theoretical point of view using quantum symmetries and the singularity structure of the moduli space. Another approach to this problem was done in [5], where derivatives to the string amplitude corresponding to the one-loop correction to the Kähler potential were calculated in order to get an explicit expression for $\partial_T^3 f$. Here we directly focus on the CP even part of string amplitudes including vertex operators associated to the gauge bosons of the internal Abelian gauge group of the torus T^2 in a background field method. The vertex operators of the gauge bosons in the zero ghost picture are

$$\tilde{V}_{A_+}^{(0)} = -\frac{1}{2} F_{\mu\nu}^T \rho(T) \bar{\partial} X^+ (X^\nu \partial X^\mu + \psi^\mu \psi^\nu) \tag{5}$$

with $\rho(T) = \sqrt{\frac{U_2}{T_2}}$ and $X^+ = \frac{1}{\sqrt{2}}\left(X_1 + iX_2\right)$ is one of the internal bosonic fields. After summing over even spin structures the only non-vanishing contributions of $\langle \tilde{V}_{A_+}^{(0)} \tilde{V}_{A_+}^{(0)} \rangle$ include fermionic contractions multiplied with contractions of internal bosons. The final answer for the two point amplitude including two gauge boson vertex operators $\tilde{V}_{A_+}^{(0)}$ is

$$\mathcal{A}(A_+, A_+) = -\frac{i}{4} F_{\mu\nu}^T F^{\mu\nu T} \Delta_{(TT)} \tag{6}$$

where the one loop threshold correction to the gauge coupling is

$$\Delta_{(TT)} = -\frac{U_2}{T_2} \int \frac{d^2\tau}{\tau_2} \sum_{P_L, P_R} \bar{P}_R^2 \, \hat{Z}_{\text{torus}} \, \bar{F}_{-2}(\bar{\tau}) \,, \tag{7}$$

with $\overline{F}_{-2} = \frac{\overline{E}_4 \overline{E}_6}{\bar{\eta}^{24}}$ the modular function of weight -2 and

$$Z_{\text{torus}}(\tau, \bar{\tau}) = \sum_{(P_L, P_R)} q^{\frac{1}{2}|P_L|^2} \bar{q}^{\frac{1}{2}|P_R|^2} := \sum_{(P_L, P_R)} \hat{Z}_{\text{torus}} \,. \tag{8}$$

where $P_{R/L}$ are Narain momenta. The $\bar{P}_R^2$ term arises from contractions of the internal bosons $\langle \bar{\partial} X^+ \bar{\partial} X^+ \rangle$. For more details the reader is referred to [7]. Similarly we determine the threshold correction $\Delta_{(UU)}$ by exchanging T_2 with U_2 and replacing $\bar{P}_R$ with its complex conjugate P_R. For the string amplitude $\langle \tilde{V}^{(0)}_{A_+} \tilde{V}^{(0)}_{A_-} \rangle$ we get the modular invariant result:

$$\Delta_{(TU)} = -\int \frac{d^2\tau}{\tau_2} \sum_{(P_L,P_R)} \left(|P_R|^2 - \frac{1}{2\pi\tau_2} \right) \hat{Z}_{\rm torus}\, \bar{F}_{-2}(\bar{\tau}) \,. \tag{9}$$

In order to compare the calculated string quantities (7) and (9) with $N = 2$ supergravity, i.e. with formulae like (4), we have to perform the τ integration and express the integrals in terms of derivatives of the underlying one loop prepotential f. The amplitudes transform with specific weights under $SL(2,\mathbf{Z})_T \times SL(2,\mathbf{Z})_U$, namely $\mathcal{A}(A_+, A_+)$ with weights $(w_T, w_U) = (2, -2)$, $(w_{\bar{T}}, w_{\bar{U}}) = (0, 0)$ and $\mathcal{A}(A_+, A_-)$ with $(w_T, w_U) = (w_{\bar{T}}, w_{\bar{U}}) = (0, 0)$.

Let us consider more general integrals which involve up to four 'charge' insertions from internal bosonic contractions.

$$\mathcal{I}_{w_T,w_U} := (T-\overline{T})^m (U-\overline{U})^n \int \frac{d^2\tau}{\tau_2^k} \sum_{(P_L,P_R)} P_L^\alpha P_R^\beta \overline{P}_L^\gamma \overline{P}_R^\delta\, \hat{Z}_{\rm torus}(\tau,\overline{\tau})\, \overline{F}_l(\overline{\tau}) \,. \tag{10}$$

Such integrals appear e.g. as threshold gauge coupling corrections for general T^2 compactification in any dimensions. The modular function $\bar{F}_l$ can be identified with the elliptic genus $\mathrm{Tr}_R\big((-1)^F q^{L_0 - c/24} \bar{q}^{\bar{L}_0 - \bar{c}/24}\big)$ which is defined as the trace in the Ramond sector of the internal CFT. The elliptic genus only gets contributions from the right moving internal sector. According to the unbroken gauge group before T^2 compactification $\bar{F}_l$ may take different forms: in $d = 6$ (with only one tensor multiplet) it becomes $\bar{F}_{-2}$ for $K3$ compactification with the gauge group $E_8 \times E_8$ being completely Higgsed away (so–called STU–model). We want this integral to have modular weights (w_T, w_U) and $(w_{\overline{T}}, w_{\overline{U}}) = (0, 0)$, which restricts the parameters to satisfy the following relations

$$\begin{aligned} m &= -\frac{w_T}{2} \\ n &= -\frac{w_U}{2} \\ \gamma &= \alpha - \frac{1}{2}(w_T + w_U) \\ \delta &= \beta + \frac{1}{2}(w_T - w_U) \,. \end{aligned} \tag{11}$$

There is also a relation for k and l which follows from modular invariance of the integrand, which is easily deduced after a Poisson resummation. Choosing the expression $\bar{F}_{-2}$ we can use the method of orbit decomposition introduced in [8] to get the final answer. For more details see Appendix B of [7].

For the two point amplitude (6) including two A_μ^+ gauge bosons we get the threshold correction:

$$\Delta_{(TT)} = -8\pi^2 i\partial_T D_T f + 8\pi^2 i\frac{(U-\overline{U})^2}{(T-\overline{T})^2}\partial_{\overline{U}} D_{\overline{U}}\overline{f} \,. \tag{12}$$

with $D_{\bar{U}} = \left(\partial_{\overline{U}} + \frac{2}{U-\overline{U}}\right)$ and where f is the $N=2$ heterotic prepotential in the chamber $T_2 > U_2$ is [9]

$$if(T,U) = \frac{i}{12\pi}U^3 + \frac{1}{(2\pi i)^4}\sum_{(k,l)>0} c_1(kl)\;\mathcal{L}i_3\left[e^{2\pi i(kT+lU)}\right] + \frac{1}{2(2\pi)^4}c_1(0)\zeta(3)\,. \tag{13}$$

The numbers $c_1(n)$ are coefficients of $\bar{F}_{-2}(\bar{\tau}) = \sum_{n\geq -1} c_1(n)\bar{q}^n$ and $\mathcal{L}i_3(x) = \sum_{p>0}\frac{x^p}{p^3}$. The prepotential $f(T,U)$ is a modular function of weight $(w_T, w_U) = (-2,-2)$. Acting with the covariant derivative $D_T^n = \partial_T - \frac{2n}{T-\bar{T}}$ on a modular function of weight $-2n$ yields a modular function of weight $-2n+2$. Thus we see that the expression of $\Delta_{(TT)}$ in terms of derivatives of the prepotential f indeed has the correct modular weights. The result for $\Delta_{(UU)}$ can be obtained from $\Delta_{(TT)}$ by exchanging T with U. Finally for $\Delta_{(TU)}$ we get

$$\Delta_{(TU)} = -8\pi^2 i\left(D_T D_U f - \bar{D}_T\bar{D}_U\bar{f}\right)\,. \tag{14}$$

The appropriate linear combination of string amplitudes which corresponds to a two point function including vertex operators of gauge bosons

$$A_\mu^T = \frac{1}{2}[A_\mu^+ - \frac{(U-\bar{U})}{(T-\bar{T})}A_\mu^-] \tag{15}$$

finally reproduces the one loop correction to the gauge coupling g_{22}^{-2} which is determined by special geometry:

$$\begin{aligned}\left[g_{22}^{-2}\right]^{1-\text{loop}} &= \frac{1}{4}\left[\frac{\Delta_{(TT)}}{8\pi^2 i} - 2\frac{(U-\bar{U})}{(T-\bar{T})}\frac{\Delta_{(TU)}}{8\pi^2 i} + \frac{(U-\bar{U})^2}{(T-\bar{T})^2}\frac{\Delta_{(UU)}}{8\pi^2 i}\right] - \frac{1}{8\pi^2 i}\frac{U_2}{T_2}G^{(1)} \\ &= -\frac{1}{32\pi^2 i}\left[\frac{U-\bar{U}}{T-\bar{T}}\int\frac{d^2\tau}{\tau_2}\sum_{(P_L,P_R)}(\bar{P}_R - P_R)^2\hat{Z}_{\text{torus}}\bar{F}_{-2}(\bar{\tau})\right]\end{aligned} \tag{16}$$

where $G^{(1)} = 4\pi^2\left(\frac{4f}{(T-\bar{T})(U-\bar{U})} - \frac{2\partial_U f}{(T-\bar{T})} - \frac{2\partial_T f}{(U-\bar{U})} - c.c.\right)$ is the Green-Schwarz term [4].

Using the same techniques we can calculate three point amplitudes including one modulus and two gauge bosons. The vertex operator of the T modulus contributes with the zero modes of an additional term $:\bar{\partial}X^+\partial X^+:$ to the correlator of the internal bosons. Summarizing we find:

$$\begin{aligned}\mathcal{A}(T, A_+, A_+) &= -2\,F_{\mu\nu}^T F^{\mu\nu\,T}\pi^3 f_{TTT} \\ \mathcal{A}(T, A_+, A_-) &= -2\,F_{\mu\nu}^T F^{\mu\nu\,U}\pi^3\left(f_{TTU} + \frac{1}{4\pi^2}G_T^{(1)}\right) \\ \mathcal{A}(T, A_-, A_-) &= -2\,F_{\mu\nu}^U F^{\mu\nu\,U}\pi^3\left(f_{TUU} + \frac{1}{4\pi^2}G_U^{(1)}\right)\end{aligned} \tag{17}$$

The holomorphic part of the above amplitudes are related to the Yukawa couplings f_{TTT}, f_{TTU} and f_{TUU} of the dual type IIA Calabi-Yau compactification.

Let us finally discuss these amplitudes from a more general point of view. The gauge kinetic terms which arise after T^2 compactification can be deduced from string amplitudes that produce the lowest non-vanishing order of gravitational one-loop corrections. The relevant correlation function one has to consider is an n-point function including graviton vertex operators where $2n$ is the number of fermionic zero modes in the considered dimension. In $d=6$ dimensions the four fermionic zero modes can be saturated by a two-point graviton correlation function:

$$\langle :\epsilon_{ij}\bar{\partial}X_1^i\left[\partial X_1^j+i(k_1\cdot\psi_1)\psi_1^j\right]e^{ik_1\cdot X_1}::\epsilon_{kl}\bar{\partial}X_2^k\left[\partial X_2^l+i(k_2\cdot\psi_2)\psi_2^l\right]e^{ik_2\cdot X_2}:\rangle\,, \quad (18)$$

which may contain both the curvature scalar $\mathcal{R}$ and the square of the Riemann tensor $\mathcal{R}_{ikjl}\mathcal{R}^{ikjl}$. Here ϵ_{ij} is the gravitational polarization tensor in six dimensions. The amplitude is determined by expanding the elliptic genus $\mathcal{A}$ of $K3$ w.r.t. $\mathcal{R}^2$. In general, in N=1, $d=6$ dimensional heterotic string theories only the 4–form part of the elliptic genus gives rise to modular invariant one–loop corrections [11, 10]. To saturate the fermionic zero modes one has to contract the four fermions which is already of the order $\mathcal{O}(k^2)$. Since the worldsheet integral of the bosonic contraction $\langle\bar{\partial}X_1^i\bar{\partial}X_2^k\rangle$ gives a zero result and thus the $\mathcal{O}(k^2)$ term of the effective action vanishes, the next to leading order arises from contractions of $\bar{\partial}X^i$ with the exponential $e^{ik\cdot X}$ which contributes another $\mathcal{O}(k^2)$ term to the amplitude. For the $\mathcal{O}(k^4)$ contribution we thus obtain [10]:

$$\Delta_{\mathcal{R}^2}^{6d,N=1}\sim\int\frac{d^2\tau}{\tau_2^2}\left(\bar{E}_2-\frac{3}{\pi\tau_2}\right)\frac{\bar{E}_4\bar{E}_6}{\bar{\eta}^{24}}=-8\pi\,. \quad (19)$$

where $\bar{E}_w$ are Eisenstein functions of weight $w=2,4,6$ and η is the Dedekind η–function. After compactification on T^2 the coupling (19) is no longer a constant but turns out to be the moduli dependent gravitational threshold correction [10, 6, 9]. However if we want to deduce the gauge kinetic term of the internal gauge bosons we keep both i and k as internal indices + which has been defined before and extract the $\mathcal{O}(k^2)$–piece of (18). What changes is that $\left(\bar{E}_2-\frac{3}{\pi\tau_2}\right)$ of (19) is replaced by

$$\int d^2z_{12}\langle\bar{\partial}X_1^+\bar{\partial}X_2^+\rangle=2\pi^2\tau_2\bar{P}_R^2\,. \quad (20)$$

This way we end up with $\Delta_{(TT)}$ of (7).

This argument can be applied for heterotic torus compactification from $d=10,8,4$ dimensions to $d=8,4,2$ dimensions. In fact, similarly to the methods established here for $d=4$ $N=2$ heterotic vacua, relations like (12) and (14), which express a certain physical quantity –given as τ-integral– in terms of a function f should also hold for all possible insertions of internal momenta like (10) in $d=2$ and $d=8$. The latter are associated to gauge fields of the internal gauge bosons. World–sheet τ-integrals as (10) for general k and l, however *without* momenta insertions, have been considered in [12, 13] in the context of heterotic/typeI duality in

$d = 10, 9, 8$ dimensions. It is the topic of [14], to discuss this issue with including momenta insertions.

Acknowledgments: K.F. would like to thank the organizers of this workshop for a stimulating atmosphere and for allowance to present this talk.

References

[1] A. Strominger, *Comm. Math. Phys.* **133** (1990) 163

[2] J. Bagger and E. Witten, *Nucl. Phys.* **B 222** (1983) 1

[3] A. Ceresole, R. d' Auria, S. Ferrara and A. van Proeyen, *Nucl. Phys.* **B 444** (1995) 92

[4] B. de Wit, V. Kaplunovsky, J. Louis and D. Lüst, *Nucl. Phys.* **B 451** (1995) 53

[5] I. Antoniadis, S. Ferrara, E. Gava, K. Narain and T. Taylor, *Nucl. Phys.* **B 447** (1995) 35

[6] I. Antoniadis, E. Gava and K.S. Narain, *Nucl. Phys.* **B 383** (1992) 92; *Phys. Lett.* **B 283** (1992) 209

[7] K. Förger, S. Stieberger, hep-th/9709004

[8] L. Dixon, V. Kaplunovsky and J. Louis, *Nucl. Phys.* **B 355** (1991) 649

[9] J.A. Harvey and G. Moore, *Nucl. Phys.* **B 463** (1996) 315

[10] W. Lerche, *Nucl. Phys.* **B 308** (1988) 102

[11] W. Lerche, B.E.W. Nilsson and A.N. Schellekens, *Nucl. Phys.* **B 289** (1987) 609; W. Lerche, B.E.W. Nilsson, A.N. Schellekens and N.P. Warner, *Nucl. Phys.* **B 299** (1988) 91; W. Lerche, A.N. Schellekens and N.P. Warner, *Phys. Rept.* **177** (1989) 1

[12] C. Bachas, C. Fabre, E. Kiritsis, N.A. Obers and P. Vanhove, hep–th/9707126

[13] E. Kiritsis, N.A. Obers, hep-th/9709058

[14] K.Förger, S. Stieberger, in preparation

Asymptotic Supergraviton States in Matrix Theory

Jan Plefka and Andrew Waldron

NIKHEF
P.O. Box 41882, 1009 DB Amsterdam
The Netherlands

Abstract: We study the Matrix theory from a purely canonical viewpoint. In particular, we identify free particle asymptotic states of the model corresponding to the 11D supergraviton multiplet along with the split of the matrix model Hamiltonian into a free and an interacting part. Elementary quantum mechanical perturbation theory then yields an effective potential for these particles as an expansion in their inverse separation. We discuss how our scheme can be used to compute the Matrix theory result for the 11D supergraviton S matrix and briefly comment on non-eikonal and longitudinal momentum exchange processes.

1 The model.

Matrix theory [1] is the conjectured description of M theory in terms of a supersymmetric matrix model. At low energies and large distances M theory, by definition, reduces to 11D supergravity. In this talk we explicitly construct asymptotic particle states in Matrix theory to be identified with the 11D supergraviton multiplet and study the scattering of these states.

The Hamiltonian of the Matrix theory is that of ten dimensional $U(N)$ super Yang-Mills dimensionally reduced to $0+1$ dimensions [2] and

arises from two disparate viewpoints. On the one hand, it is the regulating theory of the eleven dimensional supermembrane in light cone gauge quantization [3] and on the other, it is the effective Hamiltonian describing the short distance properties of $D0$ branes [4, 5, 6]. Employing the conjecture of [7], the finite N model is to be identified with the compactification of a null direction of M theory (henceforth called the $-$ direction). The quantized total momentum of the $U(N)$ system in this direction is then given by $P_- = N/R$, where R denotes the compactification radius.

We shall be primarily interested in the $U(2)$ theory, studying the Hilbert space of two supergravitons with momentum $P_- = 1/R$ each. The coordinates and Majorana spinors of the transverse nine dimensional space then take values in the adjoint representation of $U(2)$, i.e.

$$X_\mu = X^0_\mu \, i \, 1\!\!1 + X^A_\mu \, i\sigma^A \qquad \mu = 1, \ldots, 9 \tag{1.1}$$

$$\theta_\alpha = \theta^0_\alpha \, i \, 1\!\!1 + \theta^A_\alpha \, i\sigma^A \qquad \alpha = 1, \ldots, 16 \tag{1.2}$$

where σ^A are the Pauli matrices. We shall often employ a vector notation for the $SU(2)$ part in which $\vec{X}_\mu = (X^1_\mu, X^2_\mu, X^3_\mu) \equiv (X^A_\mu)$ and similarly for $\vec{\theta}$.

The Hamiltonian is then given by

$$H = H_{\mathrm{CoM}} + \frac{1}{2}\vec{P}_\mu \cdot \vec{P}_\mu + \frac{1}{4}(\vec{X}_\mu \times \vec{X}_\nu)^2 + \frac{i}{2}\vec{X}_\mu \cdot \vec{\theta}\gamma_\mu \times \vec{\theta} \tag{1.3}$$

where $H_{\mathrm{CoM}} = \frac{1}{2} R P^0_\mu P^0_\mu$ is the $U(1)$ centre of mass Hamiltonian. Note that we are using a real, symmetric representation of the $SO(9)$ Dirac matrices in which the nine dimensional charge conjugation matrix is equal to unity.

The Hamiltonian (1.3) is augmented by the Gauss law constraint

$$\vec{L} = \vec{X}_\mu \times \vec{P}_\mu - \frac{i}{2}\vec{\theta} \times \vec{\theta} \quad , \qquad [L^A, L^B] = i\,\epsilon^{ABC} L^C \tag{1.4}$$

whose action is required to vanish on physical states.

The task is now to identify the free asymptotic two-particle states of the Hamiltonian (1.3) which describe the on-shell supergraviton multiplet of eleven dimensional supergravity. This problem manifestly factorises into a $U(1)$ centre of mass state and an $SU(2)$ invariant state describing the relative dynamics of the particles.

2 The centre of mass theory.

The eigenstates of the free $U(1)$ centre of mass Hamiltonian H_{CoM} are

$$|k_\mu; h_{\mu\nu}, B_{\mu\nu\rho}, h_{\mu\hat{\alpha}}\rangle_0 = e^{ik_\mu X^0_\mu} |h_{\mu\nu}, B_{\mu\nu\rho}, h_{\mu\hat{\alpha}}\rangle_0 \tag{2.5}$$

and possess transverse $SO(9)$ momentum k_μ and on-shell supergraviton polarisations[1] $h_{\mu\nu}$, $B_{\mu\nu\rho}$ and $h_{\mu\hat{\alpha}}$ (graviton, antisymmetric tensor and gravitino, respectively). The state $|h_{\mu\nu}, B_{\mu\nu\rho}, h_{\mu\hat{\alpha}}\rangle_0$ is the $\underline{44} \oplus \underline{84} \oplus \underline{128}$ representation of the centre of mass spinor θ^0 degrees of freedom. The construction of this state is carried out in detail in [8] and allows the explicit calculation of the spin dependence of Matrix theory supergraviton amplitudes. In order to define the fermionic vacuum and creation and annihilation operators one performs a decomposition of the $SO(9)$ Lorentz algebra with respect to an $SO(7) \otimes U(1)$ subgroup [3]. This is done as follows. Firstly split vector indices $\mu = (1, \ldots, 9)$ as $(m = 1, \ldots, 7; 8, 9)$ so that an $SO(9)$ vector V_μ may be rewritten as (V_m, V, V^*) where $V = V_8 + iV_9$ and $V^* = V_8 - iV_9$. For an $SO(9)$ spinor the

[1] Note that the polarisation tensors $h_{\mu\nu}$, $B_{\mu\nu\rho}$ and $h_{\mu\hat{\alpha}}$ correspond to *physical* polarisations. The Matrix theory does away with unphysical timelike and longitudinal polarisations at the price of manifest eleven dimensional Lorentz invariance.

same decomposition is made by complexifying, in particular, for the canonical spinor variables we have

$$\lambda = \frac{\theta^0_+ + i\theta^0_-}{\sqrt{2}}\,,\ \lambda^\dagger = \frac{\theta^0_+ - i\theta^0_-}{\sqrt{2}}\,, \tag{2.6}$$

where the subscript $\pm$ denotes projection by $(1 \pm \gamma_9)/2$. The canonical anticommutation relations are now $\{\lambda_\alpha, \lambda^\dagger_\beta\} = \delta_{\alpha\beta}$ where $\alpha, \beta = 1, \ldots, 8$ and we define the fermionic vacuum $|-\rangle$ by $\lambda|-\rangle = 0$. We denote the completely filled state by $|+\rangle = \lambda^\dagger_1 \ldots \lambda^\dagger_8 |-\rangle$. One finds then the following expansion for the supergraviton polarisation state

$$\begin{aligned} |h_{\mu\nu}, B_{\mu\nu\rho}, h_{\mu\hat{\alpha}}\rangle_0 \;=\; & h|-\rangle + \frac{1}{4}h_m|-\rangle_m + \frac{1}{16}h_{mn}|\pm\rangle_{mn} + \frac{1}{4}h^*_m|+\rangle_m + h^*|+\rangle \\ & -\frac{\sqrt{3}\,i}{8}\left(B_{mn}|-\rangle_{mn} + \frac{i}{6}B_m|\pm\rangle_m + \frac{1}{6}B_{mnp}|\pm\rangle_{mnp} - B^*_{mn}|+\rangle_{mn}\right) \\ & \frac{i}{\sqrt{2}}\left(h_\alpha|-\rangle_\alpha - \frac{1}{2}h_{m\alpha}|-\rangle_{m\alpha} + \frac{1}{2}h^*_{m\alpha}|+\rangle_{m\alpha} - h^*_\alpha|+\rangle_\alpha\right)\,. \end{aligned} \tag{2.7}$$

The states in (2.7) transform covariantly with respect to $SO(7) \otimes U(1)$ and are defined in [8].

3 Asymptotic states.

Relative motions are described in the Matrix theory by the constrained $SU(2)$ quantum mechanical matrix theory defined above. However, spacetime is only an asymptotic concept in this theory. In particular diagonal matrix configurations, i.e., those corresponding to Cartan generators of $SU(N)$, span flat directions in the matrix model potential and describe spacetime configurations [1]. Transverse directions are described by supersymmetric harmonic oscillator degrees of freedom, as we will see below.

Due to the gauge constraint (1.4) quantum mechanical wavefunctions must be invariant under $SU(2)$ rotations so that there is no preferred Cartan direction. To find asymptotic states corresponding to supergraviton (i.e., spacetime) excitations in a gauge invariant way we proceed as follows. Let us suppose we wish to study states describing particles widely separated in the (say) ninth spatial direction, then we may simply declare the $SU(2)$ vector $\vec{X}_9$ to be large. The limit $|\vec{X}_9| = \sqrt{\vec{X}_9 \cdot \vec{X}_9} \to \infty$ is $SU(2)$ rotation (and therefore gauge) invariant. We search for asymptotic particle-like solutions in this limit.

To this end it is convenient to employ the (partial) gauge choice [9] in which one chooses a frame where $\vec{X}_9$ lies along the z-axis,

$$X^1_9 = 0 = X^2_9\,. \tag{3.8}$$

Calling $X_9 = (0, 0, x)$ and $\vec{X}_a = (Y_a^1, Y_a^2, x_a)$ (with $a = 1, ..., 8$) the Hamiltonian in this frame then is $H = H_{\rm V} + H_{\rm B} + H_{\rm F} + H_4$ where[2]

$$\begin{aligned} H_{\rm V} &= -\frac{1}{2x}(\partial_x)^2 x - \frac{1}{2}(\partial_{x_a})^2 & (3.9) \\ H_{\rm B} &= -\frac{1}{2}(\frac{\partial}{\partial Y_a^I})^2 + \frac{1}{2} r^2 Y_a^I Y_a^I & (3.10) \\ H_{\rm F} &= r\,\tilde{\theta}^\dagger \gamma_9 \tilde{\theta} & (3.11) \\ H_4 &= \text{“rest”}. & (3.12) \end{aligned}$$

The sum of the Hamiltonians $H_{\rm B}$ and $H_{\rm F}$ is that of a supersymmetric harmonic oscillator with frequency r and describes excitations transverse to the flat directions. Particle motions in the flat directions correspond to the Hamiltonian $H_{\rm V}$ whereby we interpret the Cartan variables $x_\mu = (x_a, x)$ asymptotically as the $SO(9)$ space coordinates.

The Hilbert space may be treated as a "product" of superoscillator degrees of freedom and Cartan wavefunctions depending on x_μ and the third component of $\vec{\theta}$ via the identity

$$H = \sum_{m,n} |m\rangle \langle m|H|n\rangle \langle n| \tag{3.13}$$

where $\{|n\rangle\}$ denote the complete set of eigenstates of $H_{\rm B}$ and $H_{\rm F}$. Since the frequency r of the superoscillators is coordinate dependent, operators $\partial/\partial x_\mu$ do not commute with $|n\rangle$ so that this "product" is not direct. This construction allows us to study an "effective" Hamiltonian $H_{mn}(x_\mu, \partial_{x_\mu}, \theta^3) = \langle m|H|n\rangle$ for the Cartan degrees of freedom pertaining to asymptotic spacetime. In particular the free Hamiltonian is given by the diagonal terms[3]

$$H_0 = \sum_n |n\rangle \langle n| \left(H_{\rm V} + H_{\rm B} + H_{\rm F} - \frac{c_n}{r^2}\right) |n\rangle \langle n| \tag{3.14}$$

and the interaction Hamiltonian then reads $H_{\rm Int} = H - H_{\rm CoM} - H_0$. Since supersymmetric harmonic oscillator zero point energies vanish, eigenstates of (3.14) are

$$|k_\mu; h_{\mu\nu}, B_{\mu\nu\rho}, h_{\mu\hat{\alpha}}\rangle = \frac{1}{x} e^{ik_\mu x_\mu} |h_{\mu\nu}, B_{\mu\nu\rho}, h_{\mu\hat{\alpha}}\rangle \otimes |0_B, 0_F\rangle \tag{3.15}$$

where $|0_B, 0_F\rangle$ is the supersymmetric harmonic oscillator vacuum. These states satisfy the correct free particle dispersion relation

$$H_0 |k_\mu; h_{\mu\nu}, B_{\mu\nu\rho}, h_{\mu\hat{\alpha}}\rangle = \frac{1}{2} k_\mu k_\mu \, |k_\mu; h_{\mu\nu}, B_{\mu\nu\rho}, h_{\mu\hat{\alpha}}\rangle \, . \tag{3.16}$$

[2]The spinors $\tilde{\theta}$ are built from θ^1 and θ^2 by complexification and a spin(9) rotation (see [8]). Note that $r^2 \equiv x_a x_a + x^2$.

[3]We subtract terms c_n/r^2 to ensure the correct asymptotic behaviour of the interaction Hamiltonian. A detailed explanation of this point may be found in [8].

Here, the supergraviton polarisation multiplet $|h_{\mu\nu}, B_{\mu\nu\rho}, h_{\mu\hat{\alpha}}\rangle$ is built from the $\underline{44} \oplus \underline{84} \oplus \underline{128}$ representation of θ^3 as in (2.7).

Therefore, upon taking the direct product of an asymptotic state (3.15) with a centre of mass eigenstate (2.5)

$$|1,2\rangle = e^{ik_\mu^{\rm tot} X_\mu^0} \frac{1}{x} e^{ik_\mu^{\rm rel} x_\mu} \otimes |0_B, 0_F\rangle \otimes |h^1_{\mu\nu}, B^1_{\mu\nu\rho}, h^1_{\mu\hat{\alpha}}\rangle_{\theta^0+\theta^3} \otimes |h^2_{\mu\nu}, B^2_{\mu\nu\rho}, h^2_{\mu\hat{\alpha}}\rangle_{\theta^0-\theta^3} \tag{3.17}$$

one obtains a state describing a pair of supergravitons widely separated in the ninth spatial direction. Its interactions, which die off as $x \to \infty$, are governed by $H_{\rm Int}$.

4 Scattering amplitudes.

The $2 \longrightarrow 2$ supergraviton scattering amplitude is then obtained by elementary quantum mechanical scattering theory as

$$\begin{aligned}\lim_{T\to\infty} \langle 1', 2'| e^{-iHT} |1,2\rangle &= \delta(k_\mu'^{\rm tot} - k_\mu^{\rm tot}) \int 4\pi x^2 \, d^9 x_\mu \qquad (4.18)\\ & \frac{e^{-ik_\mu'^{\rm rel} x_\mu}}{x} \langle \mathcal{H}^{1'}, \mathcal{H}^{2'} |\, H_{\rm Eff}(x_\mu, \partial_\mu, \theta^3_{\hat{\alpha}}) \, |\mathcal{H}^1, \mathcal{H}^2\rangle \, \frac{e^{ik_\mu^{\rm rel} x_\mu}}{x}\end{aligned}$$

where we have introduced $|\mathcal{H}^1, \mathcal{H}^2\rangle = |h^1_{\mu\nu}, B^1_{\mu\nu\rho}, h^1_{\mu\hat{\alpha}}\rangle_{\theta^0+\theta^3} \otimes |h^2_{\mu\nu}, B^2_{\mu\nu\rho}, h^2_{\mu\hat{\alpha}}\rangle_{\theta^0-\theta^3}$ and similarly for $|\mathcal{H}^{1'}, \mathcal{H}^{2'}\rangle$. The leading (Born) approximation to the "effective" Cartan Hamiltonian $H_{\rm Eff}$ is given by

$$H_{\rm Eff}{}^{(1)}(x_\mu, \partial_{x_\mu}, \theta^3_{\hat{\alpha}}) = \langle 0_B, 0_F|\, H_{\rm Int} |0_B, 0_F\rangle \tag{4.19}$$

and higher order contribution are obtained from the Lippman-Schwinger expansion

$$\begin{aligned}H_{\rm Eff}(x_\mu, \partial_\mu, \theta^3_{\hat{\alpha}}) &= \langle 0_B, 0_F| H_{\rm Int} |0_B, 0_F\rangle + \langle 0_B, 0_F| H_{\rm Int} \tfrac{1}{E-H_0+i\epsilon} H_{\rm Int} |0_B, 0_F\rangle \\ &\quad + \langle 0_B, 0_F| H_{\rm Int} \tfrac{1}{E-H_0+i\epsilon} H_{\rm Int} \tfrac{1}{E-H_0+i\epsilon} H_{\rm Int} |0_B, 0_F\rangle + \ldots \qquad (4.20)\end{aligned}$$

which due to the scaling behaviours $H_{\rm Int} \sim \mathcal{O}(x^{-1/2})$ and $H_0 \sim \mathcal{O}(x)$ turns out to be an expansion in $1/x$ the inverse separation of the two supergravitons [8].

The leading term of $H_{\rm Eff}$ is on dimensional grounds of order $1/r^2$ and receives contribution at first and second order perturbation theory in the sense of (4.20). An explicit computation [8] shows that these two contributions precisely cancel[4]. This supersymmetric cancellation is in accordance with the two loop semiclassical background field path integral calculation of [11] and yields a strong test of our proposal. Higher order contributions should then capture the revered v^4/r^7 potential for $D0$ particles [12].

Let us stress at this point, however, that the amplitudes (4.19) are restricted to the eikonal kinematical regime (i.e. high energy, straight line), as the asymptotic in- and outgoing supergraviton pairs are widely separated in the same (in this case 9th)

[4]Similar computations have been performed in a different context in[10].

spatial direction. Scattering amplitudes at arbitrary angles ($\sigma_{\mu\nu}$) may be obtained by performing an $SO(9)$ rotation of the outgoing state, i.e.

$$\langle 1', 2'|\exp(i\,\sigma_{\mu\nu}\,L^{\mu\nu}_{\mathrm{SO(9)}})\,\exp(-iH\,T)|1,2\rangle \tag{4.21}$$

where $L^{\mu\nu}_{\mathrm{SO(9)}}$ denotes the generator of $SO(9)$.

The $2 \longrightarrow 1$ supergraviton scattering channel of the $U(2)$ Matrix theory hinges on the knowledge of the zero-energy groundstate $|\mathrm{GS}\rangle$ of the $SU(2)$ supersymmetric quantum mechanics (which exists, according to [10, 13]). The "1" supergraviton state with $P_- = 2/R$ is then given by the direct product of the $U(1)$ centre of mass state $|k^{1'}_\mu, \mathcal{H}^{1'}\rangle_0$ with $|\mathrm{GS}\rangle$. Therefore the $2 \longrightarrow 1$ amplitude reads

$$\begin{aligned} \lim_{T\to\infty} \langle 1'|e^{-iHT}|1,2\rangle &= {}_0\langle k^{1'}_\mu, \mathcal{H}^{1'}| \otimes \langle \mathrm{GS}|\,\exp(-iHT)\,|k^1_\mu, k^2_\mu; \mathcal{H}^1, \mathcal{H}^2\,\rangle \\ &= {}_0\langle k^{1'}_\mu, \mathcal{H}^{1'}| \otimes \langle \mathrm{GS}|\, k^1_\mu, k^2_\mu; \mathcal{H}^1, \mathcal{H}^2\,\rangle \end{aligned} \tag{4.22}$$

since $H|\mathrm{GS}\rangle = 0$. Exact knowledge of the state $|\mathrm{GS}\rangle$ would yield us the complete non-perturbative answer for this process involving longitudinal momentum exchange. Recently there has been some progress towards uncovering the structure of the ground state [14, 15].

Acknowledgements

J. Plefka thanks the organizers for a stimulating symposium.

References

[1] T. Banks, W. Fischler, S.H. Shenker and L. Susskind, Phys. Rev. D55 (1997) 5112, hep-th/9610043.
For recent reviews see:
T. Banks, "Matrix Theory", hep-th/9710231;
D. Bigatti and L. Susskind, "Review of Matrix Theory", hep-th/9712072.

[2] M. Claudson and M.B. Halpern, Nucl. Phys. B250 (1985) 689;
R. Flume, Ann. Phys. 164 (1985) 189;
M. Baake, P. Reinicke, and V. Rittenberg, J. Math. Phys. 26 (1985) 1070.

[3] B. de Wit, J. Hoppe and H. Nicolai, Nucl. Phys. B305 (1988) 545.

[4] J. Polchinski, Phys. Rev. Lett. 75 (1995) 4724, hep-th/9510017.

[5] E. Witten, Nucl. Phys. B460 (1996) 335, hep-th/9510135.

[6] U. Danielson, G. Ferretti and B. Sundborg, Int. J. Mod. Phys. A11 (1996) 5463, hep-th/9603081;
D. Kabat and P. Pouliot, Phys. Rev. Lett. 77 (1996) 1004, hep-th/9603127.

[7] L. Susskind, "Another Conjecture about M(atrix) Theory", hep-th/9704080.

[8] J. Plefka and A. Waldron, "On the quantum mechanics of M(atrix) theory.", hep-th/9710104, to appear in Nucl. Phys. B.

[9] B. de Wit, M. Lüscher and H. Nicolai, Nucl. Phys. B320 (1989) 135.

[10] S. Sethi and M. Stern, "D-Brane Bound States Redux", hep-th/9705046.

[11] K. Becker and M. Becker, Nucl. Phys. B506 (1997) 48, hep-th/9705091; K. Becker, M. Becker, J. Polchinski and A. Tseytlin, Phys. Rev. D56 (1997) 3174, hep-th/9706072.

[12] M.D. Douglas, D. Kabat, P. Pouliot and S. Shenker, Nucl. Phys. B485 (1997) 85, hep-th/9608024.

[13] M. Porrati and A. Rozenberg, "Bound States at Threshold in Supersymmetric Quantum Mechanics", hep-th/9708119.

[14] J. Hoppe, "On the Construction of Zero Energy States in Supersymmetric Matrix Models", hep-th/9709132; "On the Construction of Zero Energy States in Supersymmetric Matrix Models II", hep-th/9709217.

[15] M.B. Halpern and C. Schwartz, "Asymptotic Search for Ground States of $SU(2)$ Matrix Theory", hep-th/9712133.

Six-Dimensional Fixed Points from Branes

Ilka Brunner, Andreas Karch

Institut für Physik, Humboldt-Universität,
Invalidenstr. 110, D-10115 Berlin

Abstract: We review the construction of six dimensional $N = 1$ fixed points in a brane picture involving D6 branes stretching between NS 5 branes.

Introduction: The classical Hanany-Witten brane setup.
Hanany and Witten [2] invented a brane configuration in IIB theory to study 3 dimensional $N = 4$ supersymmetric gauge theories. The basic ingredients they used are IIB NS 5 branes, D5 branes and D3 branes in flat space. The worldvolumes of the branes are in the following directions

	x^0	x^1	x^2	x^3	x^4	x^5	x^6	x^7	x^8	x^9
NS 5	0	1	2	3	4	5	-	-	-	-
D 5	0	1	2	-	-	-	-	7	8	9
D 3	0	1	2	-	-	-	6	-	-	-

It can be checked that this configuration preserves $1/4$ of the supersymmetries, that is 8 supercharges, as required for $N = 4$ in $d = 3$. The 3 branes are suspended between the 5 branes in the x_6 direction. The 3d, $N = 4$ theory is realized in the directions which are common to all branes. The point of view we take is that the 5 branes are much heavier than the 3 branes because they have two extra dimensions. The low energy dynamics is determined by the lowest dimensional brane in the setup. Moving around the 3 brane corresponds to changing the moduli of our theory. If we move around 5 branes this corresponds to changing parameters like masses, coupling constants, FI-terms.

What is the field theory on a 3 brane suspended between two NS 5 branes? On an infinite 3 brane there is a theory of a vector multiplet $V_{N=8}$ with 16 supercharges. The multiplet contains a vector and scalars corresponding to the fluctuations of the 5 brane in the transversal directions. In our setup, the x_6 direction is finite, therefore we are left with a $2 + 1$ dimensional field theory. The effect of the NS 5 branes is that SUSY is broken to $N = 4$. The 3 brane motion in the $7, 8, 9$–direction is locked because the 3 brane positions in these directions have to agree with the 5 brane positions. Also, the scalar A_6 coming from the dimensional reduction of the vector is projected out by the boundary conditions on the NS 5 brane. Altogether, we are left with an $N = 4$ vector multiplet. To enhance the gauge group to $U(N_c)$ we can put N_c 3 branes on top of each other. The gauge coupling of the theory is related to the distance between the NS 5branes:

$$\frac{1}{g_{YM}^2} = \frac{\Delta x_6}{g_s}$$

Here, g_s is the string coupling and $1/g_s$ would be the gauge coupling on an infinite D3. We can include matter multiplets in the fundamental representation by putting D5 branes in between the NS branes. Strings can then stretch between the D5 and D3 branes yielding matter in the fundamental representation. The mass of these multiplets corresponds to the distance (in the 345 direction) between the D5 and D3 branes. An alternative way to include matter is to add semi-infinite 3 branes to the left and right of the NS branes. Matter arises from strings stretching between a semi-infinite and finite piece of the 3 brane. The two descriptions of matter multiplets are related by the Hanany Witten effect: We move the D5 branes off to infinity. When they cross an NS brane a new D3 brane is created, which ends on the NS brane.

The Hanany Witten setup can be T-dualized along the directions 3,4,5. The dimensions of the D brane stretched between the two NS branes increases in each step. This enables us to study theories with 8 supercharges in various dimensions.

Bending and RR charge conservation.

In higher dimensions it becomes very important to take into account the disturbance caused by the D branes ending on the NS branes [12]. The end of a Dd brane looks like a magnetic monopole in the worldvolume of the NS 5 brane or as a charged particle on the $6 - d$ dimensional subspace transverse to the Dd brane. Dd branes ending from different sides on the NS brane contribute with opposite charge. The consequence is that the NS branes do not have a definite x_6 position, but the x_6 coordinate obeys a Laplace equation:

$$\Delta x_6(y) = 0,$$

where y parametrizes the transversal space. The "true" x_6 coordinate of the NS brane is the x_6 value far away from the disturbance. We can analyze the behaviour in various dimensions by looking at the solutions to the Laplace equation in various dimensions. In three dimensions, the case analyzed in the previous paragraph, the solution behaves as

$$x_6 = \frac{1}{|y|} + \mathit{constant},$$

such that for $|y| \to \infty$ we get a definite value, which we can call the x_6 position of the 5 brane.

In $d = 4$ we obtain a logarithmic behaviour. The distance in x_6 between two NS branes is proportional to the 4 dimensional gauge coupling, which is known to diverge logarithmically in 4 dimensions. This is reproduced in the brane picture.

In 5 dimensions, the transversal space is one dimensional and we obtain a linear bending of the NS brane. This can also be seen from the fact that RR charge has to be conserved at the vertices where different 5 branes come together. If we characterize a 5 brane by its charge under the NS and RR 2 forms, then a D brane has charge (0,1) and an NS brane (1,0). If they end on each other, a (1,1) brane emerges from the vertex.

The basic 6d brane setup.

We now want to study what happens in 6 dimensions in some more detail [5, 10, 11]. We consider D6 branes stretching between NS branes.

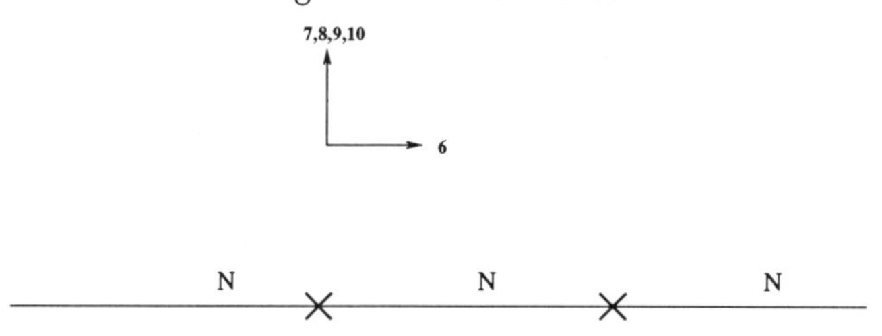

Figure 1: The brane configuration under consideration, giving rise to a 6 dimensional field theory. Horizontal lines represent D6 branes, the crosses represent NS 5 branes.

The configuration is shown in figure 1 . The worldvolume of the NS brane lies completely inside the worldvolume of the D6 which ends on it. We include matter by semi-infinite D6 branes extending to both sides of the NS branes and will discuss D8 branes later on. Because there are no transversal directions of the NS 5 brane left, there is no room for bending. The RR charge has to cancel exactly at each vertex. The net charge is given by the number of D6 branes ending from one side minus the number of D6 branes ending from the other side. Thus, we only get a consistent picture if:

$$N_c = n_l = n_r,$$

where n_l (n_r) denotes the number of D6 ending from the left (right). the total number of flavor giving semi infinite D6 is therefore

$$N_f = n_l + n_r = 2N_c$$

The low energy field theory.

What is the low energy field theory interpretation of this brane setup? According to the above philosophy, we have to look for the lowest dimensional brane in the setup. In our setup, the NS branes are as light as the finite D6 brane pieces. Only the semi infinite D6 branes are heavy. This is different from brane configurations leading to lower dimensional field theories, where the NS branes could always be considered as heavy and their motions determined parameters in the theory. The theory on a IIA NS 5 brane is the theory of a (0,2)– tensor multiplet. This multiplet consists of a tensor and 5 scalars (and fermions). Because of the presence of the

D6 branes, one half of the SUSY is broken and we are left with a (0,1) theory. The tensor multiplet decomposes into a (0,1) tensor, which only contains one scalar, and a hypermultiplet, which contains 4 scalars. The hypermultiplet is projected out from the massless spectrum because the position of the semi-infinite D6 branes fixes the position of the NS branes, so that fluctuations in the transversal directions are suppressed. The scalar in the tensor multiplet corresponds to motions of the 5 branes in the x_6 direction. We have two NS 5 branes and therefore two tensor multiplets, but effectively we keep only one of them because one of the scalars can be taken to describe the center of mass motion of the system. The vev of the other scalar gives us the distance between the NS branes. It is therefore related to the coupling of the theory. If the two 5 branes come together we arrive at a strong coupling fixed points. This theory contains tensionless strings coming from virtual membranes stretching between the 5 branes.

Altogether, the branes describe an SU(N_c) theory with a tensor and N_f hypers. The brane analysis gives the result that for a consistent theory the number of fundamentals has to be $N_f = 2N_c$. It predicts a strong coupling fixed point with this matter content.

Inclusion of D8 branes.

So far, we included fundamental matter multiplets by semi-infinite D6 branes. It should also be possible to describe the matter content by higher dimensional D branes between the NS branes [6]. In our case, these are D8 branes. D8 branes are charged under a RR nine form potential. The dual field strength of it is a constant. This constant is related to th cosmological constant appearing in massive IIA supergravity. The D8 branes divide space-time into different regions with different cosmological constant. Whenever we cross a D8 brane, the cosmological constant jumps by one unit. This is important for our brane configuration because there is a term in the action of massive IIA supergravity which is proportional to the IIA mass parameter and which couples the NS two form field B to the field strength of the 7 form potential under which the D6 is charged. The coupling reads:

$$-m \int d^{10}x B \wedge *F^{(2)}$$

($F^{(2)}$ is the field strength of the dual one form potential.) This modifies the equations of motion for the 7 form potential and therefore the RR charge cancellation condition. If a D6 brane ends on an NS brane, the equations of motion for the 7 form potential, or equivalently the Bianchi identity for the dual two form field strength is

$$dF^{(2)} = d * F^{(8)} = \theta(x_7)\delta^{(456)} - mH,$$

where $H = dB$. The δ term is the source term coming from the D6 brane ending on the NS brane. In the presence of m D8 branes the number of D6 branes ending from left and right on the NS branes should differ by m. In this way, a D8 placed in between the two NS branes has the same effect as a semi infinite D6 ending on the NS brane.

The field theoretical point of view.
We have seen that the brane construction leads via RR charge conservation to a restriction on the number of vectors and hypers in the theory. This restriction can be reproduced from a field theory point of view by an analysis of the gauge anomaly. In six dimensions the anomaly arising from vectors and hypers is

$$I = \mathrm{tr}_{adj} F^4 - \sum_R n_R \mathrm{tr} F^4$$

R denotes the representation of the matter multiplets. In our case we only have N_f fundamental matter multiplets. We can convert the trace in the adjoint to a trace in the fundamental representation and obtain:

$$I = (a - N_f)\mathrm{tr}_f F^4 + c(\mathrm{tr}_f F^2)^2$$

where a and c are group dependent factors. If the theory is consistent, the anomaly has to be cancelled. There are various options: In the simplest case, both the prefactor of the F^4 term and c vanish. In this case the theory is certainly consistent. If only the prefactor of the F^4 term vanishes but c does not vanish, the anomaly can be cancelled by introducing a Green Schwarz tree-level counterterm. This counterterm involves an antisymmetric tensor field. In the case $c > 0$ the anomaly can be cancelled without introducing gravity, whereas in the case $c < 0$ we can only get a consistent theory if we introduce gravity. Note that in 6d a tensor can be divided into a self dual and an anti self dual part. One piece is contained in the gravity multiplet and the other in a tensor multiplet. The tensor multiplet furthermore contains a scalar ϕ, whereas the gravity multiplet does not contain scalars. In the remaining case that the prefactor of the F^4 term does not vanish, the anomaly cannot be cancelled. Our gauge group is $SU(N_c)$. Here, $a = 2N_c$ and $c = 6$, so that we are in the situation, where the anomaly can be cancelled by introducing a tensor, if $N_f = 2N_c$. This is precisely the result obtained from the brane picture. For $N_c \leq 3$, SU(N_c) does not have an independent fourth order Casimir. In this case, the condition $c > 0$ only imposes an upper bound on the number of flavors. Global anomalies restrict the possible matter content further in these cases [3]. The only additional possibilities for SU(2) is 10 flavors, which can be realized by introducing an orientifold and using SU(2) $\sim$ Sp(2). The additional possibilities in the SU(3) case are 0 and 12 flavors and can not be seen in the brane picture.

From the brane picture we have predicted a strong coupling fixed point, when the expectation value of the scalar in the tensor multiplet vanishes. To verify this from a field theory point of view, we look at the following part of the action:

$$\frac{1}{g^2}\mathrm{tr}F_{\mu\nu}^2 + \sqrt{c}\phi\,\mathrm{tr}F_{\mu\nu}^2$$

We see, that one can absorb the bare gauge coupling into the expectation value of the tensor to get an effective coupling

$$\frac{1}{g_{eff}^2} = \sqrt{c}\phi$$

This is the effective coupling we see in the brane picture. At $\phi = 0$ there is a strong coupling fixed point [1].

Modifications of the basic setup.
We can introduce further building blocks into our basic brane setups to obtain other 6 dimensional fixed points. Of course, we can build a chain of k NS branes connected by D6 branes, leading to a product of SU gauge groups. This leads to a gauge group $SU(N_c)^k$ with bifundamentals and $2N_c$ fundamentals. We can also introduce orientifold 6 planes parallel to the D6 branes, leading to Sp or SO gauge groups, depending on the sign of the orientifold projection. The charge of the orientifold is twice or minus twice the charge of a D6 brane. Furthermore, it changes sign when it passes the NS branes. Taking this into account, RR charge conservation leads for one group factor to the condition $N_f = N_c - 8$ for SO and $N_f = N_c + 8$ for Sp. This is in agreement with anomaly cancellation. Combining these two options we get gauge groups $\{Sp(2N_c) \times SO(2N_c + 8)\}^{k+1/2}$ Furthermore, we can introduce O8 planes parallel to flavor giving D8 branes. If an NS 5 is stuck to the D8, we get SU gauge groups with an antisymmetric or symmetric tensor (depending on the charge of the O8).

Many such possibilities have been studied in [10, 11]. The resulting theories can also be obtained from 5 branes at ALE singularities [4].

References

[1] N. Seiberg, *Phys. Lett.* **B390** (1997) 169, hep-th/9609161.

[2] A. Hanany and E. Witten, *Nucl. Phys.* **B492** (1997) 152, hep-th/9611230.

[3] M. Bershadsky and C. Vafa, hep-th/9703167.

[4] J. Blum and K. Intriligator, hep-th/9705044.

[5] I. Brunner and A. Karch, *Phys.Lett.* **B409** (1997) 109, hep-th/9705022.

[6] A. Hanany and A. Zaffaroni, hep-th/9706047.

[7] U. Danielsson, G Ferretti, J. Kalkkinen and P. Stjernberg, *Phys. Lett.* **B405** (1997) 265, hep-th/9703098.

[8] N. Evans, C. Johnson and A. Shapere, hep-th/9703210.

[9] K. Landsteiner and E. Lopez, hep-th/9708118.

[10] I. Brunner and A. Karch, hep-th/9712143.

[11] A. Hanany, A. Zaffaroni, hep-th/9712145.

[12] E. Witten, *Nucl. Phys.* **B500** (1997) 3, hep-th/9703166.

Matrix Theory and the Six Torus

Ilka Brunner , Andreas Karch

Institut für Physik, Humboldt-Universität,
Invalidenstr. 110, D-10115 Berlin

Abstract: We review the problems associated with Matrix compactificati on T^6.

1 Introduction

Matrix theory claims to be a non-perturbative formulation of M-theory. However it is formulated in a background dependend way and hence compactifications pose a serious problem.

At the heart of the matrix construction lies the fact, that in the special reference frame chosen several degrees of freedom decouple [1, 2]. Basically one considers a compact light-like circle [2] of radius R with N units of momentum $1/R$ around the circle. In the limit that N and R both go to infinity this goes over to the so called infinite momentum frame (IMF) of [1], which loosely speaking is the reference frame of an observer boosted infinitely with respect to the experiment along the 11 direction.

Due to the infinite boost, in this frame p_{11} of all the constituents is much bigger than any other scale in the problem. The energy of a particle becomes

$$E = \sqrt{p_{11}^2 + p_{perp}^2 + m^2} \rightarrow |p_{11}| + \frac{p_{perp}^2 + m^2}{2|p_{11}|}.$$

The Hamiltonian responsible for propagation in this frame is $H = E - p_{11}$. ¿From this we see, that only modes with positive p_{11} contribute to the dynamics. In this case the Hamiltonian is governed by the finite piece

$$H = \frac{p_{perp}^2 + m^2}{2|p_{11}|}$$

while all the other modes lead to a Hamiltonian which goes to infinity with growing p_{11}. Therefore they lead to very rapid oscillations and can be neglected. This decoupling mechanism is still valid when we compactify more spacelike directions. The main task is to identify these remaining degrees of freedom.

2 The matrix description

To analyse the dynamics of M-theory in the discrete lightcone gauge, one can use IIA/M-theory duality to map this system to a better understood IIA setup. One then identifies the degrees of freedom that survive as carriers of positive p_{11} as D0 branes [1]. The same philosophy suggest, that using T-duality Matrix theory on the T^p is described by the dynamics of Dp branes. We will review this kind of construction in the following sections, especially in view of the still puzzling T^6 compactification. Recently it was shown by Seiberg [3] that this identification can basically be derived from the assumptions of M-theory/IIA duality and Lorentz invariance of M-theory by just noting that the compact lightlike circle is Lorentz equivalent to a particular limit of a compactification on a vanishing space-like circle.

2.1 The inifinite boost limit

In order to identify the surviving degrees of freedom a la [3] one proceeds in two steps. One notes that the funny frame we are interested is really equivalent to a vanishing *spacelike* circle R and one rescales all length and energy scales involved in order to keep the relevant Hamiltonian H for the surviving degrees of freedom finite. These scales are the 11d Planck length l_p, the radii of the transverse compact directions L_i and of course R. In terms of these quantities we can express p_{11}, m and p_{perp} in the expression for the energy of the relativistic excitation as

$$\begin{aligned} m &= \tilde{m}/l_p \\ p_{11} &= N/R \\ p_{perp} &= M_i/L_i \end{aligned}$$

where $\tilde{m}$, N and M_i are fixed dimensionless numbers. The first relation just expresses the fact that we want to measure energies in terms of the 11d Planck scale, while the other two are just quantization conditions for momenta along the compact directions.

Now one wants to take the limit in such a way that for $R \to 0$

$$H = E - p_{11} = \frac{p_{perp}^2 + m^2}{2|p_{11}|} \sim \frac{1/L_i^2 + 1/l_p^2}{1/R}$$

is constant. Thus the limit of infinite boost parameter amounts to taking

$$R \to 0, \quad L_i \to 0, \quad l_p \to 0 \tag{1}$$

$$l_p^2/R = f = \text{fixed}, \quad L_i/l_p = D_i = \text{fixed}$$

2.2 The type IIA string theory perspective

After one identified the limit that corresponds to an infinite boost one can analyze this limit from the IIA perspective. By considering the dynamics of D0 branes (which after all are the carriers of positive p_{11}) in the IIA setup in this particular limit one therefore derives the matrix description for various string compactifications. The parameters of the corresponding IIA theory are

$$g_s^2 = R^3/l_p^3 \quad l_s^2 = l_p^3/R \quad L_i = L_i.$$

We see that the limit (1) corresponds to $g_s \to 0$, $M_s = 1/l_s \to \infty$ and $L_i \to 0$. If we have no compact transverse dimensions we recover the original DLCQ version of the matrix description. The full quantum theory of the sector with N units of momentum p_{11} is described by the dynamics of N D0 branes in the limit that the string coupling is zero, the Planck and the string scale go to infinity while the gauge coupling of the quantum mechanics on the D0 wordvolume is fixed.

Since (1) involves taking L_i to zero the original IIA picture is not a good description anymore once we compactify transverse dimensions. We should perform a T-duality transformation along the transverse compact directions, thereby mapping the D0 partons into Dd branes, where d denotes the number of compact transverse dimensions. Since the generalization of T-duality to arbitrary manifolds is problematic, we will restrict ourselves in the rest of this work to the case of tori. This way we will derive the matrix description of the DLCQ of M-theory on T^d. It is given by Dd branes wrapping a torus with radii Σ_i in IIB (IIA) string theory for d odd (even), in the following limit for the 10d string coupling g_s, string scale M_s and 10d Planck scale M_p. g_{YM} denotes the coupling constant of the effective SYM on the Dd worldvolume. The following formulas can be simply obtained by applying the usual T-duality.

$$\begin{aligned} g_s^2 &= \frac{l_p^{3d-3}}{R^{d-3}V^2} & \Sigma_i &= \frac{l_p^3}{RL_i} \\ M_s^2 &= \frac{R}{l_p^3} & g_{YM}^2 &= \frac{l_p^{3d-6}}{R^{d-3}V} \\ M_P^8 &= \frac{V^2R^{d+1}}{l_p^{3d+9}} & & \end{aligned}$$

where $V = \prod_{i=1}^d L_i$. ¿From this we can read off how the various parameters behave in the limit (1).

$$g_s^2 \sim \begin{cases} 0 & \text{for } d < 3 \\ \text{finite} & \text{for } d = 3 \\ \infty & \text{for } d > 3 \end{cases} \qquad M_s^2 \sim \infty \text{ for all } d$$

$$M_P^8 \sim \begin{cases} \infty & \text{for } d < 7 \\ \text{finite} & \text{for } d = 7 \\ 0 & \text{for } d > 7 \end{cases} \qquad \begin{aligned} \Sigma_i &\sim \text{finite for all } d \\ g_{YM}^2 &\sim \text{finite for all } d \end{aligned}$$

Note that this time the sides of the torus stay finite, so we actually found a good string theory description. Also the gauge coupling of the theory on the brane is kept at a finite value, so we are really left with an interacting theory to describe the dynamics of the M-theory setup. For $d \leq 3$ the string coupling is finite or vanishes, while Planck and string scale go to infinity. So we can identify the wordvolume theory on the Dd by the usual string theory techniques. This way one obviously reproduced and hence derived the usual SYM on the dual torus description.

For $d = 4, 5, 6$ the string coupling blows up while the Planck scale still goes to infinity. One may hope that one can use strong/weak coupling dualities to find an appropiate description in a weakly coupled theory. We will review them in the following section on a case by case basis.

At least starting from the T^7 one won't be able to decouple the bulk gravity since the Planck scale stays finite. In the infinite boost limit several degrees of freedom decouple. The method outlined above allowed us to determine which dynamical degrees of freedom remain. It turns out that for small tori indeed all the bulk modes decouple, so that the matrix description of M-theory on T^d is described by a $d+1$ dimensional theory. Even though for $d \geq 7$ still many bulk degrees of freedom decouple (for example all the massive string modes, since M_s is sent to infinity) at least bulk gravity remains! The matrix description of M-theory in the DLCQ on those higher tori is given by a 10 dimensional theory! The same turned out to be the case for $d = 6$ [4].

3 The four-, five- and sixtorus

The fourtorus:

According to our derivation above the DLCQ of M-theory on the T^4 is given by a IIA D4 brane where we take the string coupling to infinity, the string scale to infinity and the 10d Planck scale to infinity, keeping $g_{YM}^2 = l_s g_s$, the gauge coupling of the 4+1 SYM on the D4 fixed. The D4 at infinite coupling is better thought of as the M-theory 5 brane. Translating the above limit in M-theory language we find that we are interested in the worldvolume theory of the M5 in the limit that we take the radius of the 11th dimension and also the 11d Planck scale to infinity. This worldvolume theory is known to be given by the (2,0) fixed point in 6 dimensions. One thus derived the Berkooz-Rozali-Seiberg [5] description of M-theory on the T^4.

The fivetorus:

For the fivetorus we find the theory of D5 branes of IIB at infinite string coupling, sending M_s and M_p also to infinity, where as shown above the gauge coupling of the D5 worldvolume theory stays fixed. IIB string theory offers us the possibility to map this to the theory of NS5 branes at zero string coupling. The above limit maps to $M_p \to \infty$, $g_s \to 0$, keeping M_s and hence the gauge coupling of the NS5 wordvolume gauge theory fixed. This is precisely Seiberg's realization of the 6 dimensional little string theory, which he used as the matrix description of M-theory on T^5 [6].

At this point a comment about the BPS solutions of the theory is in order. Seiberg identified the 16 BPS states preserving 1/2 of the original 32 supercharges, transforming under the $SO(5,5,Z)$ U-duality group of M-theory on T^5. They are bound states of the NS5 brane with D1, D3 or D5 branes. The energy of these states is given by

$$E = \sqrt{(T_{NS5} V_{NS5})^2 + (T_D V_D)^2}$$

where T_{brane} and V_{brane} denote the volume and the tension of the brane respectively. This is the T-dual version of the formula $E = \sqrt{p_{11}^2 + m^2}$ for the case without transverse momentum. The p_{11} momentum modes got mapped to $T_{ND5}V_{NS5}$, since this is what the D0 branes, the carriers of positive p_{11}, are mapped to under the ST^5-duality we used. m^2 is the mass of the wrapped D branes. Boosting to the IMF now amounts to taking the $g_s \to 0$ limit and switching to a light cone Hamiltonian $H = E - p_{11}$. In our case this amounts to taking

$$H = \lim_{g_s \to 0} \left(\sqrt{(T_{NS5}V_{NS5})^2 + (T_D V_D)^2} - T_{NS5}V_{NS5} \right)$$

for the light cone energy of these objects, which is the relation Seiberg used to get the excitations of the little string theory. We see again that this procedure is just equivalent to identifying the degrees of freedom that survive the infinite boost limit.

Since all the time we are working at finite R we should really see the U-duality group of M-theory on T^6, which is the discrete version of E_6. Under the $SO(5,5,Z)$ subgroup the 27 of E_6 decomposes as $27 \to 10 + 16 + 1$. In addition to the 16 states identified above, there are the 10 states corresponding to longitudinal membranes and fivebranes (that is branes wrapping R), which are momentum and winding modes in the little string theory. They correspond to bound states at threshold, so in this case $E = p_{11} + m$ and hence their lightcone energy $H = E - p_{11} = m$ is equal to their mass in the IIB string theory. These states preserve only one quarter of the supersymmetry. This is due to the fact that in the particular reference frame we chose only half of the supercharges are linearly realized. While the transverse branes break the half that's not visible anyway, they still appear as 1/2 SUSY bound states, while the longitudinal states break another half and we are left with only 1/4 linearly realized supercharges.

The 27th state is the fundamental carrier of p_{11} itself, the wrapped NS5 brane. Its space-time mass by construction is $1/R$. Indeed the full U-duality multiplet for the T^6 can be found in accordance with the fact that R is finite.

The sixtorus:

Applying the above procedure to M-theory on the T^6, one can similarly derive that the DLCQ matrix model of this theory is given by the worldvolume theory of IIA D6 branes at infinite coupling (sending M_p, M_s to infinity, holding the gauge coupling on the D6 and hence the eleven dimensional Planck scale fixed). This is the limit proposed in [7]. There it was also found that this description yields the right moduli space and BPS states.

It was however shown by Seiberg and Sethi [4] that the worldvolume theory of the D6 does not decouple from the bulk fields in this limit. Since we sent the 10d Planck scale M_p to infinity we decoupled all the 10d gravitons, but the gravitons associated with the 11th dimension of M-theory that opens up in the infinite coupling limit (the D0 branes from the IIA point of view) become massless in this limit. The coupling of these excitations to the D6 worldvolume is governed by the 11d Planck scale, which is kept fixed in the limit that's forced upon us by (1). We thus find that similar as in the case for $T^{\geq 7}$ the matrix description of the DLCQ is not a d+1 dimensional theory but involves some of the bulk modes of the full type II theory.

This conlusion can also be reached by looking again at the BPS states we should see. We are still working at finite R and thus should find the discrete E_7 of T^7

compactifications. The 56 of E_7 decomposes under the E_6 as $56 \to 27 + \bar{27} + 1 + 1$ In [7] the 27 longitudinal and 27 transverse branes were again identified as bound states of the p_{11} carriers (in this case the D6) and other branes. One of the singlets is again the wrapped D6 brane itself and correponds to the space-time state with energy $1/R$. The 56th state has space-time mass $\frac{R^2V}{l_p^9}$ and corresponds to the KK6 associated with the compact R. But this state is precisely mapped to a D0 brane in the IIA setup describing the T^6 compactification. These thus can not decouple from the matrix theory, since they are the required missing BPS state.

By now there seem to be two lines of thought about how to live with this puzzle. On one hand one might just appeal to the magic of large N. The DLCQ description of M theory on higher tori has all the problems mentioned. But once we go back to the original proposal of [1], that is go to infinite N, all these problems might go away. The D0 and the D6 repell each other. For infinite N this might be strong enough to decouple the bulk. Since we also got rid of the hidden compact dimension, we no longer would expect to see the full U-duality of the T^7 and thus the D0 branes no longer have to be there. This option was for example suggestd in [8].

The other option would be that what we've seen here is an indication that M-theory doesn't want to be compactified on $T^{\geq 6}$. This interpretation recently got some support by the observation that by replacing the T^6 by a Calabi-Yau [9], the states that correspond to those that in our case come from the D0 do not lead to any interaction with gravity.

References

[1] T. Banks, W. Fischler, S.H. Shenker, L. Susskind, hep-th/9610043.

[2] L. Susskind, hep-th/9704080.

[3] N. Seiberg, *Phys.Rev.Lett.* **79** (1997), 3577, hep-th/9710009; A. Sen, hep-th/9709220.

[4] N. Seiberg and S. Sethi, hep-th/9708085.

[5] M. Berkooz, M. Rozali, N. Seiberg, hep-th/9704089; M. Rozali, hep-th/9702136.

[6] N. Seiberg, hep-th/9705221.

[7] I. Brunner, A. Karch, hep-th/9707259; A. Hanany, G. Lifschytz, hep-th/9708037.

[8] T. Banks, 'Matrix Theory', hep-th/9710231

[9] S. Kachru, A. Lawrence, E. Silverstein, hep-th/9712223.

Nonpolynomial gauge invariant interactions of 1-form and 2-form gauge potentials

Friedemann Brandt [1] **and Norbert Dragon** [2]

[1] Departament ECM, Facultat de Física Universitat de Barcelona
and Institut de Física d'Altes Energies, Diagonal 647, E-08028 Barcelona, Spain
[2] Institut für Theoretische Physik, Universität Hannover,
Appelstraße 2, D-30167 Hannover, Germany

Abstract: A four dimensional gauge theory with nonpolynomial but local interactions of 1-form and 2-form gauge potentials is constructed. The model is a nontrivial deformation of a free gauge theory with nonpolynomial dependence on the gauge coupling constant.

Gauge invariant interactions of ordinary gauge fields (= 1-form gauge potentials) are very well known, the most famous one being undoubtedly the Yang–Mills interaction. Much less is known about the possible gauge invariant couplings between 1-form and 2-form gauge potentials, although an important coupling of that type is known for a long time: it is the celebrated coupling of a 2-form gauge potential to Chern–Simons forms which underlies among others the Green–Schwarz anomaly cancellation mechanism [1]. Here we shall construct a rather different interacting gauge theory for 1-form and 2-form gauge potentials in four dimensions. As the model has local but nonpolynomial interactions and gauge transformations, its structure is in some respect more reminiscent of gravitational interactions than of Yang–Mills theory or couplings of Chern–Simons forms to a 2-form gauge potential.

The same model was found by different means as a particular example in a more general class of theories (Eq. (17) of [2]) where it is formulated in a polynomial first-order form.

Although we will not study supersymmetric field theories here, our construction was partly motivated by the aim to gauge the central charge of the rigid N=2 supersymmetry algebra realized on the vector-tensor multiplet [3] which arises naturally in string compactifications [4]. This problem was investigated already in [5] and could be relevant among others in order to classify the still unknown couplings of the vector-tensor multiplet to N=2 supergravity.

The central charge occurs in the anticommutator of two supersymmetry transformations with the same chirality and is called central because it commutes with all generators of the supersymmetry algebra. This rigid bosonic symmetry acts non-trivially only on the 1- and 2-form gauge fields of the vector-tensor multiplet, as it is on-shell trivial on the remaining component fields of the vector-tensor multiplet. Therefore one can ask already in the nonsupersymmetric case whether it can be gauged in a reasonable way. This question is interesting in its own right and underlies our construction.

Our starting point is the standard free action for two abelian 1-form gauge potentials $A = dx^\mu A_\mu$ and $W = dx^\mu W_\mu$ and a 2-form gauge potential $B = (1/2)dx^\mu \wedge dx^\nu B_{\mu\nu}$ in flat four dimensional spacetime. The Lagrangian reads

$$\mathcal{L}_0 = -\frac{1}{4}\left(G_{\mu\nu}G^{\mu\nu} + F_{\mu\nu}F^{\mu\nu}\right) - H_\mu H^\mu \tag{1}$$

where

$$\begin{aligned} G_{\mu\nu} &= \partial_\mu W_\nu - \partial_\nu W_\mu\,, \\ F_{\mu\nu} &= \partial_\mu A_\nu - \partial_\nu A_\mu\,, \\ H^\mu &= \frac{1}{2}\,\varepsilon^{\mu\nu\rho\sigma}\partial_\nu B_{\rho\sigma}\,. \end{aligned} \tag{2}$$

The action with Lagrangian (1) has, among others, a rigid symmetry generated by

$$\delta_z A_\mu = 2H_\mu,\ \delta_z B_{\mu\nu} = \frac{1}{2}\,\varepsilon_{\mu\nu\rho\sigma}F^{\rho\sigma},\ \delta_z W_\mu = 0. \tag{3}$$

This rigid symmetry coincides indeed on-shell with the central charge of the N=2 supersymmetry algebra for the vector-tensor multiplet, cf. [3].

Our aim will now be to gauge the rigid symmetry (3). With this end in view, we look for appropriate extensions $\Delta_z A_\mu$ and $\Delta_z B_{\mu\nu}$ of $\delta_z A_\mu$ and $\delta_z B_{\mu\nu}$ transforming covariantly under the sought gauge transformations generated by

$$\begin{aligned} \delta_\xi W_\mu &= \partial_\mu \xi\,, \\ \delta_\xi A_\mu &= g\,\xi\,\Delta_z A_\mu\,, \\ \delta_\xi B_{\mu\nu} &= g\,\xi\,\Delta_z B_{\mu\nu} \end{aligned} \tag{4}$$

where ξ is an arbitrary field and g is a gauge coupling constant. Following a standard recipe in gauge theories, we try to covariantize partial derivatives of A_μ and $B_{\mu\nu}$ by means of a covariant derivative

$$\mathcal{D}_\mu = \partial_\mu - gW_\mu\Delta_z \tag{5}$$

where Δ_z is the sought extension of δ_z. We now try to covariantize (3) by replacing there $\delta_z A_\mu$ and $\delta_z B_{\mu\nu}$ with $\Delta_z A_\mu$ and $\Delta_z B_{\mu\nu}$ respectively, and ∂_μ with $\mathcal{D}_\mu$. Explicitly this yields

$$\begin{aligned} \Delta_z A_\mu &= \varepsilon_{\mu\nu\rho\sigma}(\partial^\nu B^{\rho\sigma} - gW^\nu\Delta_z B^{\rho\sigma}), \\ \Delta_z B_{\mu\nu} &= \varepsilon_{\mu\nu\rho\sigma}(\partial^\rho A^\sigma - gW^\rho\Delta_z A^\sigma). \end{aligned} \tag{6}$$

(6) determines $\Delta_z A_\mu$ and $\Delta_z B_{\mu\nu}$. Indeed, inserting the second equation (6) in the first one, we get an equation for $\Delta_z A_\mu$,

$$(E\delta_\mu^\nu + 2g^2 W_\mu W^\nu)\Delta_z A_\nu = 2Z_\mu\,, \tag{7}$$

where

$$E = 1 - 2g^2 W_\mu W^\mu,\ Z_\mu = H_\mu + gF_{\mu\nu}W^\nu. \tag{8}$$

To solve (7) for $\Delta_z A_\nu$, we only need to invert the matrix $E\delta_\mu^\nu + 2g^2 W_\mu W^\nu$. The inverse is

$$V_\mu{}^\nu = E^{-1}(\delta_\mu^\nu - 2g^2 W_\mu W^\nu). \tag{9}$$

(7) and (6) yield now

$$\Delta_z A_\mu = 2\mathcal{H}_\mu\,,\ \Delta_z B_{\mu\nu} = \frac{1}{2}\varepsilon_{\mu\nu\rho\sigma}\mathcal{F}^{\rho\sigma} \tag{10}$$

where

$$\begin{aligned}\mathcal{H}_\mu &= E^{-1}(Z_\mu - 2g^2 W_\mu W_\nu H^\nu),\\ \mathcal{F}_{\mu\nu} &= F_{\mu\nu} - 4gE^{-1}W_{[\mu}Z_{\nu]}\,.\end{aligned} \tag{11}$$

Recall that our goal was to find gauge transformations (4) under which $\Delta_z A_\mu$ and $\Delta_z B_{\mu\nu}$ transform covariantly. We can now examine whether we have reached this goal. This amounts to check whether $\mathcal{H}_\mu$ and $\mathcal{F}_{\mu\nu}$ transform covariantly, i.e. without derivatives of ξ, under the gauge transformations

$$\begin{aligned}\delta_\xi A_\mu &= 2g\,\xi\,\mathcal{H}_\mu\,,\\ \delta_\xi B_{\mu\nu} &= \frac{1}{2}g\,\xi\,\varepsilon_{\mu\nu\rho\sigma}\mathcal{F}^{\rho\sigma},\\ \delta_\xi W_\mu &= \partial_\mu\xi\,.\end{aligned} \tag{12}$$

The answer is affirmative: neither $\delta_\xi\mathcal{H}_\mu$ nor $\delta_\xi\mathcal{F}_{\mu\nu}$ contain derivatives of ξ. Indeed, an elementary, though somewhat lengthy calculation yields

$$\delta_\xi\mathcal{H}_\mu = g\,\xi\,\Delta_z\mathcal{H}_\mu,\ \delta_\xi\mathcal{F}_{\mu\nu} = g\,\xi\,\Delta_z\mathcal{F}_{\mu\nu} \tag{13}$$

with

$$\begin{aligned}\Delta_z\mathcal{H}_\mu &= V_\mu{}^\nu(\partial^\rho\mathcal{F}_{\rho\nu} - 4gW^\rho\partial_{[\rho}\mathcal{H}_{\nu]}),\\ \Delta_z\mathcal{F}_{\mu\nu} &= 4(\partial_{[\mu}\mathcal{H}_{\nu]} - gW_{[\mu}\Delta_z\mathcal{H}_{\nu]}).\end{aligned} \tag{14}$$

To construct an action which is invariant under the gauge transformations (12), it is helpful to realize that the transformations (14) are nothing but

$$\Delta_z\mathcal{H}_\mu = \mathcal{D}^\nu\mathcal{F}_{\nu\mu},\ \Delta_z\mathcal{F}_{\mu\nu} = 4\mathcal{D}_{[\mu}\mathcal{H}_{\nu]} \tag{15}$$

where the second identity is obvious from (14), whereas the verification of the first one is slightly more involved. Combining (13) and (15) it is now easy to verify that

$$\begin{aligned}
&\delta_\xi\left(-\frac{1}{4}\mathcal{F}_{\mu\nu}\mathcal{F}^{\mu\nu}-\mathcal{H}_\mu\mathcal{H}^\mu\right)\\
&= 2g\,\xi\,\mathcal{D}_\nu(\mathcal{H}_\mu\mathcal{F}^{\mu\nu})\\
&= 2g\,\xi\,\partial_\nu(\mathcal{H}_\mu\mathcal{F}^{\mu\nu})-2g^2\xi\,W_\nu\Delta_z(\mathcal{H}_\mu\mathcal{F}^{\mu\nu})\\
&= \partial_\nu(2g\,\xi\,\mathcal{H}_\mu\mathcal{F}^{\mu\nu})-\delta_\xi(2gW_\nu\mathcal{H}_\mu\mathcal{F}^{\mu\nu}).
\end{aligned}$$

This implies immediately that the Lagrangian

$$\mathcal{L} = -\frac{1}{4}(G_{\mu\nu}G^{\mu\nu}+\mathcal{F}_{\mu\nu}\mathcal{F}^{\mu\nu})-\mathcal{H}_\mu\mathcal{H}^\mu-2gW_\mu\mathcal{H}_\nu\mathcal{F}^{\mu\nu} \tag{16}$$

transforms under δ_ξ into a total derivative,

$$\delta_\xi\mathcal{L}=\partial_\nu(2g\,\xi\,\mathcal{H}_\mu\mathcal{F}^{\mu\nu}). \tag{17}$$

Hence, the action with Lagrangian (16) is gauge invariant under δ_ξ. Evidently it is also invariant under the following standard gauge transformations acting only on A_μ and $B_{\mu\nu}$ respectively:

$$\delta_\Lambda A_\mu=\partial_\mu\Lambda,\ \delta_\lambda B_{\mu\nu}=\partial_{[\mu}\lambda_{\nu]}\,. \tag{18}$$

Inserting finally the explicit expressions (11) in (16), the Lagrangian reads

$$\mathcal{L} = -\frac{1}{4}(G_{\mu\nu}G^{\mu\nu}+F_{\mu\nu}F^{\mu\nu})-E^{-1}Z_\mu Z^\mu+2E^{-1}(gW_\mu H^\mu)^2 \tag{19}$$

with E and Z_μ as in (8).

It is now easy to compute the Euler–Lagrange derivatives of $\mathcal{L}$ with respect to the fields. The result is

$$\frac{\hat\partial\mathcal{L}}{\hat\partial A_\mu} = \partial_\nu\mathcal{F}^{\nu\mu}, \tag{20}$$

$$\frac{\hat\partial\mathcal{L}}{\hat\partial B_{\mu\nu}} = -\varepsilon^{\mu\nu\rho\sigma}\partial_\rho\mathcal{H}_\sigma\,, \tag{21}$$

$$\frac{\hat\partial\mathcal{L}}{\hat\partial W_\mu} = \partial_\nu G^{\nu\mu}+2g\mathcal{F}^{\mu\nu}\mathcal{H}_\nu\,. \tag{22}$$

Note that the equations of motion for A_μ and $B_{\mu\nu}$ obtained from (20) and (21) are not covariant under δ_ξ, in contrast to the equation of motion for W_μ following from (22). However, from (14) and (15) it is obvious that (20) and (21) can be combined to covariant expressions too, which holds generally for the equations of motion in gauge theories [6]. The covariant form of the equations of motion reads

$$\begin{gathered}
\mathcal{D}_\nu\mathcal{F}^{\mu\nu}=0,\\
\mathcal{D}_{[\mu}\mathcal{H}_{\nu]}=0,\\
\partial_\nu G^{\nu\mu}+2g\mathcal{F}^{\mu\nu}\mathcal{H}_\nu=0.
\end{gathered}$$

To summarize, we have constructed an interacting four dimensional gauge theory with Lagrangian (16) or (19) for two ordinary gauge fields A_μ and W_μ and an antisymmetric gauge field $B_{\mu\nu}$. The key feature of this gauge theory is its gauge invariance under the transformations (12) which gauge the rigid symmetry (3) of the free action with Lagrangian (1). Both the Lagrangian and the gauge transformations (12) are nonpolynomial in the gauge coupling constant g and the gauge field W_μ. Nevertheless they are local, for the Lagrangian and the gauge transformations are still quadratic and linear in differentiated fields respectively. Note that the Lagrangian and the gauge transformations constitute a consistent deformation of the free Lagrangian (1) and its gauge symmetries in the sense of [7]. In particular one recovers the free theory and its gauge symmetries for $g = 0$. The generalization of all the above formulas to curved spacetime is obvious.

Let us finally compare our results to those of [5] where the central charge of the vector-tensor multiplet was gauged. First we have presented the action and the gauge transformations in an explicit and manifestly local form. In contrast, in [5] both the action and the gauge transformations are only implicitly defined (the formulas given in [5] result in local expressions too [8]). Furthermore our results appear to differ from those of [5], even when the latter are restricted to the particular nonsupersymmetric case studied here. In particular, neither the Lagrangian (19) nor the gauge transformations (12) contain Chern–Simons terms of the type occurring in [5]. This might signal that such terms are actually not needed in order to gauge the central charge of the vector-tensor multiplet. Of course, in contrast to [5], we did not study the supersymmetric case, and therefore we cannot clarify this issue here.

Acknowledgement:

We thank Piet Claus, Bernard de Wit, Michael Faux, Marc Henneaux and Piet Termonia for useful discussions and comments. F.B. was supported by the Spanish ministry of education and science (MEC).

Literatur

[1] M.B. Green and J.H. Schwarz, Phys. Lett. B 149 (1984) 117.

[2] M. Henneaux and B. Knaepen, All consistent interactions for exterior form gauge fields, to appear in Phys. Rev. D (hep-th/9706119).

[3] M. Sohnius, K.S. Stelle and P.C. West, Phys. Lett. B 92 (1980) 123.

[4] B. de Wit, V. Kaplunovsky, J. Louis and D. Lüst, Nucl. Phys. B 451 (1995) 53 (hep-th/9504006).

[5] P. Claus, B. de Wit, M. Faux, B. Kleijn, R. Siebelink and P. Termonia, Phys. Lett. B 373 (1996) 81 (hep-th/9512143); P. Claus, B. de Wit, M. Faux and P. Termonia, Nucl. Phys. B 491 (1997) 201 (hep-th/9612203).

[6] F. Brandt, Local BRST cohomology and covariance, to appear in Commun. Math. Phys. (hep-th/9604025).

[7] G. Barnich and M. Henneaux, Phys. Lett. B 311 (1993) 123 (hep-th/9304057).

[8] P. Claus, M. Faux and P. Termonia, private communication.

Three-dimensional solutions of M-theory on $S^1/Z_2 \times T_7$

Gabriel Lopes Cardoso

Institute for Theoretical Physics, Utrecht University
NL-3508 TA Utrecht, The Netherlands

Abstract: We review various three-dimensional solutions in the low-energy description of M-theory on $S^1/Z_2 \times T_7$. These solutions have an eleven dimensional interpretation in terms of intersecting M-branes.

1 Introduction

Eleven dimensional supergravity, the low-energy effective theory of M-theory, has four 1/2-supersymmetric solutions: the M-2-brane, the M-5-brane, the M-wave and (in the S^1 Kaluza-Klein vacuum) the M-monopole. We will refer to them collectively as M-branes. Intersections of these four basic configurations give rise to solutions preserving a smaller amount of supersymmetry (for a review see for instance [1, 2] and references therein). In the following, we will review various three-dimensional solutions in the low-energy description of M-theory on $S^1/Z_2 \times T_7$. These solutions have an eleven dimensional interpretation in terms of orthogonally intersecting M-branes.

The eleven-dimensional space-time line element describing a supersymmetric configuration of orthogonally intersecting M-branes is given in terms of harmonic functions which depend on some of the overall transverse directions [2, 3, 4]. If these overall transverse directions do not include the eleven-th dimension x_{11}, then its compactification on $S^1/Z_2 \times T_7$ down to three dimensions is also a solution of the low-energy effective field theory of ten-dimensional heterotic string theory compactified on a seven-torus. Such a solution is determined in terms of harmonic functions $H(z, \bar{z}) = f(z) + \bar{f}(\bar{z})$, where z and $\bar{z}$ denote the spatial dimensions. The static solutions presented in [5, 6, 7] are of this type. If, on the other hand, the overall transverse directions include the eleven-th dimension x_{11}, then compactifying the eleven-dimensional configuration on $S^1/Z_2 \times T_7$ gives rise to a solution which is determined in terms of harmonic functions $H(x_{11}, z, \bar{z})$. An example of such a supergravity solution has recently been given [8] in the context of M-theory compactified on T_8. There, it was also shown that this solution corresponds to a BPS state with mass that goes like $1/g^3$. Evidence for the existence of BPS states with masses that go like $1/g^3$ or higher inverse powers of the string coupling constant g in M-theory compactifications down to three (and lower) dimensions has been emerging in studies of U-duality multiplets of M-theory on tori [9, 8].

If the harmonic functions H are taken to be independent of the eleven-th dimension, then the resulting three-dimensional solutions can be turned into finite energy solutions [5, 7] by utilizing a mechanism first discussed in the context of four-dimensional stringy cosmic string solutions [10]. For instance, the solutions presented in [7] have, at spatial infinity ($z \to \infty$), an asymptotic behaviour corresponding to $f(z) \propto \ln z$. At finite distance these asymptotic solutions become ill-defined and so need to be modified. The associated corrections are all encoded in $f(z)$. They can be determined by requiring the solutions to have finite energy. This requirement, together with the appropriate asymptotic behaviour, determines $f(z)$ to be given by $f(z) \propto j^{-1}(z)$. In section 2 we will review some of the three-dimensional solutions constructed in [7], namely those which have a ten-dimensional interpretation in terms of a fundamental string, a wave and up to three orthogonally intersecting NS 5-branes as well as up to three Kaluza–Klein monopoles. These solutions are labelled by an integer n with $n = 1, 2, 3, 4$. The energy E carried by these (one-center) solutions is given by $E = 2n\frac{\pi}{6}$ (in units where $8\pi G_N = 1$).

Another class of solutions constructed in [7] consists of solutions carrying energies $E = n\frac{\pi}{6}$, where $n = 1, 2, 3, 4$. An example of a solution with $E = \frac{\pi}{6}$ is given by a wrapped M-monopole. In section 3 we briefly review this solution. We then compare it to the solution recently discussed in [8], which can be constructed by considering a periodic array of M-monopoles along a transverse direction and by identifying this transverse direction with the eleven-th dimension. This latter solution is thus specified by a harmonic function $H(x_{11}, z, \bar{z})$ on $S^1/Z_2 \times \mathrm{R}^2$.

2 A class of finite energy solutions

The effective low-energy field theory of the ten-dimensional heterotic string compactified on a seven-dimensional torus is obtained from reducing the ten-dimensional $N = 1$ supergravity theory coupled to $U(1)^{16}$ super Yang–Mills multiplets (at a generic point in the moduli space). The massless ten-dimensional bosonic fields are the metric $G^{(10)}_{MN}$, the antisymmetric tensor field $B^{(10)}_{MN}$, the $U(1)$ gauge fields $A^{(10)I}_M$ and the scalar dilaton $\Phi^{(10)}$ with $(0 \leq M, N \leq 9, \quad 1 \leq I \leq 16)$. The field strengths are $F^{(10)I}_{MN} = \partial_M A^{(10)I}_N - \partial_N A^{(10)I}_M$ and $H^{(10)}_{MNP} = (\partial_M B^{(10)}_{NP} - \frac{1}{2} A^{(10)I}_M F^{(10)I}_{NP}) +$ cyclic permutations of M, N, P.

The bosonic part of the ten-dimensional action is

$$\mathcal{S} \propto \int d^{10}x \sqrt{-G^{(10)}} e^{-\Phi^{(10)}} [\mathcal{R}^{(10)} + G^{(10)MN} \partial_M \Phi^{(10)} \partial_N \Phi^{(10)}$$

$$-\frac{1}{12}H^{(10)}_{MNP}H^{(10)MNP} - \frac{1}{4}F^{(10)I}_{MN}F^{(10)IMN}]. \quad (1)$$

The reduction to three dimensions (see [5] and references therein) introduces the graviton $g_{\mu\nu}$, the dilaton $\phi \equiv \Phi^{(10)} - \ln\sqrt{\det G_{mn}}$, with G_{mn} the internal 7D metric, 30 $U(1)$ gauge fields $A^{(a)}_\mu \equiv (A^{(1)m}_\mu, A^{(2)}_{\mu m}, A^{(3)I}_\mu)$ $(a = 1,\ldots,30,\ m = 1,\ldots,7,\ I = 1,\ldots,16)$, where $A^{(1)m}_\mu$ are the 7 Kaluza–Klein gauge fields coming from the reduction of $G^{(10)}_{MN}$, $A^{(2)}_{\mu m} \equiv B_{\mu m} + B_{mn}A^{(1)n}_\mu + \frac{1}{2}a^I_m A^{(3)I}_\mu$ are the 7 gauge fields coming from the reduction of $B^{(10)}_{MN}$ and $A^{(3)I}_\mu \equiv A^I_\mu - a^I_m A^{(1)m}_\mu$ are the 16 gauge fields from $A^{(10)I}_M$. The field strengths $F^{(a)}_{\mu\nu}$ are given by $F^{(a)}_{\mu\nu} = \partial_\mu A^{(a)}_\nu - \partial_\nu A^{(a)}_\mu$. Finally, $B^{(10)}_{MN}$ induces the two-form field $B_{\mu\nu}$ with field strength $H_{\mu\nu\rho} = \partial_\mu B_{\nu\rho} - \frac{1}{2}A^{(a)}_\mu L_{ab}F^{(b)}_{\nu\rho} +$ cyclic permutations.

The bosonic part of the three-dimensional action in the Einstein frame is then

$$\begin{aligned}\mathcal{S} &= \frac{1}{4}\int d^3x\sqrt{-g}\{\mathcal{R} - g^{\mu\nu}\partial_\mu\phi\partial_\nu\phi - \frac{1}{12}e^{-4\phi}g^{\mu\mu'}g^{\nu\nu'}g^{\rho\rho'}H_{\mu\nu\rho}H_{\mu'\nu'\rho'}\\ &\quad -\frac{1}{4}e^{-2\phi}g^{\mu\mu'}g^{\nu\nu'}F^{(a)}_{\mu\nu}(LML)_{ab}F^{(b)}_{\mu'\nu'} + \frac{1}{8}g^{\mu\nu}\mathrm{Tr}\,(\partial_\mu ML\partial_\nu ML)\}\ ,\end{aligned} \quad (2)$$

where $a = 1,\ldots,30$. Here M denotes a matrix comprising the scalar fields G_{mn}, a^I_m and B_{mn} [11, 5].

We now construct static solutions by setting the associated three-dimensional Killing spinor equations to zero. In doing so, we restrict ourselves to backgrounds with $H_{\mu\nu\rho} = 0$ and $a^I_m = 0$. It can be checked that the resulting solutions to the Killing spinor equations satisfy the equations of motion derived from (2). The associated three-dimensional Killing spinor equations in the Einstein frame are [6]

$$\begin{aligned}\delta\chi^I &= \frac{1}{2}e^{-2\phi}F^{(3)I}_{\mu\nu}\gamma^{\mu\nu}\varepsilon,\\ \delta\lambda &= -\frac{1}{2}e^{-\phi}\partial_\mu\{\phi + \ln\det e^a_m\}\gamma^\mu\otimes\mathbf{I}_8\,\varepsilon + \frac{1}{4}e^{-2\phi}[-B_{mn}F^{(1)n}_{\mu\nu} + F^{(2)}_{\mu\nu m}]\gamma^{\mu\nu}\gamma^4\otimes\Sigma^m\varepsilon\\ &\quad +\frac{1}{4}e^{-\phi}\partial_\mu B_{mn}\gamma^\mu\otimes\Sigma^{mn}\varepsilon\ ,\\ \delta\psi_\mu &= \partial_\mu\varepsilon + \frac{1}{4}\omega_{\mu\alpha\beta}\gamma^{\alpha\beta}\varepsilon + \frac{1}{4}(e_{\mu\alpha}e^\nu_\beta - e_{\mu\beta}e^\nu_\alpha)\partial_\nu\phi\gamma^{\alpha\beta}\varepsilon\\ &\quad +\frac{1}{8}(e^n_a\partial_\mu e_{nb} - e^n_b\partial_\mu e_{na})\mathbf{I}_4\otimes\Sigma^{ab}\varepsilon - \frac{1}{4}e^{-\phi}[e^m_a F^{(2)}_{\mu\nu(m)} - e_{ma}F^{(1)m}_{\mu\nu}]\gamma^\nu\gamma^4\otimes\Sigma^a\varepsilon\\ &\quad -\frac{1}{8}\partial_\mu B_{mn}\,\mathbf{I}_4\otimes\Sigma^{mn}\varepsilon + \frac{1}{4}e^{-\phi}B_{mn}F^{(1)n}_{\mu\nu}\,\gamma^\nu\gamma^4\otimes\Sigma^m\varepsilon\ ,\\ \delta\psi_d &= -\frac{1}{4}e^{-\phi}(e^m_d\partial_\mu e_{ma} + e^m_a\partial_\mu e_{md})\gamma^\mu\gamma^4\otimes\Sigma^a\varepsilon + \frac{1}{8}e^{-2\phi}e^m_d B_{mn}F^{(1)n}_{\mu\nu}\,\gamma^{\mu\nu}\varepsilon\\ &\quad +\frac{1}{4}e^{-\phi}e^m_d e^n_a\partial_\mu B_{mn}\gamma^\mu\gamma^4\otimes\Sigma^a\varepsilon - \frac{1}{8}e^{-2\phi}[\,e_{md}F^{(1)m}_{\mu\nu} + e^m_d F^{(2)}_{\mu\nu m}\,]\gamma^{\mu\nu}\varepsilon\ ,\end{aligned} \quad (3)$$

where $\delta\psi_d \equiv e^m_d\delta\psi_m$ denotes the variation of the internal gravitini. Here we have performed a $3+7$ split of the ten-dimensional gamma matrices [6].

In the following, we review a particular class of static solutions to the Killing spinor equations (3) constructed in [6, 7]. The solutions in this class are labelled by an integer n ($n = 1, 2, 3, 4$). The associated space-time line element is given by

$$ds^2 = -dt^2 + H^{2n} d\omega d\bar{\omega} \quad , \quad H = f(\omega) + \bar{f}(\bar{\omega}) \quad , \tag{4}$$

where $\omega = a \ln z = a(\ln r + i\theta) = \hat{r} + i\hat{\theta}$. Here, $a = \frac{n+1}{2\pi}\sqrt{|\alpha_i \alpha_{i+7}|}$, where α_i and α_{i+7} denote two electric charges carried by each of the solutions in this class. The associated field strengths of the three-dimensional gauge fields are [7]

$$F^{(1)i}_{t\beta} = \eta_{\alpha_i} \sqrt{G^{ii}} \, \frac{\partial_\beta H}{H^2} \quad , \quad F^{(2)}_{t\beta\, i} = -\eta_{\alpha_i} \sqrt{G_{ii}} \, \frac{\partial_\beta H}{H^2} \quad , \quad \beta = \hat{r}, \hat{\theta} \quad , \quad \eta_{\alpha_i} = \pm \, . \tag{5}$$

The three-dimensional dilaton field is given by $e^{-\phi} = H$. The i-th component of the internal metric G_{mn} reads $G_{ii} = |\frac{\alpha_i}{\alpha_{i+7}}|$. In addition, there are (depending on the integer n) various additional non-constant background fields G_{mn} and B_{mn}, which are also determined in terms of $f(\omega)$ [7].

Solving the Killing spinor equations (3) does not determine the form of $f(\omega)$. Its form can be determined by demanding the solution to behave as $f(\omega) \approx \frac{\omega}{2(n+1)}$ at spatial infinity [6] and by requiring the solution to have finite energy, as follows [7]. The energy carried by any of the solutions in this class is compute to be (in units where $8\pi G_N = 1$)

$$E \;=\; i\,2n \int d\omega d\bar{\omega} \, \frac{\partial_\omega f \partial_{\bar{\omega}} \bar{f}}{(f + \bar{f})^2} = i\,2n \int dz d\bar{z} \, \frac{\partial_z \hat{f} \partial_{\bar{z}} \bar{\hat{f}}}{(\hat{f} + \bar{\hat{f}})^2} \quad , \tag{6}$$

where we have introduced $\hat{f} = \frac{n+1}{\pi} f$ for later convenience. There is an elegant mechanism [10] for rendering the integral (6) finite. Let us take z to be the coordinate of a complex plane. Then there is a one-to-one map from a certain domain F on the $\hat{f}$-plane (the so called 'fundamental' domain) to the z-plane. This map is known as the j-function, $j(\hat{f}) = z$. By means of this map, the integral (6) can be pulled back from the z-plane to the domain F (the z-plane covers F exactly once). Then, by using integration by parts, this integral can be related to a line integral over the boundary of F, which is evaluated to be [10]

$$E = 2n \, \frac{2\pi}{12} = 2n \, \frac{\pi}{6} \tag{7}$$

and, hence, is finite. By expanding $j(\hat{f}) = e^{2\pi \hat{f}} + 744 + \mathcal{O}(e^{-2\pi \hat{f}}) = z$ we recover $f(\omega) \approx \frac{\omega}{2(n+1)}$ at spatial infinity.

We note that the solutions discussed above represent one-center solutions. They can be generalised to multi-center solutions via $j(\hat{f}(z)) = P(z)/Q(z)$, where $P(z)$ and $Q(z)$ are polynomials in z with no common factors. These are the analogue of the multi-string configurations discussed in [10].

It can be checked that the curvature scalar $\mathcal{R} \propto \partial_\omega f \partial_{\bar{\omega}} \bar{f}$ blows up at the special point $\hat{f} = 1$ (at this point, the j-function and its derivatives are given by $j =$

$1728, j' = 0$), whereas it is well behaved at the point $\hat{f} = e^{i\pi/6}$ (at this point, the j-function and its derivatives are given by $j = j' = j'' = 0$). It would be interesting to understand the physics at this special point in moduli space further.

As mentioned in the introduction the three-dimensional solutions reviewed in this section can be given an eleven dimensional interpretation [7] in terms of orthogonally intersecting M-branes. The two gauge fields (5), in particular, arise from a wave and from a M-2-brane in eleven dimensions, respectively.

3 Wrapped M-monopoles

Another class of solutions constructed in [7] is the class of solutions carrying energy $E = n\frac{\pi}{6}$, where $n = 1, 2, 3, 4$. An example of a solution carrying energy $E = \frac{\pi}{6}$ is obtained by compactifying the M-monopole in the following way. The M-monopole in eleven dimensions is given by the metric [12]

$$ds_{11}^2 = -dt^2 + H dy_i^2 + H^{-1}(d\psi + A_i dy_i)^2 + dx_1^2 + \ldots + dx_6^2, \qquad i = 1, 2, 3, \tag{8}$$

with $H = H(y_i)$, $F_{ij} = \partial_i A_j - \partial_j A_i = c\,\varepsilon_{ijk}\partial_k H$, $c = \pm$. Here, ψ denotes a periodic variable. Identifying one of the x_i with the eleven-th cooordinate and compactifying the remaining x_i on a five-torus yields the five-dimensional line element

$$ds_5^2 = -dt^2 + H dy_i^2 + H^{-1}(d\psi + A_i dy_i)^2. \tag{9}$$

Now, consider the case that H only depends on two of the coordinates y_i, that is $H = H(z, \bar{z})$ with $z = y_2 + iy_3$. Then we can set $A_2 = A_3 = 0$, and $\partial_2 A_1 = -c\partial_3 H$, $\partial_3 A_1 = c\partial_2 H$. The metric is now

$$\begin{aligned} ds_5^2 &= -dt^2 + H dz d\bar{z} + H dy_1^2 + H^{-1}(d\psi + A_1 dy_1)^2 \\ &= g_{\mu\nu}dx^\mu dx^\nu + G_{mn}dx^m dx^n, \end{aligned} \tag{10}$$

where the off-diagonal internal metric is given by

$$G_{mn} = \begin{pmatrix} H + A_1^2 H^{-1} & A_1 H^{-1} \\ A_1 H^{-1} & H^{-1} \end{pmatrix} . \tag{11}$$

The resulting three-dimensional solution can be turned into a finite energy solution with energy $E = \frac{\pi}{6}$ by using the mechanism described in the previous section.

Let us now compare this solution to the one obtained by identifying the eleven-th dimension not with one of the x_i, but rather with one of the y_i [8]. Compactifying the eleven-dimensional line element (8) on a six-torus yields again the five-dimensional line element (9). Compactifying over ψ yields the four-dimensional line element

$$ds_4^2 = -dt^2 + H dy_i^2 \tag{12}$$

with internal metric component $G_{\psi\psi} = H^{-1}$ and magnetic gauge field $F_{ij} = c\,\varepsilon_{ijk}\partial_k H$. The Bianchi identity of F yields $\Delta H = 0$. Now, identifying one of the transverse coordinates y_i with the eleven-th coordinate (say $y_1 = x_{11}$) and setting $z = y_2 + iy_3$ yields the Laplacian as $\Delta = \partial^2_{x_{11}} + 4\partial_z\partial_{\bar{z}}$. Thus, H is now a harmonic function on $S^1/Z_2 \times \mathrm{R}^2$. In [13] a solution to the Laplace equation on $S^1 \times \mathrm{R}^2$ was constructed. It can be readily adapted to the case at hand. We take the range of x_{11} to be $x_{11} \in [-\frac{1}{2}, \frac{1}{2}]$, with a peridic identification $x_{11} \sim x_{11} + 1$ of the endpoints. The Z_2 symmetry acts as $x_{11} \to -x_{11}$. A Z_2 invariant solution to the Laplace equation on $S^1/Z_2 \times \mathrm{R}^2$ is then given by

$$H(x_{11}, z, \bar{z}) = \sum_{n \in \mathbf{Z}} \left(\frac{1}{|n - \frac{1}{2}|} - \frac{1}{\sqrt{(x_{11} - (n - \frac{1}{2}))^2 + z\bar{z}}} \right) . \tag{13}$$

The constant term in (13) is such that H is regular at $z = 0$ on the orbifold plane $x_{11} = 0$. For $|z| \to \infty$ H, on the other hand, H reduces to $H \approx \log z\bar{z}$ on the plane $x_{11} = 0$.

Acknowledgement

We would like to thank K. Behrndt, M. Bourdeau and E. Verlinde for useful discussions.

References

[1] M. J. Duff, R. R. Khuri and J. X. Lu, *Phys. Rep.* **259** (1995) 213, hep-th/9412184

[2] J. P. Gauntlett, hep-th/9705011

[3] A. A. Tseytlin, *Nucl. Phys.* **B475** (1996) 149, hep-th/9604035

[4] E. Bergshoeff, M. de Roo, E. Eyras, B. Janssen and J. P. van der Schaar, *Class. Quant. Grav.* **14** (1997) 2757, hep-th/9704120

[5] A. Sen, *Nucl. Phys.* **B434** (1995) 179, hep-th/9408083

[6] I. Bakas, M. Bourdeau and G. L. Cardoso, hep-th/9706032

[7] M. Bourdeau and G. L. Cardoso, hep-th/9709174

[8] M. Blau and M. O'Loughlin, hep-th/9712047
C. M. Hull, hep-th/9712075

[9] S. Elitzur, A. Giveon, D. Kutasov and E. Rabinovici, *Nucl. Phys.* **B509** (1998) 122, hep-th/9707217

[10] B. R. Greene, A. Shapere, C. Vafa and S.–T. Yau, *Nucl. Phys.* **B337** (1990) 1

[11] S. Hassan and A. Sen, *Nucl. Phys.* **B375** (1992) 103, hep-th/9109038
J. Maharana and J. H. Schwarz, *Nucl. Phys.* **B390** (1993) 3, hep-th/9207016

[12] R. D. Sorkin, *Phys. Rev. Lett.* **51** (1983) 87
D. J. Gross and M. J. Perry, *Nucl. Phys.* **B226** (1983) 29

[13] H. Ooguri and C. Vafa, *Phys. Rev. Lett.* **77** (1996) 3296, hep-th/9608079

Scattering branes and octonionic Kähler σ-models

K.S. Stelle

The Blackett Laboratory, Imperial College,
Prince Consort Road, London SW7 2BZ, UK
and
TH Division, CERN
CH-1211 Geneva 23, Switzerland

Abstract: By studying p-brane probes in solitonic supergravity backgrounds, one uncovers the possibility of non-trivial scattering behaviour for the relative moduli of multiple brane solutions. This analysis is confirmed by comparison to the σ-model geometries for multiple black-hole scattering found by Shiraishi. This reveals the possibility of non-trivial σ-models for an unexpected supersymmetry type, $d = 1$, $N = 8$b supersymmetry, which corresponds to octonionic Kähler σ-models with torsion.[1]

One of the remarkable features of the set of extended-object solutions to supergravity theories in dimensions $D \leq 11$ is that, for theories that descend from the maximal $D = 11$ supergravity by dimensional reduction, all the static BPS p-brane solutions may be considered to be composed of various "intersecting" [2] combinations of four basic "elemental" solutions to the $D = 11$ theory. These four elements are the pp wave and its "dual," the NUT solution, the 2-brane [3] and the 5-brane [4]. As an example of such an intersecting solution, one may have a pp wave on a 2-brane,

$$\begin{aligned} ds_{11}^2 &= H_1^{\frac{1}{3}}(y)\left[H_1^{-1}(y)\{-dt^2 + d\rho^2 + d\sigma^2 + (H_2(y)-1)(dt+d\rho)^2\} + ds^2(\mathbb{E}^8)\right] \\ A_{[3]} &= H_1^{-1}(y)dt \wedge d\rho \wedge d\sigma \qquad \nabla^2 H_1 = \nabla^2 H_2 = 0\ , \end{aligned} \tag{1}$$

where the transverse-space harmonic function $H_1(y)$ is associated to the 2-brane while $H_2(y)$ is associated to the wave (y^m are coordinates on the 8-dimensional transverse space). Since waves are purely gravitational solutions, the harmonic function $H_2(y)$ does not appear in the 3-form gauge potential $A_{[3]}$.

[1]The work presented in this article was carried out in collaboration with G.W. Gibbons and G. Papadopoulos [1].

Note that the solution (1) is independent of time and of the two spacelike coordinates σ, ρ. This permits one to make a standard Kaluza-Klein dimensional reduction to lower dimensions, and since this reduction removes simultaneously a translation-symmetric dimension of the "extended object" as well as a spacetime dimension, this falls into the class of "diagonal" dimensional reductions [5, 6]. Making first a reduction from 11 to 10 spacetime dimensions on the coordinate σ, the solution (1) becomes a "wave-on-a-string." Analyzing the conditions for supersymmetry preservation in this spacetime background, one finds that the geometrical conditions required of the background are properly satisfied, while the supersymmetry parameter ϵ is required to satisfy simultaneously the projection conditions

$$\Gamma_{01\sigma}\epsilon = -\epsilon \tag{2a}$$

$$\Gamma_{01}\epsilon = -\epsilon\,, \tag{2b}$$

which may be imposed consistently because $[\Gamma_{01\sigma}, \Gamma_{01}] = 0$. Each of these two conditions cuts down the surviving supersymmetry by a factor of $\frac{1}{2}$, so in the end the solution preserves $\frac{1}{4}$ of the maximum possible 32 rigid supersymmetry components (which would correspond, *e.g.*, to $D = 11$ flat space). As one can see from (2a), this supersymmetry is *chiral* on the $d = 1+1$ string worldsheet. In $d = 1+1$ dimensions, there are two available types supersymmetry that correspond to $\frac{1}{4}$ supersymmetry preservation: (4,4) and (8,0). Only the latter is chiral, so this appears to be the type required to describe the $D = 10$ wave-on-a-string solution descending from (1).

After one more dimensional reduction, on the ρ coordinate this time, the solution falls properly into the family of intersecting p-branes. There is in fact not much left to intersect in this case because we are left with a $D = 9$ solution comprising just two black holes, but it still falls into the family of intersecting p-brane solutions because they do still share the common time direction. Since the two harmonic functions H_1 and H_2 can be chosen independently, with unrelated charge centers, the solution now looks like a configuration of two BPS black holes supported by independent 1-form gauge potentials. Since this solution exists for arbitrary choices of the integration constants that determine the charge centers, it is clear that this static configuration satisfies a zero-force condition, *i.e.* the resulting mutual potential energy must be flat. In the present case, this occurs as a result of a competition between gravitational and dilatonic forces, since the two components couple to different 1-form gauge potentials and so do not feel each other's electromagnetic forces.

Although the interaction between the two elements in the solution (1) vanishes for static configurations, this does not mean that the forces will cancel for time-dependent generalizations of (1). In particular, one can expect velocity-dependent forces to arise. The main question then is at what order in $(\text{velocity})^2$ do these forces arise. For the well-known case of parallel and similarly-oriented $D = 11$ membranes [3], these forces continue to vanish at order $(\text{velocity})^2$, and can occur for the first time only at order $(\text{velocity})^4$. The reason for this continued cancellation of forces may be understood by considering the relative motion problem for the two membranes to be that of finding the corresponding nonlinear σ-model for the relative moduli. In other words, one promotes the relative-position integration constants of the solution to scalar fields on the worldvolume and then works out the lowest-order effective action for these modulus fields. This is necessarily some kind of nonlinear σ-model. Now, for the two-membrane scattering problem, the fraction of residual supersymmetry is $\frac{1}{2}$, so in $d = 2+1$ this corresponds to $N = 8$ supersymmetry, with 16 independent preserved components. This is too much, however, to permit a non-trivial σ-model to exist: the only option for the corresponding σ-model metric is flat space, yielding free modulus fields, *i.e.* vanishing (velocity)^2 forces. Consequently, the lowest order at which velocity-dependent forces can occur in this case is (velocity)^4 (odd powers being ruled out by time-reversal invariance).

The case of wave-membrane scattering generalizing the solution (1), or equivalently after dimensional reduction, the scattering of two $D = 9$ black holes, looks at first blush like it should go similarly to that of membrane-membrane scattering in $D = 11$, yielding a flat sigma model metric. This expectation arises because the corresponding $d = 1+1$ (8,0) supersymmetry is very strong, even though it has only half as many surviving components as in the two-membrane scattering problem. However, this $d = 1+1$ p-brane worldvolume analysis is still not completely appropriate, because in the corresponding $D = 10$ theory one is still dealing with a wave-string scattering problem, and the analysis of gravitational wave scattering is more involved than that of first-quantized brane-brane scattering.

Let us now examine more directly the scattering problem corresponding to (1) by the standard approximation technique of considering a brane-probe source in a fixed background. We shall consider that one of the two elements in (1) is heavy and situated at the origin; since it is taken to be heavy, the motion of the other one in its field will not cause it to move from the origin. The motion of the other, "light," brane may be discussed by considering a brane-source action coupled to the background fields of the heavy brane. The brane-source action should properly be added to the supergravity action in order to provide δ-function sources on the right-hand sides of the supergravity equations; these sources should give rise to the singularities at the charge centers occurring in the harmonic functions in (1). The general bosonic p-brane action that provides these supergravity-equation sources is

$$I_{\text{probe}} = -T_\alpha \int d^{p+1}\xi \left(-\det(\partial_\mu x^m \partial_\nu x^n g_{mn}(x)\right)^{\frac{1}{2}} e^{\frac{1}{2}\varsigma^{\text{pr}}\vec{a}_\alpha\cdot\vec{\phi}} + Q_\alpha \int \tilde{A}^\alpha_{[p+1]} \qquad (3a)$$

$$\tilde{A}^\alpha_{[p+1]} = (p+1)^{-1}\partial_{\mu_1}x^{m_1}\cdots\partial_{\mu_{p+1}}x^{m_{p+1}} A^\alpha_{m_1\cdots m_{p+1}} d\xi^{\mu_1}\wedge\cdots\wedge d\xi^{\mu_{p+1}} . \qquad (3b)$$

The dilaton coupling in (3a) occurs because one needs to have the correct source for the Einstein frame, *i.e.* the conformal frame in which the Einstein-Hilbert action is free from dilatonic scalar factors. Requiring that the source match correctly to the p brane solution demands the presence of the dilaton coupling $e^{\frac{1}{2}\varsigma^{\text{pr}}\vec{a}_\alpha\cdot\vec{\phi}}$, where $\varsigma^{\text{pr}} = \pm 1$ according to whether the p-brane probe is of electric or magnetic type and where $\vec{a}_\alpha$ is the dilaton vector appearing in the dilaton coupling in the kinetic term for the gauge potential $A^\alpha_{[p+1]}$.

Now apply the brane-probe discussion to the $D = 9$ case at hand, which is a standard black hole – black hole scattering problem. One firstly notes that, owing to the fact that the light probe and the heavy background brane couple to different

1-form gauge potentials $A^{\alpha}_{[1]}$, any potential energy must come purely from the kinetic term in (3). Choosing the "static gauge," which for general p-branes is $\xi^\mu = x^\mu$ and in the present case is just $\xi^0 = t$, one may expand the determinant of the induced metric in (3). In general, for a background metric $ds^2 = e^{2A(y)}dx^\mu dx^\nu \eta_{\mu\nu} + e^{2B(y)}dy^m dy^m$, one thus obtains $\det(e^{2A(y)}\eta_{\mu\nu} + e^{2B(y)}\partial_\mu y^m \partial_\nu y^m)$. Then, expanding the determinant and the square root in (3), one obtains the potential energy from the ∂y-independent part of the resulting expansion. In the present case, this gives $V_{\text{probe}} = e^A e^{-\frac{3}{2\sqrt{7}}\phi}$, but this potential is a constant here because the heavy-brane background satisfies $A = \frac{3}{2\sqrt{7}}\phi$. Thus, we confirm the expected static zero-force condition for (1).

Continuing on to the (velocity)^2 order, one now obtains a non-trivial nonlinear sigma model with a modulus metric

$$\gamma^{mn} = H_{\text{back}}(y)\delta^{mn} , \tag{4}$$

where H_{back} is the harmonic function that determines the heavy brane's background field configuration; for the case of two black holes in $D = 9$, the harmonic function H_{back} has the structure $(1 + 1/r^6)$. But in the face of this patently non-flat modulus metric, one must ask what has gone wrong with the "folk-theorem" expectation that (8,0) σ-models must be flat.

The test-brane analysis above is confirmed by a more detailed study of the low-velocity scattering of supersymmetric black holes performed by Shiraishi [7]. The procedure here is a standard one in soliton physics: one promotes the moduli of a static solution to time-dependent functions and then substitutes the resulting generalized field configuration back into the original field equations. This leads to a set of differential equations on the modulus variables which may be viewed as the effective equations for the moduli. In the general case of multiple black hole scattering, the resulting system of differential equations is generally quite complicated. This system, however, simplifies dramatically in cases corresponding to the scattering of supersymmetric black holes, including the pair of $D = 9$ black holes in the present case, where the result involves only 2-body forces. These two-body forces may be derived from an effective action involving the position vectors of the two black holes. Separating the center-of-mass motion from the relative motion, one obtains the same modulus metric (4) as that found in the brane-probe analysis, except for a rescaling that incorporates a replacement of the brane-probe mass by the reduced mass of the two-black-hole system.

Now we should resolve the puzzle of how this non-trivial scattering modulus σ-model turns out to be consistent with supersymmetry. The modulus variables of the two-black-hole system are fields in one dimension, *i.e.* time. The N-extended supersymmetry algebra in $d = 1$ is

$$\{Q^I, Q^J\} = 2\delta^{IJ}\hat{H} \qquad I = 1, \ldots, N , \tag{5}$$

where $\hat{H}$ is the Hamiltonian. A $d = 1$, $N = 1$ σ-model is specified by a triple $(\mathcal{M}, \gamma, C_{[3]})$, where $\mathcal{M}$ is the Riemannian σ-model manifold, γ is the metric on $\mathcal{M}$ and $C_{[3]}$ is a 3-form on $\mathcal{M}$ which plays the rôle of torsion in the derivative operator acting on fermions, $\nabla_t^{(+)} = \partial_t x^i \nabla_i^{(+)}$, where $\nabla_i^{(+)}\lambda^j = \nabla_i\lambda^j + \frac{1}{2}C^j{}_{ik}\lambda^k$. The σ-model action may be written using $N = 1$ superfields $x^i(t,\theta)$ (where $x^i(t) = x^i\big|_{\theta=0}$, $\lambda^i(t) = Dx^i\big|_{\theta=0}$) as

$$I = -\tfrac{1}{2}\int dtd\theta(\mathrm{i}\gamma_{ij}Dx^i\frac{d}{dt}x^j + \frac{1}{3!}C_{ijk}Dx^iDx^jDx^k)\ . \tag{6}$$

One may additionally have a set of spinorial $N = 1$ superfields ψ^a, with Lagrangian $-\frac{1}{2}h_{ab}\psi^a\nabla_t\psi^b$, where h_{ab} is a fibre metric and ∇_t is constructed using an appropriate connection for the fibre corresponding to the ψ^a [8]. However, in the present case we shall not include this extra superfield. In order to have extended supersymmetry in (6), one starts by positing a second set of supersymmetry transformations of the form $\delta x^i = \eta I^i{}_j Dx^j$, and then requires these transformations to close to form the $N = 2$ algebra (5) and one also requires that the action (6) be invariant. In this way, one obtains the equations

$$\begin{aligned}
I^2 &= -\mathbb{1} &\text{(7a)}\\
N^i_{jk} \equiv I^i_{[j,k]} &= 0 &\text{(7b)}\\
\gamma_{kl}I^k{}_i I^l{}_j &= \gamma_{ij} &\text{(7c)}\\
\nabla^{(+)}_{(i} I^k{}_{j)} &= 0 &\text{(7d)}\\
\partial_{[i}(I^m{}_j C_{|m|kl]}) - 2I^m{}_{[i}\partial_{[m}C_{jkl]]}) &= 0\ , &\text{(7e)}
\end{aligned}$$

where (7a,b) follow from requiring the closure of the algebra (5) and (7c-e) follow from requiring invariance of the action (6). Conditions (7a,b) imply that $\mathcal{M}$ is a complex manifold, with $I^i{}_j$ as its complex structure.

What may seem surprising about the conditions (7) is that their structure is more complicated than might be expected. Experience with $d = 1 + 1$ extended supersymmetry might lead one to expect, by simple dimensional reduction, the condition $\nabla_i^{(+)}I^j{}_k = 0$ [8]. Certainly, solutions of this condition also satisfy (7c–e), but the converse is not true. Thus, the $d = 1$ extended supersymmetry conditions are "weaker" than those expected by dimensional reduction from $d = 1+1$, even though the $d = 1 + 1$ minimal spinors are, as in $d = 1$, just real single-component objects. Thus, the $d = 1+1$ theory implies a "stronger" condition; the difference is explained by $d = 1 + 1$ Lorentz invariance: not all $d = 1$ theories can be "oxidized" up to Lorentz-invariant $d = 1 + 1$ theories. Note also that the $d = 1$ "torsion" $C_{[3]}$ is not required to be closed in (7). $d = 1$ supersymmetric theories satisfying (7) are *analogous* to the (2,0) chiral supersymmetric theories in $d = 1 + 1$, but the weaker conditions (7) warrant a different notation for this wider class of models; one may call them 2b supersymmetric σ-models [1]. Such models are characterized by a Kähler geometry with torsion.

Continuing on to $N = 8$ supersymmetry, one finds an 8b generalization of the conditions (7), with 7 independent complex structures built using the octonionic structure constants $\varphi_{ab}{}^c$: $\delta x^i = \eta^a I_a{}^i{}_j Dx^j$, $a = 1,\dots 7$, with $(I_a)^0{}_b = \delta_{ab}$, $(I_a)^b{}_0 = -\delta^b{}_a$, $(I_a)^b{}_c = \varphi_a{}^b{}_c$, where the octonion multiplication rule is $e_a e_b = -\delta_{ab} + \varphi_{ab}{}^c e_c$.

Models satisfying such conditions have an "octonionic Kähler geometry with torsion," and are called OKT models [1].

Now, are there any non-trivial solutions to these conditions? Evidently, from the brane-probe and Shiraishi analyses, there must be. For our $D = 9$ black holes with a $D = 8$ transverse space, one may start from the ansatz $ds^2 = H(y)ds^2(\mathbb{E}^8)$, $C_{\mu\nu\rho} = \Omega_{\mu\nu\rho}{}^{\lambda}\partial_{\lambda}H$, where Ω is a 4-form on $\mathbb{E}^8$. Then, from the 8b generalization of condition (7d) one learns $\Omega_{0abc} = \varphi_{abc}$ and $\Omega_{abcd} = -{}^*\varphi_{abcd}$; from the 8b generalization of condition (7e) one learns $\delta^{\mu\nu}\partial_{\mu}\partial_{\nu}H = 0$. Thus we recover the familiar dependence of p-brane solutions on transverse-space harmonic functions, and so one reobtains the brane-probe or Shiraishi structure of the black-hole modulus scattering metric with

$$H_{\text{relative}} = 1 + \frac{1}{|y_1 - y_2|^6}\,. \tag{8}$$

References

[1] G.W. Gibbons, G. Papadopoulos and K.S. Stelle, *HKT and OKT geometries on soliton black hole moduli spaces, Nucl. Phys.* **B508** (1997) 623, hep-th/9706207.

[2] G. Papadopoulos and P.K. Townsend, *Intersecting M-branes, Phys. Lett.* **B380** (1996) 273, hep-th/9603087;
A.A. Tseytlin, *Harmonic superpositions of M-branes, Nucl. Phys.* **B475** (1996) 149, hep-th/9604035;
J.P. Gauntlett, D.A. Kastor and J. Traschen, *Overlapping branes in M theory, Nucl. Phys.* **B478** (1996) 544, hep-th/9604179.

[3] M.J. Duff and K.S. Stelle, *Multi-membrane solutions of D = 11 supergravity, Phys. Lett.* **B253** (1991) 113.

[4] R. Güven, *Black p-brane solitons of D = 11 supergravity theory, Phys. Lett.* **B276** (1992) 49.

[5] M.J. Duff, P.S. Howe, T. Inami and K.S. Stelle, *Superstrings in D=10 from Supermembranes in D=11, Phys. Lett.* **B191** (1987) 70.

[6] H. Lü, C.N. Pope and K.S. Stelle, *Vertical versus diagonal dimensional reduction for p-branes, Nucl. Phys.* **B481** (1996) 313: hep-th/9605082.

[7] K. Shiraishi, *Nucl. Phys.* **B402** (1993) 399.

[8] R. Coles and G. Papadopoulos, *The geometry of one-dimensional supersymmetric non-linear sigma models, Class. Quantum Grav.* **7** (1990) 427.

Four-dimensional Calabi-Yau Black Holes and their Entropies

Dieter Lüst

Institut für Physik, Humboldt-Universität,
Invalidenstr. 110, D-10115 Berlin

Abstract: We consider extremal black hole solutions of $N = 2$ supergravity which arise in the context of type II superstring compactification on Calabi-Yau 3-folds. In particular we show how the entropies of these black holes depend on the topological data of the Calabi-Yau spaces; we also construct massless black holes which are relevant for the conifold transition among different Calabi-Yau vacua.

1 Introduction

One of the celebrated successes within the recent non-perturbative understanding of string theory and M-theory is the matching of the thermodynamic Bekenstein-Hawking black-hole entropy with the microscopic entropy based on the counting of the relevant D brane configurations which carry the same charges as the black hole [1]. This comparison works nicely for type-II string, respectively, M-theory backgrounds which break half of the supersymmetries, i.e. exhibit $N = 4$ supersymmetry in four dimensions. For example, consider the type IIA superstring compactified on $K3\times T^2$. Then, the intersection of three D 4-branes, whose spatial parts of their world volumes are wrapped p^A times ($A = 1, 2, 3$) around the internal 4-cycles, together with q_0 D 0-branes leads to a four-dimensional black hole with electric charge q_0 and magnetic charges p^A. This black hole has non-vanishing event horizon A, and the corresponding Bekenstein-Hawking entropy is given by the following expression:

$$\mathcal{S}_{\rm BH} = \frac{A}{4} = 2\pi\sqrt{q_0 p^1 p^2 p^3}. \tag{1}$$

In this paper we will discuss black hole solutions and corrections to this formula which arise in string backgrounds with $N = 2$ supersymmetry in four dimensions. In particular, studying IIA compactifications on a (complex) 3-dimensional Calabi-Yau space, corrections arise due to the internal geometry and topology of the Calabi-Yau manifold [2]. In the large volume limit of the Calabi-Yay space, the Calabi-Yau intersection numbers enter the entropy formula, since the D 4-branes are now wrapped around the non-trivial Calabi-Yau 4-cycles. Furthermore, in case of generic radii also world-sheet instanton corrections, encoded by the rational instanton numbers determine the black hole solutions and their entropies [3]. In addition, we will consider corrections to the $N = 2$ black hole entropies which arise from higher-derivative terms involving higher-order products of the Riemann tensor and the vector field strengths [4, 5].

The black hole solutions in Calabi-Yau string compactifications are also very important for the topology change between different Calabi-Yau vacua. One interesting transition is the conifold transition in the complex structure moduli space of type IIB superstrings on Calabi-Yau spaces. The conifold transitions occur at those points in the moduli space where certain 3-cycles shrink to zero size and then are blown up as two cycles, changing in this way the Hodge numbers. The physical understanding of the conifold transition was provided by Strominger [6]; at the conifold point a BPS hypermultiplet black hole becomes massless, being responsible for the singularity in the moduli space metric of the $N = 2$ vector multiplets at this point. We will show [7] that near the conifold points in the Calabi-Yau moduli space there exist a massless extremal black hole solution which corresponds to a 3-brane which is wrapped around the shrinking 3-cycle. Its characteristic feature is that the scalar modulus, which describes the size of the relevant 3-cycle, varies over the 3-dimensional, uncompactified space and vanishes at some special locus in the uncompactifies space.

¿From the point of view of the low energy effective lagrangian the wrapped D branes correspond to extremal charged black hole solutions of the effective $N = 2$ supergravity. Extremal, charged, $N = 2$ black holes, their entropies and also the corresponding brane configurations were discussed in several recent papers [8, 9, 2, 10, 11, 7, 4, 12, 13, 5]. One of the key features of extremal $N = 2$ black-hole solutions is that the moduli depend in general on r, but show a fixed-point behaviour at the horizon. This fixed-point behaviour is implied by the fact that, at the horizon, full $N = 2$ supersymmetry is restored; at the horizon the metric is equal to the Bertotti-Robinson metric, corresponding to the $AdS_2 \times S^2$ geometry. The extremal black hole can be regarded as a soliton solution which interpolates between two fully $N = 2$ supersymmetric vacua, namely corresponding to $AdS_2 \times S^2$ at the horizon and flat Minkowski spacetime at spatial infinity.

2 Extremal black holes in $N = 2$ supergravity

The lowest order (up to two derivatives), four-dimensional $N = 2$ supergravity action coupled to n vector multiplets can be most conveniently formulated in the context of $N = 2$ special geometry. An intrinsic definition of special Kähler manifold can be given [14] in terms of a flat $(2n + 2)$-dimensional symplectic bundle over the

$(2n)$-dimensional Kähler-Hodge manifold, with the covariantly holomorphic sections

$$V = \begin{pmatrix} X^I \\ F_I \end{pmatrix}, \quad I = 0, \ldots, n, \tag{2}$$

obeying the symplectic constraint

$$i\langle \bar{V}, V \rangle = i(\bar{X}^I F_I - \bar{F}_I X^I) = 1 \,. \tag{3}$$

Usually the F_I can be expressed in terms of a holomorphic prepotential $F(X)$, homogenous of degree two, via $F_I = \partial F(X)/\partial X^I$. The field-dependent gauge couplings can then also be expressed in terms of derivatives of F. Note that the holomorphic sections transform non-trivially under $U(1)$ Kähler transformations. Besides V, the magnetic/electric field strengths $(F^I_{\mu\nu}, G_{\mu\nu I})$ also consitute a symplectic vector. Consequently, also the corresponding magnetic/electric charges $Q = (p^I, q_I)$ transform as a symplectic vector. In terms of these symplectic vectors the stationary solutions have been discussed in [11, 5]. Here one solved the generalized Maxwell equations in terms of $2n+2$ harmonic functions, which therefore also transform as a symplectic vector $(m, n = 1, 2, 3)$,

$$F^I_{mn} = \frac{1}{2}\epsilon_{mnp}\, \partial_p \tilde{H}^I(r) \quad , \quad G_{mnI} = \frac{1}{2}\epsilon_{mnp}\, \partial_p H_I(r) \,. \tag{4}$$

The harmonic functions can be parametrized as

$$\tilde{H}^I(r) = \tilde{h}^I + \frac{p^I}{r}\,, \qquad H_I(r) = h_I + \frac{q_I}{r}\,, \tag{5}$$

and we write the corresponding symplectic vector as $H(r) = (\tilde{H}^I(r), H_I(r)) = h + Q/r$.

Next we want to find the solutions for the four-dimensional black hole metric and for the scalar moduli fields $z^A = X^A/X^0$, which break half of the $N = 2$ supersymmetries. For the metric we make the ansatz [15]

$$ds^2 = -e^{2U} dt^2 + e^{-2U} dx^m dx^m \,, \tag{6}$$

where U is a function of the radial coordinate $r = \sqrt{x^m x^m}$. As shown in [11] the Killing spinor equations follow from the symplectic covariance of the vectors V and H, namely these vectors should satisfy a certain proportionality relation. The simplest possibility is to assume that V and H are directly proportional to each other. Because H is real and invariant under U(1) transformations, there is a complex proportionality factor, which we denote by Z. Hence we define a U(1)-invariant symplectic vector (here we use the homogeneity property of the function F),

$$\Pi = \bar{Z} V = (Y^I, F_I(Y)) \,, \tag{7}$$

so that $Y^I = \bar{Z} X^I$, and set

$$\Pi(r) - \bar{\Pi}(r) = iH(r) \,. \tag{8}$$

Now the solution for the black hole metric is given by the symplectically invariant ansatz

$$e^{-2U(r)} = Z(r)\bar{Z}(r), \tag{9}$$

where $Z(r)$ is determined as

$$Z(r) = -H_I(r)\,X^I + \tilde{H}^I(r)\,F_I(X)\,, \qquad |Z(r)|^2 = i\langle\bar{\Pi}(r), \Pi(r)\rangle\,. \tag{10}$$

The equations (8), which we call the stabilization equations, also determine the r-dependence of the scalar moduli fields:

$$z^A(r) = Y^A(r)/Y^0(r). \tag{11}$$

So the constants $(\tilde{h}^I, h_I)$ just determine the asymptotic values of the scalars at $r = \infty$. In order to obtain an asymptotically flat metric with standard normalization, these constants must fullfill some constraints. Near the horizon ($r \approx 0$), (8) takes the form used in [2] and Z becomes proportional to the holomorphic BPS mass $\mathcal{M}(z) = q_I X^I(z) - p^I F_I(X(z))$. It can be shown that the solution preserves half the supersymmetries, except at the horizon and at spatial infinity, where supersymmetry is unbroken.

From the form of the static solution at the horizon ($r \to 0$) we can easily derive its macroscopic entropy. Specifically the Bekenstein-Hawking entropy is given by

$$\begin{aligned} \mathcal{S}_{\rm BH} &= \pi\,(r^2 e^{-2U})_{r=0} = \pi\,(r^2 Z\bar{Z})_{r=0} \\ &= i\pi\left(\bar{Y}^I_{\rm hor}\,F_I(Y_{\rm hor}) - \bar{F}_I(\bar{Y}_{\rm hor})\,Y^I_{\rm hor}\right), \end{aligned} \tag{12}$$

where the symplectic vector Π at the horizon,

$$\Pi(r) \overset{r\to 0}{\approx} \frac{\Pi_{\rm hor}}{r}\,, \qquad Y^I(r) \overset{r\to 0}{\approx} \frac{Y^I_{\rm hor}}{r}\,, \tag{13}$$

is determined by the following set of stabilization equations:

$$\Pi_{\rm hor} - \bar{\Pi}_{\rm hor} = iQ. \tag{14}$$

So we see that the entropy as well as the scalar fields $z^A_{\rm hor} = Y^A_{\rm hor}/Y^0_{\rm hor}$ depend only on the magnetic/electric charges (p^I, q_I). It is useful to note that the set of stabilization equations (14) is equivalent to the minimization of Z with respect to the moduli fields [8].

3 $N = 2$, Type II Calabi-Yau superstring vacua

3.1 The large radius limit of type IIA Calabi-Yau compactifications

As an example, consider a type-IIA compactification on a Calabi-Yau 3-fold. The number of vector superfields is given as $n = h^{(1,1)}$. The prepotential, which is purely classical, contains the Calabi-Yau intersection numbers of the 4-cycles, C_{ABC}, and, as α'-corrections, the Euler number χ[1] and the rational instanton numbers n^r. Hence

[1] The corresponding term in the effective supergravity action comes from an higher derivative R^4 term in ten dimensions.

the black-hole solutions will depend in general on all these topological quantities [2]. However, for a large Calabi-Yau volume, i.e. large values of the Kähler class moduli fields $z^A = X^A/X^0$, only the part from the intersection numbers survives and the $N = 2$ prepotential is given by

$$F(Y) = D_{ABC}\frac{Y^A Y^B Y^C}{Y^0}, \qquad D_{ABC} = -\frac{1}{6}C_{ABC} \,. \tag{15}$$

Based on this prepotential we consider in the following a class of non-axionic black-hole solutions (that is, solutions with purely imaginary moduli fields) with only non-vanishing charges q_0 and p^A $(A = 1, \dots, h^{(1,1)})$. So only the harmonic functions $H_0(r)$ and $\tilde{H}^A(r)$ are nonvanishing. This charged configuration corresponds, in the type-IIA compactification, to the intersection of three D4-branes, wrapped over the internal Calabi-Yau 4-cycles and hence carrying magnetic charges p^A, plus one D0-brane with electric charge q_0. For the configuration indicated above, solving the stabilization equations eq.(8) the four-dimensional metric of the extremal black-hole is given by

$$e^{-2U(r)} = 2\sqrt{H_0(r)\, D_{ABC}\, \tilde{H}^A(r)\, \tilde{H}^B(r)\, \tilde{H}^C(r)}. \tag{16}$$

The scalars at the horizon are determined as

$$z^A_{\text{hor}} = \frac{Y^A_{\text{hor}}}{Y^0_{\text{hor}}}, \quad Y^A_{\text{hor}} = \frac{1}{2}ip^A, \quad Y^0_{\text{hor}} = \frac{1}{2}\sqrt{\frac{D}{q_0}}, \quad D = D_{ABC}\, p^A p^B p^C \,. \tag{17}$$

Finally, the corresponding macroscopic entropy takes the form [2]

$$\mathcal{S}_{\text{BH}} = 2\pi\sqrt{q_0 D}. \tag{18}$$

3.2 Topology change at the conifold point in type IIB Calabi-Yau compactications

Now consider type IIB compactifications on a Calabi-Yau space; hence $n = h^{(2,1)}$. The symplectic vector V corresponds to the period intergrals over the Calabi-Yau 3-cycles γ_I and δ_I:

$$F_I = \int_{\gamma_I} \Omega \quad , \quad X^I = \int_{\delta^I} \Omega \tag{19}$$

The conifold transitions occur at those points in the moduli space where certain 3-cycles shrink to zero size and then blowing them up as two cycles, changing in this way the Hodge numbers. In the following we will discuss the most simple situation with periods X^0 and X^1 (together with F_0 and F_1), where X^1 vanishes at the conifold point and X^0 remains finite. Near this point the prepotential can be expanded as

$$F = -i\,(Y^0)^2 \left(c + \frac{1}{4\pi}\left(\frac{Y^1}{Y^0}\right)^2 \log i\frac{Y^1}{Y^0} + (\text{analytic terms}) \right) \,. \tag{20}$$

$(c = \chi\zeta(3)/2(2\pi)^3)$.

Let us look on the structure of the space-time solution near the conifold point [7]. For simplification, we will consider again axion-free black holes with non-vanishing charges q_0 and p^1. As solution of the stabilization equations (8) we get

$$Y^0(r) = \frac{\lambda(r)}{2} \quad , \quad Y^1(r) = -i\frac{\tilde{H}^1(r)}{2} \tag{21}$$

where the function $\lambda(r)$ is given by

$$\lambda(r) = \frac{H_0(r)}{2c} - \frac{(\tilde{H}^1(r))^2}{4\pi\, H_0(r)} + \mathcal{O}((\tilde{H}^1(r))^4) \; . \tag{22}$$

Keeping only the first correction, we obtain for the function e^{-2U} in the metric

$$e^{-2U(r)} = \frac{H_0^2(r)}{4c} + \frac{(\tilde{H}^1(r))^2}{4\pi} \log \frac{2c\, \tilde{H}^1(r)}{H_0(r)} \; . \tag{23}$$

Hence, we obtained a non-singular metric, also at the points where 3-cycles vanish ($\tilde{H}^1 = 0$).

As a special case consider the black hole solution with only non-vanishing charge p^1, i.e. $q_0 = 0$. Furthermore we also set $\tilde{h}^1 = 0$. Then we obtain a *massless* black hole with metric and scalar field z^1 as follows:

$$e^{-2U(r)} = 1 + \frac{(p^1)^2}{8\pi r^2} \log \frac{c(p^1)^2}{r^2} \pm .. \quad , \quad z^1 = -i\frac{\sqrt{c}\, p^1}{r(1 \mp ..)}, \tag{24}$$

where we used $h_0 = 2\sqrt{c}$, in order to have an asymptotic Minkowski space and $\pm \cdots$ indicate higher powers in $1/r$. This black hole solutions corresponds to a single D 3-brane which is wrapped around the shrinking 3-cycle. It is the dual to the electric massless BPS state discussed by Strominger. The main property of this massless solution is that it carries only one charge and has a shrinking internal 3-cycle at spatial infinity ($z^1 \to 0$ for $r \to \infty$). Hence the Calabi-Yau space degenerates at $r = \infty$ where the topology change can take place.

4 Outlook – Higher curvature corrections and microscopic $N = 2$ entropy

The $N = 2$ black holes together with their entropies which were considered so far, appeared as solutions of the equations of motion of $N = 2$ Maxwell-Einstein supergravity action, where the bosonic part of the action contains terms with at most two space-time derivates (i.e., the Einstein action, gauge kinetic terms and the scalar non-linear σ-model). However, the $N = 2$ effective action of strings and M-theory contains in addition an infinite number of higher-derivative terms involving higher-order products of the Riemann tensor and the vector field strengths. such as nonminimal gravitational couplings R^2. As it was discussed recently in [4, 5] the Bekenstein-Hawking entropy of $N = 2$ black holes will receive corrections due to the higher derivative terms in the effective $N = 2$ supergravity action. These corrections are essential to match the macroscopic Bekenstein-Hawking entropy with the microscopic black hole entropy, which comes from counting the corresponding D-brane degrees of freedom. In fact, for the large radius limit of type IIA, Calabi-Yau compactifications, the microscopic entropy was recently computed [4, 12]:

$$\mathcal{S}_{\text{micro}} = 2\pi\sqrt{q_0 D\Big(1 + \frac{c_{2A}\, p^A}{6D}\Big)} \,. \tag{25}$$

The c_{2A} are the second Chern class numbers of the Calabi-Yau 3-fold. As shown in [4, 5], expanding this expression to lowest order in c_{2A},

$$\mathcal{S}_{\text{micro}} = 2\pi\sqrt{q_0 D} + 2\pi\frac{1}{12}c_{2A}\, p^A\sqrt{\frac{q_0}{D}} + \cdots \,, \tag{26}$$

the correction agrees with a correction to the effective action involving R^2 type terms with a coefficient function $F^{(1)}(z^A) = -\frac{1}{24}c_{2A}z^A$. In order to match the full microscopic $N = 2$ entropy from the effective action, more work is still required.

Acknowledgements

I like to thank my collaborators for the very pleasant collaboration on the material presented here.

References

[1] A. Strominger and C. Vafa, *Phys. Lett.* **B379** (1996) 99, hep-th/9601029; C.G. Callan and J.M. Maldacena, *Nucl. Phys.* **B472** (1996) 591, hep-th/9602043.

[2] K. Behrndt, G. Lopes Cardoso, B. de Wit, R. Kallosh, D. Lüst and T. Mohaupt, *Nucl. Phys.* **B488** (1997) 236, hep-th/9610105.

[3] K. Behrndt and I. Gaida, *Phys. Lett.* **B401** (1997) 263, hep-th/9702168.

[4] J.M. Maldacena, A. Strominger and E. Witten, *Black hole entropy in M theory*, hep-th/9711053.

[5] K. Behrndt, G. Lopes Cardoso, B. de Wit, D. Lüst, T. Mohaupt and W. Sabra, *Higher-Order Black-Hole Solutions in N = 2 Supergravity and Calabi-Yau String Backgrounds*, hep-th/9801081.

[6] A. Strominger, Nucl. Phys. **B451** (1995) 96, hep-th/9504090.

[7] K. Behrndt, D. Lüst and W.A. Sabra, *Moving moduli, Calabi-Yau phase transitions and massless BPS configurations in type II superstrings*, hep-th/9708065.

[8] S. Ferrara, R. Kallosh and A. Strominger, *Phys. Rev.* **D52** (1995) 5412, hep-th/9508072; S. Ferrara and R. Kallosh, *Phys. Rev.* **D54** (1996) 1514, hep-th/9602136; *Phys. Rev.* **D54** (1996) 1525, hep-th/9603090; S. Ferrara, G.W. Gibbons and R. Kallosh, *Nucl. Phys.* **B500** (1997) 75, hep-th/9702103; R. Kallosh, M. Shmakova and W.K. Wong, *Phys. Rev.* **D54** (1996) 6284, hep-th/9607077.

[9] K. Behrndt, R. Kallosh, J. Rahmfeld, M. Shmakova and W.K. Wong, *Phys. Rev.* **D54** (1996) 6293, hep-th/9608059; G. Lopes Cardoso, D. Lüst and T. Mohaupt, *Phys. Lett.* **B388** (1996) 266, hep-th/9608099.

[10] S.-J. Rey, *Nucl. Phys.* **B508** (1997) 569, hep-th/9610157.

[11] W. A. Sabra, *Mod. Phys. Lett.* **A12** (1997) 2585, hep-th/9703101; *Black holes in $N = 2$ supergravity and harmonic functions*, hep-th/9704147; K. Behrndt, D. Lüst and W.A. Sabra, *Nucl. Phys.* **B510** (1998) 264, hep-th/9705169.

[12] C. Vafa, *Black holes and Calabi-Yau threefolds*, hep-th/9711067.

[13] A. Sen, *Black holes and elementary string states in N=2 supersymmetric string theories*, hep-th/9712150.

[14] B. de Wit and A. Van Proeyen, *Nucl. Phys.* **B245** (1984) 89; *Isometries of special manifolds*, hep-th/9505097; S. Cecotti, S. Ferrara, and L. Girardello, *Int. J. Mod. Phys.* **A4** (1989) 2475; A. Strominger, *Commun. Math. Phys.* **133** (1990) 163; S. Ferrara and A. Strominger, in Strings '89, ed. R. Arnowitt, R. Bryan, M. J. Duff, D. Nanopulos and C. N. Pope, World Scientific, Singapore, (1990) 245; P. Candelas and X. de la Ossa, *Nucl. Phys.* **B355** (1991) 455; A. Ceresole, R. D'Auria, S. Ferrara and A. van Proeyen, *Nucl. Phys.* **B444** (1995) 92; L. Andrianopoli, M. Bertolini, A. Ceresole, R. D'Auria, S. Ferrara and P. Frè, *Nucl. Phys.* **B476** (1996) 397, hep-th/9603004; B. Craps, F. Roose, W. Troost and A. Van Proeyen, *Nucl. Phys.* **B503** (1997) 565, hep-th/9703082.

[15] K.P. Tod, *Phys. Lett.* **121B** (1983) 241.

Gauge-variant propagators and zero-momentum modes on the lattice

G. Damm [1] , W. Kerler [1 2] , V.K. Mitrjushkin [3]

[1] Fachbereich Physik, Universität Marburg, D-35032 Marburg, Germany
[2] Institut für Physik, Humboldt-Universität, D-10115 Berlin, Germany
[3] Joint Institute for Nuclear Research, Dubna, Russia

Abstract: We investigate the propagators of 4D SU(2) gauge theory in Lorentz (or Landau) gauge by Monte Carlo simulations. To be able to compare with perturbative calculations we use large β values. There the breaking of the Z(2) symmetry causes large effects for all four lattice directions and doing the analysis in the appropriate state gets important. We find that the gluon propagator in the weak-coupling limit is strongly affected by zero-momentum modes.

1 Introduction

Starting with Ref. [1] there has been a number of nonperturbative lattice studies of the gluon propagator in Lorentz (or Landau) gauge. However, the impact of zero-momentum modes on the propagators has not been analyzed. Recently one of us has shown that zero-momentum modes may strongly affect gauge-dependent correlators [2]. Therefore, we have performed simulations in 4D SU2 lattice gauge theory to clarify this issue.

To be able to compare quantitatively with perturbative calculations large β values must be used. There the propagators, which gauge fixing effectively makes very nonlocal objects, become sensitive to the broken Z(2) symmetry states of the deconfinement region. For comparison with perturbative results the appropriate one of these states has to be selected.

We use the Wilson action, periodic boundary conditions, fields $\mathcal{O}_\mu(x) = \frac{1}{2i}(U_{\mu x} - U^\dagger_{\mu x})$ and the Landau gauge. The propagators considered are

$$\Gamma_\mu(\vec{p},\tau) = \frac{1}{L_4}\sum_t \mathrm{Tr}\langle \tilde{\mathcal{O}}_\mu(\vec{p},t+\tau)\tilde{\mathcal{O}}_\mu(-\vec{p},t)\rangle \tag{1.1}$$

where $\tilde{\mathcal{O}}_\mu(\vec{p},\tau) = \frac{1}{V_3}\sum_{\vec{x}} e^{i\vec{p}\cdot\vec{x}}\mathcal{O}(\vec{x},\tau)$. Choosing $\vec{p} = (0,0,p_3)$ where $p_3 = \frac{2\pi}{L_3}\rho$ the transverse propagator is defined by $\Gamma_T = \frac{1}{2}(\Gamma_1 + \Gamma_2)$.

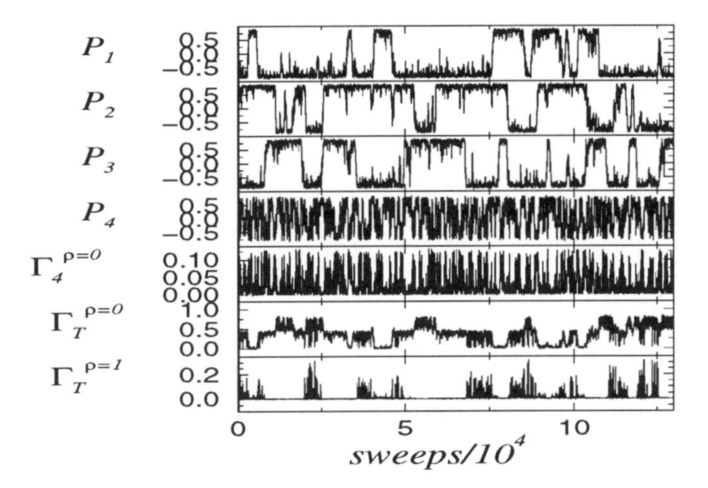

Figure 1: Time history of Polyakov loops and ($\tau = 0$) propagators at $\beta = 10$ on $4^3 \times 8$ lattice.

2 Gauge fixing procedure

The Lorentz gauge is fixed maximizing

$$F = \frac{1}{4V_4} \sum_{\mu,x} \mathrm{Tr}\, U_{\mu x} \tag{2.1}$$

by appropriate gauge transformations which implies the local condition $\sum_\mu \partial_\mu^- U_{\mu x} = 0$. Because Monte Carlo simulations in a fixed gauge would be difficult the update in the simulations is done without any constraint and solely for the purpose of measuring the observables one transforms into Lorentz gauge. This can be done because the expectation with gauge fixing $\langle P \rangle_f$ of an observable P equals the expectation without gauge fixing $\langle P^V \rangle$ of an observable P^V which is transformed with the gauge transformation maximizing (2.1). From a more general point of view [3] P^V is the gauge-invariant observable associated to P by gauge fixing. It turns out that P^V can be successively constructed which allows insight into properties to be expected.

3 Broken Z(2) symmetry

In our simulations on lattices of sizes $4^3 \times 8$, $8^3 \times 16$ and $16^3 \times 32$ we have used Polyakov loops $P_\mu = \frac{L_\mu}{V_4} \sum_{x \neq x_\mu} \mathrm{Tr} \prod_{x_\mu} U_{\mu x}$to monitor the breaking of the Z(2) symmetry. According to the two possibilities for each direction (i.e., $P_\mu > 0$ or $P_\mu < 0$) there are 16 states : $(++++), (+++-), \dots, (----)$.

The time histories of the P_μ, Γ_T and Γ_4 in Fig. 1 illustrate that the Γ_μ are, in fact, strongly affected by the indicated states. This phenomenon has been observed on all lattices considered. From the numerical analysis we find in more detail that

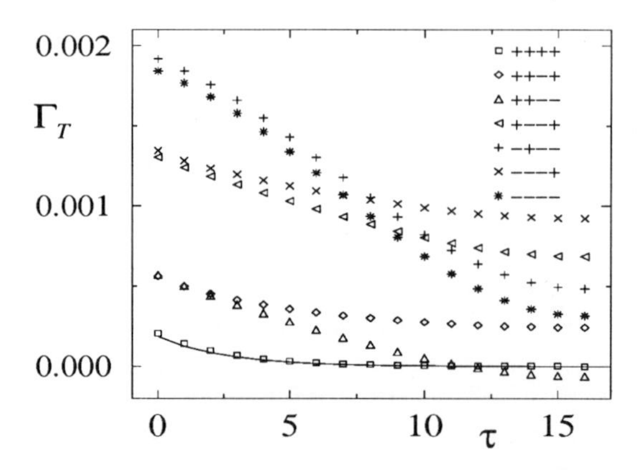

Figure 2: $\Gamma_T(\vec{p}, \tau)$ in various states for $\rho = 1$, $\beta = 10$ and $16^3 \times 32$ lattice (solid line is (4.1)).

the observables take different values in different states. As one can see in Fig. 2, $(++++)$ gives results consistent with perturbation theory while in other cases large deviations occur.

4 Forms of propagators

In the limit of large $\beta = 4/g^2$ one obtains in Gaussian approximation

$$\Gamma_T(\vec{p}, \tau) \rightarrow \frac{3g^2}{2V_4} \sum_{p_4} \frac{e^{-ip_4\tau}}{4\sin^2 \frac{p_3}{2} + 4\sin^2 \frac{p_4}{2}} \tag{4.1}$$

provided that $p_3 \neq 0$. To handle $\vec{p} = 0$ we split off the zero-four-momentum part $C_\mu = \frac{1}{V_4} \sum_x \mathcal{O}_\mu(x)$ so that the fields decompose as $\tilde{\mathcal{O}}_\mu(\vec{0}, \tau) = C_\mu + \delta\tilde{\mathcal{O}}_\mu(\tau)$ and (1.1) becomes

$$\Gamma_\mu(\vec{0}, \tau) = \mathrm{Tr}\langle C_\mu^2 \rangle + R_\mu(\tau) \; ; \tag{4.2}$$

$$R_\mu(\tau) = \frac{1}{L_4} \sum_t \mathrm{Tr}\langle \delta\tilde{\mathcal{O}}_\mu(t+\tau)\delta\tilde{\mathcal{O}}_\mu(t)\rangle \; . \tag{4.3}$$

The evaluation of (4.3) by collective-coordinate methods in Gaussian approximation leads to

$$\Gamma_\mu(\vec{0}, \tau) \rightarrow \mathrm{Tr}\langle C_\mu^2 \rangle + \frac{3g^2}{2V_4} \sum_{p_4 \neq 0} \frac{e^{-ip_4\tau}}{4\sin^2 \frac{p_4}{2}} \; . \tag{4.4}$$

which will be our reference form in the numerical study of propagators.

5 Zero-momentum modes

Now we restrict the considerations to data of the $(++++)$–state. First we note important properties of C_μ by comparing the time histories of C_μ, $\delta\tilde{\mathcal{O}}_\mu(\tau)$ and $X_\mu(\rho) = \text{Re}\,\frac{1}{V_4}\sum_x e^{i\vec{p}\cdot\vec{x}}\mathcal{O}_\mu(x)$ with $\vec{p} = (0,0,\frac{2\pi}{L_3}\rho)$ and $\rho \neq 0$, i.e. of an example of a nonzero-momentum part of the fields. Fig. 3 shows behaviors of components C^a_μ, $\delta\tilde{\mathcal{O}}^a_\mu$ and X^a_μ (where $C_\mu = \sum_a(\sigma^a/2)C^a_\mu$ etc.). For $\delta\tilde{\mathcal{O}}^a_\mu$, $X^a_\mu(1)$ and $X^a_\mu(2)$ uniform Monte Carlo noise is seen. Its magnitude is larger if the three-momentum involved in the particular quantity gets smaller. For C^a_μ in addition to such noise, with magnitude comparable to that of $\delta\tilde{\mathcal{O}}^a_\mu$, surprisingly large variations are observed. These variations exhibit different patterns for different μ, while for different a we find essentially the same pattern.

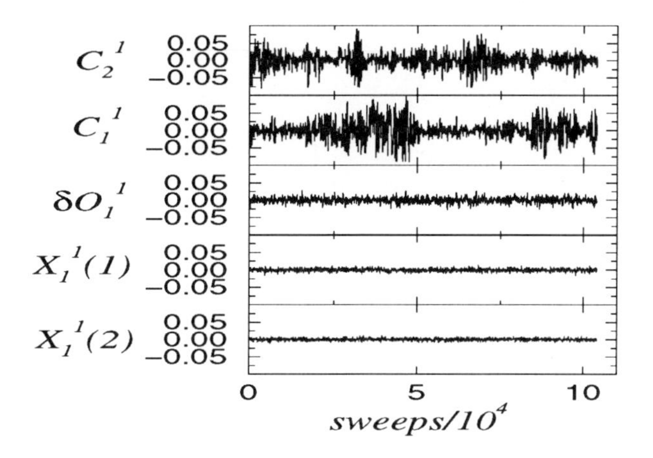

Figure 3: Time histories of C^a_μ, $\delta\tilde{\mathcal{O}}^a_\mu$, $X^a_\mu(1)$ and $X^a_\mu(2)$ for $\beta = 10$ and $16^3 \times 32$ lattice.

A very interesting aspect of these results is that in the decomposition of the fields $\tilde{\mathcal{O}}_\mu$ the parts C_μ and $\delta\tilde{\mathcal{O}}_\mu$ behave quite differently. The observed large variations of C^a_μ appear to be a characteristic consequence of zero-momentum modes. This is indeed a feature one could have expected. A similar phenomenon recently observed by various authors are the exceptional configurations which occur in quenched QCD due to zeros of the Dirac operator.

In addition to $\Gamma_T(\vec{0},\tau)$ and $R_T(\tau)$ also the quantity $\text{Tr}\langle C_T^2\rangle$ needed for the comparison with (4.4) has been determined in the simulations. From Fig. 4 it is obvious that our numerical results for the transverse propagator with $\vec{p}=\vec{0}$ agree reasonably well with that of our lowest-order calculation (4.4). It is seen that the zero-momentum part $\text{Tr}\langle C_T^2\rangle$ is large as compared to the rest, which is related to the large variations of C_μ mentioned above. It is also apparent that while $\Gamma_T(\vec{0},\tau)$ exhibits unusually large errors, which stem from the large variations of C_μ, for $R_T(\tau)$ one gets errors of

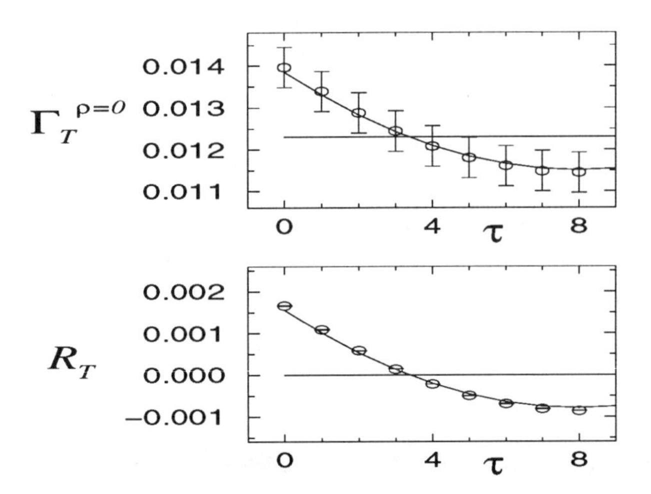

Figure 4: $\Gamma_T(\vec{0},\tau)$ and $R_T(\tau)$ compared with (4.4) (curves) and $\mathrm{Tr}\langle C_\mu^2\rangle$ (constant line in upper Figure) for $\beta = 10$ and $8^3 \times 16$ lattice.

usual size. Analogous observations as for the $8^3 \times 16$ lattice with $\beta = 10$ have been made on $4^3 \times 8$ with $\beta = 10$ and on $16^3 \times 32$ with $\beta = 10$ and $\beta = 99$. Thus our numerical results confirm the description we have given.

The importance of zero-momentum modes is quantified by the fact that $\mathrm{Tr}\langle C_\mu^2\rangle$ is large as compared to $R_\mu(0)$. To compare different lattice sizes and different values of β the respective values are to be multiplied by V_3/g^2 (which obviously cancels the extra factor implicit in our definitions). It turns out that the values of $\zeta = (V_3/g^2)\mathrm{Tr}\langle C_T^2\rangle$ are much larger than $\gamma = (V_3/g^2)(R_T(0) - R_T(L_4/2))$. From Table 1 it is seen that for increasing β this feature gets even more pronounced. The same holds for increasing lattice size. Not only ζ gets larger but there is also an increase of the ratios ζ/γ.

Table 1:

lattice	β	ζ	γ
$4^3 \times 8$	10	6.30 (16)	1.610 (5)
$8^3 \times 16$	10	15.7 (7)	3.229 (17)
$16^3 \times 32$	10	57 (10)	6.03 (22)
$16^3 \times 32$	99	375 (86)	4.98 (19)

To show also what the usual determination of effective masses would give we have

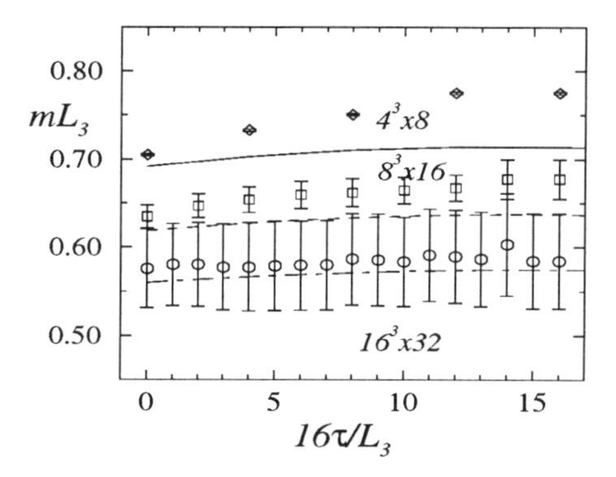

Figure 5: Effective masses from eq. (5.1) for $\vec{p} = \vec{0}$, $\beta = 10$. Curves are based on eq. (4.4).

calculated $m(\tau)$ from

$$\frac{\cosh(m(\tau)(\tau + 1 - \frac{L_4}{2}))}{\cosh(m(\tau)(\tau - \frac{L_4}{2}))} = \frac{\Gamma_T(\vec{0}, \tau + 1)}{\Gamma_T(\vec{0}, \tau)} \tag{5.1}$$

the results of which are depicted in Fig. 5. Because $m(\tau)L_\mu$ shows little dependence on lattice size, the present results may also be considered from the point of view of finite temperatures where screening masses are determined [4]. We have shown that instead of referring to effective masses the data can perfectly be described by (4.4) relying on zero–momentum modes.

To summarize, at β–values considered here the contribution of zero-momentum modes has been found to be large, to increase strongly with β, and also to increase with lattice size. Large fluctuations in the time-history have been identified as a characteristic signal for the occurrence of zero–momentum modes. The question of the role of the zero–momentum modes at smaller values of β (in the physical region) needs further study.

This research was supported in part under DFG grants Ke 250/13-1 and Mu 932/1-4, and the grant INTAS-96-370. One of us (W.K.) wishes to thank M. Müller-Preussker and his group for their kind hospitality.

References

[1] J.E. Mandula and M. Ogilvie, Phys. Lett. B 185 (1987) 127.

[2] V.K. Mitrjushkin, Phys. Lett. B 390 (1997) 293.

[3] W. Kerler, Phys. Rev. D 24 (1981)1595.

[4] U.M. Heller, F. Karsch and J. Rank, Phys. Lett. B 355 (1995) 511.

Lattice QED with Wilson Fermions: the Chiral Transition

Michael Müller-Preussker [1]

[1] Institut für Physik, Humboldt-Universität,
Invalidenstr. 110, D-10115 Berlin

Abstract: This talk presents an overview on some recent work on compact QED with Wilson fermions. We discuss the phase structure of the model and pay particular attention to the chiral limit in the Coulomb and the confinement phases. A new method for calculating the pion mass near the chiral transition is proposed avoiding the problem of exceptional configurations.

1 Introduction

We review recent results obtained in collaboration with A. Hoferichter, V.K. Mitrjushkin, H. Stüben, E. Laermann and P. Schmidt [1, 2].

Lattice QED has been discussed over the years in the literature with respect to various aspects and different kinds of lattice discretizations. Within its Coulomb phase the well established continuum perturbation theory results have to be recovered. There is the hope to find a non-trivial ultra-violet fixed point in order to solve the well-known Landau pole problem in a non-perturbative manner. This issue has been mostly studied within the non-compact $U(1)$ formulation and with staggered fermions by several groups arriving at different conclusions so far [3, 4]. Compact lattice QED is the more natural formulation, when considered to be embedded into a larger grand unifying gauge group. Very recently the pure compact $U(1)$ gauge theory with an extended Wilson action was shown likely to have a fixed point with a non-Gaussian critical exponent [5]. One has speculated that the corresponding continuum limit can be related to a supersymmetric $SU(2)$ Yang-Mills theory [6].

Here we are going to discuss compact QED with Wilson fermions. The phase structure of the theory has been discussed in [7, 8]. We concentrate on the chiral limit of this model in both the Coulomb and the confinement phases. Wilson fermions break chiral symmetry explicitly. Hopefully, the latter can be recovered by fine-tuning the bare parameters in the continuum limit. In the phase diagram a 'critical' line $\kappa_c(\beta)$ has to be related to the limit of the vanishing bare fermion mass. But, for non-vanishing lattice spacing on this line only a partial restoration of chiral symmetry can occur.

In this talk we are mainly concerned with the behavior of fermionic observables, in particular with the 'pion' norm and 'pion' correlator close to $\kappa_c(\beta)$ in the Coulomb and confinement phases of compact QED. We shall compare the full theory with its quenched approximation.

In the confinement phase near to $\kappa_c(\beta)$ the quenched approximation is obstructed by large fluctuations in the observables due to near–to–zero eigenvalues of the fermion matrix (*exceptional configurations*). We propose a variance reduction technique which allows partly to circumvent this problem and to estimate reliably the pseudoscalar mass m_π [9]. We shall demonstrate that this method works well also in the case of quenched lattice QCD [2].

2 Action, Observables and Phase Structure

We consider the standard Wilson notation for the gauge as well as the fermion part.

$$S_W = S_G(U) + S_F(U, \bar{\psi}, \psi) \tag{1}$$

$S_G(U)$ is the plaquette gauge action containing the bare coupling $\beta = 1/e^2$. The fermion contribution is

$$S_F(U, \bar{\psi}, \psi) = \sum_{f=1,2} \sum_{x,y} \bar{\psi}_x^f \mathcal{M}_{xy}(U) \psi_y^f \ , \tag{2}$$

with $\mathcal{M}(U) = \hat{1} - \kappa \mathcal{D}(U)$, $\mathcal{D}_{xy} \equiv \sum_\mu \left[\delta_{y,x+\hat{\mu}} P_\mu^- U_{x\mu} + \delta_{y,x-\hat{\mu}} P_\mu^+ U_{x-\hat{\mu},\mu}^\dagger\right]$ and $P_\mu^\pm = \hat{1} \pm \gamma_\mu$. We adopt the relation of the hopping parameter κ to the fermion mass $m_q = (1/\kappa - 1/\kappa_c(\beta))/2$.

The phase structure of the path integral quantized theory has been investigated in [1, 7, 8]. We have argued that for full QED there are four phases in the range $0 \leq \kappa < 0.3$. Whereas the Coulomb and confinement phases are quite well understood, the situation in the area $\kappa > \kappa_c(\beta)$ deserves further study (see e.g. [10]). We emphasize that the phase diagram has many similarities with that of lattice 2-flavor QCD at finite temperature.

In our investigations we mainly considered fermionic bulk variables like $< \overline{\psi}\psi >$, $< \overline{\psi}\gamma_5\psi >$ and the 'pion' norm $\langle \Pi \rangle = \left\langle \mathrm{Tr}\left(\mathcal{M}^{-1}\gamma_5\mathcal{M}^{-1}\gamma_5\right)\right\rangle_G /4V$. $\langle\ \rangle_G$ means averaging over gauge field configurations, $V = N_\tau \cdot N_s^3$ is the number of the lattice sites. Furthermore, in order to extract the 'pion' mass m_π we studied the pseudoscalar non-singlet correlator

$$\Gamma(\tau) \equiv \frac{1}{N_s^6} \cdot \sum_{\vec{x},\vec{y}} \left\langle \mathrm{Sp}\left(\mathcal{M}_{xy}^{-1}\gamma_5\mathcal{M}_{yx}^{-1}\gamma_5\right)\right\rangle_G \ . \tag{3}$$

The simulations of full QED have been carried out with the HMC method, the inversion of the matrix $\mathcal{M}$ mostly with CG and BiCGstab algorithms.

3 The Chiral Transition

For the Coulomb phase at $\beta > \beta_c \simeq 1.$ we were mainly searching for methods to determine the critical line $\kappa_c(\beta)$ [7]. We simulated for fixed $\beta = 1.1$ and varying κ with lattice sizes up to $32^3 \cdot 64$ for the quenched case and up to 12^4 for the dynamical fermion case. It turned out that $< \overline{\psi}\psi >$, $< \overline{\psi}\gamma_5\psi >$ and the 'pion' norm $\langle \Pi \rangle$ behave very smoothly. But the number of CG steps needed to achieve a definite accuracy in the fermion matrix inversion shows a clear peak at some κ_c indicating the existence of near-to-zero modes of the fermion matrix. Near to that κ_c the fermion variables become strongly fluctuating. In particular the statistical variance of the 'pion' norm exhibits a very narrow peak just at this κ value. Since this peak is increasing with the lattice size, it resembles very much a susceptibility in the neighbourhood of a smooth phase transition. The results obtained so far for the quenched approximation and for full QED look very similar. We conclude that there might be a smooth phase transition into a phase at $\kappa > \kappa_c$, which we called '3rd phase'. However, we checked that this phase is a Coulomb phase, too. The photon mass stays zero beyond κ_c .

Now let us discuss the confinement phase. As representative β–values we have considered the strong coupling limit $\beta = 0$ and two larger values $\beta = 0.6,\ 0.8$. At $\beta = 0$ for both the quenched approximation and the full theory the 'pion norm' data turned out to be compatible with a PCAC-like relation

$$\langle \Pi \rangle = \frac{C_0}{m_q} + C_1 \ , \qquad m_q \to 0 \ , \tag{4}$$

where the constant $C_0 > 0$ – up to a factor – can be identified with the subtracted chiral condensate (see [9]). The fits for $\kappa_c(0)$ provided for the quenched case $\kappa_c(0) = 0.2502(1)$ (nicely agreeing with the analytically known value $1/4$) and for the dynamical case a slightly shifted value 0.2450(6). m_π^2 has been computed as a function of κ for the full and quenched theories on an $8^3 \times 16$ lattice at $\beta = 0$. The quenched data for κ very close to κ_c were obtained by an improved statistical estimator for m_π [9] to be explained in the following paragraph. In both, quenched and dynamical cases we observe a dependence of m_π^2 on κ compatible with $m_\pi^2 \sim \left(1 - \frac{\kappa}{\kappa_c}\right)$, $\kappa \leq \kappa_c$, which in the limit $\kappa \to \kappa_c$ transforms into the PCAC–like relation $m_\pi^2 = B \cdot m_q$, $m_q \to 0+$. The linear extrapolations of m_π^2 down to zero provide estimates for $\kappa_c(0)$ compatible with the above mentioned values. Thus, we conclude that at very strong coupling a chiral limit with a vanishing pseudo-scalar mass can be defined both in the quenched and the dynamical fermion case. The chiral transition seems to be of second order.

The situation changes drastically, if we proceed to larger β–values. At $\beta = 0.8$ the quenched and dynamical cases strongly differ from each other. In the quenched case the behaviour of the 'pion norm' and of the 'pion' mass still resembles to the results obtained in the strong coupling limit $\beta = 0$. However, in the presence of dynamical fermions the fermionic bulk variables as well as gauge observables undergo a discontinuous jump at $\kappa_c(0.8) = 0.1832(3)$. The discontinuity of the 'pion norm' increases with the lattice size as has been checked for sizes $8^3 \times 16$ and $16^3 \times 32$. On top of κ_c a clear metastability behaviour was observed. The 'pion' mass passes a non-vanishing minimum value at κ_c. It only slightly decreases for increasing lattize sizes.

We conclude that at $\beta = 0.8$ the transition becomes certainly a first order transition under the influence of the fermionic determinant. As far as the pseudo-scalar mass does not vanish the standard definition of the chiral limit does not apply.

4 Exceptional Configurations and the Pion Mass

Approaching κ_c in the confinement phase the occurence of zero modes of the fermionic matrix cause strong fluctuations of the fermionic variables when computed for gauge fields within the quenched approximation. This happens in particular with the pseudo-scalar 'pion' correlator. The estimation of the 'pion' mass from the plateau in the τ–dependence of the effective mass $m_{eff}(\tau)$

$$\frac{\cosh(m_{eff}(\tau)(\tau + 1 - \frac{L_4}{2}))}{\cosh(m_{eff}(\tau)(\tau - \frac{L_4}{2}))} = \frac{\langle\Gamma(\tau + 1)\rangle}{\langle\Gamma(\tau)\rangle} . \tag{5}$$

becomes impossible, because there is no reliable estimate for the averages $\langle\Gamma(\tau)\rangle$.

In [9] it was shown that another estimator for the pseudoscalar mass can be defined which is identical to the standard estimator in the case of linear correlations

$$\overline{y}(x) = \mathrm{C} \cdot x, \tag{6}$$

where $x \equiv \Gamma(\tau)$; $y \equiv \Gamma(\tau+k)$; $k > 0$, and $\Gamma(\tau+k)$, $\Gamma(\tau)$ are calculated on individual configurations. $\overline{y}(x)$ stands for the conditional average of $y(x)$ at a fixed value of x. If (6) holds with $x \neq 0$, then

$$\frac{\langle\Gamma(\tau + k)\rangle}{\langle\Gamma(\tau)\rangle} \equiv \frac{\langle y\rangle}{\langle x\rangle} = \left\langle\frac{y}{x}\right\rangle \equiv \left\langle\frac{\Gamma(\tau + k)}{\Gamma(\tau)}\right\rangle , \tag{7}$$

and the effective masses are determined by the r.h.s. of (7). The important observation [9] is that the ratio $\Gamma(\tau + k)/\Gamma(\tau)$ calculated on individual configurations does not suffer from near–to–zero eigenvalues. The average of this ratio is statistically well-behaved, in contrast to the ratio of averages. This has been checked in quenched QED with high statistics. The linear correlation (6) was satisfied with high confidence (the correlation coefficient turned out to be in the range $0.98 \cdots 0.995$). The variance reduction technique for estimating the 'pion' mass allowed us to compute m_π for the quenched approximation for $\beta = 0$ even at $\kappa \geq \kappa_c \simeq .25$. The result is compatible with a zero mass result in the upper '4th phase'. This agrees with the expectation that there is parity-flavour violation [10].

It is interesting to ask, whether these observations also apply to quenched QCD, where the problem of exceptional configurations is well-known.

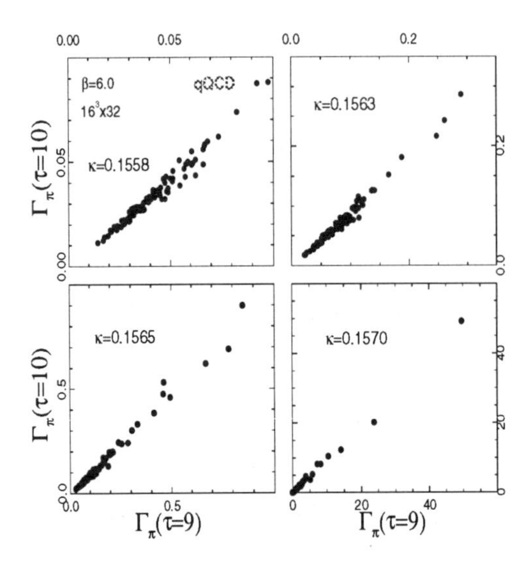

Figure 1: Scatter plots for the pion correlator in quenched QCD.

Therefore, we studied quenched QCD with Wilson fermions in the standard formulation as in QED [2]. $O(100)$ gauge fields were produced at $\beta = 6.0$ on a $16^3 \cdot 32$ lattice. For each gauge field we consider the pion propagator at different time slices. A typical situation is shown in Fig.1 where $\Gamma(\tau+1)$ over $\Gamma(\tau)$ is depicted for $\tau = 9$ at several κ's (note the different scales). As for compact QED we find again linear correlations with a correlation coefficient value $\simeq .99$. Thus, we expect the improved estimator to be applicable also for quenched QCD.

Fig.2 represents the effective mass $m^{\pi}_{eff}(\tau)$ for two different values of κ. For the smaller κ we obtain very good agreement between the two estimators. In the lower plot, very near to κ_c, the standard estimator and its error bars cannot be trusted anymore. In contrast, the improved estimator gives a very nice plateau in m^{π}_{eff} at $\tau > 5$ which still is statistically well-defined. In the meantime we have computed m_π for $\kappa \to \kappa_c$ and even beyond. There is a PCAC dependence

$$m_\pi^2 \sim m_q = \frac{1}{2}\left(\frac{1}{\kappa} - \frac{1}{\kappa_c}\right) \tag{8}$$

seen up to values very close to κ_c. At κ_c the m_π value becomes very small and certainly compatible with zero in the thermodynamic limit. For larger κ it grows again.

5 Summary

We have studied the approach to $\kappa_c(\beta)$ for several β–values within the Coulomb and the confinement phase of compact lattice QED with Wilson fermions comparing

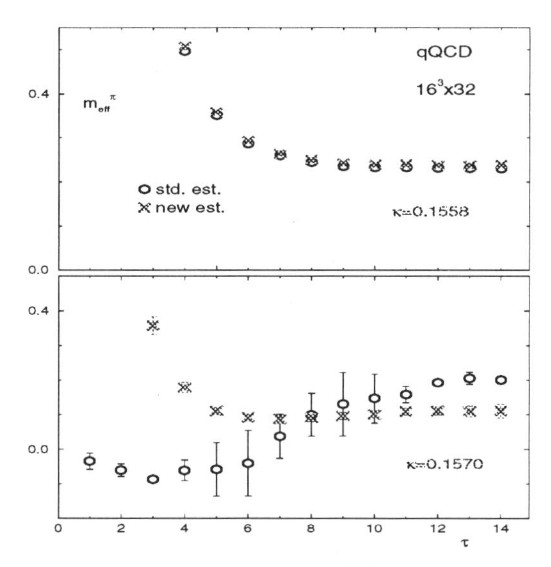

Figure 2: m^{π}_{eff} over τ for two values of κ. (o) denote the standard and (x) the improved estimator.

the full theory with its quenched approximation. We have shown the importance of vacuum polarization effects due to dynamical fermions in the context of the chiral limit. In the strong coupling limit $\beta = 0$ the only effect of dynamical fermions seems to be a renormalization of the 'critical' value κ_c, $\kappa_c^{dyn} \neq \kappa_c^{quen}$. In the quenched as well as dynamical fermion case the pseudoscalar particle becomes massless when $\kappa \to \kappa_c$. However, at larger β the presence of the dynamical ('sea') fermions drastically changes the transition. There the transition cannot be anymore associated with the zero-mass limit of a pseudoscalar particle, in sharp contrast to the quenched case. Since the transition is first order, we can speculate about the existence of tri-critical points on the line $\kappa_c(\beta)$.

We observed that due to strong linear correlations between the values of the pseudoscalar correlators at different τ the ratios $\Gamma(\tau+1)/\Gamma(\tau)$ do not suffer from near–to–zero (exceptional) eigenmodes of the fermionic matrix.

Making use of this observation we proposed another estimator of the pseudoscalar mass, which is well-defined near $\kappa_c(\beta)$ in contrast to the standard estimator.

This observation first made for quenched QED was applied to quenched QCD. It was shown that the proposed method works well also in this realistic case.

References

[1] A. Hoferichter, V.K. Mitrjushkin, M. Müller-Preussker, H. Stüben, contribution to 'LATTICE 97', Edinburgh 1997; to appear in *Nucl. Phys. B (Proc. Suppl.)* (e-print archive *hep-lat/9710047*)

[2] A. Hoferichter, E. Laermann, V.K. Mitrjushkin, M. Müller-Preussker, P. Schmidt, contribution to 'LATTICE 97', Edinburgh 1997; to appear in *Nucl. Phys. B (Proc. Suppl.)* (e-print archive *hep-lat/9710046*)

[3] M. Göckeler, R. Horsley, V. Linke, P. Rakow, G. Schierholz, H. Stüben, talk at 'LATTICE 97', Edinburgh 1997; to appear in *Nucl. Phys. B (Proc. Suppl.)* (e-print archive: *hep-lat/9801004*)

[4] see V. Azcoiti, talk at this symposium

[5] see J. Jersak, talk at this symposium

[6] see N. Sasakura, talk at this symposium

[7] A. Hoferichter, V.K. Mitrjushkin, M. Müller-Preussker, Th. Neuhaus, H. Stüben, *Nucl. Phys.* **B434** (1995) 358

[8] A. Hoferichter, V.K. Mitrjushkin, M. Müller-Preussker, H. Stüben, preprint DESY 96-207, HUB-EP-96/36 (1996) (e-print archive *hep-lat/9608053*)

[9] A. Hoferichter, V.K. Mitrjushkin, M. Müller-Preussker, *Nucl. Phys. B (Proc. Suppl.)* **42** (1995) 669; *Z. Phys.* **C74** (1997) 541

[10] S. Aoki, *Phys. Rev.* **D30** (1984) 2653; *Phys. Rev. Lett.* **57** (1986) 3136; *Phys. Lett.* **190B** (1987) 140; UTHEP-318 (e-print archive *hep-lat/9509008*)

Some radiative and topological effects in gauge field models with vacuum condensate

V. Zhukovskii [1], D. Ebert [2]

[1] Physics Faculty, Moscow State University,
119899, Moscow, Russia
[2] Institut für Physik, Humboldt-Universität,
Invalidenstr. 110, D-10115, Berlin, Germany

Abstract: After briefly reviewing some radiative effects in gauge theories with vacuum condensates the connection between the topological behaviour of 4- and 3-dimensional gauge field theories under the influence of external fields, matter density and temperature is discussed. The possibility of a Chern-Simons like term generation in the QED_4 Lagrangian by quark loops interacting with a non-abelian background field of an $SU(2)$ model of QCD is demonstrated.

One of the effective methods to study the QCD vacuum is based on the investigation of quark and gluon interactions in the background of a certain model classical gauge field. Along with instanton fields [1], efforts were undertaken to construct vacuum models, that are based upon various regular configurations of external non-abelian fields, the simplest of which is a constant chromomagnetic field of the abelian type, proposed by Savvidy [2]. As it was shown in [3], vector-potentials, by which homogeneous and constant fields are described, can only be of two types. The first one is the so called covariantly-constant field. In this case gluon fluctuations lead to a lowering of the vacuum energy [2], while the vacuum itself turns out to be unstable [4]. The vector-potentials G_μ of the second type are gauge equivalent to constant non-commuting potentials, so that the field tensor $G_{\mu\nu}$ is defined through their commutator $iG_{\mu\nu} = [G_\mu, G_\nu]$. The chromomagnetic field configuration formed by the non-abelian potentials turned out to be stable with respect to the decay into pairs of real particles [3], [5]. In a number of publications [6], [7], modelling the vacuum by the fields of the second type, a substantially nonperturbative dependence of such objects, as the photon polarization operator, on the vacuum field was obtained. Active investigations of various radiative effects in non-abelian external fields have been initiated in this connection (see, e.g.[8]).

Since the publication of the pioneer works [9], where a topological mass term for gauge fields in (2+1)-dimensional spacetime was initially investigated, effects of topology produced by the Chern-Simons modification of the gauge field Lagrangian attracted much attention. The study of radiative effects in 3-dimensional theories, carried out recently in Ref. [10], [11] demonstrated the importance of the Chern-Simons (CS) topological term for the regularization of infrared divergences in calculating higher loop diagrams even in the case when background fields are present. In particular, the one-loop electron mass operator and the photon polarization operator in QED_3 in an external magnetic field at finite temperature and matter density [10], as well as the mass operator of a quark in QCD_3 in an external chromomagnetic field [11] were calculated. A new discussion of the role of finite matter density in the effect of the dynamical Chern-Simons term generation in (2+1)–dimensional gauge theories has been initiated recently (see e.g. [12, 13]), showing that the problem of topological effects in gauge theories is far from being solved. A calculation of the photon polarization operator with exact consideration for the contribution of the non-abelian QCD vacuum condensate field was carried out in the case of scalar quarks in [7, 14, 15].

Of special interest are possible Lorentz- and parity-violating modifications of (3+1)-dimensional electrodynamics, investigated in [16]. These modifications are caused by adding to the QED Lagrangian a Chern-Simons term that couples the dual electromagnetic tensor to a four-vector. In this talk we shall demonstrate a possibility that new nontrivial topological terms of the Chern-Simons type may be generated in (3+1)–dimensional QED, when induced by the fermion loops in the non-abelian vacuum condensate of QCD.

As it is well known [9], spinor electrodynamics in (2+1)-dimensional space is described by the Lagrangian

$$\mathcal{L} = \mathcal{L}_g + \mathcal{L}_f + \mathcal{L}_{int} \tag{1}$$

with

$$\mathcal{L}_g = -\frac{1}{4}F_{\mu\nu}F^{\mu\nu} + \frac{1}{4}\theta\varepsilon^{\mu\nu\alpha}F_{\mu\nu}A_\alpha \tag{2}$$

for the gauge (electromagnetic) field, with $F_{\mu\nu} = \partial_\mu A_\nu - \partial_\nu A_\mu$,

$$\mathcal{L}_f = \bar{\psi}(i\gamma\partial - m)\psi \tag{3}$$

for the fermion field, and

$$\mathcal{L}_{int} = e\bar{\psi}\gamma^\mu\psi A_\mu \tag{4}$$

for the interaction term. The last term in $\mathcal{L}_g$ is the Chern-Simons term,

$$\mathcal{L}_{CS} = \frac{1}{4}\theta\varepsilon^{\mu\nu\alpha}F_{\mu\nu}A_\alpha, \tag{5}$$

related to the CS secondary characteristic class. Here θ is the CS parameter which has a dimension of mass. It can be generated in the effective Lagrangian by quantum corrections [9, 18], which are affected by external fields, finite temperature and the density of medium (see e.g. [19, 17, 10, 13, 12]). In the 1-loop approximation this can be seen by calculating the photon polarization operator (PO), which has two parts in QED_3, $\Pi_{\mu\nu} = \Pi^S_{\mu\nu} + \Pi^A_{\mu\nu}$, where $\Pi^S_{\mu\nu}$ is symmetric and $\Pi^A_{\mu\nu}$ is antisymmetric. The so called dynamically induced CS term in PO has the following structure:

$$\Pi^A_{\mu\nu}(q) = i\varepsilon_{\mu\nu\alpha} q^\alpha \Pi^A(q^2). \tag{6}$$

Here the value of $\Pi^A(q^2)$ at $q^2 = 0$ determines the induced CS topological photon mass $\theta_{ind} = \Pi^A(0)$ [9, 18], which leads to screening of electromagnetic fields. It should be noted that higher orders of perturbation theory do not contribute to θ_{ind} (see also [20, 21]). On the other hand, in the presence of an external magnetic field H the topological mass becomes a function of H [17, 12, 13, 10]. Moreover, in an external magnetic field the imaginary part of the amplitude of elastic scattering of a photon, related to the nonzero imaginary part of $\Pi_{\mu\nu}$, determines the decay probability of a photon [10].

In the (3+1)-dimensional isotropic space $\Pi^A_{\mu\nu} = 0$. It is evident that a non-zero value of $\Pi^A_{\mu\nu}$ means appearance of anisotropy. To demonstrate how an antisymmetric structure of the type (6) is induced in (3+1)-QED by the quark loop in a certain non-abelian background field of QCD let us consider a model with a massive quark field ψ, coupled both to an electromagnetic field A_μ and to an $SU(2)_C$ gluon field G_μ in the (3+1)-dimensional space-time, which is described by the following Lagrangian

$$\mathcal{L} = \mathcal{L}_g + \mathcal{L}_f + \mathcal{L}_{int} \tag{7}$$

with

$$\mathcal{L}_g = -\frac{1}{4} F_{\mu\nu} F^{\mu\nu} - \frac{1}{4g^2} G^a_{\mu\nu} G^{\mu\nu}_a, \tag{8}$$

$$\mathcal{L}_f = \bar{\psi}(i\gamma D - m)\psi, \tag{9}$$

and $\mathcal{L}_{int}$ is defined as in (4). Here $G^a_{\mu\nu} = \partial_\mu G^a_\nu - \partial_\nu G^a_\mu + f_{abc} G^b_\mu G^c_\nu$, and $D_\mu = \partial_\mu - iG_\mu$ ($G_\mu = G^a_\mu T^a$, T^a are the generators of the gauge group $SU(2)_C$).

If a constant non-abelian background field $G_\mu = const$ is present, then, neglecting contributions of the gluon fluctuations around this field, we may calculate the 1-loop PO of a photon, moving through the QCD vacuum, and interacting with spin 1/2-quarks in the loop. We use the quark Green's function $S(p)$ with exact consideration for the vacuum field G_μ. Taking into account, that the antisymmetric term in PO appears as a structure which is proportional to an antisymmetric tensor, we make an expansion up to a term linear in $\sigma G = \sigma_{\mu\nu} G^{\mu\nu}$. For the antisymmetric part of PO we then obtain

$$\Pi^A_{\mu\nu}(q, G) = ie^2 \int \frac{d^4p}{(2\pi)^4} \mathrm{Tr}\left[(\frac{1}{2}\sigma G)\pi_{\mu\nu} \frac{1}{(P + q/2)^2 - m^2} - (\mu \leftrightarrow \nu) \right]. \tag{10}$$

Here

$$\pi^{\mu\nu}(P,q) = [\gamma^\mu S(P+q/2)\gamma^\nu S(P-q/2)]_{\sigma G=0}\,, P_\mu = p_\mu + G_\mu. \tag{11}$$

The integral in (10) is UV finite, and after further expansion in (10), (11) up to terms linear in G_μ we obtain $\Pi^A_{\mu\nu}$ which is nonvanishing at $q^2 \to 0$

$$\Pi^A_{\mu\nu}(q,G) = \frac{5}{6\pi^2}\frac{e^2}{m^2}\,\mathrm{tr}(G_\mu G_\nu(q_\alpha G^\alpha)). \tag{12}$$

We see that if a non-abelian constant background field is present, the *QCD* vacuum corrections, like in a (2+1)-dimensional case, may induce an anisotropy in (3+1)-*QED*.

In the case of the rotationally-symmetric background field G_μ

$$G^1_1 = G^2_2 = G^3_3 = \sqrt{\lambda}, G^a_0 = 0, \quad H^a_i = \delta^a_i\lambda(i=1,2,3), \tag{13}$$

we have

$$\Pi^A_{\mu\nu}(q,\lambda) = i\varepsilon_{\mu\nu\alpha}q^\alpha\Pi^A(q^2,\lambda), \tag{14}$$

$(\mu,\nu,\alpha = 1,2,3)$ with $\Pi^A(0,\lambda) = \theta_{ind}(\lambda) = \frac{5}{24\pi^2}\frac{e^2}{m^2}\lambda^{\frac{3}{2}}$. The structure of the 4d antisymmetric PO (12), (14) is very similar to the 3d-result (6). In the field (13) one of the eigenvectors of PO has the form:

$$s_\beta = -\frac{2i}{3}\lambda^{-\frac{3}{2}}\varepsilon_{\beta\sigma\lambda\rho}\,\mathrm{tr}(G^\sigma G^\lambda G^\rho). \tag{15}$$

Then the corresponding CS structure in the *QED* action can be written in a 4-dimensional covariant way

$$S_{CS} = \frac{1}{4}\int d^4x\theta_{ind}(\lambda)\varepsilon^{\mu\nu\alpha\beta}F_{\mu\nu}A_\alpha s_\beta. \tag{16}$$

One can see that the topological current vector [9], introduced for an arbitrary varying field $G_\mu(x)$,

$$X^\beta = 4\varepsilon^{\beta\sigma\lambda\rho}\,\mathrm{tr}(G_\sigma\partial_\lambda G_\rho - \frac{2i}{3}G_\sigma G_\lambda G_\rho) \tag{17}$$

is proportional to s_β (15) in the case of a constant field. Thus one might expect that (16) after replacement $s^\beta \to X^\beta$ holds also for an arbitrary varying background field with non-zero topological charge.

The induced CS structure discussed above may provide one of the physical mechanisms for possible unusual phenomena in the propagation of light through the universe. For instance, evidence for an anisotropy in electromagnetic field propagation over cosmological distances was reported recently in [22]. They claimed to observe a new effect (extracted from Faraday rotation) of rotation of the plane of polarization of electromagnetic radiation propagating over cosmological distances, correlated with the angular position and distance of the source. It should be pointed out that this observation of anisotropy initiated many controversial comments concerning the possible interpretation of the results of [22] (see e.g. [23], and a more recent publication [24] and references therein). Such an evidence, in case it is confirmed, might signal for a non-trivial topological structure of the space over large scales, explained by microscopic reasons. The authors of [22] discussed an effective action with a gauge invariant coupling of A_μ and $F_{\mu\nu}$ to a certain background vector s_μ of the form (16). That such a term in QED_3 Lagrangian could produce an anisotropy in electromagnetic field propagation was pointed out first by Carroll, Field and Jackiw [16]. One might relate s_μ to some intrinsic "spin axis" of an anisotropic universe, associated with axion type domain walls or some other specific condensate structures of the type considered above.

As it is well known, there exists a number of nontrivial topological solutions of field equations, such as, e.g., instantons. The non-abelian constant field case, discussed above, provides just a simple idea of a possible birefringence effect in the propagation of electromagnetic or other gauge fields that may arise under the influence of a nontrivial topological background. More realistic examples are to be discussed in subsequent publications.

The authors are grateful to R.Jackiw for useful comments. One of the authors (V.Ch.Zh.) also thanks the organizers for financial support and kind hospitality.

References

[1] A. Belavin et al., *Phys. Lett.*B59 (1975) 85; E. V.Shuryak, *Nucl. Phys.* **B203** (1982) 93; D. I. Diakonov, M. V. Polyakov and C. Weiss, *Nucl. Phys.* **B461** (1996) 539

[2] G.K.Savvidy, *Phys. Lett.*B71 (1977) 133

[3] L.S.Brown and W.I.Weisberger, *Nucl. Phys.* **B157** (1979) 2857

[4] N.K.Nielsen and P.Olesen, *Nucl. Phys.* **B144** (1978) 376

[5] Sh.S.Agaev, A.S.Vshivtsev, V.Ch.Zhukovsky and O.F.Semyenov, *Izv.Vissh.Uch.Zav., Fizika,* **1** (1985) 78

[6] A.I.Milshtein and Yu.F.Pinelis, *Z.Phys.* **C27** (1985) 461

[7] A.V.Averin, A.V.Borisov and V.Ch.Zhukovsky, *Z.Phys.* **C48** (1990) 457

[8] I.M.Ternov, V.Ch.Zhukovsky, and A.V.Borisov, *Quantum processes in a strong external field (in Russian)*, Moscow, 1989; A. E. Grigoruk and V. Ch. Zhukovsky, *Vestn. Mosk. Univ., Fizika, Astron.*, **1** (1997) 20 ; A. Kyatkin, *Phys. Lett.* **B361** (1995) 105

[9] R.Jackiw and S.Templeton, *Phys. Rev.* **D23** (1981) 2291; S.Deser, R.Jackiw and S.Templeton, *Phys. Rev. Lett.* **48** (1982) 975; *Ann. Phys.(NY)* **140** (1982) 372

[10] K.V.Zhukovsky and P.A.Eminov, *Izv. Vissh. Uch. Zav.* **5** (1995) 61; *Phys. Lett.* **B359** (1995) 155; *Yad. Fiz.* **59** (1996), 1265; A.V.Borisov and K.V. Zhukovsky, *Vestn. Mosk. Univ., Fizika, Astron.*, **1** (1997) 74

[11] V.Ch.Zhukovsky, N.A.Peskov and A.Yu.Afinogenov, to appear in *Yad.Fiz.* **N5** (1998)

[12] V. Zeitlin, e-Print Archive: hep-th/9612225, e-Print Archive: hep-th/9701100

[13] A.N.Sissakian, O.Yu.Shevchenko and S.B.Solganik, e-Print Archive: hep-th/9608159, e-Print Archive: hep-th/9612140

[14] V.Ch.Zhukovskii and Sh.A.Mamedov, *Izv.Vissh.Uch.Zav., Fizika* **N10** (1990) 23

[15] V.Ch.Zhukovskii and I.V.Mamsurov, to appear in *Yad.Fiz.* **N12** (1997)

[16] S.M.Carroll, G.B.Field and R.Jackiw, *Phys. Rev.* **D 41** (1990) 1231

[17] V. Yu. Zeitlin, *Yad.Fiz.* **49** (1989) 742

[18] A.N.Redlich, *Phys. Rev. Lett.* **52** (1983) 18; *Phys. Rev.* **D 29** (1984) 2366

[19] Y.C.Kao, I.Suzuki, *Phys. Rev.* **D31** (1985) 2137; Y.C.Kao, *Phys. Rev.* **D47** (1993) 730

[20] S.Coleman, B.Hill, *Phys. Lett.* **D159** (1985) 184

[21] V.D.Spiridonov, F.V.Tkachev, *Phys. Lett.* **D260** (1991) 109

[22] B. Nodland and J. P. Ralston, *Phys. Rev. Lett.*, **78,** (1997) 3047; *Phys. Rev. Lett.*, **79,** (1997) 1958

[23] S.M.Carroll and G.B.Field, *Phys. Rev. Lett.* **79** (1997) 2394; e-Print Archive: astro-ph/9704263

[24] J. P. Ralston and B. Nodland, e-Print Archive: astro-ph/9708114

Quark and Gluon Condensates at T $\neq$ 0 within an Effective Lagrangian Approach

N. O. Agasyan, [1] D. Ebert [2] and E.–M. Ilgenfritz [2]

[1] Institute of Theoretical and Experimental Physics, Moscow, Russia
[2] Institut für Physik, Humboldt-Universität zu Berlin, Germany

Abstract: We discuss the temperature dependence of quark and gluon condensates within a conformally extended non–linear σ–model which includes broken chiral and scale invariance of QCD as well as (free) heavier hadrons. On this basis the interplay of QCD scale breaking effects and heavier hadrons in chiral symmetry restoration is studied.

1 Introduction

In this talk we will consider quark and gluon condensates of QCD [1] at finite temperature using the effective Lagrangian approach. Effective meson Lagrangians provide a compact and extremely useful method to summarize low–energy theorems of QCD [2]. They incorporate the broken global chiral and scale symmetries of QCD. In particular, to mimic the QCD scale anomaly [3] one usually introduces a scalar chiral singlet dilaton–glueball field χ [4] with an interaction potential leading to $\chi_0 = \langle 0|\chi|0\rangle \neq 0$. As discussed in [5], the kinetic and symmetry breaking mass terms of standard chiral meson Lagrangians have then to be multiplied by suitable powers of χ/χ_0 in order to reproduce the scaling behaviour of the analogous terms of the QCD Langrangian. These ideas have incited a number of investigations of the chiral and scaling behaviour of low–energy Lagrangians at finite temperature (see e.g. [6, 7, 8] and references therein).

Here, we will consider a non–linear (scaled) $O(4)$ σ–model containing one scalar (σ) and three pseudoscalar $\vec{\pi}$ fields subjected to a rescaled chiral constraint. Finally, to model the temperature effects of heavier hadrons, we have added the free Lagrangian of these particles [8]. As discussed in [9], their influence would lead to a lowering of the critical temperature T_c of restoration of chiral symmetry by about 10 percent. Therefore, a generalization of the model which includes the effect of the gluon condensate on the low–lying hadronic states (with mass, say, $\leq$ 2 GeV) is worthwile. This then requires a definite scaling prescription for hadron masses $M_h \to M_h[\chi/\chi_0]$.

In the folowing we shall quote the free energy density of the extended non-linear σ-model including pions, heavier hadrons and glueballs and then discuss the temperature dependent quark and gluon condensates $\langle\langle\overline{q}q\rangle\rangle_{|T} \propto \sigma(T)$ (including the effect of heavier hadrons) and $\langle\langle G^2\rangle\rangle_{|T} \propto \chi^4$.

2 The effective hadron Lagrangian

In order to model the approximate chiral and scale invariance of QCD we consider an extended $O(4)$ non-linear σ-model containing one scalar σ- and three pseudoscalar $\vec{\pi}$-fields as well as a scalar isoscalar dilaton (glueball) field χ. The corresponding Lagrangian is given in Euclidean notation by [8]

$$L(\sigma, \vec{\pi}, \chi) = \frac{1}{2}(\partial_\mu \sigma)^2 + \frac{1}{2}(\partial_\mu \vec{\pi})^2 + \frac{1}{2}(\partial_\mu \chi)^2 + V(\chi) + V_{SB}(\sigma, \chi) , \tag{1}$$

where the fields satisfy the rescaled chiral constraint

$$\sigma^2 + \vec{\pi}^2 = f_\pi^2 \left(\frac{\chi}{\chi_0}\right)^2 , \tag{2}$$

with $f_\pi = 93$ MeV being the pion decay constant, and where $\chi_0 = \langle 0|\chi|0\rangle$ has been introduced for dimensional reasons. This is the actual vacuum expectation value of the field χ which will slightly depend on the amount of explicit breaking of chiral symmetry.

In order to take into account the gluon contribution to the QCD scale anomaly we have included the dilaton interaction potential [4]

$$V(\chi) = K\, \chi^4 \left(\log\left(\frac{\chi}{\chi_q}\right) - \frac{1}{4} \right) \tag{3}$$

which takes its minimum value at $\chi = \chi_q$. Finally, $V_{SB}(\sigma, \chi)$ is the scaled chiral symmetry breaking term of scaling dimension 3

$$V_{SB}(\chi, \sigma) = -c\, \sigma \left(\frac{\chi}{\chi_0}\right)^2 \tag{4}$$

to be added to $V(\chi)$. The total potential $V_{tot}(\chi)$ has then its minimum shifted to χ_0. (The parameters K, χ_0 and c can be specified by considering the vacuum energy density (bag constant), the glueball mass and the pion mass.)

¿From (3,4) we get the trace anomaly

$$\Theta_{\mu\mu}^{eff} = -K\, \chi^4 - c\, f_\pi \left(\frac{\chi}{\chi_0}\right)^3 , \tag{5}$$

and the vacuum energy density

$$\begin{aligned} \epsilon_{vac} &= \frac{1}{4}\langle \Theta_{\mu\mu}^{eff} \rangle = V_{tot}(\chi_0) \\ &= -\frac{1}{4}\, K\, \chi_q^4 - c\, f_\pi + O(c^2) . \end{aligned} \tag{6}$$

It is instructive to compare ϵ_{vac} with the QCD expression for $N_f = 2$

$$\epsilon_{vac}^{QCD} = \frac{1}{4} \langle \, \Theta_{\mu\mu}^{QCD} \, \rangle = \langle \, \frac{\beta}{8 \, g} G_{\mu\nu}^a G_{\mu\nu}^a \, \rangle_{|m=0} + m \, \langle \, \overline{u}u + \overline{d}d \, \rangle \; . \tag{7}$$

This comparison suggests the following identifications:

$$\langle \, \frac{\beta}{2 \, g} \, G_{\mu\nu}^a G_{\mu\nu}^a \, \rangle_{|m=0} = - \, K \, \langle \, \chi^4 \, \rangle_{|c=0} = - \, K \chi_q^4 \tag{8}$$

and

$$m \, \langle \, \overline{u}u + \overline{d}d \, \rangle = - \, c \, f_\pi \; = \; - \; M_\pi^2 \, f_\pi^2 \; . \tag{9}$$

Equation (9) is just the well–known Gell–Mann–Oakes–Renner relation. Note that the pion mass, given by $M_\pi = \sqrt{c/f_\pi}$, scales as

$$M_\pi \to M_\pi \sqrt{\frac{\chi}{\chi_0}} \; . \tag{10}$$

For non–Goldstone particles one may derive from quark model considerations the scaling prescription

$$M_h \; \to \; M_h[\chi/\chi_0] \tag{11}$$

with some well–defined functions M_h of the ratio χ/χ_0 In order to include non–Goldstone hadrons we shall simply use a free Lagrangian $L^h(\varphi_h, M_h[\chi/\chi_0])$ for each particle type.

3 Free energy density of the extended non–linear σ–model

Let us next study the thermodynamical properties of the system of pions, glueballs and free non–Goldstone hadrons described by the Lagrangian

$$L = L(\sigma, \vec{\pi}, \chi) + \sum_h L^h(\varphi_h, M_h[\chi/\chi_0]) \; . \tag{12}$$

The partition function of the Lagrangian (12) evaluated in the saddle point approximation takes the form [8]

$$Z = \exp\left(-\beta V_3 F_{eff}(\sigma, \lambda, \overline{\chi})\right) \tag{13}$$

where the free energy density F_{eff} is given by (V_3 is the $3D$ volume)

$$\begin{aligned} F_{eff}(\sigma, \lambda, \overline{\chi}) \; &= \; V_{tot}(\sigma, \overline{\chi}) \; + \; \lambda \left(\sigma^2 - f_\pi^2 \left(\frac{\overline{\chi}}{\chi_0} \right)^2 \right) \; + \\ &+ \; F^\pi(M_{eff}^2) \; + \; F^\chi(M_\chi^2(\rho_0)) \; + \; \sum_h \; F_h(M_h[\overline{\chi}/\chi_0]) \; , \end{aligned} \tag{14}$$

and λ is an auxiliary field arising from the δ-function constraint (2). Here F^{π}, F^{χ} and F^{h} are the expressions for the thermal determinants of pions, glueballs and heavier hadrons including the zero point energy. Moreover, we have adopted a suitable subtraction procedure which leads to a *subtracted* expression of the free energy density $F_{sub}(\sigma, \lambda, \overline{\chi})$.

Then, the thermal field averages $\sigma(T)$, $\lambda(T)$ and $\chi(T)$ are obtained from solving the following saddle point equations

$$\frac{\partial F_{sub}}{\partial \sigma} = 0 \quad , \quad \frac{\partial F_{sub}}{\partial \lambda} = 0 \quad , \quad \frac{\partial F_{sub}}{\partial \overline{\chi}} = 0 \quad , \tag{15}$$

taking the derivatives at $\sigma = \sigma(T)$, $\lambda = \lambda(T)$ and $\overline{\chi} = \chi(T)$.

3.1 Quark and gluon condensates at $T \neq 0$

The quark condensate is obtained as the logarithmic derivative of the partition function with respect to the current quark mass m,

$$\langle\langle\, \overline{q}q \,\rangle\rangle_T = -\frac{T}{V}\frac{\partial \log Z}{\partial m} = \frac{\partial F_{sub}}{\partial m} \, . \tag{16}$$

Calculating $\langle\langle\, \overline{q}q \,\rangle\rangle_T$ we have to take into account also the $c(m)$–dependence of the glueball mass M_{χ}^2 and of the heavy hadron masses in (11). We obtain

$$\frac{\langle\langle\, \overline{q}q \,\rangle\rangle_T}{\langle\, \overline{q}q \,\rangle_0} = \frac{\sigma(T)}{f_\pi} \left(\frac{\chi(T)}{\chi_0}\right)^2 + \Delta \quad , \tag{17}$$

where Δ contains contributions from the glueball and the other massive hadron states.

Discarding for a moment the contributions from glueballs and heavier hadrons and considering the simplified case of massless quarks ($c = 0$, $m = 0$), the critical temperature T_c, determined from $\sigma(T_c) = 0$, is given by

$$T_c = 2\, f_\pi \,\frac{\chi(T_c)}{\chi_0} = 186\, \frac{\chi(T_c)}{\chi_0} \;\; \text{MeV} \, , \tag{18}$$

where the factor $\chi(T_c)/\chi_0$ can be expressed by the gluon condensate. As we find, the gluon condensate depends only very weakly on T in the region where $\langle\langle\, \overline{q}q \,\rangle\rangle_T \neq 0$.

4 Discussion and Summary

In Fig. 1 the thermal average $\sigma(T)$ is presented together with the quark condensate as function of temperature in the chiral limit. One finds that the gluon condensate has only a negligible influence on the behaviour of $\sigma(T)$ and on the quark condensate. The vanishing of $\sigma(T)$ at some T_c is dominated by the intrinsic chiral dynamics of pions modified by an increase of T_c by about 10 MeV (6 percent) due to glueballs, but

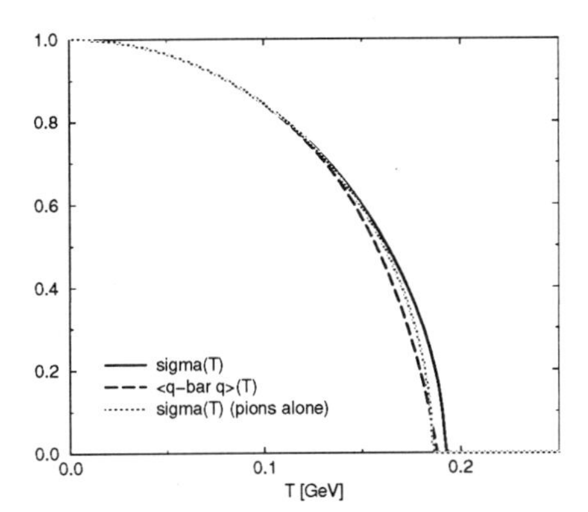

Figure 1: Temperature dependence of the thermal average $\sigma(T)$ of our model (compared with the σ–model for pions only) and the quark condensate $\langle\langle\overline{q}q\rangle\rangle_T$ for the chiral limit using a bag constant $B^{1/4} = 240$ MeV. All quantities are normalized to their values at $T = 0$.

is not related to the small reduction of $\chi(T)$ with rising temperature. This conclusion is in accord with Ref. [6] and does not confirm the corresponding conjecture of Ref. [5]. We see that heavier non–Goldstone hadrons indeed influence the quark condensate in the region of $T \leq 186$ MeVs. They yield an effective lowering by about 10 MeV of the increased value of T_c obtained for the σ–model with pions and glueballs leading to a restoration of the value $T_c \approx 2f_\pi = 186$ MeV obtained for pions alone.

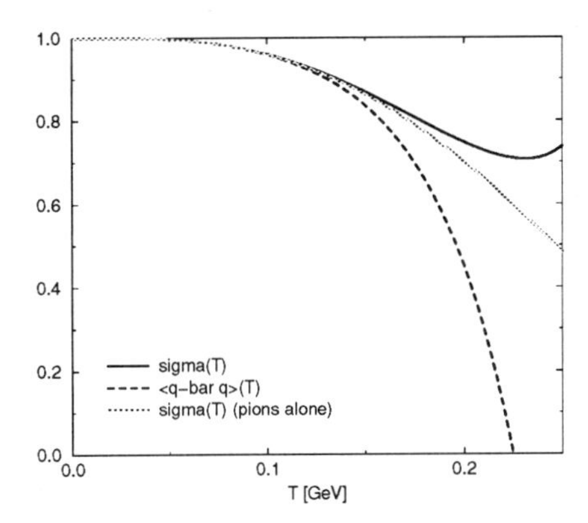

Figure 2: Same as Fig. 2 for a finite pion mass $M_\pi = 139$ MeV ($m = 7$ MeV) using the same bag constant.

Fig. 2 exhibits the same quantities for the case of finite quark masses $m = m_u =$

$m_d \approx 7$ MeV (explicit chiral symmetry breaking) which are adjusted to give the pion a mass $M_\pi = 139$ MeV. Without QCD scale breaking effects taken into account σ is now shifted upward throughout all temperatures and does not play anymore the role of an order parameter. With the coupling to the gluon condensate and, through the glueball mass, to the glueball loop, σ even bends upward for temperature $T > 220$ MeV. This can be interpreted as a kind of stabilization of σ by the gluonic degrees of freedom. A similar behaviour has been found in Ref. [7]. Note that heavier hadrons play now a decisive role in the temperature behaviour of the quark condensate. Summarizing, our model exhibits in the chiral limit, for temperatures $T \leq 200$ MeV, competing effects from glueballs (increase of T_c) and heavier hadrons (decrease of T_c) which, because of cancellation, have however only a minor influence on the chiral symmetry restoration. On the other hand, in the more realistic case of finite quark and pion masses, it is just the quark mass dependence of the heavier hadron masses which leads to a significant lowering of the quark condensate and, finally, to its vanishing.

References

[1] M. A. Shifman, A. I. Vainshtein and V. I. Zakharov, Nucl. Phys. **B 147** (1979) 385,448.

[2] H. Goldberg, Phys. Lett. **B 131** (1983) 133; R. D. Pisarski and F. Wilczek, Phys. Rev. **D 29** (1984) 338; D. Ebert and M. K. Volkov, Fortschr. Phys. **29** (1981) 35.

[3] R. Crewther, Phys. Rev. Lett. **28** (1972) 1421; M. Chanowitz and J. Ellis, Phys. Lett. **B 40** (1972) 397.

[4] J. Schechter, Phys. Rev. **D 21** (1980) 3393; H. Gomm, P. Jain, R. Johnson and J. Schechter, Phys. Rev. **D 33** (1986) 3476.

[5] B. A. Campbell, J. Ellis and K. A. Olive, Phys. Lett. **B 235** (1990) 325; Nucl. Phys. **B 345** (1990) 57.

[6] J. Dotterweich and U. Heinz, Z. Phys. **C 58** (1993) 637; J. Sollfrank and U. Heinz, Z. Phys. **C 65** (1995) 111.

[7] D. Ebert, Yu. L. Kalinovski and M. K. Volkov, Phys. Lett. **B 301** (1993) 231.

[8] N. O. Agasyan, D. Ebert and E.-M. Ilgenfritz, Prepr. HUB-EP-97/84 , hep-ph/9712344.

[9] P. Gerber and H. Leutwyler, Nucl. Phys. **B 321** (1989) 387.

Strings in the Abelian Higgs Model and QCD

D.V.Antonov [1] , D.Ebert [1]

[1] Institut für Physik, Humboldt-Universität zu Berlin
Invalidenstrasse 110, D-10115, Berlin, Germany

Abstract: We review recent progress in string representations of the Abelian Higgs Model and gluodynamics. In the first case, the partition function could be cast into the form of a sum over string world-sheets, while in the second case the string effective action follows from the Wilson loop expansion by making use of the Method of Vacuum Correlators. This enables one to express the string tension of the Nambu-Goto term and the rigidity term coupling constant via the gauge-invariant correlator of two gluonic field strength tensors, known from the lattice data. According to these data, the rigidity term coupling constant is obtained to be negative, which means the stability of strings. The Hamiltonian of the QCD string with quarks, corresponding to the obtained gluodynamics string effective action, where only the Nambu-Goto and rigidity terms are taken into account, is obtained. We also demonstrate that perturbative gluons propagating in the nonperturbative QCD vacuum lead to a correction to the rigidity term coupling constant.

One of the most challenging problems of the modern QFT is the string representation of gauge theories [1]. In the present talk, we shall briefly review the recent progress achieved in investigation of this problem in Refs. [2], [3], [4], [5], [6], [7], [8], and [9].

In Ref. [2], the partition function of the lattice Abelian Higgs Model (AHM) in the Londons' limit has been rewritten as a sum over string world-sheets, which then has been also done in the continuum case in Refs. [5] and [6]. Analogous transformations of the AHM partition function have been also performed in Refs. [3] and [4], where however the Jacobian emerging when one passes from the integration over the singular part of the phase of the Higgs field to the integration over string world-sheets has not been taken into account. Accounting for this Jacobian in Refs. [5] and [6] has led to the appearance of the Polchinski-Strominger term in the action with the coupling constant ensuring exact cancellation of the conformal anomaly in 4D. Besides accounting for the above mentioned Jacobian, in Ref. [6] external monopole currents and an additional θ-term have been also considered, and it has been demonstrated that at $\theta = 2\pi$, the resulting string effective theory is nothing else but the Neveu-Schwarz-Ramond string theory, where the problem of crumpling for the rigidity term is absent.

The mechanism of getting a string effective theory from the AHM partition function, used in all the papers [2], [3], [4], [5], and [6], is based on the application of a duality transformation to this partition function, which has been considered in Ref. [10]. This transformation replaces the integration over the regular part of the phase of the Higgs field by the integration over an auxiliary Kalb-Ramond field, which finally effectively substitutes the field of a massive gauge boson and enables one to describe the interaction of this boson with the string world sheet. Then the final string representation for the partition function has the form

$$\mathcal{Z} = \int \mathcal{D}x_\mu(\xi) \exp\left[-\int\limits_S d\sigma_{\mu\nu}(x) \int\limits_S d\sigma_{\lambda\rho}(x') D_{\mu\nu,\lambda\rho}(x-x')\right], \tag{1}$$

where S is a string world-sheet parametrized by $x_\mu(\xi)$, $\xi = (\xi^1, \xi^2)$, and $D_{\mu\nu,\lambda\rho}(x-x')$ is the propagator of the Kalb-Ramond field $h_{\mu\nu}$, following from the Lagrangian

$$\mathcal{L} = \frac{1}{12\eta^2} H^2_{\mu\nu\lambda} + \frac{e^2}{4} h^2_{\mu\nu}.$$

Here η is the v.e.v. of the Higgs field, e is the charge of this field, and $H_{\mu\nu\lambda} = \partial_\mu h_{\nu\lambda} + \partial_\lambda h_{\mu\nu} + \partial_\nu h_{\lambda\mu}$ is the field strength tensor of the Kalb-Ramond field. For simplicity, in Eq. (1) we have not taken into account the Jacobian mentioned above.

Therefore, it is the presence of a Higgs field, which gives us the possibility to get an integration over world sheets. Unfortunately, we have no similar mechanism in gluodynamics, where the only thing we are up to now able to do is to study the string effective action following from the expansion of a Wilson loop average $\langle W(C)\rangle$ defined on an arbitrary (i.e. non-flat) contour C. To this end, we shall consider $\langle W(C)\rangle$ as a statistical weight in the partition function of some effective string theory, so that the action of this theory reads

$$S_{\text{eff.}} = -\ln\langle W(C)\rangle,$$

and rewrite $\langle W(C)\rangle$ via the non-Abelian Stokes theorem and cumulant expansion. By definition, the bilocal approximation in the Method of Vacuum Correlators [11] corresponds to the case when all the cumulants higher than quadratic one can be disregarded with a good accuracy. Within this approximation, the Wilson loop average takes the form

$$\langle W(C)\rangle =$$

$$= \frac{1}{N_c} \text{tr } \exp\left[-\int\limits_S d\sigma_{\mu\nu}(w) \int\limits_S d\sigma_{\lambda\rho}(w') \langle\langle F_{\mu\nu}(w)\Phi(w,w')F_{\lambda\rho}(w')\Phi(w',w)\rangle\rangle\right], \tag{2}$$

where the bilocal cumulant could be parametrized as follows [11]

$$\langle\langle F_{\mu\nu}(w)\Phi(w,w')F_{\lambda\rho}(w')\Phi(w',w)\rangle\rangle = \frac{\hat{1}}{N_c}\Bigg\{(\delta_{\mu\lambda}\delta_{\nu\rho}-\delta_{\mu\rho}\delta_{\nu\lambda})\, D\bigg(\frac{(w-w')^2}{T_g^2}\bigg)+$$

$$+\frac{1}{2}\bigg[\frac{\partial}{\partial w_\mu}((w-w')_\lambda\delta_{\nu\rho}-(w-w')_\rho\delta_{\nu\lambda})+\frac{\partial}{\partial w_\nu}((w-w')_\rho\delta_{\mu\lambda}-(w-w')_\lambda\delta_{\mu\rho})\bigg]\times$$

$$\times D_1\bigg(\frac{(w-w')^2}{T_g^2}\bigg)\Bigg\}. \tag{3}$$

Here D and D_1 stand for two renormalization group invariant coefficient functions, $\Phi(w,w')$ denotes a parallel transporter factor from the point w' to the point w along a certain path, and T_g is the so-called correlation length of the vacuum $T_g \simeq 0.2\,\text{fm}$ [12], which is a distance at which the functions D and D_1 exponentially decrease. The surface S in Eq. (2) is an arbitrary (not necessarily minimal) surface bounded by the contour C. In such a form Eq. (2) yields a nonlocal action associated with this surface, and our aim now is to find the first few local terms in the expansion of this action in powers of $(T_g/r)^2$, where $r \simeq 1.0\,\text{fm}$ is the size of the Wilson loop in the confining regime [13].

Omitting the full derivative terms of the form $\int d^2\xi\sqrt{\hat{g}}R$, where R is a scalar curvature of the world sheet, $\hat{g} = \det\|\hat{g}_{ij}\|$ with $\hat{g}_{ij} = (\partial_i x_\mu(\xi))(\partial_j x_\mu(\xi))$, $\partial_i \equiv \partial/\partial\xi^i$, $i = 1,2$, being the induced metric tensor, we get the following string effective action in the bilocal approximation (for a detailed derivation see Ref. [7])

$$S_{\text{eff.}} = \sigma\int d^2\xi\sqrt{\hat{g}} + \frac{1}{\alpha_0}\int d^2\xi\sqrt{\hat{g}}\hat{g}^{ij}(\partial_i t_{\mu\nu})(\partial_j t_{\mu\nu}) + O\Bigg(\frac{T_g^6\alpha_s\left\langle F_{\mu\nu}^a(0)F_{\mu\nu}^a(0)\right\rangle}{r^2}\Bigg), \tag{4}$$

where

$$\sigma = 4T_g^2\int d^2z D\left(z^2\right) \tag{5}$$

is the string tension of the Nambu-Goto term,

$$\frac{1}{\alpha_0} = \frac{1}{4}T_g^4\int d^2z z^2\left(2D_1\left(z^2\right) - D\left(z^2\right)\right) \tag{6}$$

is an inverse bare coupling constant of the rigidity term, and

$$t_{\mu\nu} = \frac{1}{\sqrt{\hat{g}}}\varepsilon^{ij}\left(\partial_i x_\mu(\xi)\right)\left(\partial_j x_\nu(\xi)\right)$$

is the extrinsic curvature tensor. Eqs. (4)-(6) are the lowest terms in the derivative expansion of the nonlocal string action following from Eq. (2), which in Ref. [7] has been called "curvature expansion".

The main outcome of these equations is a negative sign of the rigidity coupling constant (6), following from the lattice data [12], according to which the nonperturbative parts of the functions D and D_1 are related to each other as $|D_1| \simeq D/3$. According to Refs. [3] and [14], this sign means the stability of strings in AHM. Namely, the negative sign corresponds to the type-II superconductor, i.e. a phase where strings are stable. Therefore, this result justifies the Method of Vacuum Correlators from the point of view of the dual Meissner mechanism of confinement [15].

As a consequence of the gluodynamics string effective action containing the rigidity term, one can find a Hamiltonian of the QCD string with quarks corresponding to this action [8]. It could be obtained from the Green function of a $q\bar{q}$-system written via the Feynman-Schwinger proper time path integral representation as follows

$$G\left(x\bar{x}|y\bar{y}\right) = \int\limits_0^\infty ds \int\limits_0^\infty d\bar{s} \int Dz D\bar{z}\; e^{-K-\bar{K}} \left\langle W(C)\right\rangle , \tag{7}$$

where $K = m_1^2 s + \frac{1}{4}\int\limits_0^s d\gamma \dot{z}^2(\gamma)$, $\bar{K} = m_2^2\bar{s} + \frac{1}{4}\int\limits_0^{\bar{s}} d\gamma \dot{\bar{z}}^2(\gamma)$ stand for the kinetic terms of a quark and antiquark with masses m_1 and m_2 respectively [16]. In particular, for the s-state of the meson in the case of large masses of a quark and antiquark, $m_1 = m_2 \equiv m \gg \sqrt{\sigma}$, we get up to the fourth power in the relative momentum of the $q\bar{q}$-system the following result

$$H = 2m + \sigma\left|\vec{r}\right| + \frac{\vec{p}^{\,2}}{m} - \left(\frac{1}{4m^3} + \frac{4\sigma\left|\vec{r}\right|}{\alpha_0 m^4}\right)\left(\vec{p}^{\,2}\right)^2 , \tag{8}$$

where $\vec{r} = \vec{z} - \vec{\bar{z}}$ is a relative distance between the quark and antiquark, and $\vec{p}$ is the corresponding relative momentum. All the terms in Eq. (8) except the last one coincide with the ones following from the $1/m$-expansion of the Hamiltonian of the so-called Relativistic Quark Model [17],

$$H_{\rm RQM} = 2\sqrt{\vec{p}^{\,2} + m^2} + \sigma\left|\vec{r}\right| ,$$

whereas the last term emerging due to the rigidity term is a new one. Therefore, we see that the rigidity term yields a nontrivial contribution to the $1/m$-expansion in the fourth power of the relative momentum. Notice also, that Eq. (8) provides us with a classical Hamiltonian, and the Weil ordering should be performed in order to make it a quantum one.

The final result of our investigations, we are going to present, is the incorporation of perturbative gluons a_μ^a, propagating in the nonperturbative confining QCD background B_μ^a, [18]. As it has been shown in Ref. [9], these gluons lead to the interaction between string world-sheet elements. In particular, we find that the string tension (5) gets no correction due to perturbative gluons, whereas the correction to the inverse bare coupling constant (6) reads

$$\left(\Delta\frac{1}{\alpha_0}\right)=g^2\int\limits_0^\infty dss\int(Du)_{00}\,e^{-\int\limits_0^s\frac{\dot{u}^2}{4}d\lambda}\left\langle\mathrm{tr}\,\mathrm{P}\,\exp\left(ig\int\limits_0^s d\lambda\dot{u}_\alpha B_\alpha(u)\right)\right\rangle_{B_\mu^a}. \qquad (9)$$

It is worth noting that since the background fields B_μ^a are strong, $gB_\mu^a=O(1)$, one should take the P-exponent on the R.H.S. of Eq. (9) as a whole. Therefore, we see that accounting for perturbative gluons leads in the lowest order to a nontrivial background dependence of the coupling constant of the rigidity term coupling constant.

References

[1] A.M. Polyakov, *Gauge Fields and Strings* (Harwood Academic Publishers, 1987);
J. Polchinski, hep-th/9210045;
A.M. Polyakov, *Nucl. Phys.* **B486** (1997) 23, hep-th/9711002.

[2] M.I. Polikarpov, U.-J. Wiese, and M.A. Zubkov, *Phys. Lett.* **B309** (1993) 133.

[3] P. Orland, *Nucl. Phys.* **B428** (1994) 221.

[4] M. Sato and S. Yahikozawa, *Nucl. Phys.* **B436** (1995) 100.

[5] E.T. Akhmedov, M.N. Chernodub, M.I. Polikarpov, and M.A. Zubkov, *Phys. Rev.* **D53** (1996) 2087.

[6] E.T. Akhmedov, *JETP Lett.* **64** (1996) 82.

[7] D.V. Antonov, D. Ebert, and Yu.A. Simonov, *Mod. Phys. Lett.* **A11** (1996) 1905.

[8] D.V. Antonov, *JETP Lett.* **65** (1997) 701.

[9] D.V. Antonov and D. Ebert, *Mod. Phys. Lett.* **A12** (1997) 1419.

[10] K. Lee, *Phys. Rev.* **D48** (1993) 2493.

[11] H.G. Dosch, *Phys. Lett.* **B190** (1987) 177;
Yu.A. Simonov, *Nucl. Phys.* **B307** (1988) 512;
for a review see Yu.A. Simonov, *Yad. Fiz.* **54** (1991) 192, *Phys. Usp.* **39** (1996) 313.

[12] A. Di Giacomo and H. Panagopoulos, *Phys. Lett.* **B285** (1992) 133.

[13] I.J. Ford et al., *Phys. Lett.* **B208** (1988) 286;
E. Laermann et al., *Nucl. Phys.* **B26** (Proc. Suppl.) (1992) 268.

[14] H. Kleinert, *Phys. Lett.* **B211** (1988) 151, *Phys. Lett.* **B246** (1990) 127;
H. Kleinert and A.M. Chervyakov, hep-th/9601030 (in press in *Phys. Lett.* **B**).

[15] G.'t Hooft, in *High Energy Physics*, ed. A. Zichichi (Editrice Compositori, 1976);
S. Mandelstam, *Phys. Lett.* **B53** (1975) 476.

[16] A.Yu. Dubin, A.B. Kaidalov, and Yu.A. Simonov, *Phys. Lett.* **B323** (1994) 41.

[17] P. Cea, G. Nardulli, and G. Preparata, *Z. Phys.* **C16** (1982) 135, *Phys. Lett.* **B115** (1982) 310;
S. Godfrey and N. Isgur, *Phys. Rev.* **D32** (1985) 189.

[18] Yu.A. Simonov, *Yad. Fiz.* **58** (1995) 113, 357.

[19] D.I. Diakonov and V.Yu. Petrov, in *Nonperturbative Approaches to QCD, Proceedings of the International Workshop at ECT*, Trento, Italy, July 10-29, 1995* (Ed. D.I. Diakonov) (PNPI, 1995), hep-th/9606104.

Topological structure in the SU(2) vacuum[1]

Thomas DeGrand, Anna Hasenfratz and Tamás G. Kovács[2]

Department of Physics, University of Colorado
Boulder, CO 80309-390, USA

Abstract: We study the topological content of the vacuum of $SU(2)$ pure gauge theory using lattice simulations. We use a smoothing process based on the renormalization group equation. This removes short distance fluctuations but preserves long distance structure. The action of the smoothed configurations is dominated by instantons, but they still show an area law for Wilson loops with an unchanged string tension. The average radius of an instanton is about 0.2 fm, at a density of about 2 fm^{-4}.

1 Introduction

Based on phenomenological models, it has been argued that instantons are largely responsible for the low energy hadron and glueball spectrum [1]. Instanton liquid models attempt to reproduce the topological content of the QCD vacuum and conclude that hadronic correlators in the instanton liquid show all the important properties of the corresponding full QCD correlators. These models appear to capture the essence of the QCD vacuum, but their derivations involve a number of uncontrolled approximations and phenomenological parameters.

Lattice methods are the only ones we presently have, which might address this connection starting from first principles. Lattice studies of instantons can suffer from several difficulties. An unambiguous topological charge can be assigned only to continuous gauge field configurations living on a continuum space-time. On the lattice, the charge can only be defined as that of an interpolated continuum field configuration. This interpolation however is non-unique on Monte Carlo generated lattice configurations.

Another problem is connected to the fact that while the continuum gauge field action is scale invariant, the lattice regularisation breaks this invariance and the action of lattice instantons typically depends on their size. This might distort the

[1]Research supported by DOE grant DE–FG02–92ER–40672
[2]Talk presented by T.G. Kovács.

size distribution of instantons and in particular can lead to an overproduction of small instantons which can even spoil the scaling of the topological susceptibility.

2 The fixed point action and charge

The framework of classically perfect fixed point (FP) actions [2] is particularly suitable to address these problems. Fixed point actions can be shown to have scale invariant instanton solutions and there is no FP-charge Q configuration with an action smaller than Q times the continuum one-instanton action. Therefore the FP procedure completely avoids the problem of dislocations. These are exact properties of the fixed point prescription [2, 3]. In this way the FP context also gives a consistent way of interpolation to define the topological charge.

The fixed point action for any given configuration V on a lattice with lattice spacing a is defined by the weak coupling saddle-point equation

$$S^{FP}(V) = \min_{\{U\}} \left(S^{FP}(U) + T(U,V) \right), \tag{1}$$

where T is the blocking kernel of a real-space RG transformation from the fine lattice (spacing $a/2$) to the coarse lattice (a) and the minimum is taken over all fine lattice configurations U. The minimising fine configuration U_{min} is the smoothest possible of those U's that block into V. It is a very special interpolating configuration which, in the weak coupling limit, gives the largest contribution to the path integral defining the RG transformation for the given coarse configuration V. Finding U_{min} for a given V will be referred to as "inverse blocking". In principle inverse blocking can be repeated several times until the resulting configuration on the finest grid becomes smooth enough that any "sensible" definition of the topological charge gives the same integer value.

In the remainder of this section we discuss the implementation of the above ideas for the 4d SU(2) gauge theory. For more details we refer the reader to [4]. Another implementation can be found in Ref. [5]. Our calculations were performed at several lattice spacings between 0.1-0.18 fm with an approximate FP action. At these values of the lattice spacing the total charge is well defined already after one step of inverse blocking but individual instantons cannot be identified at this stage. Further iteration of the inverse blocking is presently impossible due to computer memory limitations. Therefore we used a smoothing cycle based on inverse blocking followed by a blocking step but on a different coarse sublattice, diagonally shifted by $a/2$. This procedure was designed to approximate repeated steps of inverse blocking as closely as possible in order to preserve its good properties.

In order to understand how the smoothing cycle works we note that, although the inverse blocked lattice has a physical lattice spacing $a/2$ as measured by any long-distance observable, locally it is much smoother than typical Monte Carlo generated lattice configurations with the same lattice spacing. Nevertheless this locally very smooth configuration, by construction, still blocks back into the given V. This is ensured by the delicate coherence present in the U_{min} configuration on a distance scale of a. The diagonal shift by $a/2$ before blocking destroys exactly this coherence and as a result the shifted U_{min} blocks into a configuration that is locally much smoother than V was.

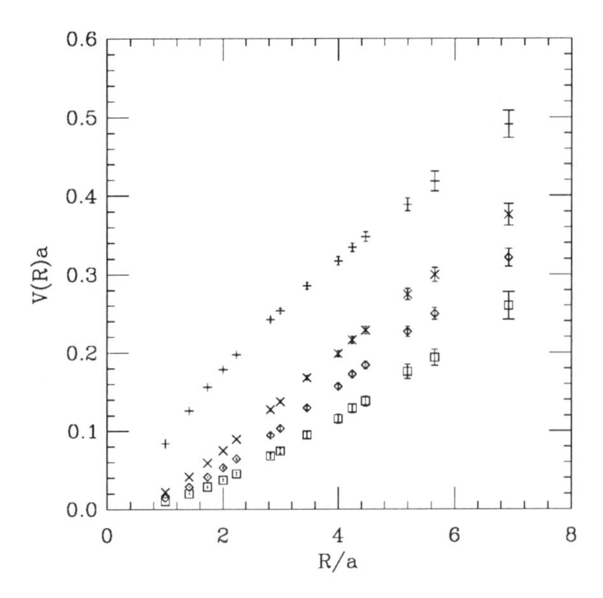

Figure 1: Potential measured on the inverse blocked 16^4 $\beta = 1.5$ configurations after performing different numbers of smoothing cycles (plusses 1 cycle, crosses 3 cycles, diamonds 5 cycles and squares 9 cycles).

After one such cycling step the shortest distance fluctuations are considerably reduced but long-distance features are preserved. Since the lattice size does not change, this step can be repeated several times. Cycling is designed to approximate repeated steps of inverse blocking and in order to see how much of the good properties of the latter it inherits, we made several checks. The stability of the long distance physical properties of the configurations can be demonstrated by the invariance of the string tension (see Fig.1). The total topological charge — as measured in each step on the fine grid — as well as artificially laid down instantons and I-A pairs also turned out to be unchanged by cycling. After about 6 cycles the locations and sizes of individual instantons could be identified.

While their locations were quite stable from the stage where we could reliably identify them, the size of some instantons kept changing slowly throughout the iteration. This was taken into account by extrapolation when measuring sizes. This might explain why on one of our configurations Ref. [6] found that the topological charge — as measured with the fermionic overlap — changed over cycling. It is very likely that over repeated cycling an already existing small instanton grew above the threshold below which the overlap cannot detect small instantons.

3 Results

The instanton size distribution obtained at different values of the lattice spacing — as defined by the Sommer parameter — is shown in Fig. 2. We obtain a topological susceptibility

$$\chi_t = \frac{\langle Q^2 \rangle}{V} = (230(10)MeV)^4, \tag{2}$$

which is about 20% larger than the values obtained with improved cooling [7] and the heating method [8]. The integrated density of topological objects (instantons plus antiinstantons) is about 2 per fm^4.

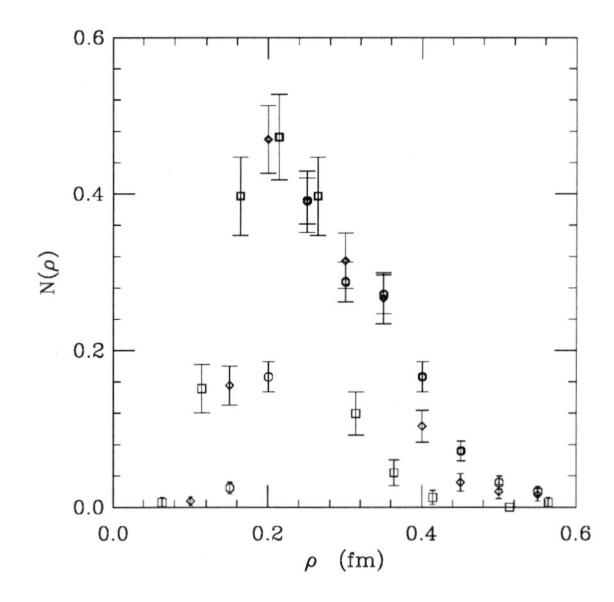

Figure 2: The density distribution of instantons. Data at $a = 0.116(2)$ fm are given by squares, $a = 0.144(1)$ fm diamonds, and $a = 0.188(3)$ fm octagons. The bold data points are ones for which the instanton radius is large compared to the lattice spacing and small compared to the simulation volume.

The smoothed configurations have essentially the same long-distance physical properties as the unsmoothed ones. On the other hand about 70% of their action can be accounted for by the instantons. This alone suggests that instantons might explain most of the long-distance features of QCD. The most straightforward way to check this is to prepare artificial configurations by laying down instantons in exactly the same way as they were found on the smoothed configurations, and then to compare the physical properties of these artificial configurations with the real smoothed ones. As an illustration in Fig. 3 we show the heavy quark potential measured with timelike Wilson loops. The comparison shows that instantons are not very likely to be responsible for confinement. Since we have no information about the relative orientation of instantons in group space, in the artificial configurations we just put them aligned.

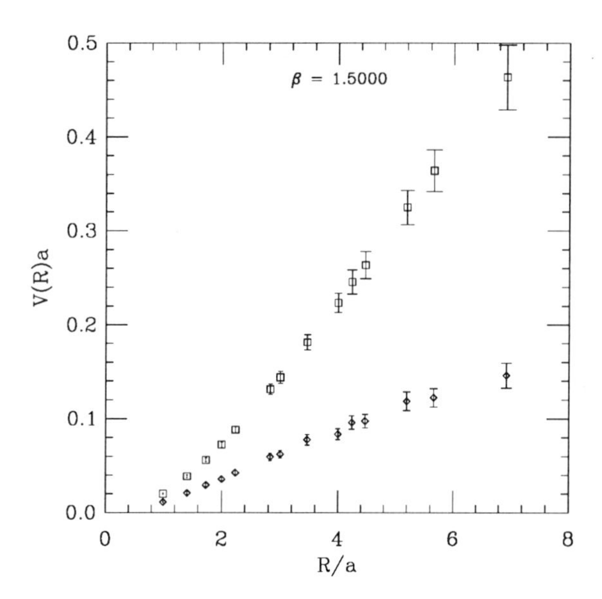

Figure 3: The heavy quark potential measured on the artificially produced instanton configurations (diamonds) and the potential measured on the corresponding 9 times smoothed "real" configurations (squares), at $\beta = 1.5$.

Finally we would like to mention further progress in this work which occured after the Buckow Conference. The heavy quark potential was computed on a randomly oriented instanton ensemble and the result turned out to be the same as on the parallel ensemble [9]. We also found that chiral symmetry breaking can be completely accounted for by the instantons [9].

Acknowledgements

TGK would like to thank the organisers of the Buckow Conference for the invitation, financial support and for putting together a very pleasant and ejoyable meeting.

References

[1] D. Diakanov, hep-ph/9602375; T. Schäfer and E. V. Shuryak, hep-ph/9610451.

[2] P. Hasenfratz and F. Niedermayer, Nucl. Phys. B414 (1994) 785; F. Niedermayer, Nucl. Phys. B (Proc. Suppl.) 53 (1997) 56; P. Hasenfratz, hep-lat/9709110.

[3] M. Blatter, R. Bulkhalter, P. Hasenfratz, and F. Niedermayer, Phys. Rev. D53 (1996) 923; R. Bulkhalter, Phys. Rev. D54 (1996) 4121.

[4] T. DeGrand, A. Hasenfratz and T.G. Kovács, Nucl. Phys. B505 (1997) 417.

[5] M. Feurstein, E.-M. Ilgenfritz, M. Müller-Preussker and S. Thurner, Berlin preprint HUB-EP-96/59, hep-lat/9611024.

[6] R. Narayanan and R.L. Singleton, hep-lat/9709014.

[7] P. de Forcrand, M. Garcia Perez and I.O. Stamatescu, Nucl. Phys. B499 (1997) 409; and talk by I.O. Stamatescu at this conference.

[8] B. Alles, M. D'Elia, A. Di Giacomo, hep-lat/9706016 and talk by A. Di Giacomo at this conference.

[9] T. DeGrand, A. Hasenfratz, and T.G. Kovács, preprint COLO-HEP-391 and hep-lat/9710078, to appear in Phys. Lett. B.

Topology in lattice QCD

A. Di Giacomo

Dipartimento di Fisica and INFN,
2 Piazza Torricelli 56100 Pisa, Italy
digiacomopi.infn.it

Abstract: The status of topology on the lattice is reviewed. Recent results show that the topological susceptibility χ can be unambigously determined. Different methods, if properly implemented, give results consistent with each other. For $SU(3)$ the Witten - Veneziano prediction is confirmed. Preliminary results for full QCD are presented. The problem there is that the usual hybrid montecarlo algorithm has severe difficulty to thermalize topology. Possible ways out are under study.

1 Introduction

Topology plays a fundamental role in QCD. The key equation is the anomaly of the $U_A(1)$ axial current[1]

$$\partial^\mu J^5_\mu(x) = 2N_f Q(x) \tag{1}$$

$J^5_\mu = \sum_{f=1}^{N_f} \bar{\psi}_f \gamma^5 \gamma_\mu \psi_f$ is the singlet axial current. N_f is the number of light flavours, and

$$Q(x) = \frac{g^2}{32\pi^2} \sum_{a,\mu,\nu} G^a_{\mu\nu} \tilde{G}^{a\mu\nu} \tag{2}$$

the topological charge density. $\tilde{G}^a_{\mu\nu} = \frac{1}{2}\varepsilon_{\mu\nu\rho\sigma} G^{a\rho\sigma}$ is the dual tensor field strength. At the classical level $\partial^\mu J^5_\mu = 0$: the right hand side of eq.(1) comes from quantum effects, whence the name of anomaly. $Q(x)$ is related to the Chern current $K_\mu(x) = \frac{g^2}{16\pi^2}\varepsilon^{\mu\alpha\beta\gamma} A^a_\alpha \left(\partial_\beta A^a_\gamma - \frac{g}{3} f^{abc} A^b_\beta A^c_\gamma\right)$ as follows

$$\partial^\mu K_\mu = Q(x) \tag{3}$$

Eq.(1) can then be written

$$\partial^\mu (J^5_\mu - 2N_f K_\mu) = 0 \tag{4}$$

from which Ward identities are derived[2].

A number of physical consequences follow from eq.'s (1), (4).

1) The $U_A(1)$ problem. This problem goes back to Gellmann's free quark model, from which the symmetries of hadron physics were abstracted. In that model $U_A(1)$ is a symmetry, i.e. $\partial^\mu J^5_\mu = 0$. This implies either the existence of parity doublets in the hadron spectrum, if the symmetry is Wigner, or, if it is Goldstone, Weinberg's inequality[3] for the η' mass $m_{\eta'} \leq \sqrt{3} m_\pi$. Neither is true in nature.

In QCD $U_A(1)$ is not a symmetry, due to the anomaly (1) and in principle there is no $U_A(1)$ problem. The anomaly can however explain the high value of the η' mass, as suggested by a $1/N_c$ expansion. The idea is that in the limit $N_c \to \infty$, with $g^2 N_c$ fixed the theory preserves all the main physical features of QCD, like confinement[4]. Non leading terms act as a perturbation on this limiting model. As $N_c \to \infty$ the anomaly disappears, being $\mathcal{O}(g^2)$ [eq.(2)], and the η' is the Goldstone boson of the spontaneously broken $U_A(1)$. The effect of the anomaly is to shift the Goldstone pole from zero mass. This shift can be computed by use of the Ward identities (4) giving[2]

$$\frac{2N_f}{f_\pi^2}\chi = m_\eta^2 + m_{\eta'}^2 - 2m_K^2 \tag{5}$$

$\chi = \chi(q^2)\big|_{q=0}$ is the topological susceptibility of the vacuum at $N_c = \infty$, or

$$\chi(q^2) = \int d^4x\, e^{iqx} \langle 0|T\left(Q(x)Q(0)\right)|0\rangle_{quenched} \tag{6}$$

$N_c = \infty$ implies quenced approximation, fermion loops being $\mathcal{O}(g^2 N_f)$.

Eq.(5) predicts, within $\mathcal{O}(1/N_c)$ accuracy

$$\chi = (180\,\mathrm{MeV})^4 \tag{7}$$

To be definite a prescription must be given for the singularity in the product $Q(x)Q(0)$ in eq.(6) as $x \to 0$. The prescription which brings to eq.(5)is[2, 5]

$$\chi = \int d^4(x-y)\partial^x_\mu \partial^y_\nu \langle 0|T(K_\mu(x)K_\nu(y))|0\rangle \tag{8}$$

By the prescription (8) δ like singularities in the product $K_\mu(x)K_\nu(y)$ as $x \to y$ are eliminated after integration. In any regularization scheme only the multiplicative renormalization of K_μ is thus left. Eq.(5) can be verified on the lattice, where $\chi_{quenched}$ can be computed. This computation provides a check of the $1/N_c$ expansion, which is a fundamental issue.

2) The behaviour of χ across the deconfining phase transition is relevant to understanding the structure of QCD vacuum[6].

3) The $U_A(1)$ Ward identities predict that in full QCD χ behaves in the chiral limit linearly in the quark mass

$$\chi \simeq \frac{1}{N_f}\langle \sum_f m_f \bar\psi_f \psi_f \rangle \tag{9}$$

4) Another important quantity to determine on the lattice is $\chi' = \frac{d}{dq^2}\chi(q^2)\Big|_{q^2=0}$ This determination in full QCD is relevant to understand the spin content of the proton[7].

5) The spin content of the proton can be determined by use of Eq.(1), as will be discussed in detail in the following.

A lattice regulator $Q_L(x)$ of the operator $Q(x)$, eq.(2) is needed. The regularized matrix elements are then determined by numerical simulations, and out of them the continuum physical quantities can be extracted by proper renormalization.

2 Defining $Q(x)$ on the lattice.

According to the general rule any gauge invariant operator on the lattice $Q_L(x)$, such that in the formal limit $a \to 0$

$$Q_L(x) = Q(x)a^4 + \mathcal{O}(a^6) \tag{10}$$

is a possible regulator of topological charge density.

A large arbitrariness by higher order terms exists in the choice of Q_L, which can be used to improve the operator in the sense which will be discussed below. A prototype choice for Q_L is

$$Q_L^{(0)}(x) = -\frac{1}{32\pi^2} \sum_{\pm\mu\nu\rho\sigma} Tr\left[\Pi^{\mu\nu}_{(0)}(x)\Pi^{\rho\sigma}_{(0)}(x)\right] \tag{11}$$

where $\Pi^{\mu\nu}_{(0)}$ is the usual plaquette operator in the plane μ, ν. $Q_L^{(0)}$ obeys eq.(10), and differs by $\mathcal{O}(a^6)$ from any other choice. We will make use of a recursive improving of the operator (11)[8]. We shall define $Q_L^{(i)}(x)$ $(i = 1, 2 \ldots)$ by a formula similar to (11), with $\Pi^{\mu\nu}_{(0)}\Pi^{\rho\sigma}_{(0)}$ replaced by $\Pi^{\mu\nu}_{(i)}\Pi^{\rho\sigma}_{(i)}$. For $i = 1$ each link is replaced by a smoothed link, for $i = 2$ each link of $\Pi_{(1)}$ is smoothed again and so on[8]. An alternative definition is the so called geometrical charge[9, 10], which again obeys the constraint eq.(10).

Any Q_L, in the limit $a \to 0$, will be a mixing of all the continuum operators with the same quantum numbers and with lower or equal dimension. In quenched QCD the only pseudoscalar with dimension ≤ 4 is $Q(x)$ itself and therefore[11]

$$Q_L(x) \underset{a\to 0}{\simeq} Z\, Q(x) \tag{12}$$

Z can be determined by a non perturbative procedure known as "heating"[12]. The idea is to measure Q_L on a state with a definite value of Q, which then determines Z. To do that an istanton is put by hand on the lattice, and quantum fluctuations at a given value of $\beta = 2N/g^2$ are produced numerically by the usual updating algorithm. A plateau after a few heating steps signals thermalization of these fluctuations, while the topological content of the configuration takes a much longer time to be changed, as can be directly tested: the values of Z for different operators Q_L at different β's are shown in Table 1.

	Op	$\chi_L \cdot 10^{-5}$	$M \cdot 10^{-5}$	M/χ_L	Z	$\chi^{1/4}$ MeV
$N_c = 3$	Q_0	2.72(6)	2.50(15)	.9	.12(4)	167(36)
$\beta = 5.9$	Q_1	2.48(5)	.88(6)	.35	.36(2)	178(6)
$a^{-1} = 1.74(4)$ GeV	Q_2	3.51(7)	.71(6)	.2	.48(2)	175(6)
$N_c = 3$	Q_0	2.14(4)	1.97(10)	.9	.18(4)	196(40)
$\beta = 6.1$	Q_1	1.12(2)	.47(3)	.42	.41(2)	178(6)
$a^{-1} = 2.17(5)$ GeV	Q_2	1.39(2)	.33(2)	.24	.54(2)	175(6)
$N_c = 2$	Q_0	3.38(8)	2.93(7)	.87	.18(6)	208(23)
$\beta = 2.44$	Q_1	3.26(8)	1.45(4)	.44	.405(13)	197(8)
$a^{-1} = 1.92$ GeV	Q_2	4.92(14)	1.34(5)	.27	.568(16)	197(8)
	Q_{Geo}	69.7 ± 1.1	52(2)	.75	.918(57)	231(13)
$N_c = 2$	Q_0	2.32(5)	2.20(3)	.95	.240(26)	197(27)
$\beta = 2.57$	Q_1	1.01(5)	.44(2)	.45	.507(9)	200(8)
$a^{-1} = 2.91(9)$ GeV	Q_2	1.16(6)	.117(5)	.16	.675(8)	198(7)
	Q_{Geo}	16.6(3)	13.26(23)	.80	.937(26)	228(12)

Table 1. χ_L, M, χ, Z, a for quenched $SU(3)$ and $SU(2)$ at various β.

3 Measuring χ

The lattice susceptibility is defined as

$$\chi_L = \sum_x \langle Q_L(x) Q_L(0) \rangle = \frac{\langle Q_L^2 \rangle}{V} \tag{13}$$

χ_L is a positive quantity by definition. However in the euclidean region, for $x \neq 0$, $\langle Q(x)Q(0)\rangle$ is a negative quantity, by reflection positivity, Q being odd under time reflection. This can be checked on the lattice and the result is shown in fig.1. If the operator is smeared over a region of size s, one expects $\langle Q(x)Q(0)\rangle$ to be negative at distances $|x| \geq s$. The peak at $x = 0$ is essential to make χ positive, and hence the prescription for the product at $x = 0$ is essential. In general, as a consequence of operator product expansion around $x = 0$, χ_L will mix with the continuum operators with the same quantum numbers and lower or equal dimension[13]

$$\chi_L = Z^2 \chi a^4 + M(\beta) + \mathcal{O}(a^6) \tag{14}$$

with

$$M(\beta) = \overline{Z}(\beta) a^4 \langle \frac{\beta(g)}{g} G^a_{\mu\nu} G^a_{\mu\nu} \rangle + \overline{\overline{Z}}(\beta) a^4 \langle m \bar{\psi}(x) \psi(x) \rangle + P(\beta) \langle I \rangle \tag{15}$$

In the quenched approximation the second term in eq.(15) will be absent.

The first term in eq.(14) corresponds to the prescription (8) and renormalizes multiplicatively. Since $\int d^4x\, \partial_\mu K^\mu = Q$ the prescription (8) implies that χ must be zero on the sector $Q = 0$. $M(\beta)$ can be obtained by measuring χ_L on that sector. This is done by the same heating technique[12] used to determine Z: a sample of configurations belonging to the $Q = 0$ sector is produced by heating the flat (zero field) configuration, in such a way that the topological charge is not changed. From eq.(14) then

$$\chi = \frac{\chi_L - M(\beta)}{a^4(\beta) Z^2(\beta)} \tag{16}$$

All the quantities in the r.h.s. of eq.(16) depend on the choice of the operator, and/or on the choice of the action, but the result must be independent of them. This appears from Table 1, and from fig.2,3.

A good operator is such that $|M(\beta)|/|\chi_L| \ll 1$ so that most of the observed signal is physcal. $Z \simeq 1$ is also desirable. The quality of different Q_L's can be appreciated from Table 1.

The behaviour of χ at deconfinement is shown in fig.2 [14], where $SU(2)$ and $SU(3)$ can be directly compared.

The existence on lattice of a nontrivial Z was first realized in ref.13. Eq.(16) was first introduced in ref.11, where, however, the renormalization constants were determined by perturbation theory.

The heating technique[12] finally allowed a non perturbative determination of them.

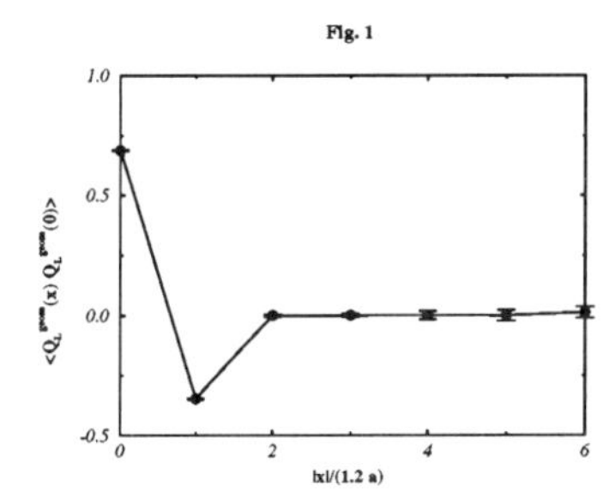

fig.1 Correlation $\langle Q(x)Q(0)\rangle$ for the geometric topological charge density.

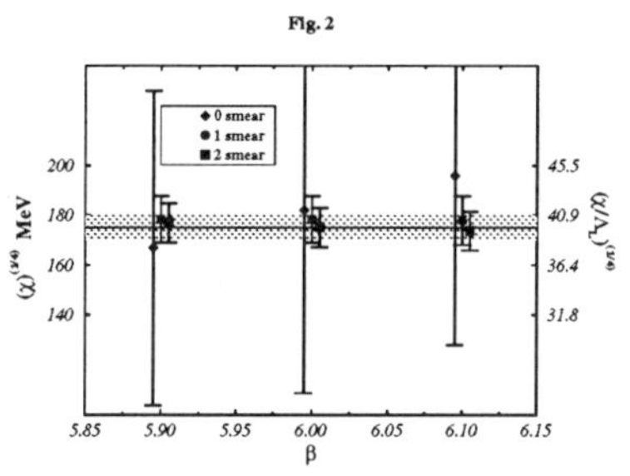

fig.2 χ for quenched $SU(3)$. Diamonds, circles and squares corrspond to 0,1 and 2 - smeared operators.

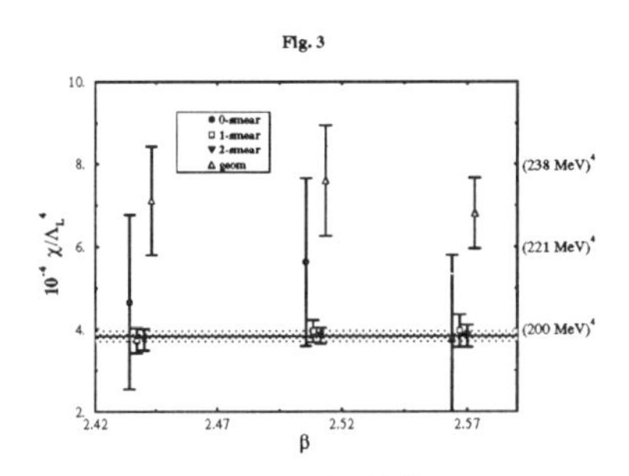

fig.3 χ for quenched $SU(2)$.

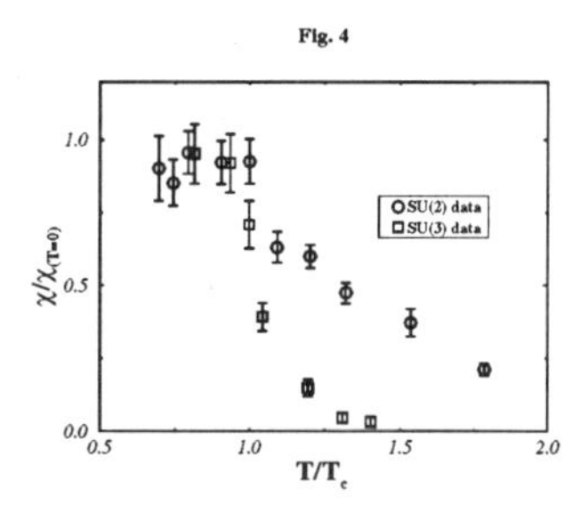

fig.4 χ across the deconfinement transition for $SU(2)$ (circles) and $SU(3)$ (squares).

4 Full QCD[17].

We have used the same technique to determine χ in full QCD with staggered fermions at $m_Q = 0.01$ and $m_Q = 0.02$ at $\beta = 5.35$. The lattice was $16^3 \times 24$. At this β value $a = 0.115(2)$ fm.

We do not have enough precision to test the dependence on m_Q [eq.(9)]. At $am = 0.01$ we get $\chi = (110 \pm 6 \pm 2)^4\,\mathrm{MeV}^4$ The first error is obtained by jake-knife technique. The second comes from the error in the determination of the scale. The expected value is $\chi \simeq (109\,\mathrm{MeV})^4$.

As we will soon discuss these errors are higly underestimated.

A similar analysis can be done for χ'. On general grounds

$$\chi' = Z^2\chi' a^2 + M'(\beta) + \mathcal{O}(a^4) \qquad M'(\beta) = P'(\beta)\langle I\rangle \tag{17}$$

χ' only mixes to the identity operator, since no other gauge invariant operator of dimension ≤ 2 exists.

The technique used to determine χ' is to improve the operator enough, so that $M'(\beta)$ will be negligeable as compared to the first term. Z is known from the analysis of χ.

A preliminary result is

$$\chi' = (258 \pm 100)\,\mathrm{MeV}^2 \qquad \sqrt{\chi'} = (16 \pm 3)\,\mathrm{MeV} \tag{18}$$

to be compared with the value computed from SVZ rules[7] $\sqrt{\chi'} = (22.3 \pm 4.8)\,\mathrm{MeV}$ Here the error coming from the normalization is much smaller than the statistical error.

Again here the error is underestimated. The reason can be in the history of the topological charge along the updating process is shown. The hybrid montecarlo algorithm is very slow in thermalizing topology[16]. The sampling is very bad and corresponds in fact to a much smaller number of independent configurations.

The same incovenience affects the determination of the spin content of the proton which will be discussed in the next section. The real error is then larger than the one estimated on the present ensemble of configurations.

5 The spin content of the proton.

The matrix element of the singlet axial current between proton states can be written as

$$\langle \vec{p}'s'|J^5_\mu(0)|\vec{p}s\rangle = \bar{u}(\vec{p}'s')\left\{G_1(k^2)\gamma^5\gamma_\mu + G_2(k^2)\gamma^5 k_\mu\right\}u(\vec{p}s) \tag{19}$$

$k = p - p'$ is the momentum transfer. $G_1(0) \equiv \Delta\Sigma$ is related to the integral of the spin dependent structure function $g_1(x, q^2)$ of deep inelastic scattering of leptons off protons. In naive parton picture $\Delta\Sigma = \Delta u + \Delta d + \Delta s$ is the fraction of theproton spin carried by the quarks and is expected to be $\Delta\Sigma = 0.7$. Experiments[18] show that it is much smaller, $\Delta\Sigma = 0.21 \pm 0.10$, and this fact is usually referred as spin crisis. By use of the anomaly equation

$$\langle \vec{p}'s'|Q(0)|\vec{p}s\rangle = \frac{1}{2N_f}\langle \vec{p}'s'|\partial^\mu J^5_\mu(0)|\vec{p}s\rangle = \frac{m_N}{N_f}A(k^2)\bar{u}(\vec{p}'s')i\gamma^5 u(\vec{p}s)$$

$A(k^2) = G_1(k^2)+(k^2/m_N)G_2(k^2)$ can be extracted from a measurement of the matrix element in the left hand side of eq.(19).

In quenched approximation $G_2(k^2)$ has a pole at $k^2 = 0$[19]. With dynamical quarks this is ot the case and, for sufficiently small k^2 $G_1(0)$ or $\Delta\Sigma$[20] can be determined.

Our preliminary result is $\Delta\Sigma = 0.05 \pm 0.05$ Again the error is underestimated due to the bad sampling of topological charge discussed above.

6 Discussion.

Our conclusions are the following.

A reliable determination of the topological susceptibility χ is possible on the lattice. All existing definitions of lattice topological charge density give results consistent with each other if proper renormalizations are performed. The value of χ for $SU(2)$ is larger than the prediction[2] but the value for $SU(3)$ is consistent with it. The value of $m_{\eta'}$ is well explained by the anomaly. Above the deconfining transition χ drops to zero, more rapidly for $SU(3)$ than for $SU(2)$[14]. In full QCD preliminary results are in agreement with expectations for χ and χ' and for the spin content of the proton. However a better sampling of topological sectors is needed. In fact the incapability of the hybrid montecarlo to thermalize topology could affect not only the measurement of quantities directly related to it, like the one we considered, but any other measurement on the lattice. Work is in progress to overcome this difficulty.

I am grateful to my collaborators B. Alles, G. Boyd, M. D'Elia, E. Meggiolaro, H. Panagopoulos, with whom most of the results presented where obtained.

References

[1] G. 't Hooft: *Phys. Rev. Lett.* **37** (1976) 8.

[2] G. Veneziano: *Nucl. Phys.* **B159** (1979) 213; E. Witten: *Nucl. Phys.* **B156** (1979) 269.

[3] S. Weinberg: *Phys. Rev.* **D11** (1975) 3583.

[4] G. t'Hooft: *Nucl. Phys.* **B72** (1974) 461.

[5] R.J. Crewter: *Riv. Nuovo Cimento* **2** (1979) 8.

[6] E. Shuryak: *Comm. Nucl. Part. Phys.* **21** (1994) 235.

[7] S. Narison, G.M. Shore, G. Veneziano: *Nucl. Phys.* **B433** (1995) 209.

[8] C. Christou, A. Di Giacomo, H. Panagopoulos, E. Vicari: *Phys. Rev.* **D53** (1996) 2619.

[9] M. Luscher: *Comm. Math. Phys.* **85** (1982) 39.

[10] A.S. Kronfeld, M.L. Laursen, G. Schierholz, U.J. Wiese, *Nucl. Phys.* **B292** (1987) 330.

[11] M. Campostrini, A. Di Giacomo, H. Panagopoulos: *Phys. Lett.* **B212** (1988) 206.

[12] A. Di Giacomo, E. Vicari: *Phys. Lett.* **B275** (1992) 429.

[13] M. Campostrini, A. Di Giacomo, H. Panagopoulos, E. Vicari: *Nucl. Phys.* **B329** (1990) 683.

[14] B. Allés, M. D'Elia, A. Di Giacomo: hep-lat 9706016, to appear in Phys. Lett. B.

[15] B. Allés, M. D'Elia, A. Di Giacomo: *Nucl. Phys.* **B494** (1997) 281.

[16] B. Allés, G. Boyd, M. D'Elia, A. Di Giacomo, E. Vicari: *Phys. Lett.* **B**.

[17] B. Allés, G. Boyd, M. D'Elia, A. Di Giacomo: In preparation.

[18] EMC collaboration: *Nucl. Phys.* **B328** (1989) 1; SMC collaboration: *Phys. Lett.* **B329** (1994) 399; E143 collaboration: *Phys. Rev. Lett.* **74** (1995) 346.

[19] R. Gupta, J.F. Mandula: *Phys. Rev.* **D50** (1994) 6931.

[20] Altmeyer et al.: *Phys. Rev.* **D49** (1994) R 3087.

Topological Properties of the QCD Vacuum at $T = 0$ and $T \sim T_c$

Philippe de Forcrand [1] , Margarita García Pérez [2], James E. Hetrick [3] , Ion-Olimpiu Stamatescu [4]

[1] SCSC, ETH-Zürich, CH-8092 Zürich, Switzerland
[2] Theory Division, CERN, CH-1211, Geneve 23, Switzerland
[3] Physics Dept., University of the Pacific, Stockton, CA 95211-0197, USA
[4] FESt, Schmeilweg 5, D-69118, Heidelberg
and
Institut für Theoretische Physik, Philosophenweg 16, D-69120, Heidelberg, Germany

Abstract: We study on the lattice the topology of $SU(2)$ and $SU(3)$ Yang-Mills theories and QCD.

1 Introduction

The vacuum of the 4-dimensional $SU(N_c)$ gauge theories has a non-trivial topological structure generated by different coverings of the spatial sphere at ∞ with gauge transformations from the $SU(2)$ subgroups. The tunneling field configurations between these different vacua, the instantons, are self-dual solutions of the classical, euclidean equations of motion of finite euclidean action and integer topological charge: $S_0 = \int d^4x s(x) = 8\pi^2$, $Q_0 = \int d^4x q(x) = \pm 1$. The so called *'t Hooft ansatz*

$$s(x) = F^2(x) = \frac{48}{\rho^4}\left[1 + \sum_{\mu=1}^{4}\left(\frac{x_\mu - x_\mu^0}{\rho}\right)^2\right]^{-4}, \quad q(x) = \frac{Q_0}{8\pi^2}s(x) \tag{1}$$

describes an isolated instanton of size ρ, centered at x^0. Obviously these are scale invariant solutions. Superpositions of N instantons *or* anti-instantons also corresponds to (higher) minima of the action: $S = NS_0$, however a pair instanton - anti-instanton is not a minimum and its total action depends, among other, by the amount of "overlap" $\omega = \rho_I\rho_A/|x_I^0 - x_A^0|^2$.
Effects associated with instantons are:

1. $U_A(1)$ symmetry breaking. This effect is succinctly described by the Witten - Veneziano formula[1] relating the topological susceptibility χ and the η' mass:

$$\chi \equiv \lim_{V\to\infty} \frac{1}{V} \int dx^4 < \mathrm{T}(q(x)q(0)) >= \frac{f_\pi}{2N_f}(m_\eta^2 + m_{\eta'}^2 - 2m_K^2). \tag{2}$$

Here χ has to be obtained in a quenched simulation. Since the instantons are essentially an $SU(2)$ phenomenon, the result should not depend strongly on N_c. The empirical data for the RHS of (2) give $\chi = (180\text{MeV})^4$. Notice that the presence of light fermions inhibits the topological susceptibility (the Dirac operator for massless fermions has zero modes for $Q \neq 0$).

2. Chiral symmetry breaking. By inducing zero modes of the Dirac operator instantons may control the chiral symmetry breaking, due to the Banks-Casher relation $< \psi\bar{\psi} >= \frac{1}{\pi} < \rho(0) >$ in which $\rho(\lambda)$ is the spectral density of the Dirac operator. Spontaneous symmetry breaking then results from a nonzero density of eigenvalues at $\lambda = 0$.

3. Dynamical effects. Instantons are believed to influence the dynamics of the intermediate distances (at the scale of $\frac{1}{2}$ fm, say) and therefore the hadron properties. Their infrared effects are, however, less certain, in particular there is an open question whether they lead to confinement.

The $U_A(1)$ symmetry breaking involves only the integrated correlation χ. This can be tested directly in numerical simulations. The effects in *2.* and *3.* depend on details of the instanton population. Predictions can be made with help of models. Numerical simulations are then asked to test the ingredients on which these models are based. The features relevant for our discussion are:

(i) The instantons can be in a gas, liquid, or crystalline phase depending on their density and overlap. The diluteness of the ensemble is expressed by the "packing fraction" $f = \pi^2/2 < \rho^4 > N/V$.

(ii) The size distribution controls the diluteness and the I-R properties. In the dilute gas approximation the estimation of the fluctuation determinant [2] permits to predict a power law for small sizes. In case of a dense population of instantons, by just integrating in a finite volume and assuming convergence of the thermodynamic limit one can derive some rather general features of the size distribution [3]. This leads to a different power law behavior for the small size distribution. We can write:

$$P(\rho) \sim \rho^p, \quad p_{dilute} = -5 + b, \quad p_{dense} = -1 + 4(b-4)/b, \quad b = 11N_c/3 \tag{3}$$

(iii) Denoting by $N_I(N_A)$ the number of instantons (anti-instantons) in a configuration, and with $N = N_I + N_A$, we write:

$$c \equiv \left(\langle N^2\rangle - \langle N\rangle^2\right)/\langle N\rangle. \tag{4}$$

The quantity c tests the Poisson character of the N - distribution. For a poissonian distribution $c = 1$, while low energy sum rules [4] suggest $c = \frac{4}{b}$.

(iv) In the "standard" model [5] the QCD instanton ensemble below the critical temperature is an interacting instanton liquid with density about N/V =1 fm^{-4}. The

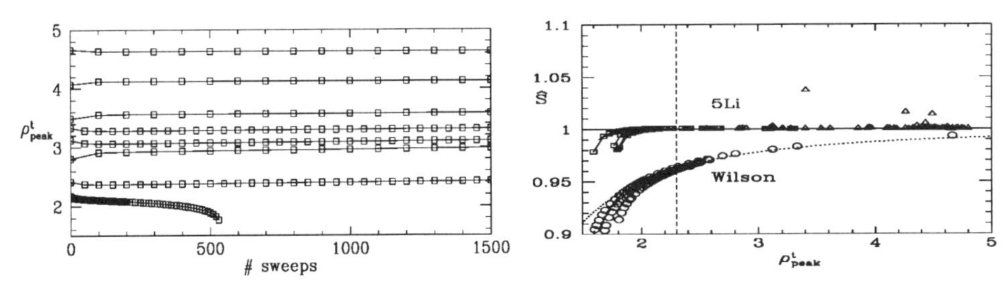

Figure 1: Evolution of the size of instantons under improved cooling (left) and the (improved: 5Li or Wilson) action of instantons of various sizes (right). For the determination of ρ_{peak} see [6].

instanton size distribution has an infrared cut-off and an average instanton size of $\sim 1/3$ fm. The dynamics of the chiral phase transition is driven by a rearrangement of the instanton ensemble at T_c – I-A molecules are formed, with a tendency to orientation along the euclidean time axis (this effect is due to light fermions).

2 Method

Topology analyses for lattice regularized fields have the usual two sources of systematic errors: cut-off effects and finite size effects. Short range topological fluctuations, at the level of the cut-off, are unphysical. They typically have an action lower than S_0 (dislocations) and can spoil the determination of the susceptibility and distort the properties of the instanton ensemble. The finite size of the lattice, on the other hand, distorts the properties of the ensemble at sizes beyond $l/2$ (l: the physical lattice size). To cope with this latter problem we must use a large enough lattice compared with the range of distances we are interested in and check the dependence on the boundary conditions. The short distance problem is more involved. The method we have adopted for obtaining sensible densities is to cool the Monte Carlo configurations to smooth out the UV-noise, and in particular to eliminate the dislocations. However, the usual plaquette action has, strictly speaking, no instanton solutions since for them the action is always smaller than S_0 and becomes lower with decreasing size. Under Wilson cooling the instantons therefore shrink and after some time disappear. We have used instead an improved action which has true, scale invariant instanton solutions for sizes $\rho > \rho_0$ with a threshold $\rho_0 \simeq 2.3a$ – see Fig. 1. This action involves 5 Wilson loops and a similar construction yields an improved charge operator which is practically an integer already on rather rough configurations [6].

Cooling proceeds by local minimization of the action and leads to the global minimum, which is determined by the topological sector Q. From the point of view of our analysis it has two effects. On the one hand it smoothes out the short distance fluctuations, including dislocations and makes apparent the physical distance structure. On the other hand it annihilates I - A pairs depending on the overlap between the opposite charges.

Since the (improved) charge operator Q stabilizes to an integer value within 1 - 2% already after very few cooling sweeps (5 - 10) the effects of type *1.* above are well defined in our method without further tuning, monitoring, etc. The susceptibility and charge distributions show good scaling behavior which implies that they reveal a physical structure. The ensemble properties, on the other hand will depend, at least in part, on the degree of cooling. This affects the discussion of points *2.* and *3.* above and the corresponding tests for instanton models.

3 Analysis

We have analyzed the $SU(2)$ and $SU(3)$ Yang-Mills theories at $T = 0$. We varied the cut-off and the lattice size such as to have the same physical volume at different discretization scales. For the $SU(2)$ analysis we used twisted boundary conditions [6]. We have studied QCD with two flavors of staggered fermions of mass $ma = 0.008$ at 3 temperatures around T_c on configurations kindly provided to us by the MILC collaboration. The lattices and the main results are given in Table 3.

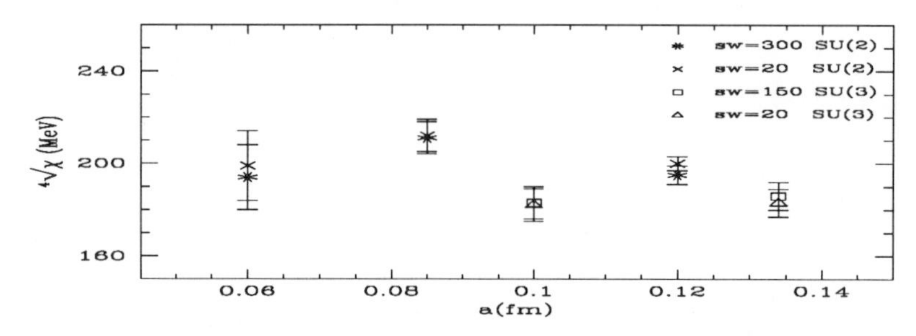

Figure 2: Topological susceptibility for quenched $SU(2)$ and $SU(3)$.

Figure 3: Isosurfaces for the charge density of a typical configuration of lattice (6), Table 1., after 20 sweeps of improved cooling; the two circles indicate a "molecule" (left) and a "caloron" (right).

TOPOLOGICAL SUSCEPTIBILITY

As can be seen from the Fig. 2 the $T = 0$ topological susceptibility is practically independent of cooling and scales correctly. $\chi^{1/4}$ shows a value of $195 - 200$ MeV

Table 1: Lattices and cooling results

Lattice	Sw.	$\langle N \rangle$	c	$\langle N \rangle / fm^4$
(1) $SU(2), \beta = 2.4$	20	10.4	0.4	2.4
$a = 0.12 fm$	50	5.0	0.5	1.2
$12^4, 217 conf.$	150	2.45	0.7	0.6
	300	2.12	1.0	0.5
(2) $SU(2), \beta = 2.6$	20	63.0	0.6	14.
$a = 0.06 fm$	50	15.6	0.65	3.6
$24^4, 115 conf.$	150	6.5	0.65	1.5
	300	3.8	0.5	0.9
(3) $SU(3), \beta = 5.85$	20	12.10	0.36	1.81
$a = 0.134 fm$	50	5.93	0.5	0.89
$12^4, 120 conf.$	150	2.52	1.0	0.38
(4) $SU(3), \beta = 6.0$	20	21.63	0.5	3.30
$a = 0.1 fm$	50	9.38	0.5	1.43
$16^4, 120 conf.$	100	5.33	0.5	0.81
(5) $QCD, \beta = 5.65$	20	50.68	0.41	1.75
$a = 0.115 fm$	50	14.90	1.	0.51
$24^3 12, 31 conf.$	100	7.77	0.8	0.27
	150	5.68	1.	0.20
(6) $QCD, \beta = 5.725$	20	42.95	0.57	2.30
$a = 0.103 fm$	50	8.86	0.82	0.47
$24^3 12, 43 conf.$	100	2.81	0.98	0.15
	150	1.84	1.03	0.10
(7) $QCD, \beta = 5.85$	20	35.0	0.4	3.95
$a = 0.0855 fm$	50	5.55	0.84	0.63
$24^3 12, 20 conf.$	100	0.50	0.9	0.06

for $SU(2)$ and $180 - 185$ MeV for $SU(3)$ with a typical error of about 5%, in very good agreement with experiment following (2) – see also Table 3. This coincides with the results obtained using improved operators [8], underrelaxed Wilson cooling [9], constrained smearing [10] and with the fermionic overlap formalism [11] (where the charge density itself has been tested). The results of [12] are about 10% higher, still compatible with (2) inside 2 standard deviations (see [13] for a possible explanation). We can conclude therefore that at present the $T = 0$ topological susceptibility is under control and in good agreement with the Witten-Venetiano formula.

The finite temperature deconfining transition of the pure Yang-Mills theory has been shown to produce a strong drop in the susceptibility [8, 10]. We have analyzed here the finite temperature QCD. Below T_c we found as expected a reduction in the susceptibility (by a factor 3), indicative of the dynamical fermion effects. Above

Table 2: Average charge and topological susceptibility

Lattice	(1)	(2)	(3)	(4)	(5)	(6)	(7)
$\langle \lvert Q \rvert \rangle$	1.6(1)	1.5(2)	1.8(2)	1.7(2)	2.0(4)	0.9(1)	0
$\chi^{1/4}$ MeV	195(4)	194(14)	184(6)	182(7)	134(10)	102(5)	0

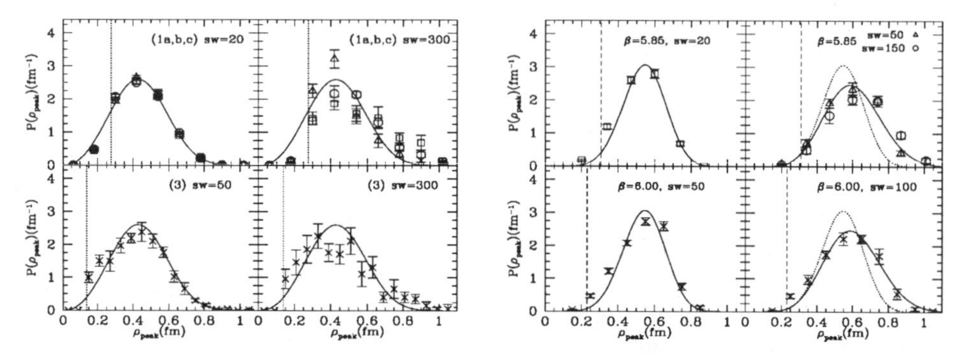

Figure 4: Size distributions for quenched $SU(2)$ (left) and $SU(3)$ (right).

T_c (which is at $\sim 160MeV$ for the MILC simulation) we found a pronounced drop in χ, similar to the quenched case. As already observed before for other staggered fermion simulations [14] also the MILC configurations show long range correlations in Q and explore few topological sectors. The results quoted in Table 3 are obtained after symmetrizing per hand the charge distribution (since this is a symmetry of the action), the errors given are only the statistical ones. We have noticed such metastabilities also in the pure $SU(3)$ simulation, there, however, one can afford to wait until the configurations decorrelate. [1]

THE INSTANTON ENSEMBLE

To observe not merely the global charge of a configuration but its structure describable in terms of instantons and anti-instantons a certain degree of smoothing must be attained (see Fig. 3 for an illustration). Using our improved cooling we achieve this smoothness with a number of cooling sweeps between 20 and 50, depending on the cut-off a. We did not try to tune a rescaling law for the amount of cooling [9], since for many of the results this was not necessary due to inherent stability properties of our cooling.

As can be seen from the tables, when the configurations are just smooth enough for us to see instantons and anti-instantons we typically have a ratio $N/|Q| \simeq 6-10$. This ratio is reduced to $N/|Q| \simeq 1-2$ in the further cooling by pair annihilation, leaving only the (anti-)instantons corresponding to the topological sector. The first remark is that the size distribution shows only little variation during this process – see Fig. 4 for the quenched simulation. This implies that all the (anti-)instantons which we observe from the start obey the same distribution. The average sizes are in the region of 0.4fm for $SU(2)$ and 0.6fm for $SU(3)$ and QCD. The small size part of the distribution departs from the dilute gas power law and shows tendency toward agreement with [3, 15] ($p \simeq 0.8$ for $SU(2)$ and 1.5 for $SU(3)$ – compare eq. (3)).

The density of (anti-)instantons is high to start with. Because of this high density and of the rather large sizes, the packing fraction here is also high. The overall distribution is here non-poissonian and in agreement with the low energy sum rules, see eq. (4) and Table 3 (while it approaches the poissonian distribution for instantons *or* anti-instantons after long cooling). This suggests a dense ensemble with a spatially random distribution biased toward small $|Q|$. Similar results at $T = 0$ have been reported in [9].

To the extent we could follow the structure described by this ensemble back to shorter cooling, it does not change in character but becomes more and more obscured by the accumulation of short range ripples and dislocations. This corroborates with the stability of distributions under further cooling (Fig. 4) and their correct scaling behavior to suggest the physical relevance of this ensemble. Since the short cooling needed before we can observe the instantons and anti-instantons might have also annihilated some pairs in the physically relevant size region, the figures we obtain for the density of (anti-)instantons are lower bounds.

For QCD at high T molecule formation with preferred orientation should be revealed by an anisotropy in the density - density correlations along space and along (euclidean) time directions. The anisotropy we observe at very short cooling is compatible with zero at the level of two standard deviations and very low in relative value (compared to the correlation itself) [7]. This signal is too weak to support the scenario of [5], however the quark mass might not be yet small enough to promote a strong effect. See also Fig. 3.

We conclude that we observe a dense ensemble randomly distributed in space, but with a bias toward small $|Q|$. This ensemble is revealed after a short cooling which essentially has filtered out dislocations and strongly overlapping IA pairs. We remark as a question of principle that any separation

$$q(x) = \text{instantons} + \text{trivial fluctuations}$$

is inherently ambiguous in the case of such strongly overlapping pairs. It is therefore not clear to us whether this latter structure is reasonably described as instantons and anti-instantons or a description in terms of other kind of excitations is more suitable for it.

ACKNOWLEDGMENTS: We wish to thank D. Diakonov, A. Di Giacomo, T. De Grand, A. Hasenfratz, M. Ilgenfritz, M. Mueller-Preussker, G. Münster, R. Narayanan and T. Kovács for discussions. MGP and IOS thankfully acknowledge partial support from the DFG.

References

[1] E. Witten, Nucl. Phys. B156 (1979) 269; G. Veneziano, Nucl. Phys. B159 (1979) 213.

[2] G. 't Hooft, Phys. Rev. Lett.37 (1976) 8; Phys. Rev. D14 (1976) 3432.

[3] G. Münster, Z. Phys. C, Particles and Fields 12 (1982) 43.

[4] E.-M. Ilgenfritz and M. Mueller-Preussker, Phys. Lett. B99 (1981) 128; D. Diakonov and V. Petrov, Nucl. Phys. B245 (1984) 259.

[5] for a review see T. Schäfer and E.V. Shuryak, hep-ph/9610451.

[6] Ph. de Forcrand, M. García Pérez and I.-O. Stamatescu , Nucl. Phys. B (Proc. Suppl.) 47 (1996) 777; Nucl. Phys. B499 (1997) 409.

[7] Ph. de Forcrand, M. García Pérez, J.E. Hetrick and I.-O. Stamatescu, hep-lat/9710001.

[8] B. Allés, M. D'Elia and A. Di Giacomo, Nucl. Phys. B494 (1997) 281.

[9] D. A. Smith and M. J. Teper, hep-lat/9801008.

[10] M. Feuerstein, E.-M. Ilgenfritz, M. Müller-Preussker and S. Thurner, hep-lat/9611024.

[11] R. Narayanan and P. Vranas, Nucl. Phys. B506 (1997) 373.

[12] T. De Grand, A. Hasenfratz and T. Kovács, hep-lat/9705009, hep-lat/9711032.

[13] R. Narayanan and R. L. Singleton Jr., hep-lat/9709014.

[14] B. Allés, G. Boyd, M. D'Elia, A. Di Giacomo and E. Vicari, hep-lat/9607049.

[15] Ph. De Forcrand, M. García Pérez, J. Hetrick, G. Münster and I.-O. Stamatescu, in preparation.

Properties of Abelian Monopoles in $SU(2)$ Lattice Gluodynamics

B.L.G. Bakker [1], M.N. Chernodub [2], F.V. Gubarev [2], M.I. Polikarpov [2] and A.I. Veselov [2]

[1] Department of Physics and Astronomy, Vrije Universiteit, De Boelelaan 1081, NL-1081 HV Amsterdam, The Netherlands
[2] ITEP, B.Cheremushkinskaya 25, Moscow, 117259, Russia

Abstract: We discuss some properties of abelian monopoles in the Maximal Abelian projection of the $SU(2)$ lattice gluodynamics. We show that in the maximal abelian projection abelian monopoles carry fluctuating electric charge and that the monopole currents are correlated with the magnetic and the electric parts of the $SU(2)$ action density.

1 Introduction

Abelian monopoles play a key role in the dual superconductor mechanism of confinement [1] in non-abelian gauge theories. Abelian monopoles appear after the so called abelian projection [2]. Condensation of abelian monopoles gives rise to the formation of an electric flux tube between the test quark and antiquark. Due to a non-zero string tension the quark and the antiquark are confined by a linear potential. There are many numerical facts [3] which show that the abelian monopoles in the Maximal Abelian (MaA) projection are responsible for the confinement. The monopole condensation in the confinement phase of gluodynamics has been established by the investigation of various monopole creation operators [4] in the MaA projection [5]. The $SU(2)$ string tension is well described by the contribution of the abelian monopole currents [6]; these currents satisfy the London equation for a superconductor [7].

Below we discuss several recently found properties of the abelian monopole currents. In Section 2 we show that in the vacuum of the $SU(2)$ lattice gluodynamics the abelian monopoles currents are correlated with the electric currents. In Section 3 we show that the abelian monopoles are locally correlated with electric and magnetic parts of the $SU(2)$ action density. All numerical calculations are performed in the MaA projection.

2 Abelian Monopoles Carry Electric Charge

Consider a (anti-) self–dual configuration of the $SU(2)$ gauge field:

$$F_{\mu\nu}(A) = \pm\frac{1}{2}\varepsilon_{\mu\nu\alpha\beta}F_{\alpha\beta}(A) \equiv \pm\tilde{F}_{\mu\nu}\,, \tag{1}$$

where $F_{\mu\nu}(A) = \partial_{[\mu,}A_{\nu]} + i[A_\mu, A_\nu]$. In the MaA projection the commutator term $\mathrm{Tr}(\sigma^3[A_\mu, A_\nu])$ of the field strength tensor $F^3_{\mu\nu}$ is suppressed, since the MaA projection is defined [5] by the minimization of the functional $R[A] = \int d^4x[(A^1_\mu)^2 + (A^2_\mu)^2]$ over the gauge transformations. Thus, in the said projection, the fields $A_\mu(x)$ are as close to abelian (diagonal) fields as possible. Therefore, in the MaA projection eq.(1) yields [8]: $f_{\mu\nu}(A) = \partial_\mu A^3_\nu - \partial_\nu A^3_\nu \approx \pm\tilde{f}_{\mu\nu}(A)$. Thus, the abelian monopole currents must be accompanied by the electric currents: $J^e_\mu = \partial_\nu f_{\mu\nu}(A) \approx \pm\partial_\nu \tilde{f}_{\mu\nu}(A) = \pm J^m_\mu$. Therefore, in the MaA projection the abelian monopoles are dyons for (anti) self-dual $SU(2)$ field configurations [8]. Below we show that in the real (not cooled) vacuum of lattice gluodynamics the abelian monopole currents are correlated with the electric currents [9].

In order to study the relation of electric and magnetic currents, we have to calculate connected correlators of these currents. The simplest correlator $\ll J^m_\mu J^e_\mu \gg \equiv \; < J^m_\mu J^e_\mu > - < J^m_\mu >< J^e_\mu >$ is zero, since $< J^m_\mu J^e_\mu >= 0$ due to the opposite parities of the operators J^m and J^e, and $< J^{m,e}_\mu >= 0$ due to the Lorentz invariance. The simplest non–trivial (normalized) correlator is

$$\bar{G} = \frac{1}{\rho^e\rho^m} < J^m_\mu(y) J^e_\mu(y) q(y) >\,, \tag{2}$$

where $q(x)$ is the sign of the topological charge density at the point x and $\rho_{m,e} = \sum_l < |J^{m,e}_l| > /(4V)$ are the densities of the magnetic and the electric charges, V is the lattice volume (total number of sites).

We perform a numerical calculation of the correlator (2) in the $SU(2)$ lattice gauge theory on the 8^4 lattice with periodic boundary conditions. We use 100 statistically independent gauge field configurations for each value of β.

The dependence of the correlator $\bar{G}$ on β is shown in Fig. 1(a). This correlator is positive for all values of β. Therefore, the abelian monopoles in the MaA projection carry an electric charge, too. According to definition (2), the sign of the electric charge of the monopole coincides with the product of the magnetic charge and the topological charge. Thus, in the gluodynamic vacuum the abelian monopoles become abelian dyons due to a non-trivial topological structure of the vacuum gauge fields.

3 Abelian Monopole Currents are Correlated with $SU(2)$ Action Density

Abelian monopoles appear as singularities in the gauge transformations [2, 3]. On the other hand, the monopole currents reproduce the $SU(2)$ string tension [6]. Thus,

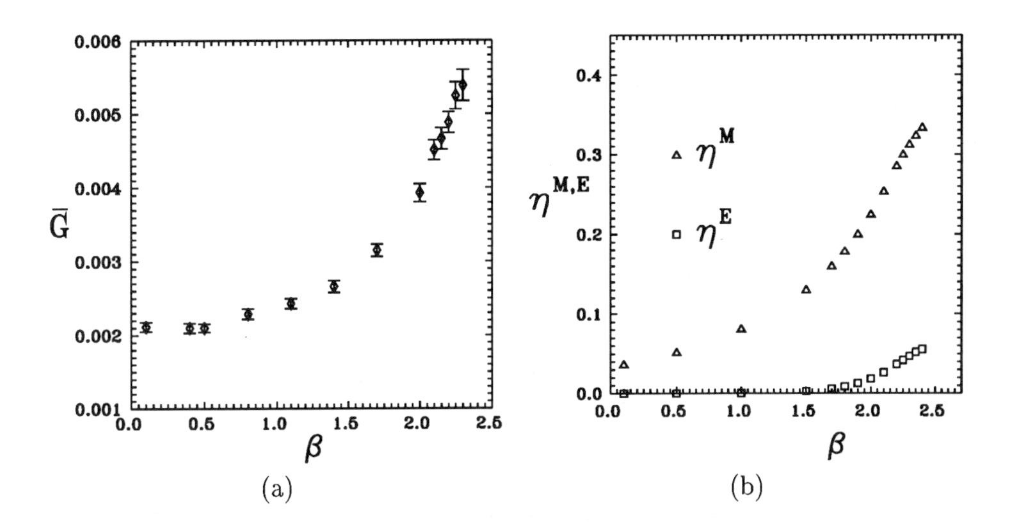

Figure 1: (a) The dependence of the correlator $\bar{G}$ on β; (b) The relative excess of the magnetic (circles, from Refs. [11]) and the electric (boxes) action density near the monopole current. The data are extrapolated to the infinite lattice size.

monopoles are likely to be related to some physical objects. A physical object is something which carries action. Below we study the local correlations of the abelian monopoles with the density of the magnetic and the electric parts of the $SU(2)$ action (the global correlation was found in Ref. [10]). We show that the monopoles are physical objects but it does not mean that these have to propagate in the Minkowski space; a chain of instantons can produce a similar effect: an enhancement of the action density along a line in Euclidean space. The simplest quantities which can show this correlation are the relative excess of the magnetic and the electric action densities $\eta^{M,E} = (S^{M,E}_m - S)/S$ in the region near the monopole current. Here S is the expectation value of the lattice plaquette action, $S_P =< (1 - \frac{1}{2} Tr\, U_P) >$. The quantities $S^{M,E}_m$ are, respectively, the magnetic and the electric parts of the $SU(2)$ action density, which are calculated on plaquettes closest to the monopole current.

In the continuum notation, the quantities $S^{M,E}_m$ have the following form:

$$S^M_m = \frac{1}{2} < \mathrm{Tr}(n_\mu(x)\, \tilde{F}_{\mu\nu}(x))^2 > , \quad S^E_m = \frac{1}{2} < \mathrm{Tr}(n_\mu(x)\, F_{\mu\nu}(x))^2 > , \tag{3}$$

$n_\mu(x)$ is the unit vector in the direction of the current: $n_\mu(x) = j_\mu(x)/|j_\mu(x)|$, if $j_\mu(x) \neq 0$, and $n_\mu(x) = 0$ if $j_\mu(x) = 0$. It is easy to see that for a static monopole ($j_0 \neq 0$; $j_i = 0, i = 1, 2, 3$) S^M_m (resp., S^E_m) corresponds to the chromomagnetic action density $(B^a_i)^2$ (resp., chromoelectric action density $(E^a_i)^2$) at the monopole current.

We calculate the quantities η^M and η^E on symmetric lattices L^4 of different lattice size $L = 8, 10, 12, 16, 20, 24, 30$ with periodic boundary conditions. In Fig. 1(b) we show the quantities $\eta^{M,E}$ extrapolated to the infinite lattice size, $(L \to \infty)$ *vs.* β. The monopole currents are calculated in the MaA projection. In Fig. 1(b) the statistical errors are smaller than the size of the symbols. It is clearly seen that the abelian monopoles are correlated with both the magnetic and the electric parts of the $SU(2)$ action density. Note that the correlation of the monopole charge with the magnetic action density is larger than the correlation with the electric part of the $SU(2)$ action.

Conclusion and Acknowledgments

Our results show that the abelian monopoles in the MaA projection of the $SU(2)$ gluodynamics *i)* have a fluctuating electric charge; *ii)* carry the $SU(2)$ action.

This work was supported by the grants INTAS-RFBR-95-0681, RFBR-96-1596740 and RFBR-96-02-17230a.

References

[1] S. Mandelstam, *Phys. Rep.*, 23C (1976) 245; G. 't Hooft, "High Energy Physics", A. Zichichi, Editrice Compositori, Bolognia, 1976.

[2] G. 't Hooft, *Nucl. Phys.*B190 [FS3], 455 (1981).

[3] T. Suzuki, *Nucl. Phys.*B *(Proc. Suppl.)* **30** (1993) 176; M. I. Polikarpov, *Nucl. Phys.*B *(Proc. Suppl.)* **53** (1997) 134; M.N. Chernodub and M.I. Polikarpov, *preprint ITEP-TH-55/97*, hep-th/9710205.

[4] L. Del Debbio *et al.*, *Phys. Lett.*B355, 255 (1995); N. Nakamura *et al.*, *Nucl. Phys. Proc. Suppl.* **53**, 512 (1997); M.N. Chernodub, M.I. Polikarpov and A.I. Veselov, *Phys. Lett.*B399, 267 (1997).

[5] A.S. Kronfeld *et al.*, *Phys. Lett.*198B, 516 (1987); A.S. Kronfeld, G. Schierholz and U.J. Wiese, *Nucl. Phys.*B293, 461 (1987).

[6] H. Shiba and T. Suzuki, *Phys. Lett.*B333, 461 (1994); J.D. Stack, S.D. Neiman and R.J. Wensley, *Phys. Rev.* D **50**, 3399 (1994); G.S. Bali *et al.*, *Phys. Rev.* D **54**, 2863 (1996).

[7] V. Singh, D. Browne and R. Haymaker, *Phys. Lett.*B306 115 (1993); C. Schlichter, G.S. Bali and K. Schilling, hep-lat/9709114.

[8] V.Bornyakov, G.Schierholz, *Phys. Lett.*B **384** (1996) 190;

[9] M.N. Chernodub, F.V. Gubarev and M.I. Polikarpov, *preprint ITEP-TH-44/97*, hep-lat/9709039; *preprint ITEP-TH-70/97*, hep-lat/9801010.

[10] H. Shiba and T. Suzuki, *Phys. Lett.*B351, 519 (1995).

[11] B.L.G. Bakker, M.N. Chernodub and M.I. Polikarpov, *Phys. Rev. Lett.*80 (1998) 30; *preprint ITEP-TH-43-97*, hep-lat/9709038.

Center dominance, Casimir scaling, and confinement in lattice gauge theory *

L. Del Debbio [1], M. Faber [2], J. Greensite [3], Š. Olejník [4]

[1] Dept. of Physics and Astronomy, University of Southampton, Southampton SO17 1BJ, UK
[2] Institut für Kernphysik, Technische Universität Wien, A–1040 Vienna, Austria
[3] The Niels Bohr Institute, DK–2100 Copenhagen Ø, Denmark
[4] Institute of Physics, Slovak Academy of Sciences, SK–842 28 Bratislava, Slovakia

Abstract: We present numerical evidence that supports the theory of quark confinement based on center vortex condensation. We introduce a special gauge ("maximal center gauge") and center projection, suitable for identification of center vortices. Main focus is then put on the connection of vortices in center projection to "confiners" in full, unprojected gauge-field configurations. Topics briefly discussed include: the relation between vortices and monopoles, first results for SU(3), and the problem of Casimir scaling.

1 Introduction

The most popular model of colour confinement in QCD relies on the idea of "dual superconductivity", due to 't Hooft and Mandelstam. A realization of the idea is the abelian-projection theory of 't Hooft [1]: he suggested to fix to an "abelian projection" gauge, reducing the SU(N) gauge symmetry to $U(1)^{N-1}$, and identifying abelian gauge fields (with respect to the residual symmetry) and magnetic monopoles. Abelian electric charges then become confined due to monopole condensation. In 1987, Kronfeld et al. [2] suggested testing 't Hooft's theory in lattice simulations, in a special gauge that makes SU(N) link variables as diagonal as possible. If one computes various physical observables using the diagonal parts of the links only, one observes "abelian" dominance [3]: the expectation values of the physical quantities in the full non-abelian theory (often) coincide with the ones in the

*Invited talk presented by Š. Olejník. His work was supported in part by the Slovak GAS, Grant No. 2/4111/97.

abelian theory obtained by the abelian projection in the maximal abelian gauge. Much evidence has been obtained for the abelian-projection picture and the model of dual superconductivity, but there remain problems to be solved. We have underlined its inability to explain approximate Casimir scaling of the linear potential between higher-representation colour sources at intermediate length scales [4, 5].

Another picture of confinement was quite popular before the advent of dual superconductivity, namely the Z_N vortex condensation theory, proposed, in various forms, by many authors [6]. According to this model, the QCD vacuum is filled with vortices, having the topology of tubes (in 3 Euclidean dimensions) or surfaces (in 4 dimensions) of finite thickness, which carry magnetic flux quantized in terms of elements of the center of the gauge group. Center vortices are assumed to condense in the QCD vacuum. The area-law fall-off of large Wilson loops comes from fluctuations in the number of center vortices linked to the loops.

The Z_N vortex condensation theory apparently suffers from the same "Casimir-scaling disease" as the abelian projection does: there seems no way of accommodating the existence of a linear potential between adjoint sources to the idea of vortices dominating the QCD vacuum. There exists a simple solution to this controversy [7], and we will discuss it at the end of this paper.

With perhaps only one exception [8], the ideas behind the vortex-condensation picture have not in the past been subjected to lattice tests. The aim of our investigation is to study the vortex theory in numerical Monte Carlo simulations, by methods and approaches inspired to some extent by earlier work of many authors in the abelian projection theory.

2 Maximal center gauge, projection and dominance

The maximal abelian gauge underscores the role of the largest abelian subgroup of the gauge group. In much the same way, one can choose a gauge condition in which the gauge group *center* is given prominent importance. In SU(2) lattice gauge theory we proposed [5, 9] to fix to *maximal center gauge* (MCG) by making link variables U as close as possible to its center elements $\pm I$. There are many (in fact infinitely many) ways how to do it; we implemented two simple choices:

1. The *indirect maximal center gauge* (IMCG) [5, 9–11]: We first fix to maximal abelian gauge (MAG) in the usual way, by maximizing the quantity

$$\sum_x \sum_\mu \mathrm{Tr}\,[U_\mu(x)\sigma_3 U_\mu^\dagger(x)\sigma_3], \tag{1}$$

then extract from $U_\mu(x)$ their diagonal parts $A_\mu(x) = \exp[i\theta_\mu(x)\sigma_3]$, and use the remnant U(1) symmetry to bring $A_\mu(x)$ as close as possible to the center elements by maximizing

$$\sum_x \sum_\mu \cos^2\theta_\mu(x). \tag{2}$$

2. The *direct maximal center gauge* (DMCG) [10,12]: To fix this gauge one looks directly for the maximum of

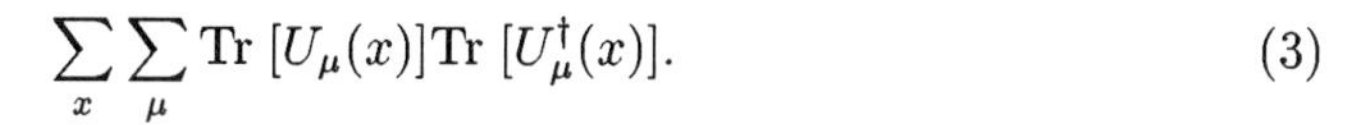

$$\sum_x \sum_\mu \text{Tr } [U_\mu(x)] \text{Tr } [U_\mu^\dagger(x)]. \tag{3}$$

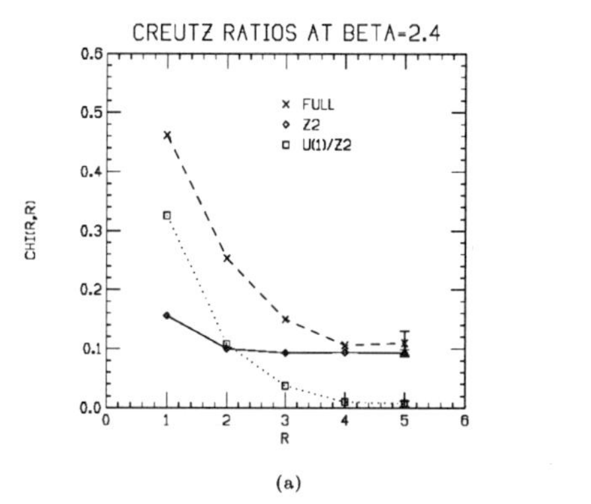

(a)

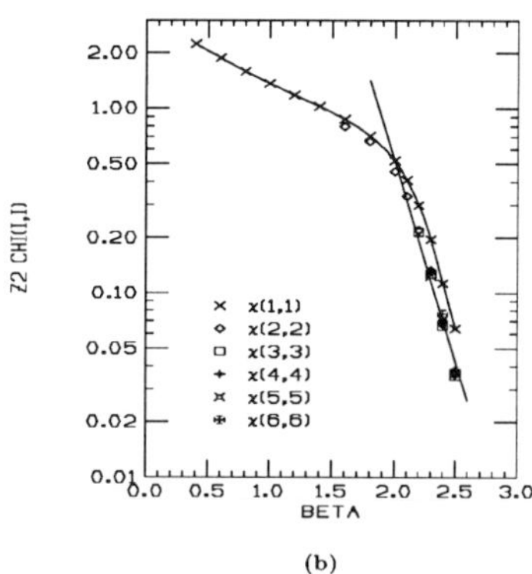

(b)

Figure 1: Creutz ratios: (a) vs. R for full, center-projected, and U(1)/Z_2-projected lattice configurations at $\beta = 2.4$ (IMCG), (b) vs. β from center-projected configurations (DMCG). The straight line is the asymptotic freedom prediction: $\sigma a^2 = (\sigma/\Lambda^2)\,(6\pi^2\beta/11)^{102/121} \exp\,(-6\pi^2\beta/11)$.

In both cases, after gauge fixing one is left with the remnant Z_2 gauge symmetry.[†]

The next step is *center projection*, i.e. replacing full link matrices U (in a particular MCG) by center elements Z, which are defined to be

$$Z \equiv \text{sign }(\cos(\theta))\, I \quad \text{(in IMCG)} \qquad \text{or} \qquad Z \equiv \text{sign }(\text{Tr }(U))\, I \quad \text{(in DMCG)}, \tag{4}$$

and to compute various physical quantities of interest, e.g. Wilson loops and Creutz ratios, using the Z links.

Figure 1a compares Creutz ratios $\chi(R,R)$ at $\beta = 2.4$ computed from full lattice configurations and from center-projected configurations. We clearly see *center dominance*: Creutz ratios computed from Z links agree with full Creutz ratios at large enough distances, the asymptotic values of the string tension almost coincide. On the contrary, Creutz ratios computed from links with the Z variable factored out (dotted line in Figure 1a) show no string tension at all. Another interesting observation is that the Creutz ratios computed from center-projected configurations almost do not depend on R; center projection removes Coulombic contributions.

In Figure 1b we plot Creutz ratios vs. β, extracted from center-projected configurations in DMCG. The straight line is the asymptotic freedom prediction with the value of $\sqrt{\sigma}/\lambda = 58$, which very well agrees with "state-of-the-art" asymptotic string tension computations [13].

[†]Qualitatively, the same physical results are obtained in both gauges (cf. [9,12]).

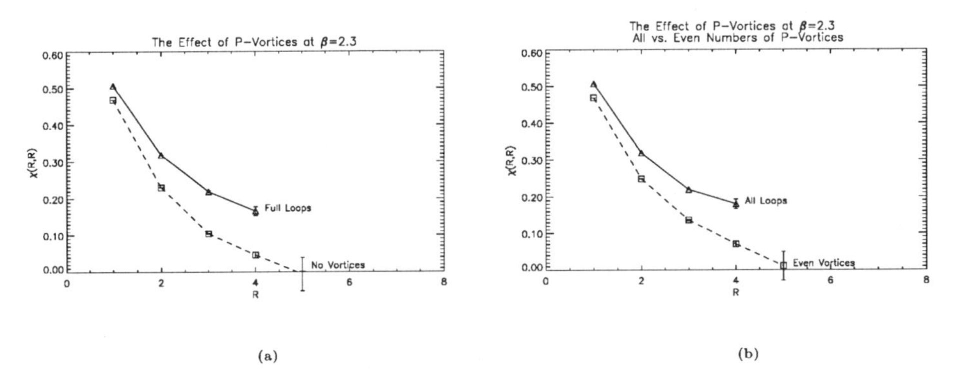

Figure 2: Creutz ratios extracted from loops with (a) no P-vortices, (b) even number of P-vortices piercing the loop, compared with the usual Creutz ratios at $\beta = 2.3$.

Our data show that the Z center variables are crucial parts of the U links in MCG, in particular they carry most of the information on the string tension. This phenomenon of *center dominance* gives rise to a whole series of questions on the role and nature of center vortex configurations in the QCD vacuum. We will list a few of those questions here and sketch our tentative answers.

3 Questions and answers

3.1 Vortices and confinement?

Question 1: Has center dominance anything to do with confinement? What, if any, is the relation of Z_2 vortices seen after center projection to "confiners" in full, unprojected configurations?

To answer the question, we first introduce the notion of a *P(rojection)-vortex*. The excitations of a Z_2 lattice gauge theory with non-zero action are "thin" vortices, having the topology of a surface, one lattice spacing thick. We will call such vortices in center projected Z-link configurations P-vortices. A plaquette is pierced by a P-vortex if, after maximal center gauge fixing and center projection, the corresponding projected plaquette has the value of -1.

However, we want to emphasize that center projection, and abelian projection as well, represents an uncontrollable truncation of full lattice configurations. Therefore we will not base our following arguments on measurements in center projected configurations. Instead, we will use center projection mainly for selecting sub-ensembles of configurations on which physical quantities are evaluated. In particular, we will compute $W_n(C)$, Wilson loops evaluated on such a sub-ensemble of configurations that precisely n P-vortices, in the corresponding center-projected configurations, pierce the minimal area of the loop. Though the data set is selected in center projection, the Wilson loops themselves are evaluated using the full, unprojected link variables.

From the computed vortex-limited Wilson loops $W_n(C)$ one can determine Creutz ratios χ_n. A simple test of a relation of P-vortices to confinement is then the following: if the presence/absence of P-vortices is *irrelevant* for confinement, then we would expect $\chi_0(I,J) \approx \chi(I,J)$ for large loops. The result of the test is shown in Figure 2a. The string tension vanishes if P-vortices are excluded from Wilson loops; it also vanishes if only odd numbers of P-vortices are excluded (Fig. 2b). The presence/absence of P-vortices seems strongly correlated with the presence/absence of "confiners" in unprojected field configurations.

However, the true "confiners" do not necessarily have to be any sort of Z_2 vortices. The natural question is, whether the objects identified using center projection tend to carry Z_2 magnetic flux. A simple argument [9, 10] leads to the expectation that $W_n(C)/W_0(C) \to (-1)^n$ for large loops.

Figure 3 shows our data for W_1/W_0 and W_2/W_0 at $\beta = 2.3$. They are consistent with the expectation and thus indicate that the confining gauge field configurations are center vortices. However, the values of $(-1)^n$ are reached for relatively *large* loop areas. The objects corresponding to P-vortices then appear to be rather "thick" Z_2 vortices. In Section 3.4 will this fact be related to a simple explanation of approximate Casimir scaling.

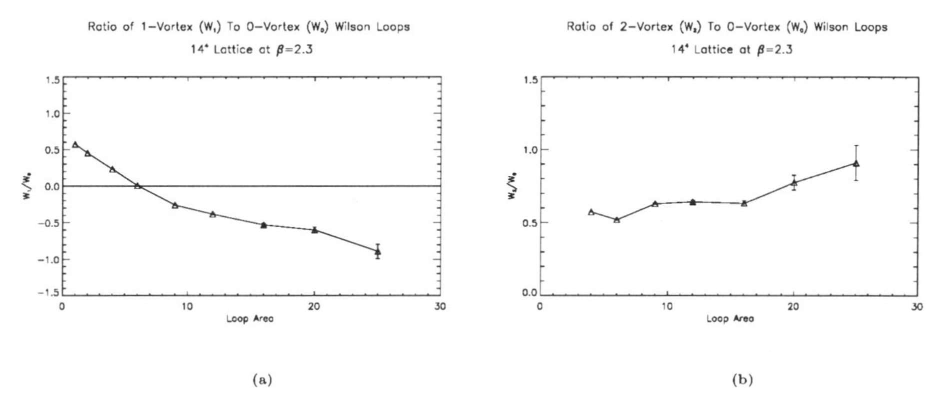

Figure 3: (a) $W_1(C)/W_0(C)$, (b) $W_2(C)/W_0(C)$ vs. loop area at $\beta = 2.3$.

Now we are ready to formulate

Answer 1: We found evidence that center dominance in maximal center gauge is a reflection of the presence of thick center vortices in the unprojected configurations. Those vortices are identified as thin P-vortices in center projection.

3.2 Vortices and/or monopoles?

Question 2: If the vacuum is dominated, at long wavelengths, by Z_2 vortex configurations, then how do we explain the numerical successes of abelian projection in maximal abelian gauge?

We believe that the question is answered in the following way:

(Probable) Answer 2: A center vortex configuration, transformed to maximal abelian gauge and then abelian-projected, will appear as a chain of monopoles alternating with antimonopoles. These monopoles essentially arise because of the projection; they are condensed because the long vortices from which they emerge are condensed.

The support for the answer was given in much detail in [10]. It is clear, however, that this question deserves further study.

3.3 SU(3)?

Question 3: In nature quarks appear in three colours. Do the observed phenomena survive transition from SU(2) to SU(3)?

The maximal center gauge in SU(3) gauge theory is defined as the gauge which brings link variables U as close as possible to elements of its center $Z_3 = \{e^{-2i\pi/3}I,\ I,\ e^{2i\pi/3}I\}$. This can be achieved e.g. by maximizing the quantity

$$\sum_x \sum_\mu \mathrm{Re}\left([\mathrm{Tr}\, U_\mu(x)]^3 \right) . \tag{5}$$

Fixing to the maximal center gauge in SU(3) gauge theory turns out to be much more difficult and CPU-time consuming than in the case of SU(2). Therefore our simulations have until now been restricted to small lattice sizes and to strong coupling.

Our strong coupling results for the SU(2) and SU(3) lattice gauge theory are compared in Figure 4. In SU(2) Monte Carlo data agree with the strong coupling expansion up to almost $\beta = 1.5$. Figure 4b shows center-projected Wilson loops in SU(3) together with results of strong-coupling expansion to leading and next-to-leading order. The agreement extends up to $\beta = 4$.

(Partial) Answer 3: The situation at strong coupling looks much the same in SU(2) and SU(3): in both cases full Wilson loops are well reproduced by those constructed from center elements alone in MCG. Thus, center dominance is seen in SU(3) gauge theory at strong coupling.

3.4 Casimir scaling?

Question 4: Is there any way of accommodating the approximate Casimir scaling of higher-representation potentials to the vortex dominated QCD vacuum?

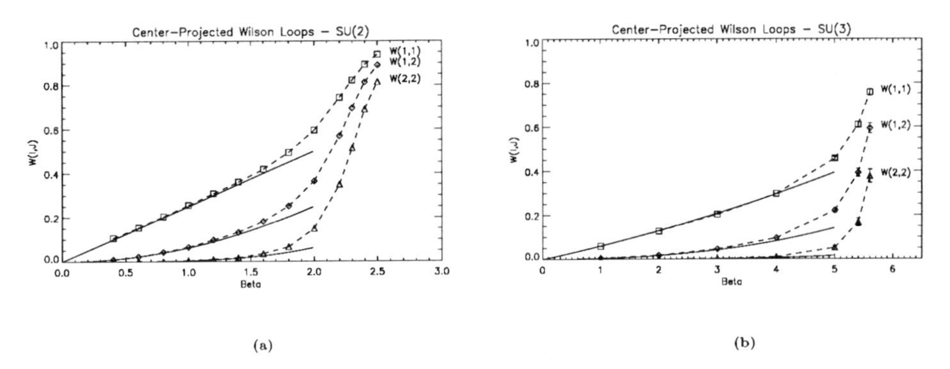

Figure 4: Center-projected Wilson loops vs. the strong-coupling expansion (solid lines) in SU(2) and SU(3) lattice gauge theory. SU(3) values were obtained on an 8^4 lattice.

At the first sight, there is not. The adjoint representation transforms trivially under the gauge group center, large adjoint Wilson loops are unaffected by center vortices. As a result, the adjoint string tension vanishes.

There is however a loophole in the above statements: Adjoint loops are unaffected, unless the core of the vortex happens to overlap with the perimeter of the loop. If, then, the vortex thickness is quite large, on the order or exceeding the typical diameter of low-lying hadrons – and our data seem to indicate the presence of rather "thick" center vortices, – the Wilson loops can be influenced by vortices up to relatively large loop sizes.

A phenomenological model of the "thick" center vortex core has been worked out in a recent paper of three of us [7]. We cannot discuss the model in detail here, we just mention that it leads to potentials between colour sources that show approximate Casimir scaling at small and intermediate distances, and colour screening of integer-representation sources at large distances.

Answer 4: The Casimir scaling of the string tensions of higher-representation Wilson loops is an effect due to the finite (and large) thickness of the center vortex cores (see also [14]).

4 Conclusions

We subjected the picture of quark confinement based on center vortices to simple tests on the lattice. We proposed a method of localizing center vortices in thermalized lattice configurations, and found vortices to be responsible for the asymptotic string tension in SU(2) lattice gauge theory. The same holds also for SU(3) at strong coupling.

Further, we posed the lattice a lot of simple questions on the nature of vortex configurations. The tests indicate that the "confiners" in QCD are center vortices, and monopoles appear along vortices as artifacts of abelian projection. It is tempting to believe that monopole condensation might be just a manifestation of the underlying vortex condensation.

Finally, the vortices observed in our simulations possess a thick core; the thickness of the core is the cause for approximate Casimir scaling of potentials at intermediate distances. This solves a long-standing problem of the center vortex theory.

The "spaghetti vacuum" picture, believed to be dead for more than a decade, returns to the stage, in quite a good health.

References

[1] G. 't Hooft, *Nucl. Phys.* **B190** (1981) 455

[2] A. Kronfeld, M. Laursen, G. Schierholz, U.-J. Wiese, *Phys. Lett.* **B198** (1987) 516

[3] T. Suzuki, I. Yotsuyanagi, *Phys. Rev.* **D42** (1990) 4257

[4] L. Del Debbio, M. Faber, J. Greensite, Š. Olejník, *Phys. Rev.* **D53** (1996) 5891

[5] L. Del Debbio, M. Faber, J. Greensite, Š. Olejník, *Nucl. Phys.* **B** (Proc. Suppl.) **53** (1997) 141.

[6] G. 't Hooft; G. Mack; J. Ambjørn, H. B. Nielsen and P. Olesen; J. M. Cornwall; R. P. Feynman; T. Yoneya; and others. See references in [7, 9, 10]

[7] M. Faber, J. Greensite, Š. Olejník, hep-lat/9710039, *Phys. Rev.* **D** (in press)

[8] E. Tomboulis, *Nucl. Phys.* **B** (Proc. Suppl.) **34** (1994) 192; **B** (Proc. Suppl.) **30** (1993) 549; *Phys. Lett.* **B303** (1993) 103

[9] L. Del Debbio, M. Faber, J. Greensite, Š. Olejník, *Phys. Rev.* **D55** (1997) 2298

[10] L. Del Debbio, M. Faber, J. Greensite, Š. Olejník, hep-lat/9708023

[11] L. Del Debbio, M. Faber, J. Greensite, Š. Olejník, hep-lat/9709032

[12] L. Del Debbio, M. Faber, J. Greensite, Š. Olejník, in preparation

[13] G. Bali, C. Schlichter, K. Schilling, *Phys. Rev.* **D51** (1995) 5165

[14] J. M. Cornwall, in Proc. of the *Workshop on Non-Perturbative Quantum Chromodynamics*, edited by K. A. Milton, M. A. Samuel (Birkhäuser, Boston, 1983) 119; also hep-th/9712248

Spectrum of the gauge Ising model in three dimensions

M. Caselle [1] , M. Hasenbusch [2] and P. Provero [1]

[1] Dip. di Fisica Teorica, Università di Torino
Via P. Giuria 1, 10125 Torino, Italy
[2] Institut für Physik, Humboldt-Universität,
Invalidenstr. 110, D-10115 Berlin

Abstract: We present a high precision Monte Carlo study of the spectrum of the Z_2 gauge theory in three dimensions in the confining phase. Using state of the art Monte Carlo techniques we are able to accurately determine up to three masses in a single channel. We compare our results with the SU(2) spectrum and with the prediction of the Isgur-Paton model. Our data strongly support the conjecture that the glueball spectrum is described by some type of flux tube model. We also compare the spectrum with some recent results for the correlation length in the 3d spin Ising model.

1 Introduction

The infrared regime of Lattice Gauge Theories (LGT) in the confining phase displays a large degree of universality. The main evidences in favour of this universality are given by the behaviour of the Wilson loop and of the adimensional ratio T_c^2/σ (where T_c denotes the deconfinement temperature) which are roughly independent from the choice of the gauge group, and show a rather simple dependence on the number of space-time dimensions. Both these behaviours are commonly understood as consequences of the fact that the relevant degrees of freedom in the confining regime are string-like excitations. The phenomenological models which try to keep into account this string-like picture are usually known as "flux-tube" models and turn out to give a very good description of the Wilson loop behaviour (see for instance [1] and references therein). The reason of this success is that in the interquark potential we have a natural scale, the string tension, which allows to define in a rather precise way a large distance ("infrared") regime in which the adiabatic approximation for the string-like excitations can be trusted. This regime can be reached by considering interquark distances large in units of the string tension.

Besides the Wilson loop and the deconfinement temperature, another important set of physical observables in LGT is represented by the glueball spectrum. In this case it is less obvious that a string like description could be used to understand the data. However, a string-inspired model exists also for the glueball spectrum: the Isgur Paton model [1] (IP in the following).

For this reason it would be very interesting to test if the same universality (which, as mentioned above, should manifest itself as a substantial independence from the choice of the gauge group) displayed by the Wilson loops also holds for the glueball spectrum. In this case we do not have the equivalent of the interquark distance, *i.e.* a parameter which can be adjusted to select the infrared region: the role of large Wilson loops is played by the higher states of the spectrum which, being localized in larger space regions, are expected to show more clearly a string-like behavior. A major problem in this respect is the lack of precise and reliable data for these higher states. An obvious proposal to overcome this problem is to begin with the (2+1) dimensional case, for which some relevant simplifications occur in the spectrum and a much higher precision can be achieved in the Monte Carlo simulations.

Following this suggestion we have studied in [2] the glueball spectrum in the case of the (2+1) Ising gauge model obtaining a high precision estimate of the first 11 states of the spectrum. These can be compared with some results for the (2+1) dimensional SU(2) model obtained with Monte Carlo simulations [3] and with variational techniques [4]. The comparison between the SU(2) and Ising spectra shows that, not only the pattern of the states is the same in the two models, but also the values of the masses (except for the lowest state) are in remarkable agreement. This is a strong evidence in favour of the above mentioned universality, and suggests that the higher states of the glueball spectrum of any LGT, (as it happens for the behaviour of large enough Wilson loops) can be predicted by some relatively simple flux-tube inspired model. We shall give in the next section few general information on the model. In Sect. 3 we give some information on the algorithm that we used and describe how the glueball spectrum was determined (for further details we refer to [2]). Sect 4. is devoted to a discussion of the glueball spectrum and to a comparison with the SU(2) results and with the IP predictions. Finally in the last section we shall make some concluding remark on the duality transformation of the glueball spectrum.

2 The Ising gauge model

The Ising gauge model is defined by the action

$$S_{gauge} = -\beta \sum_{n,\mu<\nu} g_{n;\mu\nu} \tag{1}$$

where $g_{n;\mu\nu}$ are the plaquette variables, defined in terms of the link fields $g_{n;\mu} \in -1, 1$ as:

$$g_{n;\mu\nu} = g_{n;\mu}\, g_{n+\mu;\nu}\, g_{n+\nu;\mu}\, g_{n;\nu} \quad . \tag{2}$$

where $n \equiv (\vec{x}, t)$ denotes the space-time position of the link and μ its direction. For the Ising model, as for the SU(2) model, we cannot define a charge conjugation operator. The glueball states are thus labelled only by their angular momentum J and by their parity eigenvalue $P = \pm$. The standard notation is J^P. An important simplification due to the fact that we are working in (2+1) dimensions is that in this case all the states with angular momentum different from zero are degenerate in parity. Namely J^+ and J^- (with $J \neq 0$) must have the same mass. This result holds in the continuum limit. The lattice discretization breaks this degeneration, since in this case the symmetry group is only the D_4 (dihedral) group. In particular it can be shown that the degeneration still holds on the lattice for all the odd J states, and is lifted for all the even J states (see [2] for details). An important test of the whole analysis is to see if the degeneration of the even part of the spectrum is recovered in the continuum limit. There are two other important consequences of the fact the symmetry group is reduced to D_4. The first one is that only operators with angular momentum J $(mod(4))$ can be observed and the second is that all the odd J states are grouped together in the same irreducible representation of D_4. This means that we cannot distinguish among them on the basis of the lattice symmetries. Hence in the following we shall denote the states belonging to this channel as $J = 1/3$ states.

3 The simulation

The simulations were performed with a local demon-algorithm in Multi-spin-coding technique. For the relatively small correlation length ($\xi < 5$) that we studied this algorithm should be faster than the cluster-algorithm.

A very important technical aspect of our study is the accurate determination of the mass-spectrum. Therefore we will give detailed discussion in the following. correlation functions with the separation in time.

$$\begin{aligned} G(t) &= \langle A(0)B(t)\rangle = \frac{\langle 0|AT^tB|0\rangle}{\langle 0|T^t|0\rangle} \\ &\sim \frac{1}{\lambda_0^t}\sum_i \langle 0|A|i\rangle\langle i|T^t|j\rangle\langle j|B|0\rangle = \sum_i c_i \left(\frac{\lambda_i}{\lambda_0}\right)^t = \sum_i c_i \exp(-m_i t) \quad ,(3) \end{aligned}$$

where $|i\rangle$ denotes the eigenstates of the transfer matrix and

$$c_i = \langle 0|A|i\rangle\langle i|B|0\rangle \quad . \tag{4}$$

The main problem in the numerical determination of masses is to find operators A and B that have a good overlap with a single state $|i\rangle$; i.e. that c_i is large compared with c_j, $j \neq i$. The first, important, step in this direction is to realize that, by choosing the symmetry properties of the operators A and B properly, we can select channels; i.e. we can choose, for example, A and B such that all the c_i's vanish except those corresponding to, say, a given value of the angular momentum. By using so called "zero momentum" operators, namely operators obtained by summing over a slice orthogonal to the time direction, all c_i's that correspond to nonvanishing momentum vanish.

A systematic way to further improve the overlap is to study simultaneously the correlators among several operators A_α that belong to the same channel. This is indeed a natural prescription in the context of the glueball physics since the glueballs are expected to be extended objects and choosing several extended operators on the lattice one can hope to find a better overlap with the (unknown) glueball wave function. One must then measure all the possible correlations among these operators and construct the crosscorrelation matrix defined as:

$$C_{\alpha\beta}(t) = \langle A_\alpha(t) A_\beta(0) \rangle - \langle A_\alpha(t) \rangle \langle A_\beta(0) \rangle \tag{5}$$

By diagonalizing the crosscorrelation matrix one can then obtain the mass spectrum.

This method can be further improved[6, 7] by studying the generalized eigenvalue problem

$$C(t)\psi = \lambda(t, t_0) C(t_0) \psi \tag{6}$$

where t_0 is small and fixed (say, $t_0 = 0$). Then it can be shown that the various masses m_i are related to the generalized eigenvalues as follows [6, 7]:

$$m_i = \log \left(\frac{\lambda_i(t, t_0)}{\lambda_i(t+1, t_0)} \right) \tag{7}$$

where both t and t_0 should be chosen as large as possible, $t >> t_0$ and as t is varied the value of m_i must be stable within the errors. Practically one is forced to keep $t_0 = 0, 1$ to avoid too large statistical fluctuations and at the same time t is in general forced to stay in the range $t = 1$ to 7, depending on β and the channel, to avoid a too small signal to noise ratio. This method is clearly discussed in [6, 7] and we refer to them for further details. All the results that we shall list below for the glueball spectrum have been obtained with this improved method.

4 Results and comparison with SU(2) and with the IP model.

Simulations were performed for $\beta = 0.74057, 0.74883, 0.75202$ and 0.75632. The inverse of the lightest mass at these β's is $\xi_0 = 1.864(5)$, $2.592(5)$, $3.135(9)$ and $4.64(3)$ in lattice units respectively.

In the following we normalize the masses in units of the string tension $\sqrt{\sigma}$. The continuum limit is then obtained from the extrapolation to $\xi = \infty$ where we assume that corrections are proportional to $\xi^{-\omega}$ with $\omega \approx 0.8$.

The results are listed in tab.1, where they are compared with the corresponding results for SU(2) obtained in [3] and with the predictions of the IP model. A first unambiguous result of the simulation is that the parity degeneration is indeed recovered in the continuum limit also for the even J sector. Taking into account this

Table 1: *Comparison between the Ising, SU(2) and IP spectra.*

J^P	Ising	SU(2)	IP
0^+	3.08(3)	4.763(31)	2.00
$(0^+)'$	5.80(4)		5.94
$(0^+)''$	7.97(11)		8.35
$2^\pm$	7.98(8)	7.78(10)	6.36
$(2^\pm)'$	9.95(20)		8.76
0^-	10.0(5)	9.90(27)	13.82
$(0^-)'$	13.8(6)		15.05
$(1/3)^\pm$	12.7(5)	10.75(50)	8.04

degeneration we end up with 8 independent states in the continuum limit. Looking at tab.1 we see that the biggest discrepancy in the mass values is for the lowest state, which is predicted to be too light in the IP model, and turns out to be very different in the Ising and SU(2) cases. This is due first to the lack of validity of the adiabatic approximation at small scales and second to the fact that in the IP model an "ideal" picture of string (without self repulsion terms) is assumed for the flux tube.

Apart from this state, in the remaining part of the spectrum we immediately see an impressive agreement between the Ising and SU(2) spectra. This agreement is further improved by looking at the excited states in the (0^+) channel for the SU(2) model. In [4] a variational estimate for these masses can be found (up to our knowledge no Monte Carlo estimate exists for them). In tab.2 we compare these values with the Ising ones. While the two sets of excited states disagree if measured in units of $\sqrt{\sigma}$, they agree if measured in units of 0^+. Moreover a better and better agreement is observed if ratios of higher masses are considered.

Table 2: *The 0^+ channel.*

ratio	Ising	SU(2)
$(0^+)'/0^+$	1.88(2)	1.77(2)
$(0^+)''/0^+$	2.59(4)	2.50(5)
$(0^+)''/(0^+)'$	1.37(4)	1.41(4)

We can conclude from these data that the qualitative features of the glueball spectrum are largely independent from the gauge group and well described by a flux tube effective model. While the higher states of the spectrum show a remarkable independence from the gauge group, for the lowest state the flux tube picture breaks down and the gauge group becomes important. The IP model, which is the simplest possible realization of such a flux tube, seems able to catch (at least at a qualitative level) some of the relevant features of the glueball spectrum.

5 Duality.

Another important reason of interest in the gauge Ising model is that it is related by duality to the ordinary $3D$ spin Ising model. As a consequence, one expects the glueball spectrum to be mapped in the spectrum of massive excitations of the spin model. For example lowest state 0^+ is mapped into the (inverse) correlation length of the spin model (see [2, 5] for details).

The correlation length extracted form the spin-spin correlation function of the spin model indeed reproduces consistent values. However it turns out to be extremely difficult to obtain accurate results for higher states. One particular problem is to disentangle contibutions from higher states and the cut. A field-theoretic calculation of the contributions of the cut to the spin-spin correlation function [9] turns out to be helpful [8].

References

[1] N.Isgur and J.Paton, Phys. Rev. **D31** (1985) 2910.

[2] V.Agostini, G.Carlino, M.Caselle and M. Hasenbusch, Nucl. Phys. **B484** (1997) 331.

[3] M. Teper Phys. Lett. **289B** (1992) 115, T. Moretto and M. Teper, hep-lat/9312035

[4] H.Arisue, Prog. Theor. Phys. **84** (1990) 951.

[5] M.Caselle and M.Hasenbusch, J.Phys. **A 30** (1997) 4963.

[6] A.S. Kronfeld Nucl. Phys. **B17** (Proceeding Supplement) (1990) 313

[7] M. Lüscher and U. Wolff, Nucl. Phys. **B 339** (1990) 222.

[8] M.Caselle, M.Hasenbusch and P.Provero, in preparation

[9] P.Provero, cond-mat/9709292

$U(1)$ lattice gauge theory and $N = 2$ supersymmetric Yang-Mills theory*

Jan Ambjørn[1], Domènec Espriu[2] and Naoki Sasakura[1†]

[1] The Niels Bohr Institute,
University of Copenhagen,
Blegdamsvej 17,
DK-2100 Copenhagen Ø,
Denmark
[2] Department of Physics,
University of Barcelona,
Diagonal 647,
E-08028, Barcelona,
Spain

Abstract: We discuss the physics of four-dimensional compact $U(1)$ lattice gauge theory from the point of view of softly broken N=2 supersymmetric $SU(2)$ Yang-Mills theory. We also show that the J^{PC} assignment of some of the lowest lying states can be naturally explained.

1 Introduction

The pure compact $U(1)$ lattice gauge theory is known to have a confinement phase and a Coulomb phase separated by a phase transition. Recent computer simulations indicate that the phase transition is second order and that the critical exponents associated with the transition are non-trivial [2, 3, 4, 5]. This is a remarkable situation since it, according to ordinary folklore, implies that there exists a non-trivial continuum field theory with these critical exponents. But, on the other hand, a kind of no-go theorem was argued [6], which requires that, if it is non-trivial, the associated continuum field theory must have both electrically charged fields and magnetically charged ones. One can transform the lattice system to a dual gauge field coupled

*This talk is based on [1].

†The speaker. The present address: Department of Physics, Kyoto University, Kyoto 606-8502, Japan

to a monopole field [7, 8], but one cannot find any electrically charged fields in the lattice system. Thus it seems that there are no obvious candidates for the continuum field theory associated to the second order phase transition.

The compact $U(1)$ lattice gauge theory in three dimensions has no other fixed points than the gaussian one [9], where we recover the familiar three-dimensional electrodynamics‡. However, there exists a three-dimensional non-Abelian $SU(2)$ gauge-Higgs theory, the Georgi-Glashow model, which has monopoles with finite action and where the $SU(2)$ gauge theory is spontaneously broken to $U(1)$. The physics of this model, confinement of the $U(1)$ charge and a corresponding non-vanishing string tension and a massive dual photon, is qualitatively the same as in the three-dimensional compact $U(1)$ lattice gauge theory [7], and the lattice theory describes the similar long distance physics as the continuum model, but it cannot be used to *define* in a rigorous way the full quantum field theory by approaching a fixed point.

We will thus take a pragmatic attitude and simply ask whether there exists a continuum field theory which has qualitatively the same features as observed for the compact $U(1)$ lattice theory in four dimensions. Such a question would have a meaning, even if it may turn out that the $U(1)$-transition is a weak first order phase transition§, since in this case the $U(1)$ lattice theory and the underlying continuum theory should still possess analogous long distance properties: a phase where the $U(1)$ charge is confined, and a Coulomb phase with a massive monopole field. It is clear that this is a non-trivial task for the continuum quantum theory, irrespectively of whether or not a genuine scaling limit can be defined for the lattice model.

2 The model from $N = 2$ SYM theory

An obvious candidate for a continuum field theory which may describe the same physics as the $U(1)$ lattice gauge theory is a softly broken $N = 2$ supersymmetric $SU(2)$ gauge theory. Before the soft breaking it describes at low energies a $U(1)$ theory in Coulomb phase, which consists of, in a certain strong coupling region, a light monopole hyper-multiplet interacting with a dual photon multiplet [12]. After breaking to $N=1$ supersymmetry by the introduction of the mass term superpotential of the $N = 1$ chiral multiplet in the $N = 2$ vector multiplet, it describes, at low energies, a $U(1)$ theory in confinement phase with a monopole condensate and a massive dual photon [12]. Thus the mass term superpotential triggers the phase transition from the $N = 2$ Coulomb phase to the $N = 1$ confinement phase. In order to make contact to the lattice $U(1)$ theory we have to induce further soft breaking to $N=0$ supersymmetry and find a theory with a $U(1)$ confinement-Coulomb phase transition lying entirely in the $N = 0$ sector.

A general scheme of soft breaking of the Seiberg-Witten solution, still respecting

‡Another continuum limit was suggested in [10].

§The most recent computer simulation [11] shows some evidences of a weak first order phase transition.

the monodromy properties of the singularities, was obtained in [13]. They introduced a spurion $N=2$ vector multiplet, the dilaton spurion, and the scale parameter Λ is expressed as $\exp(is)$, where s denotes the lowest scalar component of the dilaton spurion S. Thus the effective Lagrangian of the softly broken $N=2$ supersymmetric Yang-Mills theory is given by

$$\begin{aligned} \mathcal{L}_{soft} &= \frac{1}{4\pi}\mathrm{Im}\left[\int d^2\theta d^2\bar{\theta}\,\frac{\partial \mathcal{F}}{\partial A^i}\bar{A}^i + \frac{1}{2}\int d^2\theta\,\frac{\partial^2 \mathcal{F}}{\partial A^i \partial A^j}\,W^{i\alpha}W^j{}_\alpha\right], \\ & i = 0,1;\;\; A^0 = S,\; A^1 = A, \end{aligned} \tag{1}$$

where $\mathcal{F}$ is the prepotential obtained by Seiberg and Witten [12] with the substitution $\Lambda = \exp(iA^0)$ in this case. The spurion fields A^0 and W^0_α are frozen to constant values, and the auxiliary field components of them break the $N=2$ supersymmetry directly down to $N=0$. This softly broken model was shown to be in the confinement phase in the same way as the original $N=1$ model of Seiberg and Witten, the dynamics of the confinement being monitored by a monopole condensation dictated by the freezing of S [13]. For our purpose this model has the same undesirable feature that there is no $N=0$ Coulomb phase.

To have an $N=0$ theory on both sides of the transition, we add an additional $N=1$ Lagrangian

$$\mathcal{L}_z = \int d^2\theta d^2\bar{\theta}\; z^\dagger z + \left(\int d^2\theta\; l\, z\left(w - \mathrm{Tr}\,(\Phi^2)\right) + h.c.\right). \tag{2}$$

Here Φ denotes the $N=1$ chiral multiplet in the original $N=2$ vector multiplet, z is an $N=1$ chiral multiplet without any gauge charges, and l and w are free complex parameters. In the $l \to 0$ limit, z decouples from the original system and will go back to the model above. In the $l \to \infty$ limit, the kinetic term of z is negligible, and z is an auxiliary field.

First consider the $N=1$ case with vanishing auxiliary field components of the spurion fields. After elimination of the D and F components, we obtain the classical potential

$$V_{bare+z} = |l|^2|w - \mathrm{Tr}\,\phi^2|^2 + 4g^2|lz|^2\mathrm{Tr}\,(\phi^\dagger\phi) + \frac{1}{g^2}\mathrm{Tr}\,([\phi,\phi^\dagger]^2), \tag{3}$$

where ϕ denotes the scalar component of Φ, and similarly for z. The first term of (3) will constrain the value of $\mathrm{Tr}\,\phi^2$ to be close to w. When w is very large, the system can be treated semi-classically, and the gauge group will be broken at the scale of order $\sqrt{w}$ and the system will be in the $U(1)$ Coulomb phase at low energy. Taking the value w smaller, the effective coupling of the system becomes stronger, but precisely as for the original $N=2$ case, the holomorphy argument [14] ensures that the system should stay in the $U(1)$ Coulomb phase.

On the other hand, if the soft-breaking terms are introduced and the supersymmetry is broken to $N=0$, the delicate cancellation of the large fluctuations of the gauge field in the strong coupling region will be lost, and so the system will be in the confinement phase for small w. Hence we may expect that the $N=0$ system of $\mathcal{L}_{soft}+\mathcal{L}_z$ has a phase transition line of a confinement and a Coulomb phase in the parameter space of w.

We now analyze the vacuum structure of the softly broken model in the dual description, following the analysis in [13]. In this analysis we assume that the auxiliary field limit of z, i.e. the limit $l \to \infty$, simplifies the analysis in such a way that the potential obtained by the elimination of the F-component of z simply provides a constraint which relates the free parameter w and the dynamical parameter a_D, the vacuum expectation value of the scalar component of the dual photon multiplet. Thus now a_D is like a free parameter instead of w (See [1] for details.). Then, after the elimination of the D and F components of the dual photon and the monopole fields, we obtain the potential

$$\begin{aligned} V_{soft+z} &= \frac{1}{b_{11}}\left| b_{01}\bar{F}_0+\sqrt{2}m\tilde{m}+zl\frac{\partial u_q}{\partial a_D^1}\right|^2+\frac{1}{2b_{11}}\left(b_{01}D_0+|m|^2-|\tilde{m}|^2\right)^2 \\ &+ 2|a_D|^2\left(|m|^2+|\tilde{m}|^2\right)-b_{00}|F_0|^2-\frac{b_{00}}{2}D_0^2+\left(F_0 zl\frac{\partial u_q}{\partial a_D^0}+h.c.\right). \end{aligned} \quad (4)$$

Here $(m,\tilde{m})$ denote the scalar components of the monopole hyper-multiplet and

$$u_q = u-\frac{\Lambda^4}{w}f\left(\frac{a_D^2}{w},\frac{\Lambda^4}{w^2}\right) \quad (5)$$

with the order parameter $u=<\mathrm{Tr}\phi^2>$ of Seiberg-Witten [12] and an undetermined function f, while

$$b_{ij} \equiv \frac{1}{4\pi}\mathrm{Im}\ \tau_{ij}=\frac{1}{4\pi}\mathrm{Im}\ \frac{\partial^2\mathcal{F}}{\partial a_D^i \partial a_D^j}. \quad (6)$$

Finally F_0 and D_0 denote the frozen F and D components of the spurion multiplet, respectively.

In [13] the choice of parameters $F_0 \lesssim \Lambda,\ D_0=0$ was studied. In our case we find that this choice leads to an unbounded potential in a neighborhood of $a_D=0$. This implies that we have to take into account the contributions to the potential from higher derivative terms in order to understand the dynamics in this region. Since we have little control over these higher derivative terms we will not consider this case any further and turn to $F_0=0,\ D_0\neq 0$. Then one can easily see that, if

$$|a_D|^2<\frac{|b_{01}D_0|}{2b_{11}}, \quad (7)$$

the potential (4) has the following non-trivial minima:

$$\begin{aligned} |\tilde{m}|^2 &= -2b_{11}|a_D|^2+b_{01}D_0, \quad m=0 \quad \text{for } b_{01}D_0>0, \\ |m|^2 &= -2b_{11}|a_D|^2-b_{01}D_0, \quad \tilde{m}=0 \quad \text{for } b_{01}D_0<0. \end{aligned} \quad (8)$$

Outside the region (7), the minimum is given by the trivial one $m=\tilde{m}=0$. The non-zero vacuum expectation values of the monopole fields give a mass to the dual photon. Following the general folklore this leads to a confined electric charge by the dual Meissner effect. Outside the region (7), the monopole fields do not condensate, and the system is in the Coulomb phase. Thus we have shown in the dual picture that by adding a term (2) there may be a confinement-Coulomb phase transition line in the parameter space of a_D (or w), while the system on both sides of the phase transition line is an effective $N{=}0$ theory.

To regard this phase transition line as that of the pure compact $U(1)$, it is important that the light degrees of freedom are the dual photon and one of the monopole fields. One can show this by explicitly checking the mass matrices or by the global $U(1) \times U(1)_R$ symmetry of the system (See [1] for details.)

In order to substantiate further that the physics of compact lattice $U(1)$ is the same as broken $N{=}2$ supersymmetric Yang-Mills theory, we will discuss below the parity and charge conjugation quantum numbers of the one-particle state in the confinement phase. The pure $U(1)$ gauge theory $\int d^4x F_{\mu\nu}F^{\mu\nu}$ is obviously invariant under the parity and charge conjugation transformation defined by $P: A_0 \to A_0,\ A_i \to -A_i,\ \ (x_0 \to x_0,\ x_i \to -x_i)$ and $C: A_\mu \to -A_\mu$, respectively, where A_μ is the gauge potential. Defining the duality transformation by the exchange of the electric and magnetic fields $E^D = B$ and $B^D = E$, one obtains the parity and charge conjugation transformation of the dual gauge field as $P: A_0^D \to -A_0^D,\ \ A_i^D \to A_i^D$ and $C: A_\mu^D \to -A_\mu^D$, respectively. Thus the dual photon field has $J^{PC} = 1^{+-}$. Similarly one can show that the lowest physical excitation of the monopole field has $J^{PC} = 0^{++}$.

This quantum number assignment agrees with the lattice simulation. The lattice spectroscopy shows that there are one-particle states with quantum numbers 0^{++} and 1^{+-} in the confinement phase [3]. We identify these states as the lowest monopole field excitation and the dual photon one-particle state, respectively.

3 Summary

The recent remarkable lattice simulations of compact $U(1)$ show very interesting physics with a confinement-Coulomb phase transition which is reported to be of second order with non-trivial critical mass exponents. If the transition is second order, we have a new and very interesting situation in quantum field theory since there should be an underlying continuum field theory of a new kind. Presently we have nothing more to say about such a revolutionary scenario.

In this talk we have tried something more modest, namely to find a continuum four-dimensional field theory which at a qualitative level has the same low energy physics as the lattice model. Thus this continuum model will retain its explanatory value even if the transition is not really a second order transition but merely a weak first order transition.

The physics of compact lattice $U(1)$ in the Coulomb phase seems quite interesting near the phase transition: There seems to be a universal value of the "renormalized" coupling constant in the theory [15]. We have not yet made any progress in that direction.

Acknowledgment: We would like to thank E. Witten for pointing out some references and useful discussions. We are also grateful to J. Jersak for discussions at the conference. The speaker (N.S.) was supported by DANVIS grant No. 1996-145-0003 from the Danish Research Academy.

References

[1] J. Ambjørn, D. Espriu and N. Sasakura, Mod. Phys. Lett. **A12** (1997) 2665.

[2] J. Jersak, C.B. Lang and T. Neuhaus, Phys. Rev. Lett. **77**, (1996) 1933; Phys. Rev. **D54**, (1996) 6909.

[3] J. Cox, W. Franzki, J. Jersak, C.B. Lang, T. Neuhaus and P.W. Stephenson, Nucl. Phys. **B499** (1997) 371.

[4] W. Kerler, C. Rebbi and A. Weber, Phys. Lett. **B392** (1997) 438.

[5] J. Cox, W. Franzki, J. Jersak, C.B. Lang and T. Neuhaus, *Strongly coupled compact lattice QED with staggered fermions*, hep-lat/9705043.

[6] P.C. Argyres, M.R. Plesser, N. Seiberg and E. Witten, Nucl. Phys. **B461** (1996) 71.

[7] A.M. Polyakov, Phys. Lett. **B59** (1975) 82.

[8] T. Banks, R. Myerson and J. Kogut, Nucl. Phys. **B129** (1977) 493.

[9] M. Gopfert and G. Mack, Commun. Math. Phys. **82** (1981) 545.

[10] H. Neuberger, *Three Dimensional Periodic $U(1)$ Gauge Theory and Strings*, hep-th/9111055.

[11] I. Campos, A. Cruz and A. Tarancón, *First order signatures in 4D pure compact $U(1)$ gauge theory with toroidal and spherical topologies*, hep-lat/9711045.

[12] N. Seiberg and E. Witten, Nucl. Phys. **B426** (1994) 19; Nucl. Phys. **B431** (1994) 484.

[13] L. Alvarez-Gaumé, J. Distler, C. Kounnas and M. Mariño, Int. J. Mod. Phys. **A11** (1996) 4745. L. Alvarez-Gaumé and M. Mariño, Int. J. Mod. Phys. **A12** (1997) 975.

[14] N. Seiberg, *The Power of Holomorphy - Exact Results in 4D SUSY Field Theories*, hep-th/9408013.

[15] J.L. Cardy, Nucl. Phys. **B170 [FS1]** (1980) 369.
J.M. Luck, Nucl. Phys. **B210 [FS6]** (1982) 111.

Hot Electroweak Matter Near to the Critical Higgs Mass[1]

M. Gürtler[1], E.-M. Ilgenfritz[2 3], A. Schiller[1], C. Strecha[1]

[1] Institut für Theoretische Physik, Universität Leipzig, D-04109 Leipzig, Germany
[2] Institut für Physik, Humboldt-Universität, D-10115 Berlin, Germany
[3] Institute for Theoretical Physics, Kanazawa University, Kanazawa 920, Japan

Abstract: We discuss the end of the first order phase transition and the bound state spectrum, both at weak transition and at the crossover.

1 Introduction

Two years ago [1], we have presented here first results of our group obtained within the $3D$ approach, comparing the properties of the thermal phase transition in the $SU(2)$ Higgs model at small coupling (Higgs mass) $M_H^* = 35$ GeV with a more realistic $M_H^* = 70$ GeV.[2] The endpoint of the phase transition and the physics near to the corresponding temperature $T_c(m_H^{crit})$ is the topic of this talk.

In the meantime, due to efforts of three groups [3, 4, 2] using the $3D$ approach, various aspects of the high temperature electroweak phase transition (as latent heat and interface tension) have been explored in this model within this span of Higgs masses where the character of the transition changes drastically. The accuracy is inaccessible to $4D$ Monte Carlo simulations [5], although the results are consistent with each other where they can be compared [6, 7].

The interest in the properties of this phase transition resulted from the hope to work out - within the standard model - a viable mechanism for the generation of baryon asymmetry of the universe at the electroweak scale (see Z. Fodor [8]). It turned out, however, that in the case of the standard model, taking the lower bound of the Higgs mass into account, it is at most weakly first order. Together with the small amount of CP violation in the standard model this has ruled out the most economic scenario of BAU generation.

Using dimensional reduction [9] the simplest effective $3D$ $SU(2)$ Higgs theory has the action

$$S_3 = \int d^3x \Big(\frac{1}{4} F^b_{\alpha\beta} F^b_{\alpha\beta} + (D_\alpha \phi)^+ (D_\alpha \phi) + m_3^2 \phi^+ \phi + \lambda_3 (\phi^+ \phi)^2 \Big) \tag{1}$$

[1] Talk given by E.-M. Ilgenfritz, supported by DFG under grant Mu932/1-4

[2] M_H^*, together with the $3D$ gauge coupling, is labelling our lattice data. For relation to the zero-temperature Higgs mass, see Ref. [2].

with dimensionful, renormalisation group invariant couplings g_3^2 and λ_3 and a running mass squared $m_3^3(\mu_3)$. It is related to the corresponding lattice model with the action

$$S = \beta_G \sum_p (1 - \frac{1}{2}\mathrm{tr}U_p) - \beta_H \sum_{x,\alpha} \frac{1}{2}\mathrm{tr}(\Phi_x^+ U_{x,\alpha}\Phi_{x+\alpha}) + \sum_x (\rho_x^2 + \beta_R(\rho_x^2 - 1)^2) \quad (2)$$

and $\rho_x^2 = \frac{1}{2}\mathrm{tr}(\Phi_x^+\Phi_x)$ through the following relations (clarifying the meaning of M_H^*)

$$\beta_G = \frac{4}{ag_3^2}, \quad \beta_R = \frac{\lambda_3}{g_3^2}\frac{\beta_H^2}{\beta_G} = \frac{1}{8}\left(\frac{M_H^*}{80\ \mathrm{GeV}}\right)^2 \frac{\beta_H^2}{\beta_G}, \quad \beta_H = \frac{2(1-2\beta_R)}{6 + a^2 m_3^2}. \quad (3)$$

The coupling parameters in the lattice action can be expressed in terms of $4D$ couplings and masses. The bare mass squared is related to the renormalised $m_3^2(\mu_3)$ (we choose $\mu_3 = g_3^2$) through a lattice two-loop calculation [10]. Most of the $3D$ numerical investigations have been done for the $SU(2)$ Higgs model. The results obtained in the $3D$ approach indicate the validity of the dimensional reduction near to the transition temperature for Higgs masses between 30 and 100 GeV.

Our simulations have been done on the DFG-sponsored Quadrics QH2 in Bielefeld, on a CRAY-T90 and a Quadrics Q4 at HLRZ in Jülich. Algorithmic details are described in Ref. [2]. In the course of time, we have explored the parameter range $M_H^* = 35, 57, 70, 74, 76, 80$ GeV, looking for the strength of the transition, checking the approach to the continuum limit by varying $\beta_G = 8, 12, 16 \propto 1/a$ and comparing with perturbation theory. The simulations at the last three parameter values M_H^* have been devoted exclusively to locate the endpoint of the transition line [11] in the $m_H - T$ plane. The lattice size usually varied from 20^3 to 80^3. In order to unambiguously locate the critical Higgs mass we had to increase the lattice to 96^3 simulating at $M_H^* = 74$ GeV, the nearest to the critical value.

It has been essential to make extensive use of the Ferrenberg-Swendsen multihistogram technique. Data obtained at various M_H^* and β_H have been subject to global analysis by reweighting to other values, eventually extending the hopping parameter to complex values. The method constructs an optimal estimator of the spectral density of states (at fixed L and β_G) occurring in the representation of the partition function

$$Z(L, \beta_G, \beta_H) = \int dS_1 dS_2 D_L(S_1, S_2, \beta_G) \exp\left(L^3(\beta_H S_1 - \beta_H^2 S_2)\right) \quad (4)$$

from the *measured* double-histograms in the two variables

$$S_1 = 3\sum_{x,\alpha} \frac{1}{2}\mathrm{tr}(\Phi_x^+ U_{x,\alpha}\Phi_{x+\alpha}), \quad S_2 = \frac{\lambda_3}{g_3^2 \beta_G}(\sum_x \rho_x^4 - 2\sum_x \rho_x^2). \quad (5)$$

2 Finding m_H^{crit}

We have employed two methods to localise the critical Higgs mass. First we tried an interpolation of the discontinuity

$$\Delta\langle\phi^+\phi\rangle/g_3^2 = 1/8\ \beta_G\beta_H\Delta\langle\rho^2\rangle, \quad (6)$$

based on reweighting simulation data taken at $M_H^* = 70, 74$ and 76 GeV, in order to see where this turns to zero. This discontinuity is directly related to the latent heat. The pseudocritical $\beta_{H\ c}$ has been defined in two ways, by the maximum of the ρ^2 susceptibility and by the minimum of the ρ^2 Binder cumulant. For each M_H^*, an infinite volume extrapolation must be performed. For this purpose we have expressed the *physical* lattice size by the dimensionless variable $Lag_3^2 = 4L/\beta_G$. We have collected data from different lattice volumes L^3, from different β_G (= 12 and 16) as well as from different definitions of $\beta_{H\ c}$ and found them scattered along a unique function of this variable (Fig. 1). We tried to fit the finite volume scaling behaviour

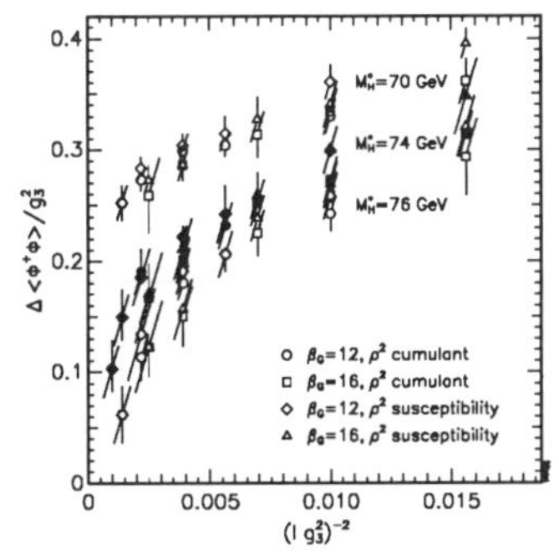

Figure 1: Thermodynamical limit for $\Delta\langle\phi^+\phi\rangle$ at three M_H^* values (here $l = aL$)

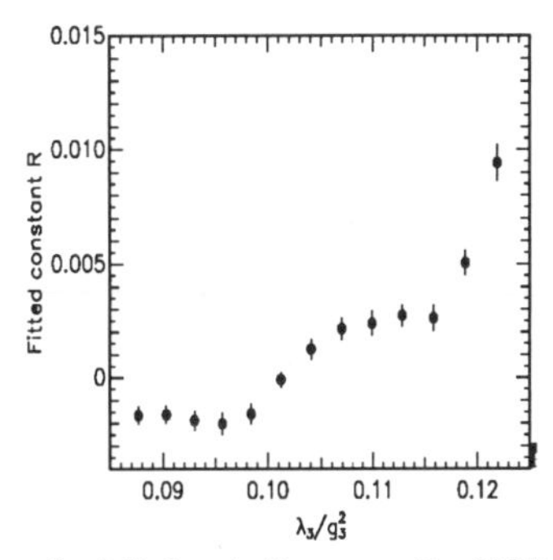

Figure 2: Minimal distance R of LY zeroes from real axis as function of λ_3/g_3^2

as $\Delta\langle\phi^+\phi\rangle_\infty - \Delta\langle\phi^+\phi\rangle_L \propto (L\ a\ g_3^2)^{-2}$ suggested by the Potts model. Using only the largest lattice volumes available, the criterion of vanishing latent heat gives an upper bound $\lambda_3^{crit}/g_3^2 < 0.107$ for the existence of a first order transition.

The other method detects the end of the transition line using the finite size analysis of the Lee–Yang zeroes. A genuine phase transition is characterised by *non-analytical* behaviour of the infinite volume free energy density. This is caused by zeroes of the partition function (in our case as function of β_H extended to complex values) clustering along a line nearest to the real axis and pinching in the thermodynamical limit. If there is a first order phase transition the first few zeroes are expected at

$$\mathrm{Im}\beta_H^{(n)} = \frac{2\pi\beta_{H\ c}}{L^3(1+2\beta_{R\ c})\Delta\langle\rho^2\rangle}\left(n - \frac{1}{2}\right), \quad \mathrm{Re}\beta_H^{(n)} \approx \beta_{H\ c}\,. \tag{7}$$

The partition function for complex couplings is obtained by reweighting from measurements at real couplings. The first zeroes can be well localised using the Newton-Raphson algorithm. We fit the imaginary part of the first zero for each available length Lag_3^2 according to $\mathrm{Im}\beta_H^{(1)} = C(Lag_3^2)^{-\nu} + R$. The scenario suggested is the change of the first order transition into an analytic crossover above M_H^{*crit}. A *positive* R signals that the first zero does not approach the real axis in the thermodynamical limit. A similar investigation has been performed recently at smaller gauge coupling in Ref. [12]. The fitted R crosses zero (Fig. 2) at $\lambda_3^{crit}/g_3^2 = 0.102(2)$. This corresponds to a zero temperature Higgs mass $m_H^{crit} = 72.4(9)$ GeV, and the phase transition line ends at a temperature of $T_c = 110(1.5)$ GeV. This refers to the experimental top mass. For the purely bosonic Higgs model we get $m_H^{crit} = 67.0(8)$ GeV and $T_c = 154.8(2.6)$ GeV.

3 Bound states below and above m_H^{crit}

To study the ground *and* excited bound states with given J^{PC} one has to use cross-correlations between operators $\mathcal{O}_i$ forming a complete set in that channel. They describe a spectral decomposition $\Psi_i^{(n)} = \langle \mathrm{vac}|\mathcal{O}_{\mathrm{i}}|\Psi^{(n)}\rangle$ with $|\Psi^{(n)}\rangle$ being the (zero momentum) energy eigenstates. According to the transfer matrix formalism, the connected correlation matrix

$$C_{ij}(t) = \langle \mathcal{O}_i(t)\mathcal{O}_j(0)\rangle = \sum_{n=1}^{\infty} \Psi_i^{(n)}\Psi_j^{(n)*}e^{-m_n t} \tag{8}$$

at time separation t gives the masses *and* wave functions.

Practically, only a *truncated* set of operators $\mathcal{O}_i(i = 1, \ldots, N)$ is used assuming that this allows to find the *lowest* states from the eigenvalue problem for $C_{ij}(t)$. However, truncation errors are not small. Solving instead the generalised eigenvalue problem

$$\sum_j C_{ij}(t)\Psi_j^{(n)} \quad = \quad \lambda^{(n)}(t, t_0)\sum_j C_{ij}(t_0)\Psi_j^{(n)} \tag{9}$$

these errors can be kept minimal ($t > t_0, t_0 = 0, 1$) [13].

The wave function of state n at fixed small time $t > t_0$ is found to be $\Psi_i^{(n)}(t) = \langle \mathrm{vac}|\ \mathcal{O}_{\mathrm{i}}\ \mathrm{e}^{-\mathrm{Ht}}\ |\Psi^{(n)}\rangle$. The components in the operator basis characterise the coupling of $\mathcal{O}_i$ to the (ground or excited) bound state n in the J^{PC} channel. The masses of these states are obtained fitting the diagonal elements $\mu^{(n)}(t)$

$$\mu^{(n)}(t) = \sum_{ij} \Psi_i^{(n)*} C_{ij}(t)\Psi_j^{(n)} \tag{10}$$

to a cosh form with t in some plateau region. In contrast to a blocking procedure used in [14], our base is built by a few types of gauge invariant operators $\mathcal{O}_i$, properly chosen with respect to lattice symmetry and quantum numbers, having different well-defined transverse extensions. In the Higgs channel (0^{++}) we use the operators ρ_x^2 and $S_{x,\mu}(l) = \frac{1}{2}\mathrm{tr}(\Phi_x^+ U_{x,\mu} \ldots U_{x+(l-1)\mu,\mu}\Phi_{x+l\mu})$ as well as quadratic Wilson loops of size $l \times l$, in the W-channel (1^{--}) the operators $V_{x,\mu}^b(l)) = \frac{1}{2}\mathrm{tr}(\tau^b\Phi_x^+ U_{x,\mu} \ldots U_{x+(l-1)\mu,\mu}\Phi_{x+l\mu})$ and in the 2^{++} channel $S_{x,\mu}(l) - S_{x,\nu}(l)$ where l expresses the operator extension in lattice units. In our procedure, the operator $\Phi^{(n)}$ projecting maximally onto state n is a optimised superposition $\Phi^{(n)} = \sum_1^N a_i^{(n)}\mathcal{O}_i$ with coefficients provided by the (properly normalised) solutions $\Psi_i^{(n)}$ of the generalised eigenvalue equation. The coefficients give direct access to the spatial extension of the states under study. This construction, at different β_G, should reveal the same wave function in position space as function of lag_3^2.

As examples we present in Figs. 3,4 squared wave functions in the 0^{++} channel vs. lag_3^2 near the end of the phase transition at $M_H^* = 70$ GeV (from a 30^3 lattice). In the symmetric phase the second excitation is a pure W-ball state in agreement

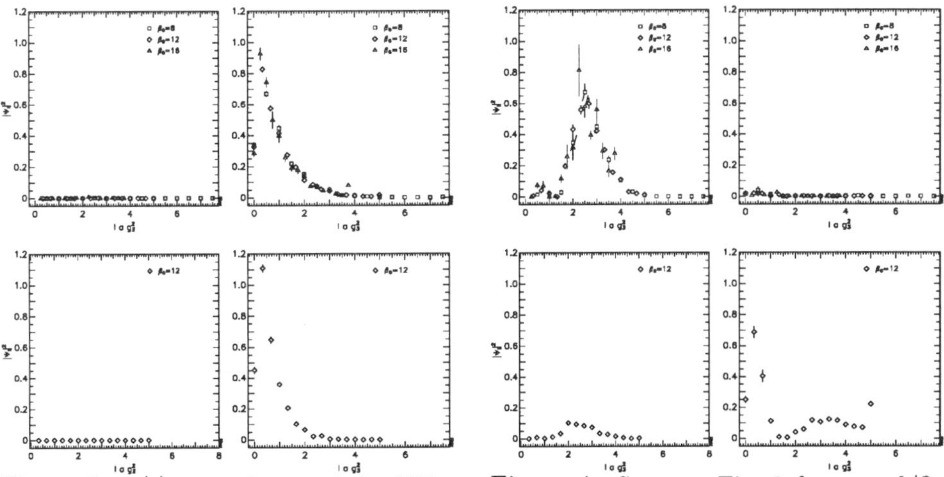

Figure 3: 0^{++} ground state, left: Wilson loop projection, right: $S_{x,\mu}(l)$ projection, upper/lower plot: symmetric/Higgs phase

Figure 4: Same as Fig. 3 for second/first excited state

with [14] (the first excitation is a Higgs state, too). In the broken phase, the first excited Higgs state still contains some admixture of pure gauge matter which vanishes only deeply inside this phase. This seems to be a precursor of the near end of the transition. More details, statistics and results for other quantum numbers will be discussed elsewhere [15].

Since Wilson loops have different projection properties in both phases we have studied the flow of the mass spectrum with β_H or $m_3^2(g_3^2)/g_3^4$ at $M_H^* = 100$ GeV ($\lambda_3/g_3^2 = 0.15625$) safely above the critical Higgs mass, too, passing there an analytic crossover. In Fig. 5 we present the 0^{++} spectrum in the vicinity and far from the crossover. The wave functions are similar to those discussed before. Here the spectrum on the "symmetric" side shows decoupling of Higgs and W-ball states, too, the lowest W-ball mass is roughly independent on β_H. Passing the crossover (from "symmetric" to the "broken" side) the first Higgs excitation contains also excited glue. These contributions vanish at lower temperature as indicated by the right figure. The minimum of the ground state mass remains finite, therefore the correlation length fits into the lattice.

The phenomenological interest is now concentrated on extensions of the standard model, with MSSM being the most promising candidate. Concerning non-perturbative physics in general, however, the lattice version of the standard Higgs model remains interesting as a laboratory for investigating the behaviour of hot gauge fields coupled to scalar matter, for the characterisation of possible bound states, for the understanding of real time topological transitions and as a cross-check for analytical approximation schemes.

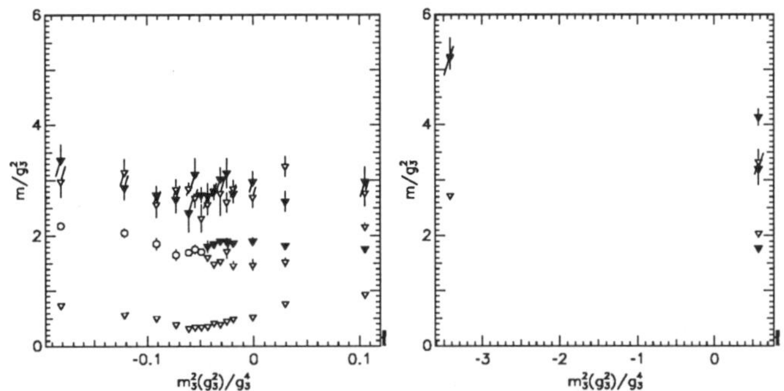

Figure 5: 0^{++} spectrum beyond the critical Higgs mass; left: near to crossover, right: far from crossover. Triangles denote Higgs states, full symbols W-ball states and circles Higgs states with admixture of pure gauge matter

References

[1] M. Gürtler *et al.*, *Nucl. Phys.* **B (Proc. Suppl.) 49,** (1996) 312

[2] M. Gürtler *et al.*, *Nucl. Phys.* **B483** (1997) 383

[3] K. Kajantie *et al.*, *Nucl. Phys.* **B466** (1996) 189

[4] F. Karsch *et al.*, *Nucl. Phys.* **B474** (1996) 217

[5] Z. Fodor *et al.*, *Nucl. Phys.* **B439** (1995) 147; F. Csikor *et al.*, *Nucl. Phys.* **B474** (1996) 421; Y. Aoki, *Phys. Rev.* **D56** (1997) 3860; F. Csikor, Z. Fodor, J. Heitger, *KEK-TH-541*(1997), hep-lat/9709098

[6] K. Rummukainen, *Nucl. Phys.* **B (Proc.Suppl.) 53** (1997) 30

[7] M. Gürtler, E.-M. Ilgenfritz, A. Schiller, *Eur. Phys. J.* **C1,** (1998) 363

[8] Z. Fodor, these proceedings

[9] K. Kajantie *et al.*, *Nucl. Phys.* **B458** (1996) 90

[10] M. Laine, *Nucl. Phys* **B451** (1995) 484

[11] M. Gürtler, E.-M. Ilgenfritz, and A. Schiller, *Phys. Rev.* **D56,** (1997) 3888

[12] F. Karsch *et al.*, *Nucl. Phys.* **B (Proc. Suppl.) 53** (1997) 623

[13] M. Lüscher, U. Wolff, *Nucl. Phys.* **B339** (1990) 222; C.R. Gattringer, C.B. Lang, *Nucl. Phys.* **B391** (1993) 463

[14] O. Philipsen, M. Teper,H. Wittig, *Nucl. Phys.* **B469** (1996) 445; *OUTP-97-44-P*, hep-lat/9709145

[15] M. Gürtler, E.-M. Ilgenfritz, A. Schiller, C. Strecha, in preparation

Gauge-Fixing Approach to Lattice Chiral Gauge Theories

Wolfgang Bock [1], Maarten F.L. Golterman [2], Yigal Shamir [3]

[1] Institute of Physics, Humboldt University,
Invalidenstr. 110, 10115 Berlin, Germany
[2] Department of Physics, Washington University
St. Louis, MO 63130, USA
[3] School of Physics and Astronomy, Tel-Aviv University
Ramat Aviv, 69978 Israel

Abstract: We review the status of our recent work on the gauge-fixing approach to lattice chiral gauge theories. New numerical results in the reduced version of a model with a U(1) gauge symmetry are presented which strongly indicate that the factorization of the correlation functions of the left-handed neutral and right-handed charged fermion fields, which we established before in perturbation theory, holds also nonperturbatively.

1 Introduction

The nonperturbative formulation of chiral gauge theories on the lattice is a long-standing and, to date, still unsolved problem. The local chiral gauge invariance on the lattice is broken for non-zero values of the lattice spacing, even in models with an anomaly-free spectrum. The failure of many proposals in the past was connected to the fact that the strongly fluctuating longitudinal gauge degrees of freedom alter the fermion spectrum leading to vector-like instead of chiral gauge theories. Already several years ago the Rome group proposed to use perturbation theory in the continuum as guideline and to transcribe the gauge-fixed continuum lagrangian of a chiral gauge theory to the lattice [1]. The hope is that, for a smooth gauge-fixing condition (like e.g. the Lorentz gauge), preferably smooth gauge configurations are selected from a gauge orbit, whereas rough gauge field configurations are suppressed by a small Boltzmann weight. The gauge-fixed model in the continuum is invariant under BRST symmetry. This symmetry is broken on the lattice, but the hope is that it can be restored by adding all relevant and marginal counterterms which are allowed by the exact symmetries of the lattice theory. Concrete lattice implementations of the nonlinear gauge $\sum_\mu \{\partial_\mu A_\mu + A_\mu^2\} = 0$ and of the Lorentz gauge $\sum_\mu \partial_\mu A_\mu = 0$ were first given in refs. [2] and [3].

As a first step we have studied a model with U(1) gauge symmetry. The important advantage of the abelian case is that the ghost sector drops out from the path integral and no Fadeev–Popov term needs to be included in the action. As a second simplification we included only a gauge-boson mass counterterm which is the only dimension-two counterterm and ignored all dimension-four counterterms. We notice already here that the lattice action can be formulated such that a fermion-mass counterterm (which is the only dimension-three counterterm) needs not to be added to the action. The coefficient of the gauge-boson mass counterterm has to be tuned such that the photons are massless in the continuum limit (CL). This value of the coefficient corresponds, for sufficiently small values of the gauge coupling, to a continuous phase transition between a "ferromagnetic" (FM) and a novel, so-called ferromagnetic "directional" (FMD) phase. The CL has to be taken from the FM side of the phase transition where the photon mass is larger than zero and the expectation value of the vector field vanishes. The FMD phase is characterized by a non-vanishing condensate of the vector field and a broken hypercubic rotation invariance.

Motivated by previous investigations of lattice chiral gauge theories we first restricted the gauge fields to the trivial orbit, where only the dynamics of the longitudinal gauge degrees of freedom is taken into account [4]. We shall refer to this model in the following as the "reduced model." In perturbation theory and by a high-statistics numerical simulation we could show in this reduced model that

1. the $U(1)_{L,\mathrm{global}} \otimes U(1)_{R,\mathrm{global}}$ is restored on the FM-FMD phase transition which is a central prerequisite for the construction of a chiral gauge theory on the lattice, and

2. the fermion spectrum in the CL contains only the desired left-handed charged fermion and a right-handed neutral "spectator" fermion.

The first statement applies also to the strongly coupled symmetric phase of the Smit-Swift model in which the unwanted species doublers were shown to decouple. The fermion spectrum in this phase however contains only a neutral Dirac fermion which decouples completely from the gauge fields when they are turned on back again. Later it was argued that a lattice chiral gauge theory can indeed not be defined within a symmetric phase or on its boundaries [5].

The outline of the rest of the paper is as follows: In Sect. 2. we introduce the fully gauged U(1) model and its reduced version. In Sect. 3 we review our previous results for the phase diagram and the fermion spectrum in the reduced model. Our new numerical results which further substantiate the above statement about the fermion spectrum are presented in Sect. 4. In the last section, we briefly summarize our results and give a brief outlook to future projects.

2 The Model

The fully gauged U(1) lattice model is defined by the following action

$$S_V = \frac{1}{g^2} \sum_{x\mu\nu} \{1 - \mathrm{Re}\, U_{\mu\nu x}\} + \frac{1}{2\xi g^2} \left\{ \sum_{x,y,z} \Box_{xy}(U)\Box_{yz}(U) - \sum_x B_x^2(V(U)) \right\}$$

$$-\kappa \sum_{\mu x} (U_{\mu x} + U_{\mu x}^\dagger) + \sum_{x,y} \left\{ \overline{\psi}_x \gamma_\mu \left([D_\mu(U)]_{xy} P_L + [\partial_\mu]_{xy} P_R \right) \psi_y - \frac{r}{2} \overline{\psi}_x \Box_{xy} \psi_y \right\} \quad (1)$$

$$B_x(V(U)) = \tfrac{1}{4} \sum_\mu (V_{\mu x - \hat{\mu}} + V_{\mu x})^2 , \quad V_{\mu x} = \mathrm{Im}\, U_{\mu x} . \quad (2)$$

The action in eq. (1) includes the following terms (from the left to the right): the usual plaquette term ($\propto 1/g^2$), the Lorentz gauge-fixing term ($\propto 1/2\xi g^2$), the gauge-boson mass counterterm ($\propto \kappa$), the "naive" kinetic term for the fermions and the Wilson term ($\propto r$) which we use to remove species doublers. $U_{\mu x} = \exp(igA_{\mu x})$ is the lattice link variable, $U_{\mu\nu x}$ the plaquette variable, g is the gauge coupling, ξ is the gauge-fixing parameter, r is the Wilson parameter and $P_{\mathrm{L,R}} = \frac{1}{2}(1 \mp \gamma_5)$ are the left-and right-handed chiral projectors. ∂_μ and $D_\mu(U)$ designate the free and covariant antihermitian nearest-neighbor lattice derivatives, and $\Box$ and $\Box(U)$ the free and covariant nearest-neighbor lattice laplacians. The gauge-fixing action on the lattice was constructed such that it reduces in the classical CL to $\frac{1}{2\xi}\int d^4x(\partial_\mu A_\mu)^2$, and has an absolute minimum at $U_{\mu x} = 1$, validating weak coupling perturbation theory [2, 3]. We also notice that the fermionic part of (1) is invariant under the shift symmetry $\psi_R \rightarrow \psi_R + \epsilon_R$, $\overline{\psi}_R \rightarrow \overline{\psi}_R + \overline{\epsilon}_R$ [6]. This symmetry implies that a fermion-mass counterterm does not need be added to the action. We ignore here all dimension four counterterms, which we believe to be less important [3]. The lattice model is defined by the following path integral

$$Z = \int \mathcal{D}U\mathcal{D}\overline{\psi}\mathcal{D}\psi\, e^{-S_V(U;\psi_L,\psi_R)} = \int \mathcal{D}\phi\mathcal{D}U\mathcal{D}\overline{\psi}\mathcal{D}\psi\, e^{-S_H(\phi;U;\psi_L,\psi_R)} . \quad (3)$$

In the second equation we have made the gauge degrees of freedom ϕ explicit. These gauge degrees of freedoms are nothing but group-valued scalar (Higgs) fields. The action S_H is obtained from S_V by performing in eq. (1) a gauge rotation $U_{\mu x} \rightarrow \phi_x^\dagger U_{\mu x}\phi_{x+\mu}$, $\psi_{\mathrm{L}x} \rightarrow \phi_x^\dagger \psi_{\mathrm{L}x}$. The scalar fields ϕ_x emerge in all those terms which are gauge non-invariant, i.e. in the gauge-fixing term, the gauge-boson mass counterterm and the Wilson term. S_H is invariant under a $\mathrm{U(1)_{L,local}}\otimes\mathrm{U(1)_{R,global}}$ symmetry: $\psi_{\mathrm{L}} \rightarrow g_{\mathrm{L}x}\psi_{\mathrm{L}x}$, $\psi_{\mathrm{R}} \rightarrow g_{\mathrm{R}}\psi_{\mathrm{R}x}$, $U_{\mu x} \rightarrow g_{\mathrm{L}x} U_{\mu x} g_{\mathrm{L}x+\hat{\mu}}^\dagger$, $\phi_x \rightarrow g_{\mathrm{L}x}\phi_x g_{\mathrm{R}}^\dagger$.

Next, we introduce the "reduced" model, which is obtained by setting $U_{\mu x} = 1$ in S_H. The $\mathrm{U(1)_{L,local}}\otimes\mathrm{U(1)_{R,global}}$ symmetry turns into a $\mathrm{U(1)_{L,global}}\otimes\mathrm{U(1)_{R,global}}$ symmetry. The action of this reduced model reads

$$S_H^{\mathrm{red}} = \tilde{\kappa} \left\{ \sum_x \phi_x^\dagger (\Box^2\phi)_x - B_x^2(V^r(\phi)) \right\} - \kappa \sum_x \phi_x^\dagger (\Box\phi)_x + \tfrac{1}{2} \sum_{x\mu} \left\{ \overline{\psi}_x \gamma_\mu \psi_{x+\mu} \right.$$

$$\left. -\overline{\psi}_{x+\mu}\gamma_\mu\psi_x - r\left((\overline{\psi}_x(\phi_{x+\mu}^\dagger P_L + \phi_x P_R)\psi_{x+\mu} + \mathrm{h.c.}) - 2\overline{\psi}_x(\phi_x^\dagger P_L + \phi_x P_R)\psi_x \right) \right\} , \quad (4)$$

where $V^r_{\mu x} = \mathrm{Im}(\phi^\dagger_x \phi_{x+\mu})$. The reason why this reduced model is of interest is that it should lead in the CL to a theory of free chiral fermions in the correct representation of the gauge group. This is a necessary condition for the construction of chiral gauge theory with unbroken symmetry. The failure of many previous proposals of chiral gauge theories, like e.g. of the Smit-Swift model, was connected to the fact that the fermion spectrum is altered by the strongly fluctuating gauge degrees [4]. In the following sections we shall reexamine this important question in our model.

3 Phase Diagram and Fermion Spectrum of the Reduced Model

Let's first consider the phase diagram of the pure bosonic part of the action (4) which only includes the gauge-fixing term and the gauge-boson mass counterterm. The $(\kappa, \widetilde{\kappa})$-phase diagram of this higher derivative scalar field theory contains at large $\widetilde{\kappa}$ a FM phase at $\kappa > \kappa_c$, where $\langle V^r_{\mu x}\rangle = 0$ and $\langle \phi \rangle > 0$, and a FMD phase at $\kappa < \kappa_c$ with $\langle V^r_{\mu x}\rangle \neq 0$ [2]. Both phases are separated by a continuous phase transition at $\kappa_c(\widetilde{\kappa})$ [7]. To one-loop order we find $\kappa_c = 0.02993 + O(1/\widetilde{\kappa})$ [7]. As explained in Sect. 1, in the fully gauged model κ has to be tuned from the FM side towards this phase transition in order to obtain massless photons. In the reduced model we have computed the order parameter $\langle \phi \rangle$ in the FM phase, both in perturbation theory in $1/\widetilde{\kappa}$ and also numerically, and find that it vanishes in the limit $\kappa \searrow \kappa_{\mathrm{FM-FMD}}$ [7, 8]. This phenomenon is associated with the $1/(p^2)^2$ propagator for the ϕ-field fluctuations. The vanishing of $\langle \phi \rangle$ implies that the $\mathrm{U}(1)_{\mathrm{L,global}} \otimes \mathrm{U}(1)_{\mathrm{R,global}}$ symmetry which is broken to its diagonal subgroup in the FM phase, is restored on the FM-FMD phase transition line, an essential prerequisite for the construction of a chiral gauge theory with unbroken gauge symmetry.

We now introduce the four fermion operators $\psi^{\mathrm{n}}_{\mathrm{R}} = \psi_{\mathrm{R}}$, $\psi^{\mathrm{n}}_{\mathrm{L}} = \phi^\dagger \psi_{\mathrm{L}}$, $\psi^{\mathrm{c}}_{\mathrm{L}} = \psi_{\mathrm{L}}$ and $\psi^{\mathrm{c}}_{\mathrm{R}} = \phi \psi_{\mathrm{R}}$. The fields with the superscripts c (charged) and n (neutral) transform nontrivially under the $\mathrm{U}(1)_{\mathrm{L,global}}$ and $\mathrm{U}(1)_{\mathrm{R,global}}$ subgroups respectively. We have calculated the neutral and charged fermion propagators both to one-loop order in perturbation theory in $1/\widetilde{\kappa}$ [8], and also numerically [9]. We could show that 1.) the unwanted species doublers decouple in all cases and 2.) only the $\psi^{\mathrm{n}}_{\mathrm{R}}$- and $\psi^{\mathrm{c}}_{\mathrm{L}}$-propagators exhibit in the CL isolated poles at $p = (0,0,0,0)$, which correspond to the desired massless fermion states. We found that non-analytic terms occur in the $\psi^{\mathrm{n}}_{\mathrm{L}}$- and $\psi^{\mathrm{c}}_{\mathrm{R}}$-propagators and that there are no poles in these channels in the CL. If the $\psi^{\mathrm{n}}_{\mathrm{R}}$- and $\psi^{\mathrm{c}}_{\mathrm{L}}$ fermions are the only free fermions that exist in the CL of the reduced model, we would expect that the $\psi^{\mathrm{n}}_{\mathrm{L}}$-and $\psi^{\mathrm{c}}_{\mathrm{R}}$-correlation functions in coordinate space *factorize* for sufficiently large separations $|x-y|$ in the following manner

$$\langle \psi^{\mathrm{n}}_{\mathrm{L},x} \overline{\psi}^{\mathrm{n}}_{\mathrm{L},y} \rangle \sim \langle \psi^{\mathrm{c}}_{\mathrm{L},x} \overline{\psi}^{\mathrm{c}}_{\mathrm{L},y} \rangle \langle \phi^\dagger_x \phi_y \rangle \, , \qquad \langle \psi^{\mathrm{c}}_{\mathrm{R},x} \overline{\psi}^{\mathrm{c}}_{\mathrm{R},y} \rangle \sim \langle \psi^{\mathrm{n}}_{\mathrm{R},x} \overline{\psi}^{\mathrm{n}}_{\mathrm{R},y} \rangle \langle \phi_x \phi^\dagger_y \rangle \, . \tag{5}$$

We were able to show in one-loop perturbation in $1/\widetilde{\kappa}$ that the two relations in eq. (5) hold both in the FM phase and also in the CL, i.e. for $\kappa \searrow \kappa_{\mathrm{FM-FMD}}$ [8]. It is important to confirm these relations also nonperturbatively.

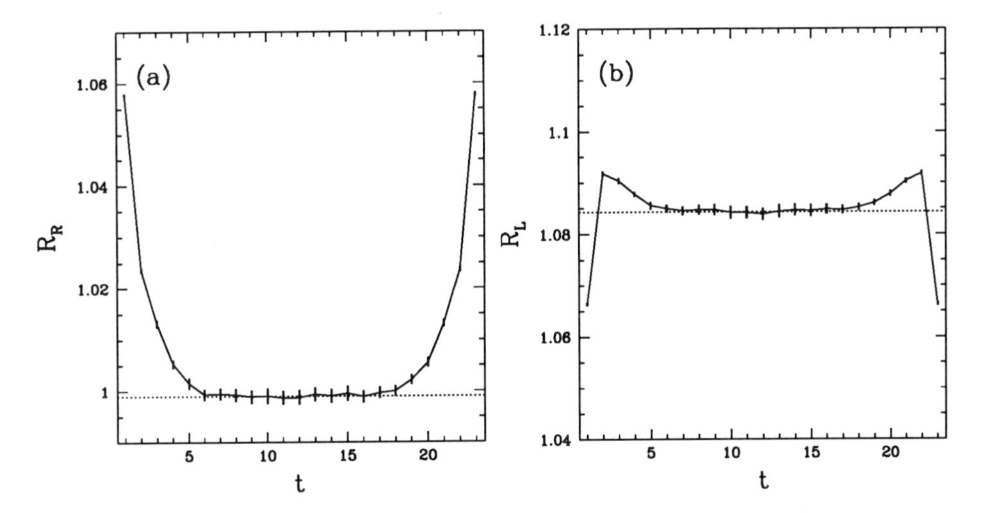

Figure 1: The ratios R_R (a) and R_L (b) as function of t at $(\tilde{\kappa}, \kappa) = (0.2, 0.3)$ ($r = 1$). The lattice size is $6^3 24$. The data points are connected by solid lines and the horizontal dotted lines are to guide the eyes.

4 Numerical Results

To this end we have performed a quenched simulation at the point $(\tilde{\kappa}, \kappa) = (0.2, 0.3)$ in the FM phase. We set $r = 1$ and determined the five correlation functions $\langle \psi^{\mathrm{n}}_{\mathrm{L},x} \overline{\psi}^{\mathrm{n}}_{\mathrm{L},y} \rangle$, $\langle \psi^{\mathrm{n}}_{\mathrm{R},x} \overline{\psi}^{\mathrm{n}}_{\mathrm{R},y} \rangle$, $\langle \psi^{\mathrm{c}}_{\mathrm{L},x} \overline{\psi}^{\mathrm{c}}_{\mathrm{L},y} \rangle$, $\langle \psi^{\mathrm{c}}_{\mathrm{R},x} \overline{\psi}^{\mathrm{c}}_{\mathrm{R},y} \rangle$ and $\langle \phi^{\dagger}_x \phi_y \rangle$ on an cylindrical lattice of size $L^3 T$ with $L = 6$, $T = 24$. For the fermion fields we used antiperiodic (periodic) boundary conditions in the temporal (spatial) directions, whereas for the scalar field we used periodic boundary conditions in all directions.

To compute the four fermionic correlation functions numerically we have to employ point sources and sinks which implies that a very high statistics is required to obtain a satisfactory signal to noise ratio. We set

$$y(x,t) = x_1 \hat{1} + x_2 \hat{2} + x_3 \hat{3} + \mathrm{mod}(x_4 + t, T) \hat{4} , \tag{6}$$

with $t = 1, \ldots, T-1$. For a given scalar field configuration we have randomly picked a source point on each time slice and averaged over the resulting T correlation functions. For the computation of the fermionic correlation functions we used in total 1300 scalar configurations which were generated with a 5-hit Metropolis algorithm and, in order to reduce autocorrelation effects, were separated by 2000 successive Metropolis sweeps.

In the case of the bosonic correlation function we summed, for a given configuration, over all lattice sites x, $\langle (1/(L^3 T) \sum_x \phi^{\dagger}_x \phi_{y(x,t)} \rangle$, where $y(x,t)$ is given again by eq. (6), and measured the bosonic correlation function on each of the 1300×2000 configurations.

To check if relation (5) holds also nonperturbatively we have computed the two ratios

$$R_L = \frac{\langle \psi^{\rm n}_{{\rm L},x} \overline{\psi}^{\rm n}_{{\rm L},y} \rangle}{\langle \psi^{\rm c}_{{\rm L},x} \overline{\psi}^{\rm c}_{{\rm L},y} \rangle \langle \phi^\dagger_x \phi_y \rangle} \,, \qquad R_R = \frac{\langle \psi^{\rm c}_{{\rm R},x} \overline{\psi}^{\rm c}_{{\rm R},y} \rangle}{\langle \psi^{\rm n}_{{\rm R},x} \overline{\psi}^{\rm n}_{{\rm R},y} \rangle \langle \phi^\dagger_x \phi_y \rangle} \,, \tag{7}$$

which should approach a constant at sufficiently large separations t. The two ratios are displayed in fig. 1 as a function of t. The two graphs clearly show that both ratios start to flatten off at $t \approx 5$ and are, within errors, indeed constant at larger separations. The fact that $R_R = 1$ (for larger t) is consistent with shift symmetry.

5 Conclusion

The quenched results presented in the last section suggest that the factorization of the $\psi^{\rm n}_L$-and $\psi^{\rm c}_R$-correlation functions (cf. eq. (5)) which we established before in 1-loop perturbation theory (cf. ref. [8]) remain valid also nonperturbatively.

As future direction of research we plan to study the U(1) model with full dynamical gauge fields. This requires the fermion representation to be anomaly free. We furthermore want to extend the gauge-fixing approach to the nonabelian gauge theories. This is a non-trivial issue, because it is not known whether the BRST formulation of gauge theories can be defined consistently beyond perturbation theory.

Acknowledgements: WB is supported by the Deutsche Forschungsgemeinschaft under grant Wo 389/3-2, MG by the US Department of Energy as an Outstanding Junior Investigator, and YS by the US-Israel Binational Science Foundation, and the Israel Academy of Science.

References

[1] A. Borelli, L. Maiani, G.-C. Rossi, R. Sisto, M. Testa, Phys. Lett. B221 (1989) 360; Nucl. Phys. B333 (1990) 335.

[2] Y. Shamir, Phys. Rev. D57 (1998) 132.

[3] M. Golterman, Y. Shamir, Phys. Lett. B399 (1997) 148.

[4] Y. Shamir, Nucl. Phys. B (Proc. Suppl.) 47 (1996) 212.

[5] Y. Shamir, Phys. Rev. Lett. 71 (1993) 2691.

[6] M. Golterman, D. Petcher, Phys. Lett. B225 (1989) 159.

[7] W. Bock, M. Golterman, Y. Shamir, hep-lat/9708019.

[8] W. Bock, M. Golterman, Y. Shamir, hep-lat/9801018.

[9] W. Bock, M. Golterman, Y. Shamir, hep-lat/9709154.

Lattice Chiral Schwinger Model: Selected Results*

V. Bornyakov [1], G. Schierholz [2 3] and A. Thimm [2 4]

[1] Institute for High Energy Physics IHEP, RU-142284 Protvino, Russia
[2] Deutsches Elektronen-Synchrotron DESY and HLRZ, D-15735 Zeuthen, Germany
[3] Deutsches Elektronen-Synchrotron DESY, D-22603 Hamburg, Germany
[4] Institut für Theoretische Physik, Freie Universität Berlin, D-14195 Berlin, Germany

Abstract: We discuss a method for regularizing chiral gauge theories. The idea is to formulate the gauge fields on the lattice, while the fermion determinant is regularized and computed in the continuum. A simple effective action emerges which lends itself to numerical simulations.

1 Introduction

The regularization of chiral fermions by means of a space-time lattice has well-known difficulties. To evade them it has been suggested [1, 2, 3] to discretize only the gauge fields and treat the fermions in the continuum by introducing an interpolation of the latter [4]. In the present talk we shall test the idea in the chiral Schwinger model. First results of this approach have been reported in [5]. For similar ideas see [6].

To compute the effective action we start from Wilson fermions. The action is

$$\begin{aligned} S_\pm &= \frac{1}{2a_f}\sum_{n,\mu}\left\{\bar\psi(n)\gamma_\mu[(1+P_\pm U^f_\mu(n))\psi(n+\mu)-(1+P_\pm U^{f\dagger}_\mu(n-\mu))\psi(n-\mu)]\right\} \\ &+ S_W(U^f), \end{aligned} \tag{1}$$

where $P_\pm = (1\pm\gamma_5)/2$, and where we have denoted the lattice spacing of the fermionic lattice by a_f. Later on we will take the limit $a_f \to 0$. In practice $a_f = a/N$, N integer, where a is the lattice spacing of the gauge field action. The link variables on the fine lattice, U^f, are obtained by a suitable interpolation [4] from the links on the original lattice, U. The usual (gauged) Wilson term S_W reads

$$S_W(U^f) = \frac{1}{2a_f}\sum_{n,\mu}\bar\psi(n)[2\psi(x) - U^f_\mu(n)\psi(n+\mu) - U^{f\dagger}_\mu(n-\mu)\psi(n-\mu)]. \tag{2}$$

*Talk given by V. Bornyakov

We will consider ungauged, $S_W(U^f = 1)$, and partially gauged Wilson terms as well.

The effective action is obtained in three steps. First one defines

$$\exp(-W_\pm) = \int \mathcal{D}\psi\mathcal{D}\bar{\psi}\exp(-S_\pm). \tag{3}$$

Then one performs the limit $\lim_{a_f\to 0} W_\pm$. One finds that this action is not invariant under chiral gauge transformations, not even in the anomaly-free model. Gauge invariance can, however, be restored by adding a local counterterm.† The counterterm can be identified analytically and its coefficient be calculated in one-loop perturbation theory [7]. It is

$$c\sum_x A_\mu^2(x), \tag{4}$$

where $A_\mu = \lim_{a_f\to 0}(1/a_f)\,\mathrm{Im}\,U_\mu^f$, and $c = -0.0202$‡ for both gauged and ungauged Wilson terms. We then arrive at the effective action

$$\widehat{W}_\pm = \lim_{a_f\to 0} W_\pm + c\sum_x A_\mu^2(x). \tag{5}$$

The effective action $W_\pm$ has been computed by means of the Lanczos method. Note that the action is non-hermitean. The Lanczos vectors were re-orthogonalized after every iteration. For smooth gauge fields it was shown that [5]

$$\mathrm{Re}\widehat{W}_\pm = \frac{1}{2}(W + W_0), \tag{6}$$

where W (W_0) is the effective action of the vector model (free theory). This result was conjectured in [1] for 'perturbative' gauge fields.

Because of lack of space we shall restrict ourselves in the written version of the talk to rough gauge fields. We assume that the reader is familiar with the problem and with previous results on the subject.

2 Effective action revisited

Before we turn to the problem of rough gauge fields, let us briefly mention some new results on the effective action.

On the L^2 torus the gauge field can be written

$$A_\mu(x) = \frac{2\pi}{L}t_\mu + \varepsilon_{\mu\nu}\partial_\nu\alpha(x) + \mathrm{i}g^{-1}(x)\partial_\mu g(x), \tag{7}$$

assuming $A_\mu(x) \in [-\pi,\pi)$, where $t_\mu \in [-1,1)$ are the zero momentum modes (torons), $\Box\,\alpha(x) = -F_{12}(x)$ and $g(x) \in U(1)$ is a gauge transformation.§ We assume periodic boundary conditions for the gauge fields and antiperiodic boundary conditions for the fermions.

†It was shown in [3] for a rather general class of gauge fields, not including compact $U(1)$ fields though, that the effective action can be made gauge invariant and finite.

‡This value can also be extracted from [8].

§As we shall see this assignment is not unique if we allow large gauge transformations.

Writing $a_\mu = (2\pi/L)\, t_\mu$ and $\widetilde{A}_\mu = A_\mu - a_\mu$, the real part of the effective action factorizes in the form

$$\mathrm{Re}\widehat{W}_\pm(A) = \mathrm{Re}\widehat{W}_\pm(\widetilde{A}) + \mathrm{Re}\widehat{W}_\pm(a) - W_0. \tag{8}$$

Formally this relation follows from the property of the Dirac operator

$$D(\widetilde{A} + a) = \exp(-\mathrm{i}ax)\, D(\widetilde{A})\, \exp(\mathrm{i}ax). \tag{9}$$

A similar expression can be derived for the imaginary part. The toron part of the effective action, $\widehat{W}_\pm(a)$, is known analytically [9]. For the fluctuating part we obtain

$$\widehat{W}_\pm(\widetilde{A}) = \sum_\alpha \frac{q_\alpha^2}{4\pi} \int \mathrm{d}^2x \left\{ \widetilde{F} \frac{1}{\Box} \widetilde{F} + \mathrm{i}\epsilon_\alpha \widetilde{F} \frac{1}{\Box} \partial_\mu \widetilde{A}_\mu \right\}, \tag{10}$$

where $\widetilde{F} = \varepsilon_{\mu\nu}\partial_\mu \widetilde{A}_\nu$, $\tilde{\partial}_\mu = \varepsilon_{\mu\nu}\partial_\nu$, and q_α and ϵ_α are the fermion charge and chirality, respectively. This result is in agreement with the well known expression for the effective action in $\mathbb{R}^2$ [10]. Note that the imaginary part of (10) vanishes in the anomaly-free model.

To check the result (10) and our method of calculation, we discretized the continuum configuration [11] $A_\mu(x) = c_\mu \cos(2\pi k x/L) + (2\pi/L)\, t_\mu$ with $c_1 = c_2 = 0.32$ and $k_1 = 1$, $k_2 = 0$ (as chosen in [11]) and computed $W_\pm$. In Fig. 1 we show $\mathrm{Im}\, W_\pm(A)$ as a function of $r = a_f/a$ for two toron fields. For the extrapolated values we obtain -0.004078 and 0.002098, respectively. This is to be compared with the analytical values -0.004074 and 0.002094, respectively. We find the same good agreement with the analytic formulae for the real part.

In [7] we shall give an analytic proof of eqs. (8) and (10).

3 Vortex-antivortex configuration

In [5] we reported numerical evidence for gauge invariance of the anomaly-free effective action under a class of gauge transformations. This did not include 'singular' gauge transformations which create a vortex-antivortex pair.¶

Let us consider a lattice gauge field configuration

$$\theta_\mu(s) = \theta_\mu^v(s - v) - \theta_\mu^v(s - \bar{v}) - \frac{2\pi}{L^2}\varepsilon_{\mu\nu}(v_\nu - \bar{v}_\nu) \tag{11}$$

where $\theta_\mu^v(s) = 2\pi\varepsilon_{\mu\nu}\partial_\nu G(s)$, $G(s)$ being the lattice inverse Laplacian. This configuration corresponds to a vortex-antivortex pair at positions v and $\bar{v}$, respectively [13]. It is gauge equivalent to the vacuum configuration $\theta_\mu(s) = 0$. This configuration gives rise to a non-zero toron field $t_\mu = -(1/L)\, \varepsilon_{\mu\nu}(v_\nu - \bar{v}_\nu)$. The imaginary part of the effective action is again given by the toron field contribution. It is non-zero. The real

¶Note that singular gauge transformations generally create a problem in the overlap approach [12].

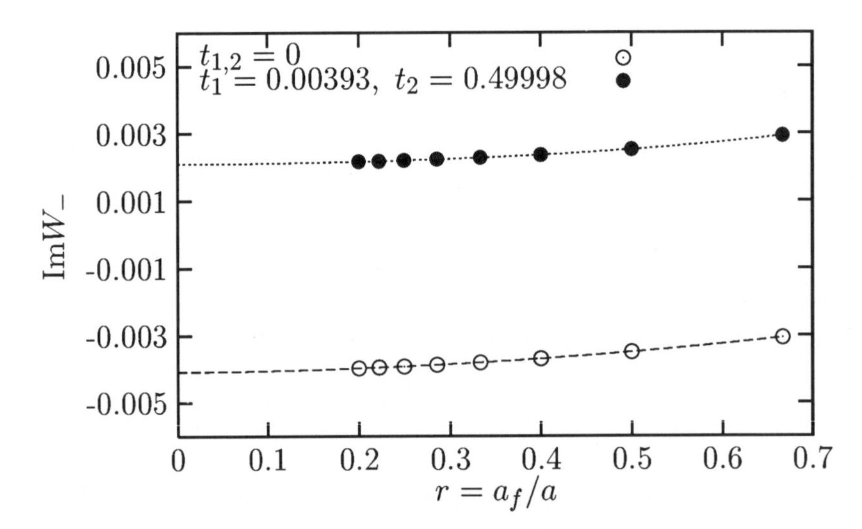

Figure 1: $\mathrm{Im}W_-$ as a function of r for the configuration described in the text and $(q,\epsilon) = (1,-1)$.

part of the effective action is found to be divergent as $a_f \to 0$, in agreement with the analytic result (10). In Fig. 2 we show $\mathrm{Re}\, W_-^{\Sigma} = \mathrm{Re}\, W_- + c \sum_{s,\mu} 2\,(1 - \cos(\theta_\mu(s)))$ as a function of r for two different distances of the vortices. The divergence is $\propto \log(1/a_f)$. Thus we conclude that these configurations have zero weight in the partition function. We also expect that they do not contribute to any observable.

4 Index theorem

The lattice action must fulfill the index theorem in order to be in the same universality class as the continuum action. In general, the index theorem states that the number of zero modes of positive chirality minus the number of zero modes for negative chirality is equal to the topological charge, $n_+ - n_- = Q$. In two dimensions it even holds that $n_+ = Q$ for $Q > 0$ and $n_- = |Q|$ for $Q < 0$. Accordingly, a right(left)-handed fermion has $Q\,\theta(Q)$ $(-Q\,\theta(-Q))$ zero modes.

We have checked that the vector model satisfies the index theorem. For the eigenvalues we find, both numerically and analytically, the asymptotic behavior $E_0 = 2\pi|Q|(a_f/aL)^2$. For the chiral model, and both gauged and ungauged Wilson terms, the index theorem is, however, broken. For the ungauged Wilson term there are no zero modes at all, while for the gauged Wilson term we find zero modes of both chiralities like in the vector model.

The index theorem can be restored by considering the following Wilson term

$$S_W^{\pm}(U^f) = \frac{1}{4a_f} \sum_{n,\mu} \bar{\psi}(n) P_{\pm}[2\psi(x) - U_\mu^f(n)\psi(n+\mu) - U_\mu^{f\,\dagger}(n-\mu)\psi(n-\mu)]$$

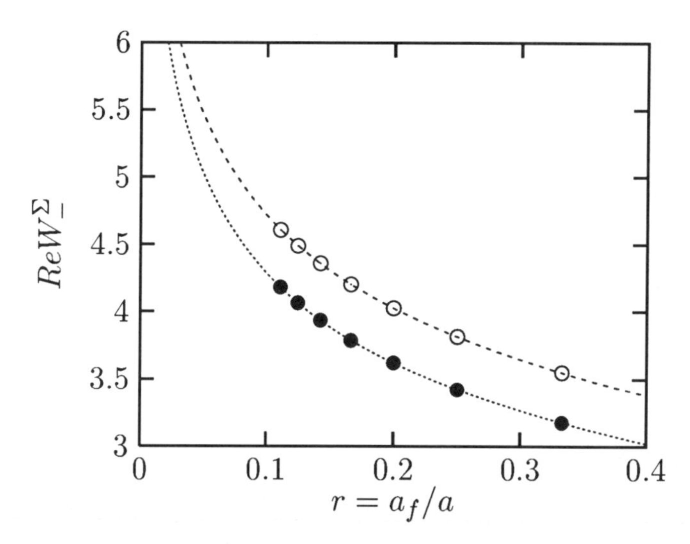

Figure 2: $\mathrm{Re}W_-^\Sigma$ as a function of r for the configuration (11) and $(q,\epsilon) = (1,-1)$.

$$+ \quad \frac{1}{4a_f} \sum_{n,\mu} \bar{\psi}(n) P_\mp [2\psi(x) - \psi(n+\mu) - \psi(n-\mu)]. \tag{12}$$

The Wilson term (12) has the property $S_W^+(U^f) + S_W^-(U^f) = S_W(U^f) + S_W(0)$. We have checked numerically that it fulfills the index theorem for $|Q| = 1$. The next step is to verify the index theorem also for higher topological charges.

In the $Q = 0$ sector the Wilson term (12) gives the same results as before. The coefficient c of the counterterm is different though. In this case we obtain $c = -0.05971$.

5 Conclusions

We may consider the problem of formulating the chiral Schwinger model on the lattice as being solved for the $Q = 0$ sector. Numerical simulations are now feasible. The real part of the effective action is effectively half the action of the corresponding vector model, while the imaginary part of the anomaly-free model can be computed analytically from the toron fields. We hope to be able to report results on the sectors with non-vanishing topological charge in the near future.

Acknowledgements

One of us (VB) would like to thank DESY-Zeuthen for its hospitality. This work has been supported in part by INTAS and the Russian Foundation for Fundamental Sciences through grants INTAS-96-370 and 96-02-17230a.

References

[1] L. Alvarez-Gaumé and S. Della-Pietra, in *Recent Developments in Quantum Field Theory*, eds. J.Ambjorn, B. J. Durhaus and J. L. Petersen (North-Holland, 1985).

[2] M. Göckeler and G. Schierholz, Nucl. Phys. B (Proc. Suppl.) 29B,C (1992) 114, *ibid.* 30 (1992) 609.

[3] G. 't Hooft, Phys. Lett. B349 (1995) 491.

[4] M. Göckeler, A.S. Kronfeld, G. Schierholz and U.-J. Wiese, Nucl. Phys. B404 (1993) 839.

[5] V. Bornyakov, G. Schierholz and A. Thimm, DESY 97-178 (1997) (hep-lat/9709037).

[6] P. Hernandez and R. Sundrum, Nucl. Phys. B455 (1995) 287; G. T. Bodwin, Phys. Rev. D54 (1996) 6497; I. Montvay, CERN-TH-95-123 (1995) (hep-lat/9505015).

[7] V. Bornyakov, G. Schierholz and A. Thimm, in preparation.

[8] A. A. Slavnov and N. V. Zverev (hep-lat/9708022).

[9] L. Alvarez-Gaumé, G. Moore and C. Vafa, Commun. Math. Phys. 106 (1986) 1; M. Göckeler and G. Schierholz, unpublished (1994); R. Narayanan and H. Neuberger, Phys. Lett. B348 (1995) 549.

[10] R. Jackiw and R. Rajaraman, Phys. Rev. Lett. 54 (1985) 1219.

[11] R. Narayanan and H. Neuberger, Nucl. Phys. B443 (1995) 305.

[12] R. Narayanan and H. Neuberger, Nucl. Phys. B477 (1996) 521; T. Aoyama and Y. Kikukawa, Nucl. Phys. B (Proc. Suppl.) 53 (1997) 638.

[13] V. Mitrjushkin, Phys. Lett. B389 (1996)713.

Chiral symmetry breaking and confinement in QCD

Yu.A.Simonov

Institute of Theoretical and Experimental Physics
117218, Moscow, B.Cheremushkinskaya 25, Russia

Abstract: The effective quark Lagrangian is derived in the limit of large N_c, containing vacuum gluon field correlators as interacting kernels. Keeping only the Gaussian field correlator one obtains nonlocal nonlinear equations for the heavy–light Green's function exhibiting the scalar confinement and hence chiral symmetry breaking.

A similar type of equations is obtained for the gluon propagator which allows to calculate the Gaussian field correlator, selfconsistently from the first principles at large N_c. As a first estimate the gluon correlation length T_g is found analytically in terms of string tension, which is numerically close to the lattice data.

1 Introduction

The chiral symmetry breaking (CSB), confinement and their interrelation in QCD belong to the unsolved complicated problems of QCD. Here I propose the novel approach based on new nonlinear gauge–invariant equations for the quark and gluon Green's function.

To write gauge–invariant equations one considers quark or gluon propagator in presence of a static source at the origin.

2 Equations for the quark propagator

We start with the quark Green's function and write the effective quark Lagrangian in presence of a static source, using the averaging of the partition function over gluonic field A_μ.

To take into account the static source we consider the generalized coordinate gauge [1] and express A_μ through $F_{\mu\nu}$ as

$$A_\mu(x) = \int_C ds \frac{dz_\alpha(s,x)}{ds} F_{\alpha\beta}(z) \frac{dz_\beta}{dx_\mu} \tag{1}$$

where the contour C starts at x_μ and is described by $z_\mu(x,s)$ (in the usual coordinate gauge $z_\mu(x,s) = sx_\mu, 0 \le s \le 1$). The effective Lagrangian is (a more extended version of this derivation see in [1] and [2])

$$\mathcal{L}_{eff}(\psi^+\psi) = \int \psi^+(x)(-i\hat{\partial} - im)\psi(x)d^4x +$$

$$\frac{1}{2N_c}\int d^4x d^4y (\psi_a^+(x)\gamma_\mu\psi_b(x))(\psi_b^+(y)\gamma_{\mu'}\psi_a(y)) \times \tag{2}$$

$$\times J_{\mu\mu'}(x,y)$$

where we have defined

$$J_{\mu\mu'}(z,w) = \int_C^z du_\alpha \int_C^w dv_\gamma (\delta_{\alpha\gamma}\delta_{\beta\delta} - \delta_{\alpha\delta}\delta_{\beta\gamma})\frac{du_\beta}{dz_\mu}\frac{dv_\delta}{dw_{\mu'}} \times$$

$$\times D(u-v) \tag{3}$$

and $D(u)$ is the Kronencker part [3] of the field correlator $< F(u)F(0) >$, yielding the string tension

$$\sigma = \frac{1}{2}\int_{-\infty}^{\infty} d^2u D(u) \tag{4}$$

Note that we have neglected in (2) higher field correlators, which can be shown to yield subdominant contribution.

The Lagrangian (2) can be used to obtain equations for the quark Green's function S in the large N_c limit, where the following rule of replacement holds

$$\psi_b(x)\psi_b^+(y) \to < \psi_b(x)\psi_b^+(y) >= N_c S(x,y), \tag{5}$$

One obtains a system of equations for the quark Green's function S and the mass operator M

$$iM(z,w) = J_{\mu\nu}(z,w)\gamma_\mu S(z,w)\gamma_\nu \tag{6}$$

$$(-i\hat{\partial}_z - im)S(z,w) - i\int M(z,z')S(z',w)d^4z' = \delta^{(4)}(z-w) \tag{7}$$

The system of equations (6-7) is exact in the large N_c limit, when higher correlators are neglected and defines unambiguously both the interaction kernel M and the Green's function S. One should stress at this point again that both S and M are not the one-particle operators but rather two–particle operators, with the role of the second particle played by the static source. It is due to this property, that S and M are gauge invariant operators, which is very important to take confinement into account properly. Had we worked with one–particle operators, as is the habit in QED and sometimes also in QCD, then we would immediately loose the gauge invariance and the string, and hence confinement.

3 Chiral symmetry breaking

The CSB can manifest itself in solutions of (6),(7) in several ways. One is the appearance of the nonzero chiral condensate

$$< \bar{\psi}(0)\psi(0) >= iN_c tr\ S(0,0) \tag{8}$$

This was estimated using the relativistic WKB method in [2] to be

$$< \bar{\psi}(o)\psi(0) >\sim -N_c\sigma/T_g const \tag{9}$$

where T_g is the gluonic correlation length [1-3] and const is of the order of unity.

Another manifestation of CSB is the scalar confinement, which is seen at large distances r. Indeed one write expansion for the Green's function S and M in the inverse powers of the string mass $M_{str} = \sigma r + m$, and the time–averaged Green's function $\bar{S}$ satisfies an equation [1]:

$$[-i\vec{\gamma}\vec{\partial} - i(m + \sigma|\vec{z}|)]\bar{S}(\vec{z},\vec{w}) = \delta^{(3)}(\vec{z} - \vec{w}) \tag{10}$$

In (10) the string term σr enters as a scalar, which signals CSB.

Equations (9) and (10) exemplify the connection between CSB and confinement in the large N_c limit. The necessary appearance of CSB in this limit was proved earlier in [4] but no hint to the possible mechanism was given. Here we demonstrate that CSB occurs due to the string formation (which is contained in factor σ in (9), (10) and the kernel J in (6),(7)), but this effect comes to the existence only due to the solution of nonlinear equations (6),(7). One should mention that these equations are similar to those of NJL [5], but are nonlocal and therefore do not need cut-off. The CSB solution is not obtained by perturbation expansion of (6),(7), but rather is an extra, nonperturbative solution existing due to the nonlinearity. A similar situation occurs in the NJL model [5].

4 Equation for the gluon propagator [6]

We now consider gluons in the field of the static (adjoint) source and use again the gauge (1) to take the source into account.

To write these equations we divide all $N_c^2 - 1$ gluonic fields into two groups: a small group of r fields $b^i_\mu, i = 1,...r$, which later be called "valence gluons", and a large group of fields $\bar{A}^a_\mu$, $a = r + 1,...N_c^2 - 1$, which form a background for valence gluons.

As will be shown below this background creates in the confining phase after vacuum averaging an adjoint string. Hence, physically the problem reduces to the motion of valence gluons in the field of the string. Since the string is a white object, the action of the string on valence gluons generates color–diagonal mass operator and the resulting equation for the valence gluon propagator simplifies considerably. We shall find out that these equations are nonlinear and selfcoupled, so that one can look for selfconsistent solutions for the valence gluon propagator and the field correlator. Inserting the resulting field correlator in the kernel, one realizes the selfcoupling mentioned above.

Our purpose now is to write the effective action for the valent gluons b_μ, which is obtained when one has averaged over "background fields" $\bar{A}_\mu$. Note at this point, that both b_μ and $\bar{A}_\mu$ contain perturbative and nonperturbative contributions, and the separation made is of technical character, rather than physical, at any distance. In the specific situation to be considered below, when distance to the static source is large, the field $\bar{A}_\mu$ creates the adjoint string and is predominantly nonperturbative, as well as b_μ.

As a result of averaging of the total action $S(A)$, one obtains

$$e^{-S_{eff}(b)} =< e^{-S(b+\bar{A})} >_{\bar{A}} \tag{11}$$

To calculate $S_{eff}(b)$ the cluster expansion theorem is used for (11) and only quadratic cumulants are retained. The leading term at large distances is $S_{eff}(b) = \int L_{eff}(b) d^4x \cong \int < S^{(2,2)} >_A d^4x$ where we keep the dominant contribution to S_{eff} at large N_c and large distance

$$< S^{(2,2)} >_A = \frac{1}{2}\chi^{ik,ab}\{b^i_\mu(x)b^k_\mu(x) < \bar{A}^a_\lambda(x)\bar{A}^b_\lambda(x) >_{\bar{A}} -$$

$$-b^i_\mu(x)b^k_\nu(x) < \bar{A}^a_\mu(x)\bar{A}^b_\nu(x) >_{\bar{A}}\}, \tag{12}$$

where

$$\chi^{ik,ab} \equiv g^2 \sum_{c=1}^{N_c^2-1} f^{cia} f^{ckb}.$$

Now using(1) one can introduce the kernel $J_{\mu\nu}$ (3) and write

$$< S^{(2,2)} >_{\bar{A}} = \frac{N_c}{2C_2^f} b^i_\mu(x) b^i_\nu(x) (\delta_{\mu\nu} J_{\lambda\lambda}(x,x) - J_{\mu\nu}(x,x)), \tag{13}$$

where $C_2^f = \frac{N_c^2-1}{2N_c}$.

The next step is the same as in section 2 made for the quark propagator, namely one is using at large N_c the formal replacement

$$b_\mu(x) b_\nu(y) \to < b^i_\mu(x) b^k_\nu(y) >_{b,\bar{a}} \equiv G^{ik}_{\mu\nu}(x,y) \tag{14}$$

and obtainsan equation of the Dyson–Schwinger–type,

$$(-\partial^2_\lambda \delta_{\mu\rho} + \partial_\mu \partial_\rho) G_{\rho\nu}(x,y) + \int M^{(g)}_{\mu\rho}(x,z) G_{\rho\nu}(z,y) d^4z = \delta^{(4)}(x-y), \tag{15}$$

where the mass operator $M^{(g)}$ is approximately equal at large distances to $M^{(2,2)}$, where

$$M^{(2,2)}_{\mu\nu}(x,y) = \frac{N_c}{C_2^f} \delta^{(4)}(x-y) [J_{\lambda\lambda}(x,y)\delta_{\mu\nu} - J_{\mu\nu}(x,y)], \tag{16}$$

For more extended treatment and derivation of equations (14),(15) the reader is referred to [6].

The kernel $J_{\lambda\mu}$ in (15) is expressed through the field correlator $< FF >$. To make equations selfconsistent one should express the latter through the gluon Green's function G (14). This is possible since one can always refer the color indices in each term of $\sum_{a=1}^{N_c^2-1} < F^a F^a >$ to the group of fields b_μ. Then one can write symbolically

$$tr < F(x)F(y) > \sim \partial_\mu \partial_\nu G(x,y) + (G(x,y))^2 + perm + ... \tag{17}$$

where ellipsis stands for higher cumulants.

The system of equations (14-16) allows for a nonperturbative solution, which violates the scale invariance present in the equations. This solution is defined by fixing one nonperturbative scale, e.g. the string tension σ. Then equations (14-16) predict that i) both field correlators $D(x), D_1(x)$ [3] exponentially decay at large x; $D(x), D_1(x) \sim exp(-x/T_g)$ in agreement with lattice data [7], and ii) the gluon correlation length T_g is connected to σ as [6]

$$1/T_g = (2.33)^{3/4} \sqrt{\frac{9\sigma}{2\pi}} \tag{18}$$

Insertion of the standard value $\sigma \approx 0.2 GeV^2$ yields $T_g \approx 0.2 fm$, which is in good agreement with lattice data [7].

5 Conclusion

We have derived equations for the gluon and quark propagators in the field of a static source. These equations possess symmetry (chiral for the quark and scale invariance for the gluon) which is violated by the nonperturbative solutions. One obtains in this way the CSB due to the confining kernel, and the confining kernel itself satisfies nonlinear equations. Properties of this kernel are in agreement with lattice data. The support from RFBR grant 97-02-16406 and the RFBR-DFG grant 96-02-00088G, 95-02-05436, 97-02-17491 is gratefully acknowledged.

References

[1] Yu.A.Simonov, Chiral symmetry breaking and confinement in the heavy–light $q\bar{q}$ system in QCD, hep-ph/9712248.

[2] Yu.A.Simonov, Theory of light quarks in the confining vacuum, hep-ph/9704301.

[3] H.G.Dosch and Yu.A.Simonov, Phys. Lett. **B205** (1988) 339.

[4] S.Coleman and E.Witten, Phys. Rev. Lett. **45** (1980) 100.

[5] Y.Nambu and G.Iona-Lasinio, Phys. Rev. **122** (1961) 345
S.P.Klevansky, Rev. Mod. Phys. **64** (1992) 649.

[6] Yu.A.Simonov, Selfcoupled equations for the field correlators, hep-ph/9712250.

[7] A.Di Giacomo and H.Panagopoulos, Phys. Lett. **B285** (1992) 133.

$O(a)$ Improvement for Quenched Wilson Fermions

S. Capitani[1], M. Göckeler[2], R. Horsley[*3], B. Klaus[4,5], H. Oelrich[4], H. Perlt[6], D. Petters[4,5], D. Pleiter[4,5], P. Rakow[4], G. Schierholz[1,4], A. Schiller[6] and P. Stephenson[7]

[1] Deutsches Elektronen-Synchrotron DESY, D-22603 Hamburg, Germany
[2] Institut für Theoretische Physik, Universität Regensburg, D-93040 Regensburg, Germany
[3] Institut für Physik, Humboldt-Universität zu Berlin, D-10115 Berlin, Germany
[4] DESY, HLRZ und Institut für Hochenergiephysik, D-15735 Zeuthen, Germany
[5] Institut für Theoretische Physik, Freie Universität Berlin, D-14195 Berlin, Germany
[6] Institut für Theoretische Physik, Universität Leipzig, D-04109 Leipzig, Germany
[7] Dipartimento di Fisica, Università degli Studi di Pisa e INFN, Sezione di Pisa, 56100 Pisa, Italy

Abstract: We briefly describe some of our recent results for the mass spectrum and matrix elements using $O(a)$ improved fermions for quenched QCD. Where possible a comparison is made between improved and Wilson fermions.

1 Introduction

Upon discretising the QCD-Lagrangian in the Wilson formulation we find that the gluon piece has $O(a^2)$ errors, but for the fermion part, the additional Wilson term, necessary to avoid the fermion 'doubling' problem gives us $O(a)$ errors. Thus we would expect that looking at any physical quantity, for example a mass ratio, we have

$$\frac{m_H}{m_{H'}} = r_0 + ar_1 + a^2 r_2 + O(a^3). \tag{1}$$

Symanzik developed a systematic perturbative progamme to $O(a^n)$ in which a basis of irrelevant operators is added to the Lagrangian to completely remove $O(a^{n-1})$ effects. Restricting this to on-shell (or physical) quantities, [1], enables the equations of motion to be used to reduce the required set of operators, in both the action and for improved operators in matrix elements. In this talk we briefly report our progress on computing the hadron spectrum and matrix elements for $O(a)$ improved fermions, where in this case we expect for a physical quantity such as a mass ratio

$$\frac{m_H}{m_{H'}} = r_0 + a^2 r_2' + O(a^3). \tag{2}$$

---

*Talk presented by R. Horsley at the 31st Ahrenshoop Symposium on the Theory of Elementary Particles, September 1997, Buckow, Germany.

The ground-work has been laid by the Alpha collaboration who succeeded in non-perturbatively calculating several improvement coefficients for $\beta \equiv 6.0/g^2 \geq 6.0$. Reviews and further references may be found in [2]. We shall here neither describe their method nor go into details of how the mass spectrum or matrix elements are computed on the lattice. A description of our spectrum results is given in ref. [3] while for Wilson matrix elements see, for example, [4]. Up to now we have generated configurations on $(16^3, 24^3) \times 32$ lattices at $\beta = 6.0$ and on $24^3 \times 48$ lattices at $\beta = 6.2$. We have used the string tension, $\sqrt{K}$, to determine the lattice spacing a at these two g^2 values. A physical value of $\sqrt{K} = 0.427\text{GeV}$ is taken to set the scale.

2 The mass spectrum

We have computed the ρ ($J^{PC} = 1^{--}$) and nucleon (N) masses together with the a_0 (0^{++}), a_1 (1^{++}) and b_1 (1^{+-}) masses. The latter three mesons being p-wave states are difficult to measure with our symmetric sources and our results should only be taken as indicative of possible trends. In Fig. 1 we plot this hadron spectrum

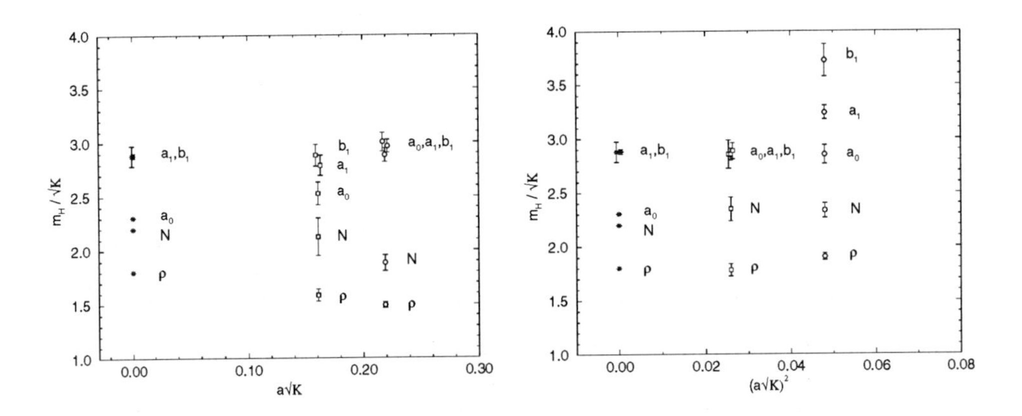

Figure 1: The hadron spectrum for light hadrons at $\beta = 6.0$ (circles), 6.2 (squares) using Wilson fermions (left picture) and using $O(a)$ improved fermions (right picture) against the lattice spacing, a. Experimental numbers are shown as stars.

both for Wilson and $O(a)$ improved fermions. Roughly the same amount of *CPU* time has gone into producing each of these pictures. From eqs. (1,2) we expect that the dominant discretisation terms are such that linear extrapolations in a for Wilson fermions and a^2 for $O(a)$ improved fermions are sufficient. Clearly with only two points we must limit ourselves here to qualitative observations†. First we note that due to the absence of the $O(a)$ term for the improved fermions, the convergence to the continuum limit is faster. On the other hand it does appear

†Presently, for $O(a)$ improved fermions, most groups have concentrated on $\beta = 6.0$, 6.2, [5], but there has been a recent interesting attempt to go to lower β values, [6].

that, especially for heavier particles, the signal fluctuates more. It also seems that at least for the ρ and N particles, $O(a)$ improved fermions results lie closer to the continuum result – the Wilson results have a noticeable gradient. An alternative way of looking at these results is given in Fig. 2. While it is not necessary that quenched QCD should reproduce exactly the continuum spectrum, in many present-

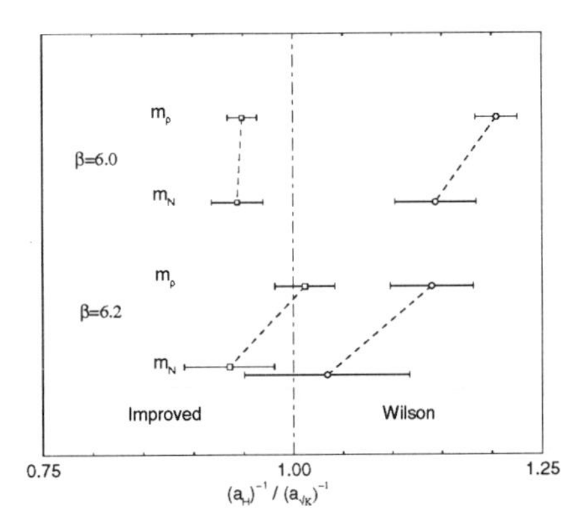

Figure 2: a_H^{-1} estimates using $H = \rho$ or N as a scale normalised to the string tension for Wilson fermions (circles) compared to $O(a)$ improved masses (squares).

day applications a hadron mass (typically the ρ or N) is used to set the scale. Potential discrepancies between these scales can be revealed by normalising them against the string tension (which has $O(a^2)$ errors). Again this confirms the previous impression: $O(a)$ improved fermions for the ρ or N perform better than Wilson fermions.

3 Matrix elements

Matrix elements, such as decay constants or moments of structure functions, are more complicated to calculate than masses as the operators must be appropriately renormalised, so that $\mathcal{O}^R = Z_O(1 + b_O a m_q)\mathcal{O}$, with $\mathcal{O} = O + \sum_I c_I a O_I$. For present-day β values, non-perturbative (NP) determinations of Z_O, b_O and $\{c_I\}$ are preferable. (Here we are mainly interested in matrix elements in the chiral limit, so we do not need b_O; the other irrelevant operators O_I are only required in the improved case to ensure complete $O(a)$ cancellation.) In Fig. 3 we show Z_V ($V_\mu = \overline{q}\gamma_\mu q$) and Z_A ($A_\mu = \overline{q}\gamma_\mu\gamma_5 q$) for both Wilson and $O(a)$ improved fermions. We see that in all cases, while first order perturbation theory lies $\geq 10\%$ away from NP computations, tadpole improvement (TI) always gives results closer to the NP line. For the Wilson case we expect to find larger discrepancies (of $O(a)$) between the various NP estimates of Z than for improved fermions (of $O(a^2)$). This seems most obvious for Z_V. We next note that both the Alpha, [10], and our non-perturbative determination of

Z_V for $O(a)$ improved fermions are in very good agreement with each other. From current conservation we expect $\langle N|\mathcal{V}_\mu^R|N\rangle = p_\mu/E_N\chi_q$ with $\chi_u = 2$, $\chi_d = 1$. This at (eg) zero momentum allows a non-perturbative determination of Z_V (and b_V), for both Wilson and $O(a)$ improved fermions. This result is exemplified in Fig. 4 where we perform a check of momentum effects and restoration of rotational invariance, by computing $\langle N|\mathcal{V}_4^R|N\rangle$, $\langle N|\mathcal{V}_1^R|N\rangle \times E_N/p_1$. While all results for $\langle N|\mathcal{V}_4^R|N\rangle$ are consistent with each other, $\langle N|\mathcal{V}_1^R|N\rangle \times E_N/p_1$ shows some spread; indeed at $\beta = 6.0$, Wilson fermions seem to perform better than the $O(a)$ improved fermions.

In the left picture of Fig. 5 we apply the Alpha result for Z_A to a determination of g_A. There seems to be an approach to the experimental value of g_A (but with rather large $O(a^2)$ corrections). For the moments of the quark distributions in the nucleon, we obtain the results shown in the right hand picture in Fig. 5. The lowest moment, $\langle x\rangle$, derived from the operator $\overline{q}\gamma_\mu D_\nu q$ is interesting, not only because it contains one derivative, but also because there are two distinct lattice representations involving diagonal or off-diagonal elements, the latter requiring for its evaluation a non-zero spatial momentum. (Again we always used the minimum possible.) This potentially allows a better study of the approach to the continuum limit. At present a non-perturbative renormalisation of $\langle x\rangle$ is not known; however no tadpole term contributes to first order perturbation theory, so that this automatically lies close to the *TI* result, so we might hope that any error in the renormalisation constant is small. A further problem for the $O(a)$ improved operators is the addition of at least one irrelevant operator. However, as discussed in [8], we hope that its effect is also small. For $\langle x\rangle$, from Fig. 5, all results appear to be reasonably consistent with each other, and for $O(a)$ improved fermions there does not seem to be any large $O(a^2)$ effect. However, there seems to be a larger deterioration of the signal (in comparison to the Wilson case) when introducing a momentum into the operator.

In conclusion, it would seem that $O(a)$ Symanzik improvement does indeed bring us closer to the goal of calculating continuum masses and matrix elements in *QCD*. The numerical calculations were performed on the Quadrics facility at *DESY-IfH*.

References

[1] M. Lüscher et al., Comm. Math. Phys. 97 (1985) 59.

[2] M. Lüscher et al., Nucl. Phys. B(Proc. Suppl.) 53 (1997) 905, hep-lat/9608049; S. Capitani et al., hep-lat/9709125; H. Wittig, hep-lat/9710013; R. Sommer, hep-ph/9711243.

[3] M. Göckeler et al., hep-lat/9707021.

[4] M. Göckeler et al., Phys. Rev. D53 (1996) 2317, hep-lat/9508004; hep-ph/9708270; C. Best et al., Phys. Rev. D56 (1997) 2743, hep-lat/9703014.

[5] A. Cucchieri et al., hep-lat/9711040; P. A. Rowland, Lat97.

[6] R. G. Edwards et al., hep-lat/9711052.

[7] S. Capitani et al., hep-lat/9709049.

[8] S. Capitani et al., hep-lat/9709036.

[9] S. Aoki et al., Nucl. Phys. B(Proc. Suppl.) 53(1997) 209, hep-lat/9608144.

[10] M. Lüscher et al., Nucl. Phys. B491 (1997) 344, hep-lat/9611015.

[11] G. M. de Divitiis et al., hep-lat/9710071.

[12] M. Göckeler et al., hep-lat/9710052.

[13] A. D. Martin et al., Phys. Lett. B354 (1995) 155, hep-ph/9502336.

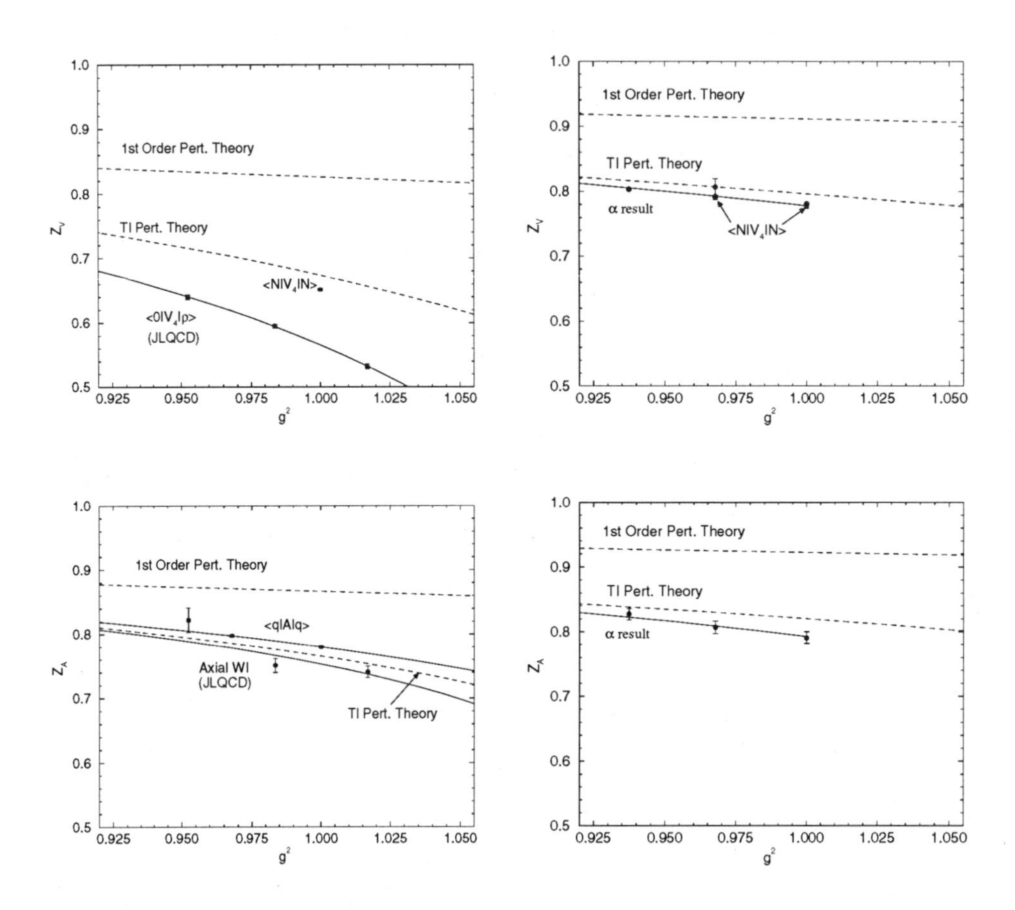

Figure 3: Various determinations of Z_V (upper row) and Z_A (lower row) for Wilson (left column) and $O(a)$ improved fermions (right column). The tadpole improved (*TI*) results have been obtained from first order perturbation results, $Z_O(g^2) = 1 + c_O g^2 + O(g^4)$, [7], using the procedure given in [3]. Also plotted are results using the local vector current between nucleon, [8] (filled squares) and ρ states, [9] (filled circles), and the local axial current between quark states, [12] (filled squares). Where appropriate, a simple Padé interpolation/extrapolation has been applied, $Z_O(g^2) = (1 + p_O g^2 + q_O g^4)/(1 + r_O g^2)$ with $p_O - r_O = c_O$. For the $O(a)$ improved fermions we also give the Alpha results (filled circles) and Padé extrapolation, [10] and from [11] (open circle).

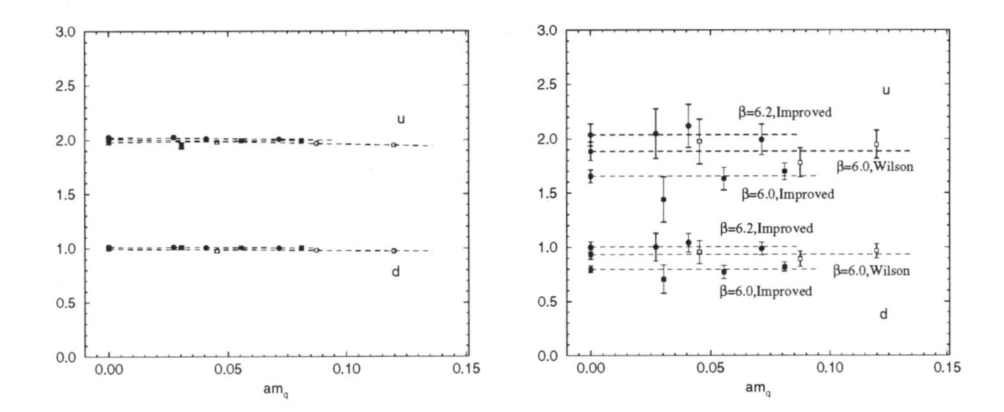

Figure 4: $\langle N|\mathcal{V}_4^R|N\rangle$, left picture, $\langle N|\mathcal{V}_1^R|N\rangle \times E_N/p_1$, right picture, at $\vec{p} = (p_1, 0, 0)$ with p_1 being the minimum possible momenta, namely $p_1 = 2\pi/16a \approx 765$MeV or $2\pi/24a \approx 694$MeV at $\beta = 6.0,\ 6.2$ respectively. The expected values are $\chi_u = 2$, $\chi_d = 1$. Filled circles, squares denote $O(a)$ improved fermions at $\beta = 6.2,\ 6.0$ respectively, while open squares are for $\beta = 6.0$ Wilson fermions. To guide the eye, constant fits are made to the chiral limit for the right hand picture.

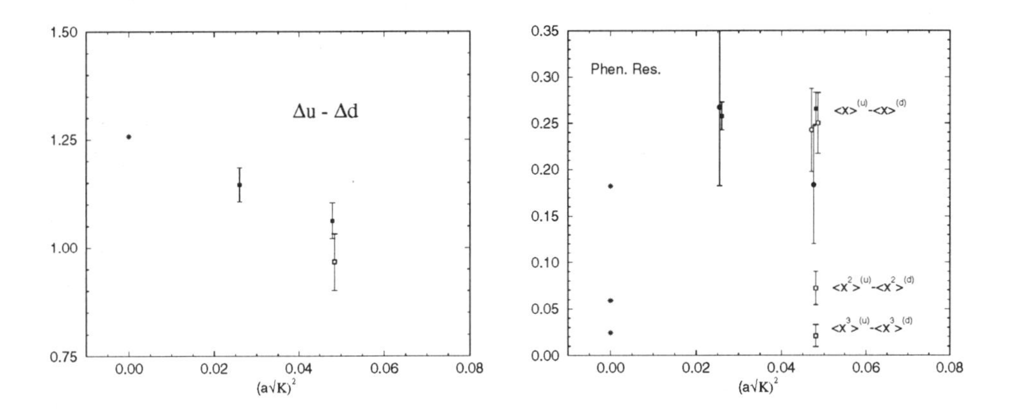

Figure 5: $g_A = \Delta u - \Delta d$, left picture, and $\langle x^n\rangle^{(u)} - \langle x^n\rangle^{(d)}$ moments at a scale of $\mu \sim 1.95$GeV, right picture. The phenomenological results are denoted with stars ($g_A = 1.26$ and $\langle x^n\rangle^{(u)} - \langle x^n\rangle^{(d)}$ by the *MRS* parametrisation, [13]). $O(a)$ improved fermion results are given by filled symbols. For $\langle x\rangle$, results using a zero-momentum diagonal matrix element are given by a square while the non-zero momentum off-diagonal results are denoted with a circle.

The role of the Polyakov loop in the Dirac operator of QCD at finite temperature

Vicente Azcoiti

Departamento de Fisica Teórica, Universidad de Zaragoza,
Plaza San Francisco, 50009 Zaragoza, Spain

Abstract: We show how all the contributions to the determinant of the Dirac-Kogut-Susskind operator of QCD at finite temperature containing a net number of Polyakov loops become irrelevant in the infinite volume limit. We discuss also on two of the most interesting physical implications of this result: i) the restoration of the Polyakov symmetry in the full theory with dynamical fermions and ii) the total suppression of baryonic thermal fluctuations in QCD at finite temperature.

1 Introduction

The fermion determinant in QCD and in general in any gauge theory is a gauge invariant non local operator, the main contributions of which being complex combinations of closed Wilson loops including the Wilson line or Polyakov loop [1] which is closed trough the boundary of the finite time direction.

In the pure gauge model the gauge action is invariant under Polyakov transformations and, because of that, the mean value of the Polyakov loop has been extensively used as order parameter in the investigations of the finite temperature phase transition. The situation however change in the full theory with dynamical fermions where the Polyakov symmetry is explicitly broken by the contribution to the integration measure of the determinant of the Dirac operator. The general wisdom in this case is that the Polyakov loop is no more an order parameter for the finite temperature phase transition.

Notwithstanding that I will show how in lattice QCD with staggered fermions all the contributions to the fermion determinant containing a net number of Polyakov loops become irrelevant in the infinite spatial volume limit. The Polyakov symmetry is recovered in this limit and we can therefore kill all these contributions from the beginning and work with a theory with dynamical fermions which preserves the Polyakov symmetry.

But this is, maybe, neither the only nor the most interesting physical implication of this result. In fact the Hilbert space of physical states can be decomposed as a direct sum of Hilbert spaces, each one of them corresponding to a fixed value of the quark number operator. The conservation of baryonic charge in QCD implies that all these spaces are invariant under the hamiltonian and we can therefore write the partition function as a sum of canonical partition functions, each one of them corresponding to a fixed value of the quark number operator. This decomposition of the partition function coincides with the integrated partition function obtained from the Polyakov loop expansion of the fermion determinant, the quark content being the net number of Polyakov loops. Since, as we will show, all the contributions with a net number of Polyakov loops are irrelevant, we conclude that thermal fluctuations of physical states with non vanishing baryonic charge are fully suppressed in QCD at finite temperature.

A method to implement in standard simulations the Polyakov symmetry and to kill baryonic thermal fluctuations in full QCD will also be developed here. The rest of the paper is organized as follows: section 2 describes some general features of the partition function of QCD at finite temperature. In section 3 we show the main result of this work with the help of an extra abelian degree of freedom. Section 4 is devoted to discuss some interesting physical consequences which follow from the results of section 3. In section 5 we discuss on the generalization of these results for Wilson fermions.

2 The Partition Function

The partition function of QCD at finite temperature T

$$Z = Tr(e^{\frac{-H}{T}}) \tag{1}$$

is the trace over the Hilbert space of physical states of minus the inverse temperature times the hamiltonian. Taking into account the conservation of baryonic charge we can write the partition function as a sum of canonical partition functions at fixed baryon number as follows:

$$Z = \sum_{k} Tr_k(e^{\frac{-H}{T}}), \tag{2}$$

where Tr_k in (2) indicates the trace over the subspace of fixed quark number k and $k = 0, +3, -3, +6, -6, ...$

The decomposition of the partition function given in (2) is the standard representation used in the investigations of QCD at finite baryon density. In this last case the partition function (2) is slightly modified by the introduction of a chemical potential μ in the following way [2], [3]

$$Z = \sum_{k} e^{\frac{\mu}{T}k} Tr_k(e^{\frac{-H}{T}}). \tag{3}$$

The previous decomposition (2) of the partition function as a sum of canonical partition functions at fixed baryon number corresponds in lattice regularized QCD to the Polyakov loop expansion of the integrated partition function which, for staggered fermions, can be written as follows

$$Z = \bar{a}_{3V_x} + \bar{a}_{3(V_x-1)} + ... + \bar{a}_0 + \bar{a}_{-3} + ... + \bar{a}_{-3V_x}, \tag{4}$$

where $\bar{a}_i$ ($\bar{a}_{-i}$) is the integral over the gauge group of all the contributions to the fermion determinant containing i forward (backward) Polyakov loops respectively, the integral being weighted with the pure gauge Boltzmann factor. The maximum number of Polyakov loops which can appear in a given coefficient of (4) for SU(3) and Kogut-Susskind fermions is three times the spatial lattice volume. $\bar{a}_0$ is the contribution with no net number of Polyakov loops and of course we have also all the symmetric contributions corresponding to backward Polyakov loops.

Since each coefficient in (4) represents the partition function at a fixed baryon number, this expression is also consistent with the fact that V_x is the maximum number of baryons allowed by the Fermi-Dirac statistics in a finite and discrete space of V_x points and staggered fermions.

All the averaged coefficients $\bar{a}_i$ in (4) are positive definite since they correspond to the decomposition of the partition function as a sum of canonical partition functions over the subspaces of fixed baryon number. Due to the Z(3) Polyakov symmetry of the pure gauge action, the only non vanishing coefficients are those containing a multiple of three times Wilson lines, which on the other hand reflects the fact that the only physical states in QCD are mesons, baryons and combinations of them.

3 The Extra-Abelian Degree of Freedom

Let us consider the determinant of the Dirac-Kogut-Susskind operator for the slightly modified gauge configuration which consists in multiplying all link variables at a fixed time-slice and pointing forward in the time direction by the global phase factor $e^{i\eta}$ and all the hermitian conjugates by $e^{-i\eta}$. Taking into account that the standard way to introduce a chemical potential in the lattice [3] is to multiply all links pointing forward (backward) in the time direction by a factor e^{μ} ($e^{-\mu}$) respectively, this is just what corresponds to consider an imaginary chemical potential $\mu = i\eta/L_t$, L_t being the lattice temporal extent.

The fermion determinant for a given gauge configuration and with the previous extra-degree of freedom can be written as follows

$$Det\Delta(i\eta) = a_{3V_x}e^{i3V_x\eta} + ... + a_1e^{i\eta} + a_0 + a_{-1}e^{-i\eta} + ... + a_{-3V_x}e^{-i3V_x\eta}. \tag{5}$$

The coefficients a_i (a_{-i}) in (5), averaged over the gauge group with the corresponding pure gauge Boltzmann factor, are the coefficients $\bar{a}_i$ ($\bar{a}_{-i}$) which appear in (4).

$$\bar{a}_i = \int [DU] a_i(U) e^{-\beta S_G(U)}. \tag{6}$$

The first interesting property of expression (5) which follows from the hermiticity and chiral properties of the fermion matrix Δ is that

$$Det\Delta(i\eta) \geq 0 \tag{7}$$

for every η.

By inverse Fourier transformation we can write

$$a_j = \frac{1}{2\pi}\int e^{-ij\eta} det\Delta(i\eta) d\eta, \tag{8}$$

$a_j = 0$ if $j > 3V_x$, $j < -3V_x$. These relations and the inequality (7) together tell us that $a_0 \geq 0$ and the absolute values $|a_j| \leq a_0$ for every j. Furthermore the same relations also hold for the integrated coefficients which appear in (4), i.e.

$$\bar{a}_k \leq \bar{a}_0, \tag{9}$$

where the absolute value in (9) disappears because of the fact that the integrated coefficients are real and positive.

Taking into account the symmetry properties of the coefficients ($\bar{a}_k = \bar{a}_{-k}$) we can write the partition function (4) as follows

$$Z = \bar{a}_0(1 + \frac{2\bar{a}_3}{\bar{a}_0} + ... + \frac{2\bar{a}_{3V_x}}{\bar{a}_0}), \tag{10}$$

which gives for the free energy density f the following expression

$$f = \frac{T}{V_x} log\bar{a}_0 + \frac{T}{V_x} log(1 + \frac{2\bar{a}_3}{\bar{a}_0} + ... + \frac{2\bar{a}_{3V_x}}{\bar{a}_0}). \tag{11}$$

This expression and the inequalities (9) imply that in the thermodynamical limit $V_x \to \infty$ the free energy density can be computed as:

$$f = \frac{T}{V_x} log\bar{a}_0, \tag{12}$$

i.e., the only relevant contribution in the determinant of the Dirac operator to the thermodynamics of QCD at finite temperature is that corresponding to a zero net number of Polyakov loops. We can therefore kill all the irrelevant contributions from the beginning and restore the Polyakov symmetry in full QCD.

4 Some Relevant Physical Implications

The main result of the analysis here developed is contained in equation (12). This section will be devoted to discuss some physical consequences which follow from it.

i) Since the only relevant contribution to the partition function of QCD at finite temperature is that with zero net number of Polyakov loops, the Polyakov symmetry is restored in the infinite volume limit of full QCD with dynamical fermions. The Polyakov loop is therefore a good order parameter for full QCD. A practical way to implement this result in numerical simulations is to include an extra-abelian degree of freedom, as done in section 3 of this paper.

ii) Another interesting conclusion which follows from the fact that the coefficient $\bar{a}_0$ in (12) does not depend on the boundary conditions for the fermion field is that periodic and antiperiodic boundary conditions give rise to the same physics.

iii) The physical meaning of equation (12) for the free energy is that the partition function of QCD at finite temperature is dominated by the canonical partition function computed over the Hilbert subspace of physical states of vanishing baryon number. In other words, baryonic thermal fluctuations are fully suppressed in QCD at finite temperature.

iv) It has been pointed out recently that in numerical simulations of quenched QCD at finite temperature and in the broken deconfined phase, the chiral condensate seems to depend crucially on the Z_3 phase in which the gauge dynamics settles [4] and the chiral symmetry restoration transition appears to occur at different temperatures depending of the phase of the Polyakov loop [5]. After the analysis here developed it is clear that the correct way to solve this puzzle and to implement the quenched approximation in QCD is to take for the chiral condensate operator its Polyakov loop invariant part. This result also suggest that a investigation of the finite size effects in finite temperature full QCD induced by the irrelevant contributions to the partition function could be of great interest.

5 Wilson Fermions

We have shown in the previous sections of this paper how the thermodynamics of QCD, when regularized in a space-time lattice and using staggered fermions, is controlled by the contribution to the fermion determinant with no net number of Polyakov loops, i.e., by the thermal fluctuations of physical states with vanishing baryon number.

The two main ingredients to get this result are the conservation of baryonic charge in QCD and the inequalities (9) of section 3 which tell us that the partition function at fixed baryon number reaches its maximum value in the Hilbert subspace corresponding to zero baryon number. The first one of the two ingredients is independent of the lattice regularization for the Dirac operator. The second one however is based on the positivity of the determinant of the Dirac operator for any gauge configuration and any value of the extra-abelian degree of freedom $e^{i\eta}$ (equation (7) of section 3).

Since we have made use of the hermiticity and chiral properties of the Dirac-Kogut-Susskind operator in order to get equation (7), it is natural to ask whether our result applies to any fermion regularization or rather it is related to the presence-absence of the chiral anomaly.

Let us say from the beginning that even if we have not yet a definite answer to this question, there are strong indications suggesting that the chiral anomaly does not play any relevant role here. These indications come from the analysis of the properties of the fermion determinant for Wilson fermions. As well known, the Dirac-Wilson operator Δ can be written as

$$\Delta = I - \kappa M, \tag{13}$$

where κ is the hopping parameter and the matrix M verifies the following chiral relation

$$\gamma_5 M \gamma_5 = M^+. \tag{14}$$

Equation (14) implies that if λ is eigenvalue of M, λ^* is also eigenvalue of M, i.e., the fermion determinant is always real. However it could be negative and in fact this unpleasant situation has been found for some gauge configurations in numerical simulations of the Schwinger model done in the unphysical strong coupling region [6]. However the unitary character of the gauge group implies that all the eigenvalues are upper bounded by the relation

$$|\lambda| \leq 8 \tag{15}$$

which implies that for $\kappa \leq 1/8$, $det\Delta \geq 0$. It is easy to verify that under the previous condition $\kappa \leq 1/8$, the positivity of $det\Delta$ also holds in the presence of the extra-abelian degree of freedom introduced in section 3.

In other words, all the results of this paper can be extended in a straightforward way to Wilson fermions if we impose the restriction $\kappa \leq 1/8$, i.e., the hopping parameter region associated to a positive bare fermion mass.

6 Aknowledgements

This work has been partially supported by CICYT (Proyecto AEN97-1680). The author thanks Giuseppe Di Carlo and Angelo Galante for discussions.

References

[1] A.M. Polyakov, *Phys. Lett.* **B72** (1977) 477

[2] J. Kogut, H Matsuoka, M. Stone, H.W. Wyld, S. Shenker, J. Shigemitsu, D.K. Sinclair, *Nucl. Phys.* **B225** (1983) 93

[3] P. Hasenfratz, F. Karsch, *Phys. Lett.* **B125** (1983) 308

[4] S. Chandrasekharan, N. Christ, *Nucl. Phys.* **B** *(Proc. Suppl.)* **47** (1996) 527

[5] M.A. Stephanov, *Phys. Lett.* **B375** (1996) 249

[6] V. Azcoiti, G. Di Carlo, A. Galante, A.F. Grillo, V. Laliena, C.E. Piedrafita, *Phys. Rev.* **D53** (1996) 5069

Resolution of the Landau pole problem in QED *

M. Göckeler [1], R. Horsley [2], V. Linke [3], P. Rakow [4], G. Schierholz [4,5], H. Stüben [6]

[1] Institut für Theoretische Physik, Universität Regensburg, D-93040 Regensburg
[2] Institut für Physik, Humboldt Universität zu Berlin, D-10115 Berlin
[3] Institut für Theoretische Physik, Freie Universität Berlin, D-14195 Berlin
[4] Deutsches Elektronen Synchrotron DESY, Institut für Hochenergiephysik und HLRZ, D-15735 Zeuthen
[5] Deutsches Elektronen Synchrotron DESY, D-22603 Hamburg
[6] Konrad-Zuse-Zentrum für Informationstechnik Berlin, D-14195 Berlin

Abstract: We present new numerical results for the renormalized mass and coupling in non-compact lattice QED with staggered fermions. Implications for the continuum limit and the role of the Landau pole are discussed.

1 Introduction

In the 1950s Landau investigated the relation between the bare charge e and the renormalized charge e_R in QED. He found [1]

$$\frac{1}{e_R^2} - \frac{1}{e^2} = \frac{N_f}{6\pi^2} \ln \frac{\Lambda}{m_R}, \tag{1}$$

where Λ is the momentum cutoff, m_R is the renormalized mass of the electron and N_f is the number of flavours (for staggered fermions $N_f = 4$). It is well known that (1) implies two potential problems when $\Lambda \to \infty$:

- When e is fixed the theory has a *trivial* continuum limit, i.e., $e_R \to 0$.
- When $e_R > 0$ is fixed e becomes singular at $\Lambda_{\text{Landau}} = m_R \exp\left(6\pi^2/N_f e_R^2\right)$. This singularity is called the *Landau pole.*

These problems do not affect the phenomenological success of QED because in practical perturbative calculations the cutoff can be chosen to be large compared with

*Talk presented by G. Schierholz at the 31st Ahrenshoop Symposium on the Theory of Elementary Particles, September 1997, Buckow, Germany.

experimental energies. Finding a solution to them is of fundamental theoretical interest and requires non-perturbative methods. We have therefore extensively studied QED on the lattice using non-compact gauge-fields and dynamical staggered fermions [2, 3].

In this talk [4] we report on new measurements of the renormalized mass and charge. This was started in [2] and is now extended to lattices of size 16^4 and bare masses am down to 0.005 (a denotes the lattice spacing). Using our measurements we have determined *functions* $am_R(e^2, am)$ and $e_R^2(e^2, am)$. These functions imply that the theory is trivial. They also give a resolution of the Landau pole problem.

2 The renormalized mass

The renormalized mass was obtained from fits to the fermion propagator as explained in [2]. To get am_R as a function of am and e we need to know the equation of state which relates the bare parameters to the chiral condensate $\sigma \equiv a^3\langle\bar{\chi}\chi\rangle$. In [3] we have found that the σ data obey a mean field equation of state with logarithmic corrections

$$am = A_0 \frac{\sigma^3}{\ln^{p_0}(1/\sigma)} + A_1 \left(\frac{1}{e^2} - \frac{1}{e_c^2}\right) \frac{\sigma}{\ln^{p_1}(1/\sigma)} . \tag{2}$$

We fitted this expression to σ data that were extrapolated to infinite lattice size and obtained $1/e_c^2 = 0.19040(9)$, $A_0 = 1.798(5)$, $p_0 = 0.324(15)$, $A_1 = 6.76(3)$, $p_1 = 0.485(7)$, $\chi^2/\text{d.o.f} = 7.6$.

We have observed that σ can be well described by a polynomial in am_R

$$\sigma = A_1 am_R + A_3 a^3 m_R^3 + A_5 a^5 m_R^5 + A_7 a^7 m_R^7 \tag{3}$$

where the first parameter $A_1 \equiv 0.6197$ can be taken from perturbation theory and a fit to data from 12^4 and 16^4 lattices gave $A_3 = -0.321(5)$, $A_5 = +0.169(13)$, $A_7 = -0.040(7)$, $\chi^2/\text{d.o.f.} = 2.1$. Because the results of both lattice sizes fall on a universal curve we conclude that the polynomial (3) is also valid on an infinite lattice.

3 The renormalized charge

The determination of the renormalized charge has been improved since [2]. The method can only be sketched here. It consists of making a global fit to *all* gauge field propagators $D(k)$ that we have measured

$$\frac{1}{e^2 D(k)} - \frac{1}{e^2} = -\Pi(k, am_R, L) \tag{4}$$

where L is the linear lattice size. The *ansatz* for the fit function Π was taken from [5] to be

$$\Pi = U - \frac{V}{U} \ln(1 - e^2 U) \tag{5}$$

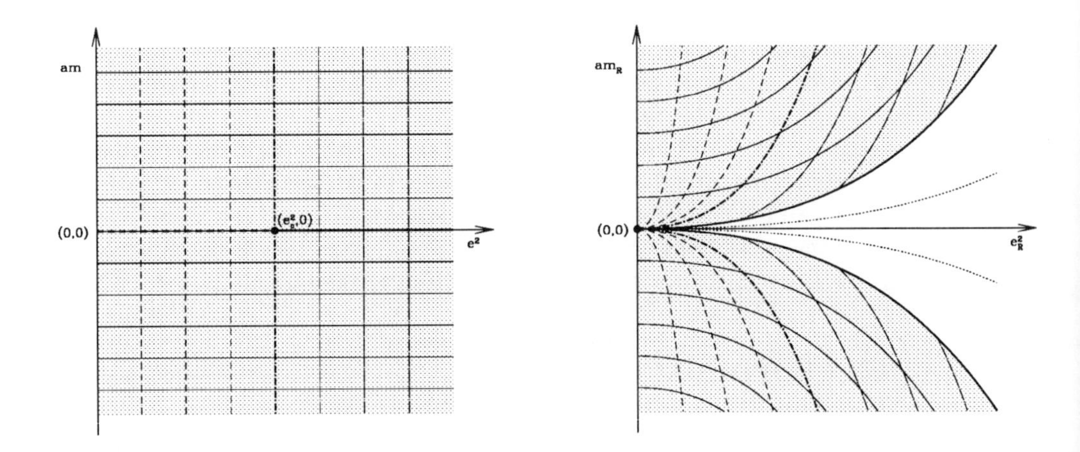

Figure 1: Sketch of the mapping $(e^2, am) \leftrightarrow (e_R^2, am_R)$.

where U is given by 1-loop lattice perturbation theory (see [2]) and where we have set

$$V(k, am_R, L) = v_0 + v_1 U(k, am_R, L)\,. \tag{6}$$

We then find $e_R^2(e^2, am_R)$ using $e_R^2 = Z_3 e^2$ and $Z_3 = \lim_{k\to 0}\lim_{L\to\infty} D(k)$ from

$$\frac{1}{e_R^2} - \frac{1}{e^2} = -\Pi(0, am_R, \infty)\,. \tag{7}$$

A simultaneous fit of (4) to the gauge field propagators at our 52 values (e^2, am, L) gave $v_0 = -0.00207(2)$, $v_1 = -0.0328(7)$, $\chi^2/\text{d.o.f.} = 1.7$. Since $U(0, am_R, \infty) \approx (N_f/6\pi^2)(-0.31 + \ln am_R)$ we only find small corrections to the old result (1).

4 Discussion

We can now discuss the mapping $(e^2, am) \leftrightarrow (e_R^2, am_R)$. A global qualitative view of this mapping is shown in Figure 1, while a quantitative plot of $(1/e^2, am) \leftrightarrow (1/e_R^2, am_R)$ is shown in Figure 2.

In both figures accessible regions are plotted in grey. The whole plane of bare parameters is accessible but this plane is mapped only onto a part of the plane of the renormalized parameters. The border of the accessible region is shown as a thick line. It is the image of the corresponding thick line on the $am = 0$ line starting/ending at the critical value of the coupling constant.

On the line $am_R = 0$ in Figure 1 the only accessible point is the origin. This reflects triviality. In Figure 2 triviality is expressed by the fact that no line of (finite) constant $1/e_R^2$ flows into the critical point.

The dotted line in both figures is the position of the Landau pole, i.e., the line of pairs (e_R^2, am_R) with $1/e^2 = 0$ from (7) with $\Pi \equiv U$. This line is well separated from the border for all finite e_R^2.

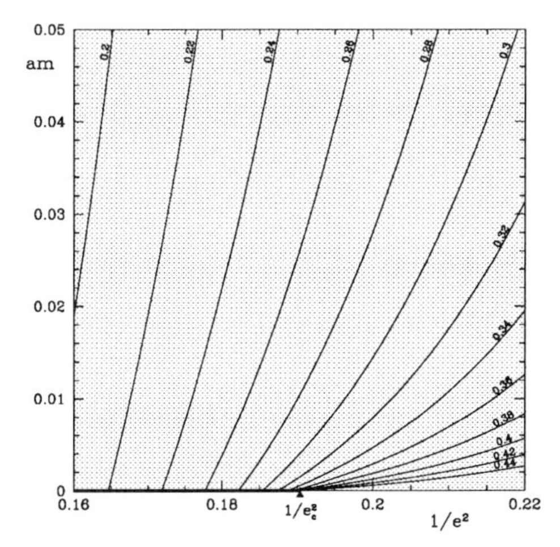

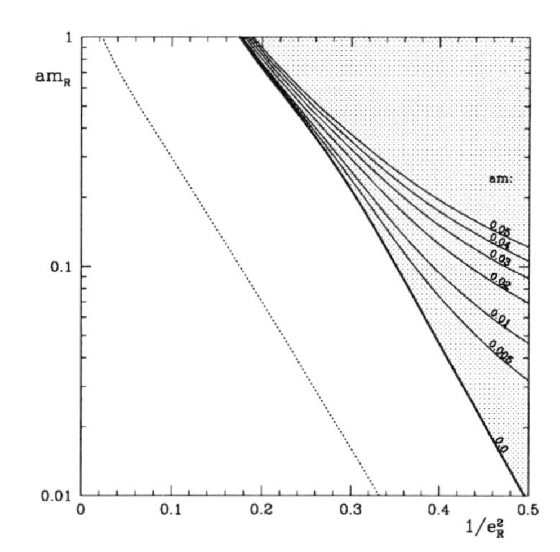

Figure 2: Quantitative picture of the mapping $(1/e^2, am) \leftrightarrow (1/e_R^2, am_R)$. In the plane of the bare parameters lines of constant $1/e_R^2$ are shown indicating the renormalization group flow of e_R^2. In the plane of the renormalized parameters the lines of constant am are shown.

5 Conclusions

From the presented analysis we conclude that non-compact lattice QED with staggered fermions has a trivial continuum limit. In addition our analysis implies a resolution of the Landau problem. The resolution is that for given e and am the theory does not allow arbitrary choices of am_R. Instead through chiral symmetry breaking the theory itself provides a minimal lattice spacing or maximal cutoff which is below Λ_{Landau}. Non-perturbatively the function Π is very close to what one finds in perturbation theory, but there is no Landau pole problem because the pole always lies in the forbidden region.

Acknowledgements

Our new measurements were done on the CRAY T3D at the Konrad-Zuse-Zentrum Berlin. Financial support by the Deutsche Forschungsgemeinschaft is gratefully acknowledged.

References

[1] L. D. Landau, in: Niels Bohr and the Development of Physics, ed. W. Pauli, Pergamon Press, London, 1955.

[2] M. Göckeler, R. Horsley, P. Rakow, G. Schierholz and R. Sommer, Nucl. Phys. **B371** (1992) 713.

[3] M. Göckeler, R. Horsley, V. Linke, P. Rakow, G. Schierholz and H. Stüben, Nucl. Phys. **B487** (1997) 313.

[4] This material is based on the talk presented by H. Stüben at Lattice '97, available at `hep-lat/9801004`, and a recent letter by the same authors, available at `hep-th/9712244`.

[5] D.V. Shirkov, Nucl. Phys. **B371** (1992) 467.

The Coulomb law in the $U(1)$ lattice gauge theory[1]

G. Cella [1], U.M. Heller [2], V.K. Mitrjushkin [3] and A. Viceré [1]

[1] INFN in Pisa and Dipartimento di Fisica dell'Universitá di Pisa, Italy
[2] SCRI, Florida State University, Tallahassee, FL 32306-4052, USA
[3] Lab.Theor.Phys., JINR, Dubna, Russia

Abstract: We study the heavy charge potential in the Coulomb phase of pure gauge compact $U(1)$ theory on the lattice. We calculate the static potential $V_W(T,\vec{R})$ from Wilson loops on a $16^3 \times 32$ lattice and compare with the predictions of lattice perturbation theory. We investigate finite size effects and, in particular, the importance of non–Coulomb contributions to the potential.

1 Introduction

The action of the pure gauge $U(1)$ theory is [1]

$$S_W(U) = \beta \sum_x \sum_{\mu>\nu} \Bigl(1 - \cos\theta_{x,\mu\nu}\Bigr) , \qquad \beta = \frac{1}{g^2} , \tag{1}$$

where g^2 is the bare coupling constant, and the link variables are $U_{x\mu} = \exp(i\theta_{x\mu})$, $\theta_{x\mu} \in (-\pi, \pi]$. The plaquette angles are given by $\theta_{x,\mu\nu} = \theta_{x,\mu} + \theta_{x+\hat{\mu},\nu} - \theta_{x+\hat{\nu},\mu} - \theta_{x,\nu}$. This action makes up the pure gauge part of the full QED action S_{QED}, which is supposed to be compact if we consider QED as arising from a subgroup of a non–abelian (e.g., grand unified) gauge theory [2]. At small enough β's the strong coupling expansion shows an area–law behaviour of the Wilson loops, while at large β's a deconfined phase exists. The two phases are separated by a phase transition at some 'critical' value β^* whose existence was assumed in [1]. The weak coupling – deconfined – phase is expected to be a Coulomb phase, *i.e.*, a phase with massless noninteracting vector bosons (photons) and Coulomb–like interactions between static charges.

There are scarcely any doubts about the existence of the Coulomb phase in the weak coupling region for Wilson's QED, though – to our knowledge – there is no

---

[1]Work supported by the Deutsche Forschungsgemeinschaft under research grant Mu 932/1-4 and EEC–contract CHRX-CT92-0051 and by the US DOE under grants # DE-FG05-85ER250000 and # DE-FG05-96ER40979.

rigorous proof. It is worthwhile to mention that for the Villain approximation [3] such a proof is available. However, a detailed study has shown that the Villain action is quantitatively a rather bad approximation to the Wilson action in the weak coupling region [4].

A perturbative expansion suggests for the lattice potential $V^{latt}(\vec{R})$ the expression $V^{latt}(\vec{R}) = \sum_{n=1}^{\infty} g^{2n} \cdot V_n(\vec{R})$, where up to an additive constant V_1 is the lattice analog of the continuum Coulomb potential $\sim 1/4\pi R$ (for an explicit expression see (3) below). One–loop corrections ($\sim g^4$) do not change the functional dependence of the potential but at the two–loop level ($\sim g^6$) non–Coulomb–like contributions appear (see below). The analytical and numerical study of these contributions, as well as of the finite volume behavior of the potential, constitutes the aim of the present work.

The behavior of the heavy charge potential in the weak coupling phase in pure $U(1)$ gauge theory was the subject of several numerical studies (see, e.g. [5, 6, 7, 8]). In all cases consistency with a Coulomb–like behavior was reported. However, we feel that a more elaborate and systematic study is necessary. In particular, Monte Carlo simulations give precise enough measurements of the potential, that finite size effects and finite lattice spacing effects, *i.e.*, deviations of the lattice Coulomb potential from the continuum $1/R$ behavior should not be neglected. In addition, finite T effects in the extraction of the potential from Wilson loops have to be taken into account. We shall find, on the other hand, that the non-Coulomb–like contributions to the potential are negligible.

2 The potential from Polyakov loop correlations

One way to define the heavy charge potential $V = V_P(\vec{R})$ is $V_P(\vec{R}) = -\frac{1}{N_4} \cdot \ln \Gamma_P(\vec{R})$, where we consider a finite lattice of size $N_s^3 \times N_4$ and $\Gamma_P(\vec{R})$ is the Polyakov loop correlator

$$\Gamma_P(\vec{R}) = \frac{1}{Z} \int_A e^{ig \sum_{x\mu} J^P_{x\mu} A_{x\mu}} \cdot e^{-S_W(A)} , \tag{2}$$

if the currents $J^P_{x\mu}$ correspond to a static heavy charge–anticharge pair. We calculated the potential $V_P(\vec{R})$ perturbatively up to order $O(g^6)$. Graphically the different contributions are shown in Figure 1.

In the tree approximation (Figure 1a) one obtains the lattice Coulomb potential

$$g^2 V_{Coul}^{(N_s)}(\vec{R}) = \frac{g^2}{N_s^3} \sum_{\vec{p} \neq 0} \frac{1 - \cos \vec{p}\vec{R}}{\vec{\mathcal{K}}^2} , \tag{3}$$

where $\mathcal{K}_i = 2 \sin \frac{p_i}{2}$. The lattice Coulomb potential defined in eq. (3) becomes close to the continuum expression when $1 \ll R \ll \frac{1}{2} N_s$ for sufficiently large lattice size N_s. For lattice sizes that are typically used in numerical simulations (i.e. $N_s \sim 16 \div 32$) the lattice Coulomb potential in eq. (3) differs considerably from the continuum expression.

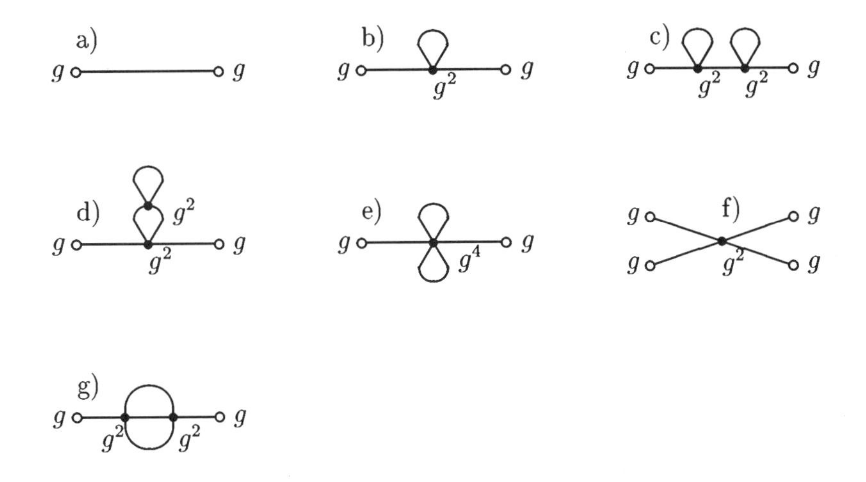

Figure 1: Different diagrams contributing to the potential $V_P(\vec{R})$.

The $O(g^4)$ contribution (Figure 1b) as well as $O(g^6)$ contributions shown in Figures 1c to 1e result in a renormalization of the coupling without changing the form of the potential. However, two diagrams at order $\sim O(g^6)$ contain non–Coulomb–like contributions. One of them is the four–prong–spider graph (Figure 1f) which gives the contribution to the potential

$$V^{(f)}(\vec{R}) \;=\; -\frac{2g^6}{3N_s^9} \sum_{\{\vec{p}^{(i)}\}} \delta_{\vec{p}^{(1)}+\ldots+\vec{p}^{(4)},0} \prod_{i=1}^{4} \sin\frac{\vec{p}^{(i)}\vec{R}}{2} \cdot \sum_{j=1}^{3}\prod_{i=1}^{4} \frac{\mathcal{K}_j(p^{(i)})}{\vec{\mathcal{K}}^2(p^{(i)})} \;. \tag{4}$$

Another non–Coulomb–like contribution comes from the two–loop bubble diagram shown in Figure 1g is $V^{(g)}(\vec{R}) = \frac{g^6}{48} V_{Coul}(\vec{R}) + V^{(g)}_{nCoul}(\vec{R})$, where

$$V^{(g)}_{nCoul}(\vec{R}) = \frac{g^6}{6N_s^3} \sum_{\vec{p}} \frac{1-\cos\vec{p}\vec{R}}{(\vec{\mathcal{K}}^2)^2} \sum_{i,j=1}^{3} \mathcal{K}_i\mathcal{K}_j \cdot \left(T_{ij}(\vec{p}) - T_{ij}(\vec{0})\right) , \tag{5}$$

and

$$T_{ij}(\vec{p}) = \frac{1}{(N_4N_s^3)^2} \sum_{qq'} D_{ij}(q)D_{ij}(q')D_{ij}(p-q-q') \,, \quad D_{ij}(q) = \frac{\mathcal{K}_i\mathcal{K}_j + \mathcal{K}_4^2\cdot\delta_{ij}}{\mathcal{K}^2}(q) \,. \tag{6}$$

Therefore, up to order $O(g^6)$ the static charge potential $V_P(\vec{R})$ is

$$V_P(\vec{R}) = g^2_{2-loop}\cdot V_{Coul}(\vec{R}) + g^6\cdot V_{nCoul}(\vec{R}) \,, \tag{7}$$

with

$$V_{nCoul}(\vec{R}) = V^{(f)}(\vec{R}) + V^{(g)}_{nCoul}(\vec{R}) \,, \tag{8}$$

and where

$$g^2_{2-loop} = g^2(1 + \frac{1}{4}g^2 + \frac{11}{96}g^4) \ . \tag{9}$$

V_{nCoul} is almost three orders of magnitude smaller than V_{Coul}, a difference that will even be enhanced when including the coupling constants for weak coupling. Furthermore V_{nCoul} approaches its asymptotic value much faster than $1/R$. Since the long distance behavior is governed by the small momentum region, we can estimate it in the infinite volume limit from power counting in the small momentum region of the integrals. For the two-loop bubble diagram, eq. (5), since the zero-momentum part corresponding to $T_{ij}(\vec{0})$, which contributes to V_{Coul}, was split off, we expect a $1/R^3$ approach to a constant. For the four–prong–spider graph, eq. (4), we expect, apart from a constant, a $1/R^5$ fall-off at large distance. Log–log plots of the numerically computed non-Coulomb contributions on finite lattices confirm these expectations.

3 The potential and g_R^2 from Wilson loops

The potential can also be obtained from Wilson loops: $V = V_W(T, \vec{R})$. It is defined as follows

$$V_W(T; \vec{R}) = \ln \frac{W(T, \vec{R})}{W(T+1, \vec{R})} \ , \tag{10}$$

where $W(T, \vec{R})$ is the Wilson loop with 'time' extension T and space extension $\vec{R}$.

The Wilson loop can be on–axis or off-axis, and the space–like parts of the loop can include the contribution of many different contours. In our calculations we have chosen planar loops and two types of non–planar contours: the 'plane-diagonal' contour with space–like part in the plane (x_2, x_3) at fixed x_1 (say, $x_1 = 0$) connecting points $x_2 = x_3 = 0$ and $x_2 = x_3 = R_0$, and the 'space-diagonal' contour in the $3d$ space (x_1, x_2, x_3) connecting points $x_1 = x_2 = x_3 = 0$ and $x_1 = x_2 = x_3 = R_0$. For the off-axis loops we average over all paths that follow the straight line between the endpoints as closely as possible. For example, for the loop through the origin and $x_2 = x_3 = R_0$ the paths considered go through all points with $x_2 = x_3$ in between. We average over all combinations of first taking a step in the 2-direction followed by a step in the 3-direction and vice versa.

Our Monte Carlo data were produced on a $16^3 \times 32$ lattice for 11 values of β. We measured planar and nonplanar Wilson loops $W(T, \vec{R})$ with $T \leq T_{max} = 15$ and $R \leq R_{max} = 8\sqrt{3}$.

In the Coulomb phase the potential obtained from Monte Carlo simulations is expected to be very similar to the perturbative lattice Coulomb potential. Since the latter can have sizable finite size and finite T effects, as discussed earlier, we decided to use as a fit formula for the numerical potential $V_W(T; \vec{R})$ the lowest order perturbative expression, with the only fit parameter being the renormalized coupling constant g_R^2. The success of such fits will indicate that the non-Coulomb contributions are small, like in perturbation theory, and that non-perturbative effects

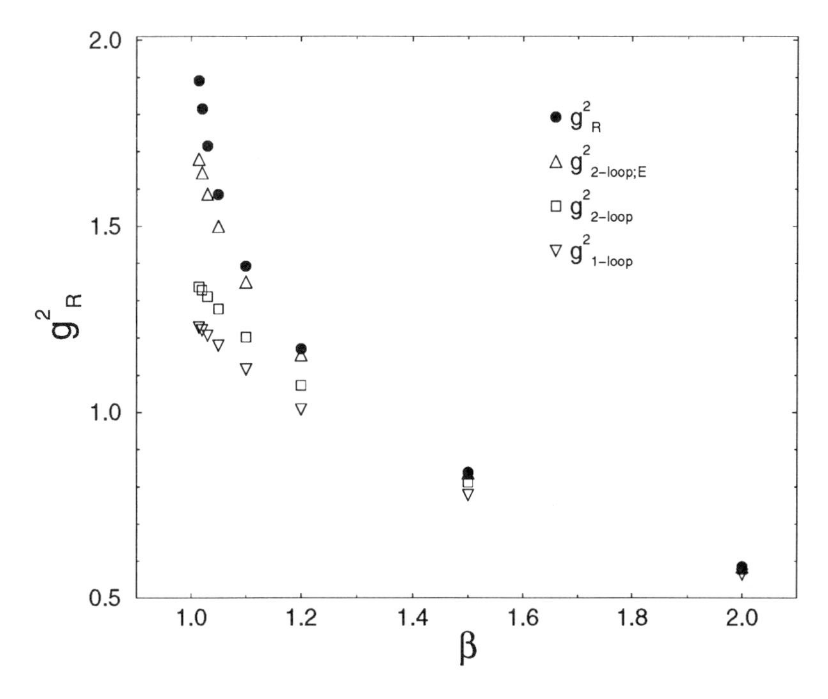

Figure 2: The renormalized coupling g_R^2 extracted from the fit on a $16^3 \times 32$ lattice as compared with 1–loop and 2–loop perturbative values.

just go into the renormalization of the coupling. The number of degrees of freedom, $N_{d.o.f.}$ in our fits varied between $N_{d.o.f.}^{max} = 344$ and $N_{d.o.f.}^{min} = 31$ with decreasing β, from $\beta = 10$ to $\beta = 1.015$. In all the cases we obtained $\chi^2/N_{d.o.f.} \lesssim 1.0$ (with the only exception being for $\beta = 1.015$ where $\chi^2/N_{d.o.f.} \simeq 2.1$).

At large values of β ($\beta \geq 5$) our one-parameter fit works well for the whole set of data points, *i.e.* all points with $1 \leq T \leq 15$ and $1 < R \leq 8\sqrt{3}$ ($N_{d.o.f.} = 344$). The non–Coulomb–like corrections are therefore negligible despite our rather small statistical errors. With decreasing β these corrections become noticeable at small values of R (recall the fast fall-off of V_{nCoul} like $1/R^3$) and small values of T. After excluding the data points corresponding to these small R and T values from the fit, we obtained fits with a high confidence level, but fewer degrees of freedom, as described above. Remarkably, the lattice Coulomb potential works, at least at large distance, really well even at β–values very close to the phase transition point, β^*, to the confined phase. This observation in fact constitutes one of the main results of this paper.

The extracted values of the renormalized coupling g_R^2 as well as the perturbative values g_{1-loop}^2 and g_{2-loop}^2 are shown in Figure 2. The difference between g_R^2 and the perturbative value g_{2-loop}^2 becomes noticeable at $\beta \lesssim 2.0$. This is the region where one expects the perturbative three–loop contributions to become important. In addition nonperturbative effects may start to significantly contribute.

For non–abelian lattice gauge theory it has proved useful, in perturbative computations, to use an the 'improved' coupling, such as $g_E^2 \equiv \frac{1}{c_1}\Big(1 - \langle U_p \rangle\Big)$ [11, 12], where U_p denotes the plaquette and $c_1 = 1/4$ is the first coefficient in its perturbative expansion. For pure gauge $U(1)$ the perturbative expansion of $\langle U_p \rangle$ is known to three loops[13]

$$\langle U_p \rangle = 1 - \frac{1}{4}g^2 - \frac{1}{32}g^4 - 0.0131185g^6 - 0.00752g^8 + \dots \,. \tag{11}$$

From this we get a perturbative expansion of g_E^2 in terms of the bare g^2. We then can express g^2 in terms of g_E^2 and substitute in the perturbative formula for g^2_{2-loop}, eq. (9),

$$g^2_{2-loop,E} = g_E^2(1 + \frac{1}{8}g_E^2 + 0.0308594g_E^4 + \dots). \tag{12}$$

$g^2_{2-loop,E}$ based on the 'improved' coupling g_E^2 is indeed a somewhat better prediction for g_R^2 than the purely perturbative g^2_{2-loop}. This fact can also be seen in Figure 2.

4 Conclusions

We have made an analytical and numerical study of compact lattice pure gauge $U(1)$ theory in the Coulomb phase. The main point of interest was the study of the *non*–Coulomb contributions to the heavy charge potential. For this purpose we calculated perturbatively the heavy charge potential $V_P(\vec{R})$ defined from the Polyakov loop correlations in the 2–loop (i.e., $\sim g^6$) approximation. This calculation shows that the non–Coulomb contribution $V_P^{nCoul}(\vec{R})$ is much smaller than the Coulomb contribution $V_P^{Coul}(\vec{R})$, at least at large distances.

The conclusions obtained within perturbation theory were confirmed by numerical calculations of the potential $V_W(T;\vec{R})$ defined from planar and nonplanar time–like Wilson loops. We used as a fit formula for $V_W(T;\vec{R})$ the lowest order perturbative expression with the only fit parameter being the renormalized coupling constant g_R^2. These fits worked well for all distances at large β, and for sufficiently big T and R even down to very close to the phase transition to the strong coupling confined phase. It is worth noting that it is impossible to obtain a good χ^2 using for the fit a continuum potential $\sim 1/R$. The values of the renormalized coupling g_R^2 are stable with respect to a smearing procedure.

Our main conclusion is that *compact* pure gauge $U(1)$ theory can serve equally well as a non–compact version to describe the physics of free photons in the weak coupling region. The only difference is a finite, R–independent renormalization of the coupling constant $g^2 \to g_R^2$ in the compact theory.

References

[1] K. Wilson, Phys. Rev. **D10** (1974) 2445.

[2] A. Polyakov, *Gauge fields and strings*, Harwood Academic Publishers, (1987).

[3] J. Villain, J. Phys. (Paris) **36** (1975) 581.

[4] W. Janke and H. Kleinert, Nucl. Phys. **B270** (1986) 135

[5] G. Bhanot, Phys. Rev. **D24** (1981) 461.

[6] T.A. DeGrand and D. Toussaint, Phys. Rev. **D24** (1981) 466.

[7] J. Jersák, T. Neuhaus and P.M. Zerwas, Phys. Lett. **133B** (1983) 103; Nucl. Phys. **B251** (1985) 299.

[8] J.D. Stack and R.J. Wensley, Nucl. Phys. **B371** (1992) 597.

[9] J.L. Cardy, Nucl. Phys. **B170** (1980) 369.

[10] J.M. Luck, Nucl. Phys. **B210** (1982) 111.

[11] G. Parisi, in *Proceedings of the XXth Conference on High Energy Physics*, Madison 1980.

[12] G. Martinelli, G. Parisi and R. Petronzio, Phys. Lett. **100B** (1981) 485.
F. Karsch and R. Petronzio, Phys. Lett. **139B** (1984) 403.

[13] R. Horsley and U. Wolff, Phys. Lett. **105B** (1981) 290.

Symmetry properties of massive gauge theories in nonlinear background gauges: Background dependence of Green functions

B. Geyer[1] **and D. Mülsch**[2]

[1] Naturwissenschaftlich-Theoretisches Zentrum and Institut für Theoretische Physik,
Universität Leipzig, Augustusplatz 10-11, D-04109 Leipzig, Germany
[2] Wissenschaftszentrum Leipzig e.V., Goldschmidtstr. 26, D-04109 Leipzig, Germany

Abstract: Nonabelian gauge theories with a generic background field A_μ in nonlinear gauges are investigated in the framework of osp(1,2) covariant quantization. Ward identities, gauge and consistency conditions of the theory are determined. The A_μ-dependence of the 1PI vertex functions as well as the complete Green functions is determined relative to the corresponding vacuum functions.

1 Introduction

Given a quantum field theory with classical background configuration the question arises how the renormalization properties of the theory with background field are possibly changed and whether their Green functions may be related to the Green functions without them. Here, this problem will be considered for nonabelian gauge theories with generic background field A_μ within the algebraic BRST approach. At first, this approach has been used by Kluberg-Stern and Zuber [1]. They also pointed out that the classical action

$$S_{\rm YM}(A) = -(2g)^{-2} \int d^4x \, {\rm Tr}(F_{\mu\nu}(A+Q)F^{\mu\nu}(A+Q)), \tag{1}$$

is invariant under two different kinds of transformations :

$$\text{type I}: \qquad \delta A_\mu = D_\mu(A)\delta\Omega, \qquad \delta Q_\mu = [Q_\mu, \delta\Omega], \tag{2}$$

$$\text{type II}: \qquad \delta A_\mu = 0, \qquad \delta Q_\mu = D_\mu(A+Q)\delta\Omega. \tag{3}$$

For *linear* background gauges also the gauge fixed action is invariant under type I transformations, whereas type II transformations have to be changed into the BRST transformations. Their Ward identities are the (local) Kluberg-Stern–Zuber(KSZ) and Slavnov–Taylor(ST) identity, respectively. Obviously, there exists another kind of transformations,

$$\text{type III}: \qquad \delta Q_\mu = D_\mu(Q)\delta\Omega, \qquad \delta A_\mu = [A_\mu, \delta\Omega], \tag{4}$$

which however cannot be required to hold unchanged for the renormalized quantum action.

With the aim to determine the A_μ–dependence of the Green functions Rouet [2] used this formalism. He could show that the on–shell amputated physical Green functions in the background field A_μ are deduced from those in the vacuum by a mere translation of the (renormalized) quantum field $z_Q Q_\mu$, i.e.

$$Z_{\text{phys}}(A|J) = \exp\left(-i(\hbar z_Q)^{-1} \int d^4x \, \text{Tr}(A^\mu J_\mu)\right) Z_{\text{phys}}(0|J). \tag{5}$$

In a recent paper [3] we have shown that, if massive gauge fields, e.g. with mass $m(s) = (1-s)m$, are considered, as it is necessary at least intermediately in the framework of the BPHZL–renormalization [4] to avoid IR-singularities, then only *nonlinear* gauges are permissible [1]. Therefore, the Curci-Ferrari model [6] in the (nonlinear) Delbourgo-Jarvis gauge [7] has been considered. This gauge has the additional advantage that the theory is invariant under BRST as well as anti-BRST transformations $\mathbf{s}$ and $\bar{\mathbf{s}}$, and, in the massive case, necessarily also under the symplectic transformations generated by $\mathbf{q}, \mathbf{d}$ and $\bar{\mathbf{d}}$ (for definitions see Chapter 2).

Here, we extend these considerations to include a generic background field A_μ. Imposing the gauge condition on the generating functional for the 1PI vertex functions $\Gamma(A|Q)$ it can be shown that a further identity, the local Lee identity, emerges which is related to type III symmetry above. Together with the KSZ identity it leads to a *linear* differential equation for the A_μ–dependence of $\Gamma(A|Q)$. Its solution relates $\Gamma(A|Q)$ and $\Gamma(0|Q)$; in addition, it can be shown that A_μ does not require any further z–factors and, therefore, the renormalization of $\Gamma(A|Q)$ is determined by that of $\Gamma(0|Q)$. By Legendre transformation this gives a *complete* solution of the problem originally considered by Rouet and, after restriction to the physical subspace, confirmes his result. Furthermore, the method used here may be extended to consider topological nontrivial background fields such as instantons and merons.

[1] The well known fact that massive gauge theories, if renormalizable, violate unitarity does not matter here since after carrying out all necessary subtractions the limit $s \to 1$ may be taken [5].

2 Massive gauge theory in nonlinear background gauge

The theory is defined by the following gauge fixed (classical) action including appropriate sources

$$S(A|g,m;\xi,\rho,\sigma) = -(2g)^{-2}\int d^4x \mathrm{Tr}(F_{\mu\nu}(A+\mathcal{Q})F^{\mu\nu}(A+\mathcal{Q})) \quad (6)$$
$$+(\mathbf{s}_m\bar{\mathbf{s}}_m+m^2)\int d^4x \mathrm{Tr}\big(\tfrac{1}{2}\mathcal{Q}_\mu\mathcal{Q}^\mu+\xi\bar{\mathcal{C}}\mathcal{C}-G_\mu\mathcal{Q}^\mu+U\bar{\mathcal{C}}-\bar{U}\mathcal{C}+\tfrac{1}{2}\sigma G_\mu G^\mu+\rho\bar{U}U\big)$$

with $\mathcal{Q}_\mu = Q_\mu+\sigma G_\mu,\quad \mathcal{C}=C+\rho U,\quad \bar{\mathcal{C}}=\bar{C}+\rho\bar{U},\quad \mathcal{B}=B+(\zeta-\bar{\zeta})\,\rho\, S,\quad (7)$

where $Q_\mu, C, \bar{C}$ and B are the gauge, ghost, antighost and auxiliary field, respectively, and $G_\mu, U, \bar{U}$ and S are associated sources having the same quantum numbers as these fields, respectively; ξ is the gauge parameter, ζ an arbitrary real parameter ($\bar{\zeta}=1-\zeta$) reflecting the freedom in defining B according to $\mathcal{B}=\bar{\zeta}\mathbf{s}_m\bar{\mathcal{C}}-\zeta\bar{\mathbf{s}}_m\mathcal{C}$, whereas σ and ρ are to be introduced since the dynamical fields and the associated sources mix under renormalization.

The mass-dependent (anti)BRST operators $\mathbf{s}_m$ (and $\bar{\mathbf{s}}_m$) are defined through their action on the fields and the sources according to:

$$\begin{aligned}
\mathbf{s}_m A_\mu = 0,\quad \mathbf{s}_m\mathcal{Q}_\mu = D_\mu(A+\mathcal{Q})\mathcal{C},\quad \mathbf{s}_m\mathcal{C} = -(1/2)\{\mathcal{C},\mathcal{C}\},\\
\mathbf{s}_m\bar{\mathcal{C}} = \mathcal{B}-\zeta\left\{\mathcal{C},\bar{\mathcal{C}}\right\},\quad \mathbf{s}_m\mathcal{B} = \zeta\left[\mathcal{B}+\bar{\zeta}\left\{\mathcal{C},\bar{\mathcal{C}}\right\},\mathcal{C}\right]-m^2\mathcal{C},\\
\mathbf{s}_m G_\mu = P_\mu,\quad \mathbf{s}_m P_\mu = 0,\quad \mathbf{s}_m U = R,\quad \mathbf{s}_m R = 0,\\
\mathbf{s}_m\bar{U} = S,\quad \mathbf{s}_m S = -m^2 U,\quad \mathbf{s}_m\bar{P}_\mu = -m^2 G_\mu,\quad \mathbf{s}_m\bar{R} = -2m^2\bar{U}.
\end{aligned} \quad (8)$$

Corresponding relations defining the action of the antiBRST operator $\bar{\mathbf{s}}_m$ are obtained by the following conjugation $\mathbf{C}$ which leaves the classical action (6) invariant and, obviously, is an extension of the ghost–antighost conjugation:

$$\begin{aligned}
&C\to\bar{C},\quad U\to\bar{U},\quad P_\mu\to\bar{P}_\mu,\quad R\to\bar{R},\quad S\to -S,\quad \zeta\to\bar{\zeta},\\
&\bar{C}\to -C,\quad \bar{U}\to -U,\quad \bar{P}_\mu\to -P_\mu,\quad \bar{R}\to R,\quad (S\equiv\bar{S}),\quad \bar{\zeta}\to\zeta;
\end{aligned} \quad (9)$$

Q_μ, G_μ and B as well as ξ are $\mathbf{C}$–even. The additional sources $\bar{P}_\mu$ and $\bar{R}$ together with S and $\bar{U}$ couple to the nonlinear parts of the BRST transformations (8) of the fields and analogously for the $\mathbf{C}$–conjugate fields and sources; G_μ couples to $\mathbf{s}_m\bar{\mathbf{s}}_m\mathcal{Q}_\mu$.

For later convenience we introduce the Delbourgo-Jarvis-Nakanishi-Ojima(DJNO) transformations [7] defined by:

$$\begin{aligned}
\mathbf{d}A_\mu = \mathbf{d}\mathcal{Q}_\mu = \mathbf{d}G_\mu = 0,\quad \mathbf{d}\bar{\mathcal{C}} = \mathcal{C},\quad \mathbf{d}\mathcal{C} = 0,\quad \mathbf{d}\mathcal{B} = (\zeta-\bar{\zeta})\mathcal{C}^2,\\
\mathbf{d}\bar{P}_\mu = P_\mu,\quad \mathbf{d}P_\mu = 0,\quad \mathbf{d}\bar{U} = U,\quad \mathbf{d}U = 0;\quad \mathbf{d}S = R,\quad \mathbf{d}R = 0,\quad \mathbf{d}\bar{R} = 2S.
\end{aligned} \quad (10)$$

Again, the anti–DJNO transformations are obtained by $\mathbf{C}$–conjugation.

As in the case $A_\mu = 0$ (see [11]) the algebra generated by $\left\{\mathbf{s}_m, \bar{\mathbf{s}}_m; \mathbf{d}, \bar{\mathbf{d}}, \mathbf{q}\right\}$ is easily verified to be the superalgebra $osp(1,2)$ [8]. It is given by the following $\mathbf{q}$–graded commutation relations

$$\mathbf{s}_m^2 = -m^2\mathbf{d}, \qquad \{\mathbf{s}_m, \bar{\mathbf{s}}_m\} = m^2\mathbf{q}, \qquad \bar{\mathbf{s}}_m^2 = -m^2\bar{\mathbf{d}}, \tag{11}$$

$$[\mathbf{d}, \mathbf{s}_m] = 0, \qquad \left[\bar{\mathbf{d}}, \mathbf{s}_m\right] = -\bar{\mathbf{s}}_m, \qquad [\mathbf{q}, \mathbf{s}_m] = \mathbf{s}_m, \tag{12}$$

$$\left[\bar{\mathbf{d}}, \bar{\mathbf{s}}_m\right] = 0, \qquad [\mathbf{d}, \bar{\mathbf{s}}_m] = \mathbf{s}_m, \qquad [\mathbf{q}, \bar{\mathbf{s}}_m] = -\bar{\mathbf{s}}_m, \tag{13}$$

$$[\mathbf{q}, \mathbf{d}] = 2\mathbf{d}, \qquad \left[\mathbf{q}, \bar{\mathbf{d}}\right] = -2\bar{\mathbf{d}}, \qquad \left[\mathbf{d}, \bar{\mathbf{d}}\right] = -\mathbf{q}. \tag{14}$$

Using these relations it is easy to check that the action $S(A|g,m;\xi,\rho,\sigma)$ is invariant under (anti)BRST as well as (anti)DJNO transformations:

$$\begin{aligned} &\mathbf{s}_m S(A|g,m;\xi,\rho,\sigma) = 0, \quad \mathbf{d}\, S(A|g,m;\xi,\rho,\sigma) = 0,\\ &\bar{\mathbf{s}}_m S(A|g,m;\xi,\rho,\sigma) = 0, \quad \bar{\mathbf{d}}\, S(A|g,m;\xi,\rho,\sigma) = 0; \end{aligned} \tag{15}$$

invariance under $\mathbf{q}$ is trivial since $S(A|g,m;\xi,\rho,\sigma)$ has ghost number zero by definition. The (anti)DJNO symmetry of the action according to (11) is a consequence of the (anti)BRST invariance!

In addition to this $osp(1,2)$–symmetry, which is related to broken gauge, i.e. type II, symmetry, the action (6) is invariant (eventually modulo a mass term) under (generalized) type I and type III transformations defined as

$$\begin{aligned} &\delta_{\mathrm{I}} A_\mu = D_\mu(A)\delta\Omega, \qquad \delta_{\mathrm{I}} \mathcal{Q}_\mu = [\mathcal{Q}_\mu, \delta\Omega], \qquad \delta_{\mathrm{I}} G_\mu = [G_\mu, \delta\Omega], \qquad \delta_{\mathrm{I}}\Phi = [\Phi, \delta\Omega],\\ &\delta_{\mathrm{III}} \mathcal{Q}_\mu = D_\mu(\mathcal{Q})\delta\Omega, \ \delta_{\mathrm{III}} G_\mu = D_\mu(G)\delta\Omega, \ \delta_{\mathrm{III}} A_\mu = [A_\mu, \delta\Omega], \ \delta_{\mathrm{III}}\Phi = [\Phi, \delta\Omega], \end{aligned}$$

where Φ, here and in the following, denotes the remaining fields and sources:

$$\delta_{\mathrm{I}} S(A|g,m;\xi,\rho,\sigma) = 0, \quad \delta_{\mathrm{III}} S(A|g,m;\xi,\rho,\sigma) = m^2\partial_\mu Q^\mu. \tag{16}$$

Whereas the first symmetry can be formulated as a Ward identity, the KSZ identity of the (renormalized) theory, the second one leads to the Lee–identity (24).

3 Ward identities for the vertex functional Γ

In order to construct the renormalized theory associated to the classical action (6) the symmetries of that action have to be expressed by corresponding Ward identities of the 1PI functional. These identities hold, first of all, at lowest perturbative order, i.e. for $\Gamma^0(A|g,m;\xi,\rho,\sigma)$; they are to be required to hold for the renormalized 1PI functional $\Gamma(A|g,m;\xi,\rho,\sigma)$. Since these identities are purely algebraic requirements they may be formulated for a generic functional, denoted also by Γ:

$$^{A}\mathbf{K}\,\Gamma = 0 \qquad \text{(Kluberg – Stern – Zuber identity)}, \tag{17}$$

$$\mathbf{S}_m(\Gamma) = 0 \qquad \text{(Slavnov – Taylor identity)}, \tag{18}$$

$$\mathbf{D}\,(\Gamma) = 0 \qquad \text{(Delduc – Sorella identity)}, \tag{19}$$

$$\mathbf{T}_\zeta(\Gamma) = 0 \qquad \text{(Bonora – Pasti – Tonin identity)}, \tag{20}$$

and, in addition, the **C**-conjugated second and third identity; the first identity and the last one (fixing the definition of $\mathcal{B}$) are **C**–even.

The above operations are defined in the following manner: The Kluberg-Stern–Zuber (KSZ) operator extends the local Ward operator $\mathbf{W} = \sum [\Phi, \delta/\delta\Phi\}$ according to ${}^A\mathbf{K} = D_\mu(A)\delta/\delta A_\mu + \mathbf{W}$ with the commutator or anticommutator depending on even or odd **q**–grading of Φ, respectively. Integrating (17) over spacetime we get the rigid Ward identity $\int d^4x \, \{[A, \delta/\delta A] + \mathbf{W}\} \, \Gamma = 0$ which is defined also for $A_\mu = 0$. The (bilinear) Slavnov-Taylor (ST) and Delduc-Sorella (DS) [9] operations as usual are given by $\mathbf{S}_m(\Gamma) = \int d^4x \sum \mathrm{Tr}\left((\mathbf{s}_m\Phi)\delta\Gamma/\delta\Phi\right)$ and $\mathbf{D}(\Gamma) = \int d^4x \sum \mathrm{Tr}\left((\mathbf{d}\Phi)\delta\Gamma/\delta\Phi\right)$, respectively, where the nonlinear parts of $\mathbf{s}_m\Phi$ and $\mathbf{d}\Phi$ are to be replaced by the functional derivative of Γ with respect to the corresponding source $\bar{P}_\mu, \bar{R}, S$ or $\bar{U}$. The Bonora-Pasti-Tonin (BPT) operation [10] is defined by $\mathbf{T}_\zeta(\Gamma) = \partial\Gamma/\partial\zeta - \int d^4x \mathrm{Tr}\left(\delta\Gamma/\delta S \cdot \delta\Gamma/\delta B\right)$.

For linear gauges the action S, if appropriately chosen, coincides with the effective action in tree approximation Γ^0. But in the case of nonlinear gauges this will not be true. Namely, the ST operation applied to $S(A|g, m; \xi, \rho, \sigma)$ leads to the expression $\mathbf{S}_m(S) = 2\zeta\bar{\zeta}\rho \int d^4x \mathrm{Tr}(\delta S/\delta B \cdot \delta S/\delta \bar{C})$ which vanishes only in the special case $\zeta\bar{\zeta} = 0$. Therefore, S has to be changed in such a manner that this unwanted term is cancelled. The explicit determination is very cumbersome, it leads to the following remarkable result:

$$\Gamma^0(A|g, m; \xi, \rho, \sigma) = S(A|g, m; \xi, \rho, \sigma) + \frac{\zeta\bar{\zeta}\rho}{1 - 4\zeta\bar{\zeta}\rho\xi} \int d^4x \mathrm{Tr}\left(\frac{\delta S}{\delta B}\frac{\delta S}{\delta B}\right). \tag{21}$$

As can be seen by an explicit computation this expression fulfills the identities (17) – (20) and is stable against small perturbations. Therefore, it constitutes the correct starting point for perturbative renormalization.

4 Gauge and consistency conditions

The auxiliary field B, despite of the fact that it does not decouple from the dynamics, is not a primary one. However, its dynamics can be constrained to be the same as for the classical action by the following gauge condition (which, in fact, is the renormalized equation of motion):

$$\begin{aligned} \tau \, \delta\Gamma/\delta B &= \xi(2B + (\zeta - \bar{\zeta}) \, \delta\Gamma/\delta S) + D_\mu(A)(G^\mu - \mathcal{Q}^\mu) + [\mathcal{Q}_\mu, G^\mu] \\ &+ \left\{U, \bar{\mathcal{C}}\right\} - \left\{\bar{U}, \mathcal{C}\right\}, \end{aligned} \tag{22}$$

where $\tau = 1 - \rho\xi$. Contrary to the common belief, (22) is a well posed condition since its nonlinear content is respected by $\delta\Gamma/\delta S$.

Applying $\delta/\delta B$ on the ST identity the antighost equation of motion obtains:

$$\begin{aligned} \tau \, \delta\Gamma/\delta\bar{C} &= \xi(2m^2 C - \delta\Gamma/\delta\bar{U}) - D_\mu(A)\left((1 - \sigma)P^\mu - \delta\Gamma/\delta\bar{P}_\mu\right) \\ &- [Q_\mu, P^\mu] + \left[\bar{C}, R\right] + \left[\bar{U}, \delta\Gamma/\delta\bar{R}\right] + \left[G_\mu, \delta\Gamma/\delta\bar{P}_\mu\right] \\ &- [C, S + \zeta\delta\Gamma/\delta B] + [U, B + \zeta\delta\Gamma/\delta S]; \end{aligned} \tag{23}$$

an analogous condition for the ghost equation of motion obtains by **C**–conjugation.

If now $\delta/\delta\bar{C}$ is applied on the anti–ST identity we finally get a further independent condition, the local Lee identity ($\bar{\sigma} = 1 - \sigma$):

$$(D_\mu(A)\mathbf{\Delta}^\mu(\sigma) + \mathbf{W})\Gamma = m^2 D_\mu(A)Q^\mu \quad \text{with} \quad \mathbf{\Delta}^\mu(\sigma) \equiv \bar{\sigma}\delta/\delta Q_\mu + \delta/\delta G_\mu. \tag{24}$$

It expresses a consistence condition related to Type III symmetry. This can be seen if the A_μ–dependence of Γ is taken into account. In the case $A_\mu = 0$, where the KSZ identity disappears, the integrated Lee identity equals the rigid Ward identity.

Together with the identities (17) – (20) the conditions (22) – (24) constitute a basic set of algebraic relations which express the symmetries to be fulfilled by the vertex functional. Additional consistency conditions cannot appear since the linearized ST– and DS–operations fulfil (anti–) commutation relations analogous to (11) – (14); the BPT-identity is not independent, but follows by the help of $\partial\mathbf{S}_m(\Gamma)/\partial\zeta = 0$.

5 Background dependence of Greens functions

Now, combining the KSZ– and the Lee–identity the following *linear* functional differential equation for Γ obtaines

$$\left(\frac{\delta}{\delta A_\mu} - \bar{\sigma}\frac{\delta}{\delta Q_\mu} - \frac{\delta}{\delta G_\mu}\right)\Gamma(A|Q,G,\Phi) + m^2 Q^\mu = 0. \tag{25}$$

Its solution is given by

$$\Gamma(A|Q,G,\Phi) = \exp(A\cdot\mathbf{\Delta})\Gamma(0|Q,G,\Phi) - m^2\{1 + \tfrac{1}{2}(A\cdot\mathbf{\Delta})\}(A\cdot Q), \tag{26}$$

with the obvious abbreviation $(X \cdot Y) = \int d^4x \mathrm{Tr}(X_\mu Y^\mu)$. ¿From this it follows that – up to m^2–dependent term – the A_μ–dependence of $\Gamma(A|Q,G,\Phi)$ is given by a shift of Q and G in $\Gamma(0|Q,G,\Phi)$, namely

$$\Gamma(A|Q,G,\Phi) = \Gamma(0|Q + \bar{\sigma}A, G + A, \Phi) - m^2\left(A\cdot(Q + \tfrac{1}{2}\bar{\sigma}A)\right). \tag{27}$$

By this approach we obtained a complete description of the background field dependence for the vertex functional. It should be noted that this was possible because (25) is a linear differential equation, and this results from the fact that a nonlinear gauge has been chosen. Let us further remark that our derivation is purely algebraic. Therefore, the solution holds to any order of perturbation theory. Furthermore, without much effort it can be shown that A_μ will not be renormalized. Therefore, it is sufficient to prove renormalizability of $\Gamma(0|Q,G,\Phi)$.

The obtained result may be transfered to the generating functional Z of the complete Green functions by applying a Legendre transformation. Therefore let us introduce $W(A|J,G,\Phi) = i\hbar \ln Z(A|J,G,\Phi)$, being defined through $\Gamma(A|Q,G,\Phi) +$

$W(A|J,G,\Phi)+(J\cdot Q)+(K\cdot B)+(L\cdot\bar{C})-(\bar{L}\cdot C)=0$. Then the differential equation corresponding to (25) is given by

$$\left(\frac{\delta}{\delta A_\mu}-\frac{\delta}{\delta G_\mu}+m^2\frac{\delta}{\delta J_\mu}-\bar{\sigma}\frac{J^\mu}{i\hbar}\right)Z(A|J,G,\Phi)=0. \tag{28}$$

In the same manner as above the solution for the Greens functions is obtained

$$Z(A|J,G,\Phi)=\exp\{\bar{\sigma}\left(A\cdot(J-\tfrac{1}{2}m^2A)\right)/i\hbar\}Z(0|J-m^2A,G+A,\Phi). \tag{29}$$

Furthermore, comparing the Lee identity for $G_\mu\neq 0;\Phi=0$,

$$\left(\frac{\bar{\sigma}}{i\hbar}\partial_\mu J^\mu-m^2\partial_\mu\frac{\delta}{\delta J_\mu}+D_\mu(G)\frac{\delta}{\delta G_\mu}+[J_\mu,\frac{\delta}{\delta J_\mu}]\right)Z(0|J,G,0)=0,$$

with the corresponding one for $G_\mu=0$ (being deduced from the rigid Ward identity)

$$\left((i\hbar z_Q)^{-1}\partial_\mu J^\mu+[J_\mu,\delta/\delta J_\mu]\right)Z(0|J,0,0)=0,$$

the G_μ–dependence of the renormalized functional $Z(0|J,G,0)$ may be determined explicitly:

$$Z(0|J,G,0)=\exp\left\{(\sigma z_\sigma/i\hbar z_Q)\left((J+\tfrac{1}{2}m^2G)\cdot G\right)\right\}Z(0|J,0,0). \tag{30}$$

Combining (29) and (30) we obtain for the physical Green functions in Landau gauge ($\xi=0$), being restricted by the gauge condition $D_\mu(A)\delta Z_{\rm phys}(A|J)/\delta J_\mu=0$ and ζ–independence $\partial Z_{\rm phys}(A|J)/\partial\zeta=0$, in the limit $m^2=0, G_\mu=0$ exactly Rouet's result equ. (5) above.

At this stage we have to comment that in paper [2] the fact has been overlooked that the verification of equ. (10) of that paper presupposes an assumption which is not proved. By our method, being completely different from the one used by Rouet, this gap has been closed and his result is now verified.

6 Concluding remarks

The Curci-Ferrari model of massive gauge fields in nonlinear gauge has been generalized to the presence of a generic background field – not necessarily being a solution of the field equations. The symmetries of the model are enlarged due to the appearance of the background field. From the KSZ– and the Lee identity a linear differential equation for 1PI vertex functional and, after Legendre transform, an analogous one for the generating functional of the Greens functions is obtained. Their solutions relate the vertex resp. Green functions with background field A_μ to the corresponding one for $A_\mu=0$. In particular, a former proof of Rouet has been strenghtened.

Furthermore, it is claimed that A_μ is not renormalized and, therefore, the problem of renormalization of the theory with background is reduced to the corresponding problem of the theory without background field. This, however, has been shown in an earlier paper [11]. There, it is proved, that only three independent z–factors appear, namely z_Q, z_B and z_g (see also [3]). A more detailed presentation of the above results will be given elsewhere.

Finally let us remark that this method of enlarging the symmetries can be applied also to the special cases of instanton and meron configurations. Despite the fact that in these cases additional zero modes appear which are to be considered in the same spirit as has been done for the gauge zero modes the formalism introduced here works with some additional technicalities.

References

[1] H. Kluberg-Stern, and J. B. Zuber, Phys. Rev. **D12** (1975) 467, 482, and 3159

[2] A. Rouet, Phys. Lett **B84** (1979) 448

[3] B. Geyer and D. Mülsch, Nucl. Phys. B (Proc. Suppl.) **56B** (1997) 253;
B. Geyer and D. Mülsch, *Symmetry properties of massive gauge theories in non-linear background gauge: Background dependence of Green functions*, Leipzig NTZ 27/97, hep-th/9711050
B. Geyer, P. M. Lavrov, and D. Mülsch, *OSp(1,2)-covariant Lagrangean Quantization of Irreducible Massive Gauge Theories*, NTZ 30/97, hep-th 9712024

[4] J. H. Lowenstein and W. Zimmermann, Nucl. Phys. **B86** (1975) 77,
J. H. Lowenstein, Commun. Math. Phys. **47** (1976) 53

[5] J. H. Lowenstein and W. Zimmermann, Commun. Math. Phys. **44** (1975) 73
J. H. Lowenstein and E. Speer, Commun. Math. Phys. **47** (1976) 43

[6] G. Curci and R. Ferrari, Nuovo Cim. **A32** (1976) 151,
I. Ojima, Zs. Phys. **C13** (1082) 173

[7] R. Delbourgo and P. D. Jarvis, J. Phys. **A15** (1082) 611,
N. Nakanishi and I. Ojima, Z. Phys. **C6** (1980) 155

[8] W. Nahm, V. Rittenberg and M. Scheunert, J. Math. Phys. **17** (1976) 1626, 1640, M. Scheunert, W. Nahm and V. Rittenberg, J. Math. Phys. **18** (1976) 146, 155; see also: L. Frappat, P. Sorba and A. Sciarrino, Dictionary on Lie superalgebras, ENSLAPP-AL-600/96 (hep-th/9607161)

[9] F. Delduc and S. P. Sorella, Phys. Lett **B63** (1976) 91

[10] L. Bonora, P. Pasti and M. Tonin, Nuovo Cim. **64A** (1081) 378

[11] B. Geyer and D. Mülsch, *Gauge (in)dependence, nonlinear gauges and the concept of extended BRST and anti–BRST symmetries*, Leipzig preprint, NTZ 29/1996

Quantum chaos in the quark matrix at finite temperature

Harald Markum[1], Rainer Pullirsch[1], Klaus Rabitsch[1], Tilo Wettig[2]

[1] Institut für Kernphysik, Technische Universität Wien,
Wiedner Hauptstr. 8-10, A-1040 Wien, Austria
[2] Institut für Theoretische Physik, Technische Universität München,
D-85747 Garching, Germany

Abstract: The eigenvalue spectrum of the staggered Dirac matrix in SU(3) gauge theory and in full QCD is investigated on a $6^3 \times 4$ lattice for various values of gluon coupling and quark masses both in the confinement and in the deconfinement phase. We study the nearest-neighbor spacing distribution $P(s)$ as a measure of the fluctuation properties of the eigenvalues. In both phases except far into the deconfinement region, the lattice data agree with the Wigner surmise of random-matrix theory which is indicative of quantum chaos. No signs of a transition to Poisson regularity at the deconfinement phase transition are found.

1 Introduction

The eigenvalue spectrum of the Dirac operator is important to understand certain features of QCD. For example, the accumulation of small eigenvalues is, via the Banks-Casher formula [1], related to the spontaneous breaking of chiral symmetry. Recently, the fluctuation properties of the eigenvalues in the bulk of the spectrum have also attracted attention. It was shown in Ref. [2] that on the scale of the mean level spacing they are described by random-matrix theory (RMT). In particular, the nearest-neighbor spacing distribution $P(s)$, i.e., the distribution of spacings s between adjacent eigenvalues on the unfolded scale (see below), agrees with the Wigner surmise of RMT. According to the so-called Bohigas conjecture [3], quantum systems whose classical counterparts are chaotic have a nearest-neighbor spacing distribution given by RMT whereas systems whose classical counterparts are integrable obey a Poisson distribution, $P(s) = e^{-s}$. Therefore, the specific form of $P(s)$ indicates the presence or absence of quantum chaos.

At the localization transition in disordered mesoscopic systems, one observes a transition in $P(s)$ from Wigner to Poisson behavior [4]. In Ref. [2], the question was raised if there is such a transition from chaotic to regular behavior in the case of the lattice Dirac operator. The present study serves to investigate this question (see Ref. [5] for first results in quenched QCD). Recently, a Wigner to Poisson transition was also studied in the context of a spatially homogeneous Yang-Mills-Higgs system [6]. The question if chaos in an N-component φ^4-theory in the presence of an external field survives quantization was investigated in Ref. [7].

In RMT, one distinguishes between several universality classes which are determined by the symmetries of the system. For the case of the QCD Dirac operator, this classification was done in Ref. [8]. Depending on the number of colors and the representation of the quarks, the Dirac operator is described by one of the three chiral Gaussian ensembles of RMT. For the fluctuation properties in the spectrum bulk, the predictions of the chiral ensembles are identical to those of the ordinary Gaussian ensembles [9]. In Ref. [2], the Dirac matrix was studied in SU(2) using both staggered and Wilson fermions corresponding to the symplectic and orthogonal ensemble, respectively. Here, we study SU(3) with staggered fermions corresponding to the chiral Gaussian unitary ensemble. We thus cover the last remaining symmetry class in the confined phase. The RMT result for the nearest-neighbor spacing distribution is rather complicated. It can be expressed in terms of so-called prolate spheroidal functions, see Ref. [10] where $P(s)$ has also been tabulated. A very good approximation to $P(s)$ is provided by

$$P(s) = \frac{32}{\pi^2} s^2 e^{-\frac{4}{\pi}s^2} . \tag{1}$$

This is the Wigner surmise for the unitary ensemble.

2 Eigenvalue analysis

We generated gauge field configurations using the standard Wilson plaquette action for SU(3) with and without dynamical fermions in the Kogut-Susskind prescription. We have worked on a $6^3 \times 4$ lattice with various values of the inverse gauge coupling $\beta = 6/g^2$ both in the confinement and deconfinement phase. The boundary conditions were periodic for the gluons, and periodic in space and anti-periodic in Euclidean time for the fermions. We typically produced 10 independent equilibrium configurations for each β. This is sufficient because of the spectral ergodicity property of RMT which allows to replace ensemble averages by spectral averages if one is only interested in bulk properties.

The Dirac operator, $\not{D} = \not{\partial} + ig\not{A}$, is anti-hermitian so that all eigenvalues are imaginary. For convenience, we denote them by $i\lambda_n$ and refer to the λ_n as the eigenvalues in the following. Because of $\{\not{D}, \gamma_5\} = 0$ the λ_n occur in pairs of opposite sign. All spectra were checked against the analytical sum rules $\sum_n \lambda_n = 0$ and $\sum_{\lambda_n > 0} \lambda_n^2 = 3V$, where V is the lattice volume.

To construct the nearest-neighbor spacing distribution $P(s)$ from the eigenvalues, one first has to "unfold" the spectra. This procedure is a local rescaling of the energy scale so that the mean level spacing is equal to unity on the unfolded scale. Ensemble and spectral averages are only meaningful after unfolding.

3 Results and Discussion

We begin the discussion of our results with the pure gluonic case. Figure 1 shows the staircase function $N(E)$ which is the number of eigenvalues with $\lambda \leq E$. The staircase function is not very sensitive to the value of β, neither in the confinement ($\beta \leq 5.6$) nor in the deconfinement ($\beta \geq 5.8$) regime. However, across the phase transition $N(E)$ is diminished for small E reflecting the suppression of small eigenvalues.

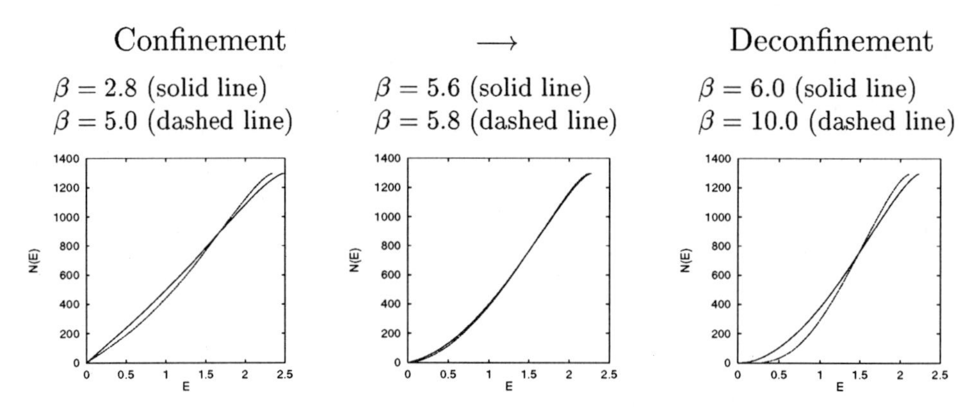

Figure 1: The staircase function $N(E)$ representing the number of positive eigenvalues $\leq E$ for typical configurations on a $6^3 \times 4$ lattice in pure SU(3). The confinement (deconfinement) phase corresponds to $\beta \leq 5.6$ ($\beta \geq 5.8$).

Figure 2 shows the nearest-neighbor spacing distribution $P(s)$ corresponding to the parameters of Fig. 1, compared with the RMT prediction. In the confinement phase, we find the expected agreement of $P(s)$ with the Wigner surmise of Eq. (1). In the deconfinement phase, we still observe agreement with the RMT result up to $\beta = 10.0$ (we have also plotted the Poisson distribution for comparison).

We continue with the case of full QCD with $N_f = 3$ degenerate flavors of staggered quarks with mass $ma = 0.1$ and $ma = 0.05$, respectively. The staircase function $N(E)$ shows no visible dependence on the masses considered. It looks unexceptional and very similar to the results of pure SU(3) in Fig 1. For lack of space we do not include the figures here. Figure 3 compares the nearest-neighbor spacing distribution $P(s)$ of full QCD with the RMT result. No mass dependence is visible. Again, in the confinement as well as in the deconfinement phase we still observe agreement with the RMT result up to $\beta = 8.0$. The observation that $P(s)$ is not influenced by the presence of dynamical quarks could have been expected from the results of Ref. [9]. Those calculations, however, only apply to the case of massless dynamical quarks.

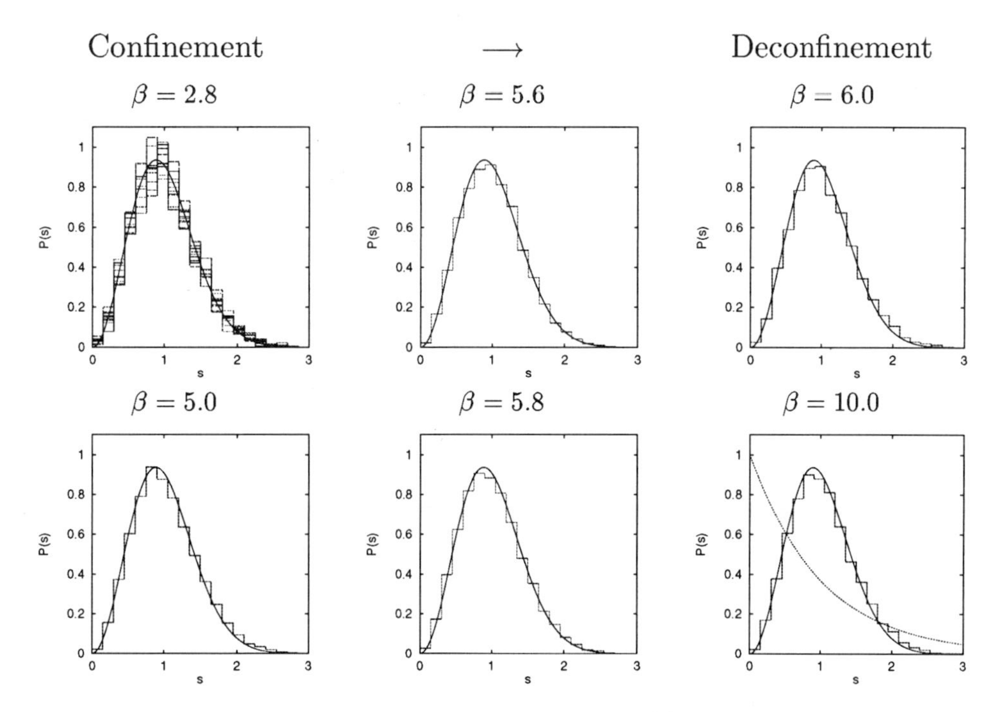

Figure 2: The nearest-neighbor spacing distribution $P(s)$ averaged over 10 independent configurations (except for $\beta = 2.8$ where all single configurations are plotted to give an indication of the deviations) on a $6^3 \times 4$ lattice in pure SU(3) (histograms) compared with the random-matrix result (solid lines). For comparison, the Poisson distribution $P(s) = e^{-s}$ is plotted for $\beta = 10.0$ (dotted line). There are no changes in $P(s)$ across the deconfinement phase transition at $\beta \approx 5.7$.

Our results, as those of Ref. [2], strongly indicate that massive dynamical quarks do not affect $P(s)$ either.

No signs for a transition to Poisson regularity are found. Thus, the deconfinement phase transition does not seem to coincide with a transition in the spacing distribution. For very large values of β far into the deconfinement region (not shown), the eigenvalues start to approach the degenerate eigenvalues of the free theory, given by $\lambda^2 = \sum_{\mu=1}^{4} \sin^2(2\pi n_\mu / L_\mu)/a^2$, where a is the lattice constant, L_μ is the number of lattice sites in the μ-direction, and $n_\mu = 0, \ldots, L_\mu - 1$. In this case, the nearest-neighbor spacing distribution is neither Wigner nor Poisson. However, it is possible to lift the degeneracies of the free eigenvalues using an asymmetric lattice where L_x, L_y, etc. are relative primes [11]. For large lattices, the nearest-neighbor spacing distribution of the non-degenerate free eigenvalues is then given by the Poisson distribution. While it may be interesting to search for a Wigner to Poisson transition on such asymmetric lattices, it is unlikely that such a transition will coincide with the deconfinement phase transition.

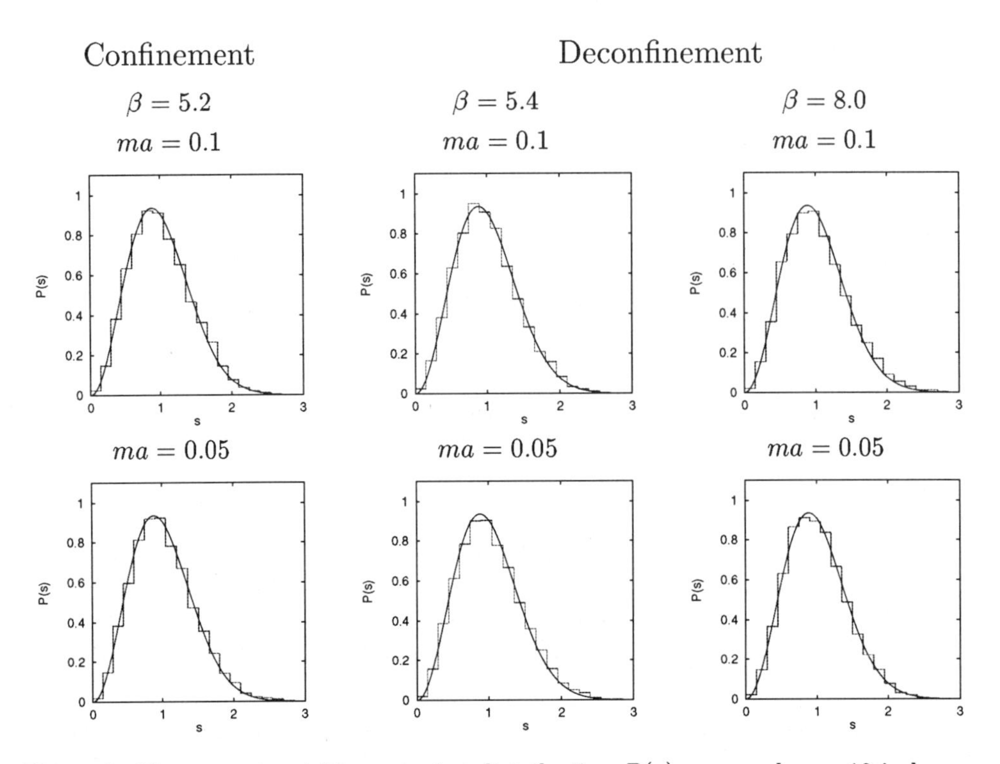

Figure 3: The nearest-neighbor spacing distribution $P(s)$ averaged over 10 independent configurations on a $6^3 \times 4$ lattice in full QCD (histograms) compared with the random-matrix result (solid lines). Again, there are no changes in $P(s)$ across the deconfinement phase transition at $\beta \approx 5.3$.

We do not believe that the absence of a signature for a transition from Wigner to Poisson behavior at the deconfinement phase transition is due to the finite lattice size. Even for the small lattice size we used, the agreement of $P(s)$ with the RMT curve is nearly perfect. This leads us to believe that we should have seen some sign of a transition if it existed in the thermodynamic limit.

4 Conclusions

We have searched for a transition in the nearest-neighbor spacing distribution $P(s)$ from Wigner to Poisson behavior across the deconfinement phase transition of pure gluonic and of full QCD. Such a transition exists, e.g., at the localization transition in disordered mesoscopic systems [4]. In a Yang-Mills-Higgs system a smooth transition along a Brody distribution was seen [6]. We found no signature of a transition in our lattice data, neither for pure SU(3) nor for full QCD. The data agree with the RMT result in both the confinement and the deconfinement phase except for extremely large values of β where the eigenvalues are known analytically. Our analysis of full

QCD with two different quark masses showed no influence on the nearest-neighbor spacing distribution.

A few words to explain our results are in order. Temporal monopole currents survive the deconfinement phase transition leading to confinement of spatial Wilson loops. Thus, even in the deconfinement phase, the gauge fields retain a certain degree of randomness. It would be interesting to try to disentangle those spatial contributions to the Dirac matrix to check the above mechanism.

Acknowledgments

This work was supported in part by FWF project P10468-PHY and by DFG grant We 655/11-2. We thank T.S. Biró, E.-M. Ilgenfritz, M.I. Polikarpov, and J.J.M. Verbaarschot for helpful discussions.

References

[1] T. Banks and A. Casher, Nucl. Phys. B 169 (1980) 103.

[2] M.A. Halasz and J.J.M. Verbaarschot, Phys. Rev. Lett. 74 (1995) 3920; M.A. Halasz, T. Kalkreuter, and J.J.M. Verbaarschot, Nucl. Phys. B (Proc. Suppl.) 53 (1997) 266.

[3] O. Bohigas, M.-J. Giannoni, and C. Schmit, Phys. Rev. Lett. 52 (1984) 1.

[4] B.L. Al'tshuler, I.Kh. Zharekeshev, S.A. Kotochigova, and B.I. Shklovskiĭ, Zh. Eksp. Teor. Fiz. 94 (1988) 343 [Sov. Phys. JETP 67 (1988) 625]; B.I. Shklovskiĭ, B. Shapiro, B.R. Sears, P. Lambrianides, and H.B. Shore, Phys. Rev. B 47 (1993) 11487.

[5] H. Markum, R. Pullirsch, K. Rabitsch, and T. Wettig, hep-lat/9709103; T.S. Biró, M. Feurstein, and H. Markum, hep-lat/9711002.

[6] L. Salasnich, Mod. Phys. Lett. A 12 (1997) 1473.

[7] L. Casetti, R. Gatto, and M. Modugno, hep-th/9707054.

[8] J.J.M. Verbaarschot, Phys. Rev. Lett. 72 (1994) 2531.

[9] D. Fox and P.B. Kahn, Phys. Rev. 134 (1964) B1151; T. Nagao and M. Wadati, J. Phys. Soc. Jpn. 60 (1991) 3298; 61 (1992) 78, 1910.

[10] M.L. Mehta, *Random Matrices*, 2nd ed. (Academic Press, San Diego, 1991).

[11] J.J.M. Verbaarschot, private communication.

Low-lying Eigenvalues of the improved Wilson-Dirac Operator in QCD

Hubert Simma[1], Douglas Smith[2] (UKQCD collaboration)

[1] DESY-Zeuthen, Platanenallee 6, D-15738 Zeuthen
[2] Dept. of Physics and Astronomy, University of Edinburgh, Edinburgh EH9 3JZ

Abstract: The spectral flow of the low-lying eigenvalues of the improved and unimproved Wilson-Dirac operator is studied on instanton-like configurations and on thermalized quenched configurations at various β-values. We also investigate the space-time localisation and chirality of the corresponding eigenvectors.

1 Introduction

The spectrum of the Dirac operator in QCD around zero is of particular physical interest, because in the continuum it is directly related to topological concepts and to the phenomenon of spontaneous breaking of chiral symmetry.

The topological charge of a smooth classical gauge field is related by the Atiyah-Singer index theorem to the difference between the number of right- and left-handed exact zero modes of the massless Dirac operator in that gauge background. After performing the ensemble average of the path integral and in the limit of an infinite volume, the Banks-Casher formula relates the chiral condensate $\langle\bar{\psi}\psi\rangle$ to the spectral density for zero eigenvalues of the Dirac operator [1]. Moreover, in the framework of an effective chiral Lagrangian one finds sum rules for moments of the spectral density in individual topological sectors, and the derivative of the spectral density at zero can be related to the decay constant of the pion [2].

The situation is substantially more complicated in numerical simulations of QCD on finite discrete lattices. Besides the restriction to a finite volume and the difficulties of defining a topological charge, the use of the Wilson action explicitly breaks chiral symmetry.

In numerical simulations the low-lying eigenvalues of the Wilson-Dirac operator are also crucial from a practical point of view. In the quenched approximation (almost-)zero modes are believed to be the origin of "exceptional" configurations, on which the numerical computation of light quark propagators is difficult and expensive, and which lead to large statistical errors of hadronic quantities at small quark masses [3]. In simulations with dynamical fermions the eigenvalues of the Wilson-Dirac operator play a central rôle for the performance of the algorithms and their scaling behaviour when approaching the chiral limit [4].

In this work we investigate the appearance and space-time localisation of zero (or low-lying) modes of the hermitian Wilson-Dirac operator $\gamma_5 M$. In particular, we are interested in the effects of the Sheikholeslami-Wohlert (SW) improvement term, the variation of the zero-mode distribution with the gauge coupling, and the connection between the eigenmodes and topological properties of the underlying gauge field.

This contribution is organised as follows: After the discussion of general properties of the spectra of M and $\gamma_5 M$ in the next section, we investigate in sect. 3 the zero modes on simple, instanton-like gauge configurations, before and after heating, and with and without improvement. In sect. 4, we present numerical data for thermalized quenched configurations on 16^4 lattices, and discuss our results in sect. 5.

2 Properties of the Eigenmodes of γ_5M

The (improved) Wilson-Dirac matrix is given by

$$M \;=\; 1-\kappa(H+\frac{i}{2}c_{SW}F_{\mu\nu}\sigma_{\mu\nu})\;, \tag{1}$$

where H is the usual hopping term, and κ is related to the bare quark mass m_0 by $\kappa = (8+2m_0)^{-1}$. The SW improvement term contains the usual clover combination $F_{\mu\nu}$ of the SU(3) gauge links. The boundary conditions for the fermion fields are anti-periodic in time direction and periodic in the space directions.

Due to the γ_5-hermiticity, $M^\dagger = \gamma_5 M \gamma_5$, the eigenvalues of M come in complex conjugate pairs μ and μ^*. For $c_{SW} = 0$, the spectrum on even lattices is also reflection symmetric along the line Re$\mu = 1$, i.e. $Mr = (1-\kappa\rho)r \;\Leftrightarrow\; Ms = (1+\kappa\rho)s$,with the eigenvectors related by the stagger transformation $s(x) = (-1)^{\sum_\mu x_\mu}\, r(x)$.

The elementary relation, $(\mu_k^* - \mu_i)\cdot r_k^\dagger \gamma_5 r_i = 0$, between eigenvalues μ_i and corresponding right eigenvectors r_i of M, suggests to distinguish two classes of eigenmodes:

- "bulk" modes with $\mu \notin \mathbf{R}$ and vanishing chirality $(r, \gamma_5 r)$
- real modes with $\mu \in \mathbf{R}$ and possibly non-zero chirality

All eigenvalues lie within the disk $\{\mu = 1+\kappa\rho : |\rho| \le 8\}$. On configurations with increasing β-values, the bulk modes depopulate the vicinity of the real axis except for five regions corresponding to the different corners of the Brillouin zone where the quark propagator has poles.

One may expect that some of the real modes are related to topological properties of the gauge field by an approximate lattice remnant of the index theorem, but of course also topologically trivial gauge fields can have real modes (e.g. $U_4(x) = diag(1,-1,-1)$ on one time slice and all other $U_\mu(x) = \mathbf{1}$).

In the following we shall investigate eigenmodes of the *hermitian* operator

$$Q \equiv \gamma_5 M/\,(1+8\kappa)\;. \tag{2}$$

The normalisation is such that the eigenvalues $\lambda(\kappa)$ of Q satisfy $0 \leq \lambda^2 \leq 1$, and we refer to the part of the spectrum closest to zero as the "low-lying" eigenvalues. They can be computed in an efficient and reliable way by using an accelerated conjugate gradient algorithm [5], which also provides approximate eigenvectors and is viable for large lattice sizes.

Recalling the relation $M^{-1} = Q^{-1}\gamma_5/(1+8\kappa)$, we note that all hadronic propagators can be constructed from Q^{-1}, which has a simple spectral representation in terms of the orthonormal eigenvectors of Q.

The eigenmodes (eigenvalues and corresponding eigenvectors) of $Q(\kappa)$ have a non-trivial κ-dependence. The real modes, $\mu = 1 - \kappa\rho \in \mathbf{R}$, of M are equivalent to zero-modes of Q at κ-values $\kappa_0 = 1/\rho$. In general, the spectral flow $Q(\kappa)$ provides indirect information about the complex spectrum of M, and in particular the flow of the low-lying eigenvalues $\lambda(\kappa)$ probes the spectrum of M close to the real axis. The five regions, where bulk modes of M populate the vicinity of the real axis, correspond to extrema of the low-lying eigenvalues of $Q(\kappa)$ at κ-values roughly of the order of 0.125, 0.25, $\pm\infty$, -0.25, and -0.125. In the special case of a flat gauge field each eigenvalue $\mu = 1 - \kappa\rho$ of M corresponds to a minimal absolute value in the spectral flow of $\tilde{\lambda} = \lambda \cdot (1+8\kappa)/\kappa$ at $\kappa = 1/\mathrm{Re}\rho$.

The chiralities $\chi_i \equiv (u_i, \gamma_5 u_i)$ of the eigenvectors of Q are related to the spectral flow of $\tilde{\lambda}_i(\kappa)$ by the identity

$$\chi_i(\kappa) = \frac{d\tilde{\lambda}_i}{d(1/\kappa)}\,, \tag{3}$$

i.e. the zeroes of the chirality correspond to extrema in the spectral flow of $\gamma_5 M/\kappa$. Eq. (3) yields the first order κ-dependence of $\lambda(\kappa)$, and from the flow of a zero modes close to the position κ_0, where λ vanishes, one can thus estimate κ_0 as

$$\kappa_0 \approx \kappa \cdot \left(1 + \frac{\chi}{\lambda - \chi}\right)\,,$$

which is very helpful for an efficient iterative search for possible zero modes.

3 Instanton-like Configurations

We first investigate to what approximation the index theorem holds for Wilson fermions on simple instanton-like configurations. These are set up on even lattice sizes in the singular gauge as described in [6], followed by a few cooling sweeps to reduce boundary mismatches. We consider configurations with up to two well-separated instantons or anti-instantons in the same SU(2) subgroup and a net topological charge $\nu = 0, \pm 1, \pm 2$.

In the physical κ-region we find $n_+ = \nu$ zero modes with chiralities of the same sign as ν, and $n_- = 0$ with opposite sign chiralities. On multiple-instanton configurations their positions are split depending on the size and distance of the instantons. In the four unphysical regions one has $(n_+, n_-) = (0, 4\nu)$, $(6\nu, 0)$, $(0, 4\nu)$, and $(\nu, 0)$, respectively. In the following we consider only the physical region $0.1 \leq \kappa \leq 0.2$.

The spectral flow of the low-lying bulk modes on cold instanton-like configurations is characterised by a value of κ_χ which is close to the free-case value of $1/8$ and essentially independent of the instanton size ρ. On the other hand, the position κ_0 and the chirality $\chi_0(\kappa_0)$ of the zero mode seem to be sensitive to discretisation effects and are strongly ρ-dependent. The following table shows the values for single instantions on a 16^4 lattice:

	$c_{SW} = 0$		$c_{SW} = 1$	
ρ	κ_0	$\chi_0(\kappa_0)$	κ_0	$\chi_0(\kappa_0)$
2	.135	.66	.1258	.986
4	.127	.90	.1250	.999
8	.126	.99	.1250	.999

We note the clear effect of the improvement term (using the tree-level value $c_{SW} = 1$): It significantly rises the chirality $\chi(\kappa_0)$ of the zero mode and moves the position κ_0 towards lower values and closer to κ_χ.

When the configurations are heated by a moderate number of update sweeps, the zero-mode remains, but the position κ_0 changes and fluctuates for different heating trajectories. The heating also moves the positions κ_χ, where the chiralities of the lowest bulk modes vanish, to values around $\kappa_{\text{crit}}(\beta)$, where β is the value at which the heating was performed. For the improved operator the variation of κ_0 is much smaller, and the central value is lower and significantly closer to κ_χ than for $c_{SW} = 0$.

Also on the heated configurations the eigenvectors are centered on the instantons (except for $\kappa \ll \kappa_\chi$), but the space-time extension of the eigenvector density at κ_0 is not as clearly scaling with the size of the instanton as in the cold case.

4 Quenched Configurations

To study the spectrum of the Wilson-Dirac operator on quenched configurations, we used 10 thermalized 16^4 configurations for each of the gauge couplings, $\beta = 5.9$, 6.0, and 6.1. The spectral flows were computed for $c_{SW} = 0$ and for the non-perturbative value $c_{SW}(\beta)$ of ref. [7].

The distribution of the zero modes for the improved operator at $\beta = 6.0$ is shown in Fig. 1. The peak of the distribution around κ_{crit} is more pronounced in the improved case than for $csw = 0$. The typical behaviour of the chiralities on one of these configurations is illustrated in Fig. 2. The circles denote the lowest (zero) modes and the lines connect points from the same bulk modes to guide the eye. One notices that the chiralities of the low-lying bulk modes vanish around the same value $\kappa_\chi \approx \kappa_{\text{crit}}(\beta)$.

The space-time localisation of the low-lying eigenmodes for the unimproved case, is known to change around κ_{crit} from a typically exponential decay of the eigenvector density to an even stronger, almost point-like, localisation [8]. For the improved operator we find in the region $\kappa \ll \kappa_\chi$ a somewhat stronger localisation of the lowest-lying modes than for $c_{SW} = 0$, but this difference tends to disappear towards κ_χ. Above κ_χ a sharp drop of the "participation ratio" indicates again a very strong

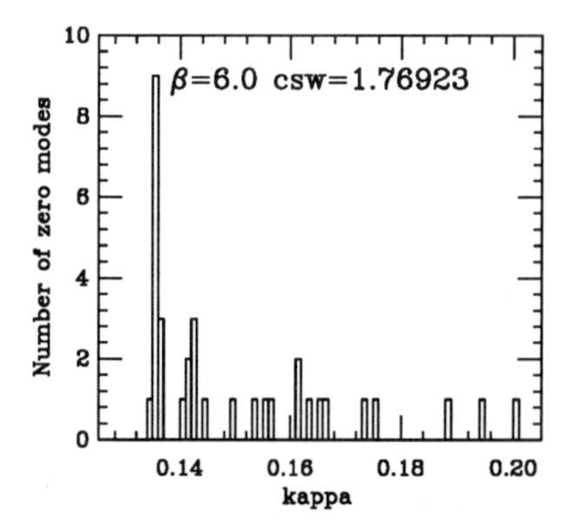

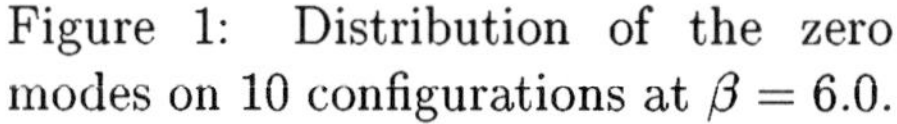
Figure 1: Distribution of the zero modes on 10 configurations at $\beta = 6.0$.

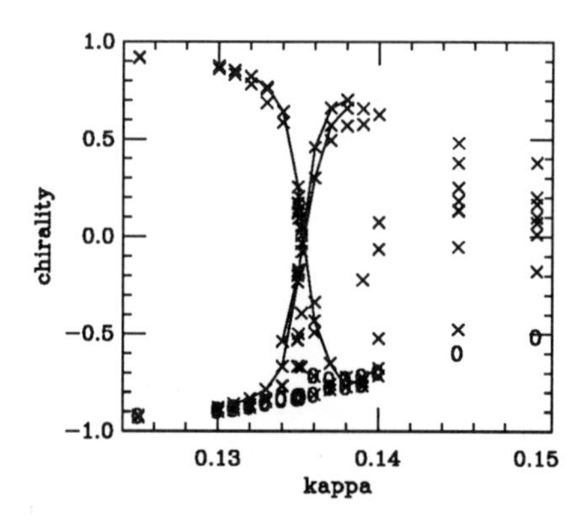

Figure 2: Typical flow of the chirality of the lowest modes.

(almost point-like) localisation similar to the unimproved case. The κ-dependence of the localisation strength is the same for the flow of zero modes and of low-lying bulk modes. Thus, zero modes appearing at very low $\kappa_0 \leq \kappa_\chi$ are in general weaker localised than those at higher κ_0.

To investigate the localisation of the zero modes in relation to topological properties of the underlying gauge field, we approximate the charge density of the cooled gauge configurations by a superposition of single-instanton charge distributions [9]. We find that the peaks closest to the maximum of the eigenvector density are in average significantly higher (i.e. correspond to smaller instantons) than the average value of the configuration. There also seems to exist a weak correlation between the chirality of the eigenmode and the sign of the topological charge of the closest peak.

5 Discussion

On simple instanton-like configurations we confirmed that the improved and unimproved Wilson-Dirac operator approximately satisfy an index theorem similar to the case of staggered fermions or QED_2 [10]. The presence of the zero modes is rather stable under the effect of roughening the gauge fields by heating.

Inclusion of the SW improvement term significantly increases the chirality of the zero modes and moves κ_0 closer to κ_{crit}. On thermalized quenched configurations the improvement term renders the zero mode distribution stronger peaked around κ_{crit}. This is in accordance with the picture that in the continuum limit the κ-region with vanishing mass gap (i.e. with zero modes of Q) shrinks to a point [11].

The number of zero modes found on the 16^4 lattices clearly decreases with increasing β-values. This, and the much lower abundance of zero modes on 8^4 lattices, indicates that the number of zero modes grows with the physical volume.

The eigenvectors of the improved operator are somewhat stronger localised than in the unimproved case, and their positions tend to lie close to peaks of the topological charge density of the (cooled) gauge field. The width of these peaks is typically smaller than the average for the configuration.

On the four exceptional configurations encountered by UKQCD at $\beta = 6.0$ on $16^3 48$ and $32^3 64$ lattices we verified the presence of a zero mode at a near κ-value $\kappa_0 < \kappa_{\text{crit}}$. In all cases the eigenvector is localised close to or on a narrow peak in the topological charge density, and the chirality has the same sign as the charge of the peak, but is not correlated with the overall topological charge of the configuration.

The spectral flow of $Q(\kappa)$ can also provide instructive information about the complex spectrum of M. For instance, the position κ_χ of the chirality zeroes of the lowest bulk modes approximately coincides with κ_{crit} (defined by an ensemble average, e.g. from the pion mass), and may be an interesting quantity to characterise the "critical" κ-value for individual gauge configurations.

Further details and numerical data shall be presented elsewhere [12].

Thanks are due to the organisers for an interesting and pleasant Symposium. We thank M. Bäker, K. Jansen, M. Lüscher and S. Pickles for helpful discussions, and R. Petronzio for generous computing time on the APE at Tor Vergata. Part of the numerical work was also performed at DESY-Zeuthen, at the University of Edinburgh, and at the APE Lab, and we are grateful for their hospitality and support.

References

[1] T. Banks and A. Casher, *Nucl. Phys.* **B169** (1980) 103

[2] H. Leutwyler, and A. Smilga, *Phys. Rev.* **D46** (1992) 5607;
A. Smilga, hep-th/9503049

[3] W. Bardeen, et al., hep-lat/9710084 and hep-lat/9705002

[4] K. Jansen, *Nucl. Phys.* **B** (Proc. Suppl.) **53** (1997) 974

[5] T. Kalkreuter and H. Simma, *Comp. Phys. Comm.* **93** (1996) 33

[6] M. Laursen, J. Smit, and J. Vink, *Nucl. Phys.* **B343** (1990) 522

[7] M. Lüscher, et al., *Nucl. Phys.* **B491** (1997) 323

[8] K. Jansen, et al. *Nucl. Phys.* **B** (Proc. Suppl.) **53** (1997) 262

[9] D. Smith and M. Teper (UKQCD Collaboration), hep-lat/9801008

[10] I. Barbour, and M. Teper, *Phys. Lett.* **B175** (1986) 445;
S. Itoh, Y. Iwasaki, and T. Yoshié, *Phys. Rev.* **D36** (1987) 527;
J. Smit, and J. Vink, *Nucl. Phys.* **B286** (1987) 485;
C. Gattringer, et al., *Nucl. Phys.* **B508** (1997) 329 and hep-lat/9712015

[11] R.G. Edwards, U.M. Heller, R. Narayanan, R.L. Singleton, hep-lat/9711029

[12] H. Simma, in preparation

String Duality and Novel Theories without Gravity

Shamit Kachru

Department of Physics
University of California, Berkeley
and
Lawrence Berkeley National Laboratory
University of California
Berkeley, CA 94720, USA

Abstract: We describe some of the novel 6d quantum field theories which have been discovered in studies of string duality. The role these theories (and their 4d descendants) may play in alleviating the vacuum degeneracy problem in string theory is reviewed. The DLCQ of these field theories is presented as one concrete way of formulating them, independent of string theory.

1 Introduction

Recent advances in string theory have led to the discovery of many new interacting theories without gravity. These theories are found by taking special limits of M-theory, in which many of the degrees of freedom decouple. In this talk we will:
I. Describe some examples of these new theories.
II. Review why it is important to fully understand these examples.
III. Propose a definition of these theories, in the light-cone frame, which is manifestly independent of M or string theory.

2 Examples

2.1 Theories with $(2,0)$ Supersymmetry

The first (and simplest) examples were found by Witten [1], in studying type IIB string theory on $K3$. He considers the situation where the $K3$ develops an $A-D-E$ singularity. In the IIA theory, one finds extra massless gauge bosons in these circumstances. These extra vectors of the (1,1) supersymmetry are required by string-string duality, and arise from D2 branes which wrap the collapsing 2-cycles and become massless in the singular limit.

In the IIB theory, there is a chiral (2,0) supergravity in six dimensions. The only massless multiplet of the (2,0) supersymmetry (other than the gravity multiplet) is the tensor multiplet, which consists of 5 scalars, some chiral fermions, and a self-dual two form $B_{\mu\nu}$ which satisfies

$$dB = *dB \tag{1}$$

The (2,0) supergravity *requires* the presence of precisely 21 tensor multiplets for anomaly freedom. Therefore, it is hard to envision a scenario where one finds extra massless particles at the singular point in moduli space. However, further compactification on an S^1 yields a theory related to the IIA theory by T-duality, so one must find (after S^1 compactification) gauge bosons of the $A - D - E$ gauge group. What is their IIB origin?

Further compactify the IIA and IIB theories on circles with radii $R_{A,B}$. Then T-duality relates the theories with $R_A = \frac{1}{R_B}$ (we are temporarily setting the string scale α' to one for simplicity). The relation between the six-dimensional string couplings $\lambda_{A,B}$ is

$$\frac{1}{\lambda_A} = \frac{R_B}{\lambda_B} \tag{2}$$

If we consider a point in IIA moduli space a distance ϵ from the singular point, then there are W-bosons coming from wrapped D2 branes whose masses go like

$$M_W = \frac{\epsilon}{\lambda_A} \tag{3}$$

So in type IIB, the mass is

$$M_W = \frac{\epsilon R_B}{\lambda_B} \tag{4}$$

This looks like the mass of a *string* wrapped around the S^1 in the IIB theory! But this string is not the critical type IIB string; from equation (4) it must have a tension

$$T = \frac{\epsilon}{\lambda_B} \tag{5}$$

Of course, this string comes from a D3 brane wrapped around a collapsing sphere in the $K3$ of area $\simeq \epsilon$.

For very small ϵ, $T << \frac{1}{\alpha'}$. So we get an $A - D - E$ series of quantum theories in six dimensions which contain light string solitons. Because the noncritical strings are very light compared to the fundamental string scale, one can decouple gravity. Then, it is believed that one is left with an interacting quantum field theory in six dimensions. As $\epsilon \to 0$, one approaches a nontrivial fixed point of the renormalization group.

In six-dimensions, strings are dual to strings. The particular light strings in question are self-dual (the $H = dB$ they produce is self-dual as in equation (1)), so the "coupling" of these quantum theories is fixed and of order one. In other words, there is no coupling constant which can serve as an expansion parameter.

By using ALE spaces instead of $K3$, one can find such interacting theories for each A_k or D_k singularity. For the A_k theories, there is another simple description due to Strominger and Townsend [2, 3]. For instance, consider two parallel M5 branes in eleven-dimensional flat spacetime. There are membranes which can end on the M5 branes, yielding a noncritical string on the fivebrane worldvolume with tension proportional to the separation.

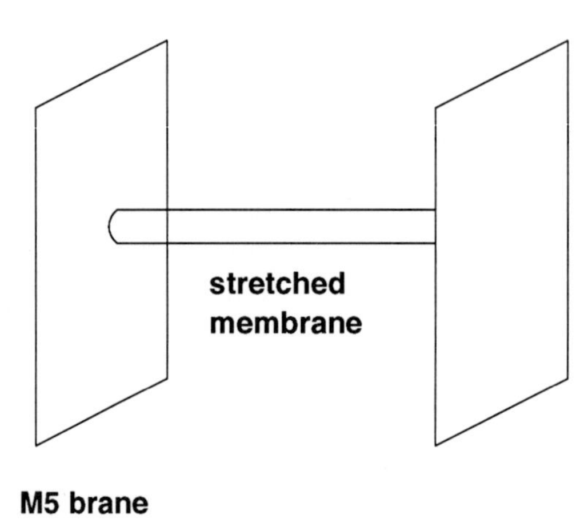

Figure 1: A membrane stretching between two M5 branes.

Each fivebrane has a tensor multiplet on its worldvolume (the five scalar components parametrize the transverse position of the fivebrane in eleven dimensions). If we denote the two five-tuples of scalars by $\vec{\phi}_{1,2}$, then the VEVs $\langle\vec{\phi}_{1,2}\rangle$ label different vacua of an effective six-dimensional theory. When $\vec{\phi}_1 \to \vec{\phi}_2$, the noncritical strings become tensionless and we find another description of the A_1 fixed point above.

More precisely, if we say the fivebranes are separated by a distance L, the limit one wishes it to take is

$$M_{pl} \to \infty, \quad L \to 0, \quad T = LM_{pl}^3 \; fixed \tag{6}$$

In this limit, gravity decouples but the noncritical strings stay light, yielding an interacting theory without gravity. The obvious generalization with k fivebranes yields the A_{k-1} (2,0) fixed point, with moduli space

$$\mathcal{M}_k = \frac{R^{5k}}{S_k} \tag{7}$$

given by the positions of the parallel fivebranes, mod permutations. At generic points on $\mathcal{M}_k$, the low energy theory has k tensor multiplets.

2.2 Theories with $(1,0)$ Supersymmetry

New interacting 6d theories with (1,0) supersymmetry have also been discovered [4, 5]. Perhaps the simplest example is the following. Consider Horava and Witten's description of the $E_8 \times E_8$ heterotic string as M-theory on S^1/Z_2 [6] . The length of the interval is related to the heterotic g_s, while E_8 gauge fields live on each of the two "end of the world" ninebranes.

We can consider a fivebrane at some point on the interval. Its position in the S^1/Z_2 is parametrized by the real scalar ϕ in a (1,0) tensor multiplet, while its other transverse positions are scalars in a (1,0) hypermultiplet. Since it is a scalar in six dimensions, ϕ naturally has dimension two; we will say the two E_8 walls are located at $\phi = 0$ and $\phi = \frac{1}{\alpha'}$. Then, one has noncritical strings on the fivebrane world volume with tensions

$$T_1 = \phi, \;\; T_2 = (\frac{1}{\alpha'} - \phi) \tag{8}$$

coming from membranes with one end on the fivebrane and one end on the ninebranes.

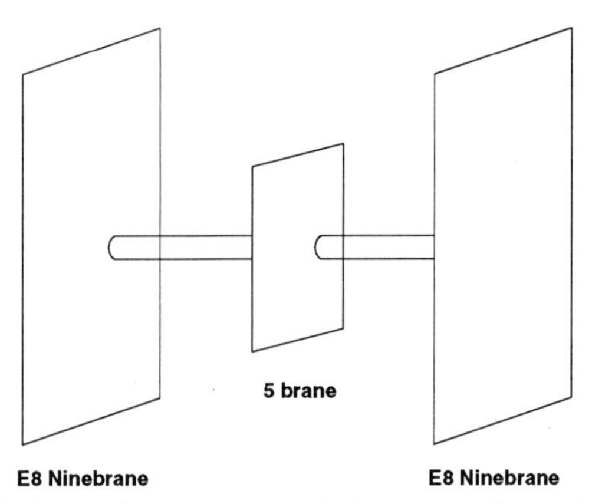

Figure 2: A fivebrane between two ninebranes and two 5-9 membranes.

As $\phi \to 0$ or $\phi \to \frac{1}{\alpha'}$, one again finds that the lightest degrees of freedom in the theory are solitonic self-dual noncritical strings. The theory on the fivebrane needs to also have an E_8 global symmetry, to couple consistently to the E_8 gauge fields on the end of the world. Therefore, one concludes that when the fivebrane hits the ninebrane, one finds a nontrivial (1,0) supersymmetric RG fixed point, with E_8 global symmetry.

For both the (2,0) A_k theories and the E_8 theory in 6d, there is no known UV free Lagrangian which flows, in the IR, to the fixed point of interest. Therefore, it is of intrinsic interest to find a definition of these quantum field theories which is independent of string or M-theory. We will propose such a definition in §4. Before doing that, it seems proper to provide some motivations for the study of these theories.

3 Why are these theories of interest?

There are at least three motivations for studying these theories:
a) These are the first examples of nontrivial fixed point quantum field theories above four dimensions. For instance, if one does a very naive analysis of gauge field theory in d dimensions

$$L = \int d^d x \, \frac{1}{g^2} F_{\mu\nu}^2 + \cdots \tag{9}$$

one finds that $[g^2] = 4 - d$ (in mass units), so the theory is *infrared free* for $d > 4$. Hence, the theories of §2 are of intrinsic interest as a new class of interacting quantum field theories.
b) These theories play a crucial role in the study of M(atrix) theory compactifications. In M(atrix) theory, one starts with the maximally supersymmetric D0 brane quantum mechanics, with N zero branes giving the DLCQ in a sector with light-like momentum N and the $N \to \infty$ limit yielding the 11d uncompactified theory [8, 9]. To study compactifications on a transverse T^n, one then T-dualizes the $U(N)$ D0 brane quantum mechanics to obtain a description of M-theory on T^n as the $n+1$ dimensional $U(N)$ Super Yang-Mills theory compactified on the dual torus, $\tilde{T}^n$.

An obvious problem with this approach is that for $n > 3$, the Super Yang-Mills is ill-defined at short distances (it is not renormalizable). Let us consider the first such case: M-theory on T^4. The U-duality group in this case is $SL(5, Z)$. This suggests that perhaps the M(atrix) definition involves some 5+1 dimensional QFT compactified on a $\tilde{T}^5$, geometrizing the $SL(5, Z)$ U-duality group as the modular group of the torus. The unique candidate which is well-defined (and has the correct supersymmetry) is the A_N (2,0) quantum field theory of §2 [10, 11]. But, how does this prescription relate to our expectation that the theory should be 4+1 $U(N)$ SYM?

The 5d SYM theory has a conserved $U(1)$ current

$$j = *(F \wedge F) \tag{10}$$

We can identify j with the Kaluza-Klein $U(1)$ symmetry of the (2,0) theory compactified on an S^1 of radius $\tilde{L}_5$, if we say

$$2\pi \tilde{L}_5 = \frac{g^2}{2\pi} \tag{11}$$

The 5d theory has particles which are 4d instantons, and whose action is given by $4\pi^2 \frac{n}{g^2}$ for the "n-instanton" particle. These in turn can be interpreted as Kaluza-Klein modes coming from the (2,0) theory with a momentum $p_5 = \frac{n}{\tilde{L}_5}$ around the hidden "extra" circle which promotes $\tilde{T}^4$ to $\tilde{T}^5$. In this way, one ends up with the prescription that M-theory on T^4 is defined by the (2,0) theory on $\tilde{T}^5$ [10, 11]. This makes the SL(5,Z) U-duality manifest.
c) These novel interacting theories play a crucial role in the unification of M-theory vacua. Consider, for instance, the heterotic $E_8 \times E_8$ theory compactified on $K3$. There is a Bianchi identity for the three-form field strength H which looks like

$$dH = Tr(R \wedge R) - Tr(F \wedge F) \tag{12}$$

where R is the curvature and F is the Yang-Mills field strength. Integrating (12) over the $K3$, we find that there should be $n_{1,2}$ Yang-Mills instantons in the two E_8s, with

$$n_1 + n_2 = 24 . \tag{13}$$

It is then natural to ask: How are vacua with different choices of $n_{1,2}$ connected to each other?

Consider the n_1 instantons in one E_8 wall. Instanton moduli space has singularities, including points where a single instanton shrinks to "zero size." In the heterotic theory, this small instanton can now be represented as a fivebrane sitting at the E_8 wall. But, now there is a new branch in the moduli space of vacua - in addition to re-expanding into a large E_8 instanton, the fivebrane can move off into the S^1/Z_2 interval! In the process, one loses 29 hypermultiplet moduli (the moduli of one E_8 instanton) and gains a single tensor multiplet (the real scalar parametrizes the position of the fivebrane in the interval). Hence, one is left with $n_1 - 1$ instantons on the E_8 wall. By moving across the interval and entering the other wall as an instanton, the fivebrane can effect a transition from a vacuum with instanton distribution (n_1, n_2) to a vacuum with instanton distribution $(n_1 - 1, n_2 + 1)$. In this way, the perturbative heterotic vacua with different numbers of instantons in the two E_8s are all connected [4, 5]. More generally, one can modify equation (13) to read $n_1 + n_2 + n_5 = 24$, where now n_5 is the number of five-branes in the interval [7].

We have glossed over an important point here: The (1,0) tensor multiplet (on the fivebrane worldvolume) contains a self-dual tensor $B^+_{\mu\nu}$. No conventional mass term is possible for the tensor, since there is no $B^-_{\mu\nu}$ that $B^+_{\mu\nu}$ can pair up with. So, how can transitions changing the number of tensor multiplets ever occur?

When the transition occurs, we are precisely in the situation described in §2.2, where there is an interacting (1,0) superconformal field theory with E_8 global symmetry. There is no weakly coupled description of this fixed point, and a phase transition can occur there. By going through this nontrivial fixed point, it is possible to connect the two branches with different numbers of tensor multiplets. So, the novel theories of §2 are of apparent use in unifying 6d (0,1) supersymmetric vacua.

In fact, related theories also seem to play an important role in connecting 4d $N = 1$ vacua. For instance, one can compactify the $E_8 \times E_8$ heterotic string on a Calabi-Yau threefold M which is a $K3$ fibration. In many examples, one finds a low-energy theory with *chiral* gauge representations. For instance, if one has embedded an $SU(3)$ bundle V with $c_1(V) = 0$ in one of the E_8s, the unbroken subgroup of E_8 is E_6. The matter fields come in the **27** and $\overline{\mathbf{27}}$ representations, and one finds for the net number $N = |\#\mathbf{27} - \#\overline{\mathbf{27}}|$ of generations:

$$N = \frac{1}{2} \left| \int_M c_3(V) \right| . \tag{14}$$

Among the singular loci in the moduli space of vacua, there are places where V develops a curve of singularities which corresponds to a small instanton in the generic $K3$ fiber of M. One can represent this small instanton as a fivebrane wrapping the base of the fibration. In certain cases, there is a new branch of the moduli space where the wrapped fivebrane can move away from the E_8 wall into the S^1/Z_2. It was argued in [12] that in many cases this changes the net number of generations of the E_6 gauge theory remaining on the wall. Related phenomena were discussed in [13, 14]. So, phase transitions through close relatives of the E_8 fixed point in six dimensions can also connect up 4d string vacua with different net numbers of generations.

4 A Proposed M(atrix) Description

In §3, we have seen several interesting applications of the new interacting 6d field theories. However, for both the (2,0) supersymmetric theories and the (1,0) theories with E_8 global symmetry, there is no obvious definition of the theory that doesn't involve an embedding in M-theory. This is a very *un – economical* way of defining a quantum field theory – one starts with far too many degrees of freedom, and must decouple most of them from the quantum field theory of interest.

An alternative way of describing the (2,0) theories was proposed in [15, 16], and extended to the (1,0) theories in [17, 18]. We will discuss the simplest case – the A_{k-1} (2,0) theory, i.e. the theory of k coincident M 5-branes. We know several suggestive facts about this theory:

- If we compactify the 6d theory on a circle with radius R, it produces a 5d $U(k)$ Super Yang-Mills theory with coupling $g_5^2 = R$.
- The Kaluza-Klein particles (with $p_5 = \frac{1}{R}$) are "instantons" of the $U(k)$ gauge theory (i.e., 4d Yang-Mills instantons which look like particles in 5d).

These facts suggest that, in analogy with the M(atrix) approach to M-theory [8], we should search for a light-cone quantization of the A_{k-1} (2,0) theory.

4.1 DLCQ of (2,0) A_{k-1} Theory

Let us take our 5+1 dimensions to be parametrized by $X^0, \cdots, X^5$. In normal light-cone quantization, one defines $X^\pm = X^0 \pm X^1$ and gives initial conditions on a surface of fixed X^+. Then, one evolves forward in light-cone "time" using the Hamiltonian $H = P_+$. The modes of quantum fields with $P_- < 0$ are canonical conjugates of modes of $P_- > 0$, so we can choose the vacuum to be annihilated by the $P_- < 0$ modes. The $P_- = 0$ modes are not dynamical (but can give rise to subtleties, which will be mentioned below).

In *discrete* light-cone quantization (or DLCQ) [9], one in addition compactifies the light-like direction

$$X^- \simeq X^- + 2\pi R \tag{15}$$

Then, P_- is quantized in units of $1/R$. For finite N, the DLCQ Fock space is very simple, since there are a finite number of modes. However integrating out the zero momentum modes can, in the DLCQ of *some* theories, lead to complicated interactions [19]. The decompactification limit (where one recovers Lorentz invariance) is taken by going to large R at fixed P_-, which is equivalent to going to large N.

Following [20], one may find it fruitful to view the compactification of X^- as the limit of a space-like compactification

$$X^- \simeq X^- + 2\pi R, \quad X^+ \simeq X^+ + \frac{R_s^2}{R} \tag{16}$$

where $R_s \to 0$. For finite R_s, one can boost this to a spatial compactification

$$X^1 \sim X^1 + R_s, \quad P_- = \frac{N}{R} \to P_1 = \frac{N}{R_s} \tag{17}$$

Then, our interest is in describing modes of momentum P_1 as $R_s \to 0$. In many cases, this can be rather complicated. But for the (2,0) superconformal theories, we get a very weakly coupled ($g^2 = R_s$) $U(k)$ Super Yang-Mills theory in 4+1 dimensions, with N "instantons" (carrying charge under $J = *(F \wedge F)$). If we want these to have finite energy in the *original* reference frame, they must have very small velocities. So for $R_s \to 0$, it seems that we should get a quantum mechanical sigma model on the moduli space of N $U(k)$ instantons. The space-time supersymmetry implies that this sigma model should have 8 supercharges.

We will now give a more direct derivation of the relevant quantum mechanics from M(atrix) theory.

4.2 Derivation from M(atrix) Theory

Following Berkooz and Douglas [21], we know that the background of k longitudinal 5-branes in M(atrix) theory can be represented by studying the theory of N zero branes in the background of k D4 branes in Type IIA string theory. The presence of the D4 branes (and the consequent 0-4 strings) break the supersymmetry of the quantum mechanics to $N = 8$.

The resulting quantum mechanics is in fact the dimensional reduction of a 6d (0,1) supersymmetric system, which is a $U(N)$ gauge theory with k fundamental hypermultiplets and an additional adjoint hyper. The quantum mechanics has a $U(N)$ gauge symmetry and an $SO(4)_{\|} \times SO(5)_{\perp} \times U(k)$ global symmetry. The bosonic fields are $X_{\perp}$, $X_{\|}$ and q with charges $(\mathbf{N^2}, \mathbf{1}, \mathbf{5}, \mathbf{1})$, $(\mathbf{N^2}, (\mathbf{2}, \mathbf{2}), \mathbf{1}, \mathbf{1})$ and $(\mathbf{N}, (\mathbf{2}, \mathbf{1}), \mathbf{1}, \mathbf{k})$ under the gauge and global symmetries. Roughly speaking, $X_{\perp}$ characterizes the positions of the 0 branes transverse to the 4-branes, while $X_{\|}$ characterizes their positions in the directions along the 4-branes.

The moduli space of vacua has various branches, but two are of particular interest to us [1]:

1) The Coulomb branch

On this branch, $X_{\perp}, X_{\|} \neq 0$ while $q = 0$. This is the branch where the 0 branes are moving around in spacetime away from the D4 branes, as depicted in Figure 3 below.

[1] Of course strictly speaking the quantum mechanics has no moduli space of vacua, but as usual in discussions of M(atrix) theory we can imagine a moduli space in the Born-Oppenheimer approximation.

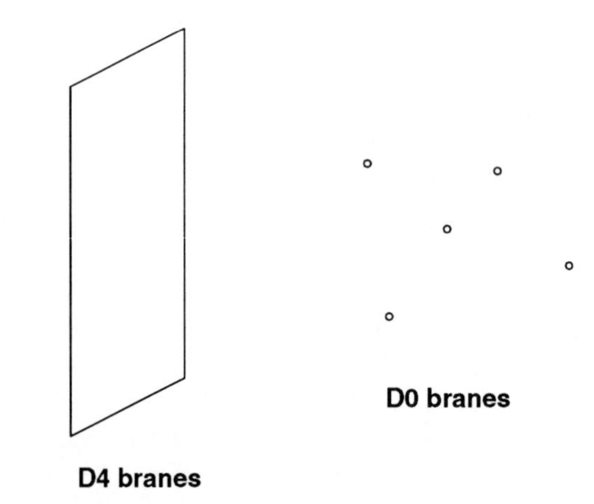

Figure 3: A picture of a point on the Coulomb branch.

2) The Higgs branch

On this branch $X_\perp = 0$ while $X_\parallel, q \neq 0$. The D0 branes are inside of the D4 branes, as shown below in Figure 4. We will denote the Higgs branch moduli space for given N and k by $\mathcal{M}_{N,k}$.

So roughly speaking, the Higgs branch is concerned with the physics on the fivebranes while the Coulomb branch captures the physics away from the fivebranes (e.g. 11d supergravity). This leads us to believe that the quantum mechanics on the Higgs branch will offer a M(atrix) description of the interacting field theories discussed in §2.1. More precisely, if we want to decouple gravity from the physics on the fivebranes, we need to take the limit $M_{pl} \to \infty$. Then in particular, the coupling in the quantum mechanics g_{QM} which is related to R and M_{pl} by $g_{QM}^2 = R^3 M_{pl}^6$ also satisfies $g_{QM} \to \infty$. This is the infrared limit of the quantum mechanics.

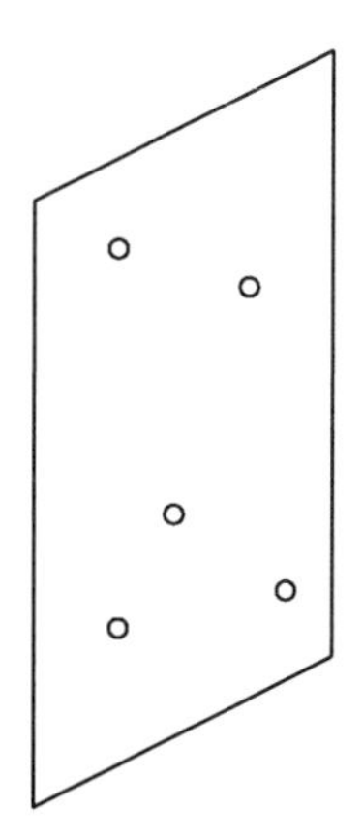

Figure 4: A picture of a point on the Higgs branch.

The surprising fact is that in this strong coupling limit, many simplifications occur:
1) For $g_{QM} \to \infty$, the Higgs branch physics decouples from the Coulomb branch! This is because the masses of the massive W bosons go off to infinity with g_{QM}, and there is a tube of infinite length in the Coulomb branch as one approaches $X_{\perp} \to 0$. So, we are left with quantum mechanics on the Higgs branch $\mathcal{M}_{N,k}$.
2) By the ADHM construction, $\mathcal{M}_{N,k}$ is actually the moduli space of N instantons in $U(k)$ gauge theory [22]! This is in accord with the fact that D0 branes are expected to behave like instantons in D4 branes [23].
3) By theorems about the absence of couplings between vector and hypermultiplets, the Higgs branch does *not* receive quantum corrections. Therefore, even the strong coupling $g_{QM} \to \infty$ limit does not correct the sigma model on $\mathcal{M}_{N,k}$.

We conclude that the A_{k-1} (2,0) theory has a description in terms of the quantum mechanics on the moduli space of N $U(k)$ instantons in the $N \to \infty$ limit. One can similarly derive a M(atrix) description for the E_8 theories with (1,0) supersymmetry [17, 18]. We will not review that here.

5 Conclusions

I have tried to emphasize in this talk the fruitful interaction that has occurred in the past year or two between three different research directions. The new interacting theories in $d \geq 4$ seem to have important uses in both the quest for a nonperturbative definition of (compactified) M-theory, and in the search for resolutions to the vacuum degeneracy problem.

Several advances have been made in areas very closely related to my talk since the conference in Buckow. I summarize some recent developments and future directions here:

- One can try to use similar M(atrix) formulations to study other interesting field theories, for instance familiar 4d field theories [24].
- One can use the quantum mechanical formulation of the 6d theories to try and compute quantities of interest in these conformal field theories, e.g. correlation functions of various local operators [25].
- One can try to use the M(atrix) descriptions of the "little string theories" of [26] to compute interesting properties of these novel theories without gravity.
- One can investigate M-theory compactification on six and higher dimensional manifolds in the M(atrix) formulation. There are problems with obtaining a simple M(atrix) description of T^6 compactifications (as discussed in e.g. [20]), while the situation seems to be better for Calabi-Yau compactifications [27]. Successful definitions of the M(atrix) compactifications will involve new theories without gravity.
- One can further pursue the study of interesting phase transitions in 4d $N = 1$ string vacua [12, 13, 14] by using new constructions (utilizing D-branes or F-theory) to find simple examples.

Acknowledgements

The results discussed in this talk were obtained in collaboration with O. Aharony, M. Berkooz, N. Seiberg, and E. Silverstein. This work was supported in part by NSF grant PHY-95-14797, by DOE contract DOE-AC03-76SF00098, and by a DOE Outstanding Junior Investigator Award.

References

[1] E. Witten, "Comments on String Dynamics," hep-th/9507121.

[2] A. Strominger, "Open P-branes," Phys. Lett. **383B** (1996) 44, hep-th/9512059.

[3] P. Townsend, "D-branes from M-branes," hep-th/9512062.

[4] O. Ganor and A. Hanany, "Small E_8 Instantons and Tensionless Noncritical Strings," Nucl. Phys. **B474** (1996) 122, hep-th/9602120.

[5] N. Seiberg and E. Witten, "Comments on String Dynamics in Six Dimensions," Nucl. Phys. **B471** (1996) 121, hep-th/9603003.

[6] P. Horava and E. Witten, "Heterotic and Type I String Dynamics from Eleven-Dimensions," Nucl. Phys. **B460** (1996) 506, hep-th/9510209.

[7] M.J. Duff, R. Minasian, and E. Witten, "Evidence for Heterotic-Heterotic Duality," Nucl. Phys. **B465** (1996) 413, hep-th/9601036.

[8] T. Banks, W. Fischler, S. Shenker and L. Sussind, "M theory as a Matrix Model: A Conjecture," Phys. Rev. **D55** (1997) 112, hep-th/9610043.

[9] L. Susskind, "Another Conjecture About Matrix Theory," hep-th/9704080.

[10] M. Rozali, "Matrix Theory and U Duality in Seven Dimensions," Phys. Lett. **400B** (1997) 260, hep-th/9702136.

[11] M. Berkooz, M. Rozali and N. Seiberg, "Matrix Description of M theory on T^4 and T^5," Phys. Lett. **408B** (1997) 105, hep-th/9704089.

[12] S. Kachru and E. Silverstein, "Chirality Changing Phase Transitions in 4d String Vacua," Nucl. Phys. **B504** (1997) 272, hep-th/9704185.

[13] G. Aldazabal, A. Font, L. Ibanez, A. Uranga and G. Violero, "Non-Perturbative Heterotic D=6, D=4, N=1 Orbifold Vacua," hep-th/9706158.

[14] I. Brunner, A. Hanany, A. Karch and D. Lüst, "Brane Dynamics and Chiral Nonchiral Transitions," hep-th/9801017.

[15] O. Aharony, M. Berkooz, S. Kachru, N. Seiberg and E. Silverstein, "Matrix Description of Interacting Theories in Six Dimensions," hep-th/9707079.

[16] E. Witten, "On the Conformal Field Theory of the Higgs Branch," hep-th/9707093.

[17] D. Lowe, "$E_8 \times E_8$ Small Instantons in Matrix Theory," hep-th/9709015.

[18] O. Aharony, M. Berkooz, S. Kachru and E. Silverstein, "Matrix Description of (1,0) Theories in Six Dimensions," hep-th/9709118.

[19] S. Hellerman and J. Polchinski, "Compactification in the Lightlike Limit," hep-th/9711037.

[20] N. Seiberg, "Why is the Matrix Model Correct," Phys. Rev. Lett. **79** (1997) 3577, hep-th/9710009.

[21] M. Berkooz and M. Douglas, "Five-branes in M(atrix) Theory," Phys. Lett. **395B** (1997) 196, hep-th/9610236.

[22] M. Atiyah, V. Drinfeld, N. Hitchin and Y. Manin, "Construction of Instantons," Phys. Lett. **65B** (1978) 185.

[23] M. Douglas,"Branes within Branes," hep-th/9512077.

[24] O. Ganor and S. Sethi, "New Perspectives on Yang-Mills Theories with Sixteen Supersymmetries," hep-th/9512071.

[25] O. Aharony, M. Berkooz and N. Seiberg, "Light-Cone Description of (2,0) Superconformal Theories in Six Dimensions," hep-th/9712117.

[26] N. Seiberg, "New Theories in Six Dimensions and Matrix Description of M Theory on T^5 and T^5/Z_2," Phys. Lett. **408B** (1997) 98, hep-th/9705221.

[27] S. Kachru, A. Lawrence and E. Silverstein, "On the Matrix Description of Calabi-Yau Compactifications," hep-th/9712223.

Critical Phenomena, Strings, and Interfaces

Gernot Münster

Institut für Theoretische Physik I, Universität Münster
Wilhelm-Klemm-Str. 9, D-48149 Münster, Germany

1 Introduction

Traditionally, the participants of the "Ahrenshoop International Symposium on the Theory of Elementary Particles" form two nearly disjoint subsets, consisting of string theorists and lattice gauge theorists. So, for a plenary speaker the question arises: is it possible to give a talk which addresses both of these species? I don't have an answer, but this contribution is meant as an attempt.

Considering the basic theoretical objects which are being studied, there is no apparent relation. The geometrical objects of string theory are world-sheets of open or closed strings. We shall not speak about the additional internal degrees of freedom here. A parameterized world-sheet is described by functions $x^\mu(\sigma,\tau)$. In lattice gauge theory, on the other hand, the basic objects are group valued variables $U(x,\mu)$ associated with the links (x,μ) of a space-time lattice. That looks very different.

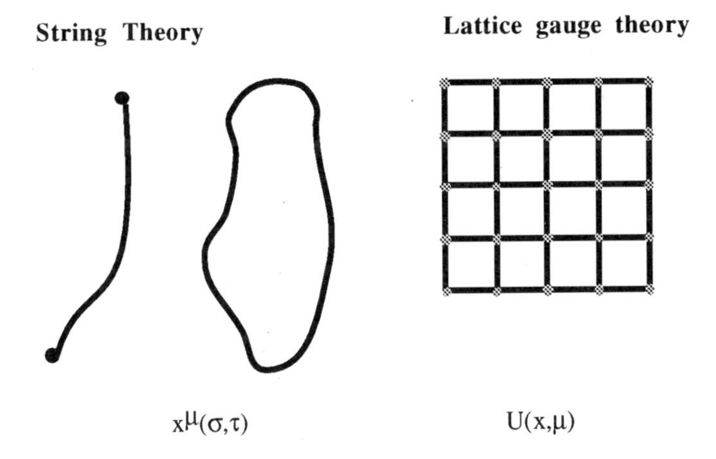

Let us consider bosonic string theory in d dimensional space-time a little bit closer. The Nambu-Goto action of a world sheet, parameterized by $x^\mu(\sigma,\tau)$, with μ

running from 1 to d, is

$$S = \alpha \cdot \text{Area} = \alpha \int d\sigma d\tau \sqrt{g}\,, \tag{1}$$

where

$$g = \det \left(\frac{\partial x_\mu}{\partial \xi_\alpha} \frac{\partial x^\mu}{\partial \xi_\beta} \right), \qquad \xi_\alpha = \sigma, \tau. \tag{2}$$

Most people would start from the Polyakov action [1] nowadays, but let us stick to the Nambu-Goto action for the time being.

To quantize string theory basically means to give meaning to functional integrals of the type

$$Z = \int D[x^\mu(\sigma,\tau)]\; e^{-S}\,. \tag{3}$$

As is known since long there are obstacles to naive quantization for any dimension d. After employing reparametrization invariance to fix the so-called conformal gauge, the remaining conformal symmetry is generated by operators L_n, which obey the Virasoro algebra

$$[L_n, L_m] = (n-m)L_{n+m} + \frac{d-26}{12}(m^3 - m)\delta_{n+m,0}\,, \tag{4}$$

where the ghost contribution to the L_n's is included. Consistent straightforward quantization requires the central extension to vanish. Therefore only

$$d = 26 \tag{5}$$

is allowed.

Now let us turn to lattice gauge theory. The link variables $U(x,\mu)$ represent the gauge field $A_\mu(x)$ in the sense of elementary parallel transporters on the lattice:

$$U(x,\mu) \simeq e^{-aA_\mu(x)} \in \mathrm{SU}(N)\,, \tag{6}$$

where a is the lattice constant. The simplest action for a lattice gauge field is the Wilson action

$$S = -\frac{\beta}{N} \sum_p \mathrm{Re}\,\mathrm{Tr}\, U(p)\,, \tag{7}$$

where p are the elementary plaquettes on the lattice, $U(p)$ is the ordered product of the four link variables belonging to the boundary of plaquette p, and β is inversely proportional to the coupling constant squared. The basic functional integral is of the type

$$Z = \int D[U(x,\mu)]\, e^{-S}\,. \tag{8}$$

One method to evaluate such integrals approximatively is the strong coupling expansion, i.e. an expansion in powers of β, analogous to high temperature expansions in statistical mechanics. For example, the strong coupling expansion for Wilson loops leads to diagrams, which are (more or less) surfaces on the lattice bounded by the loop. These surfaces look like world-sheets of strings. Indeed, it turns out that

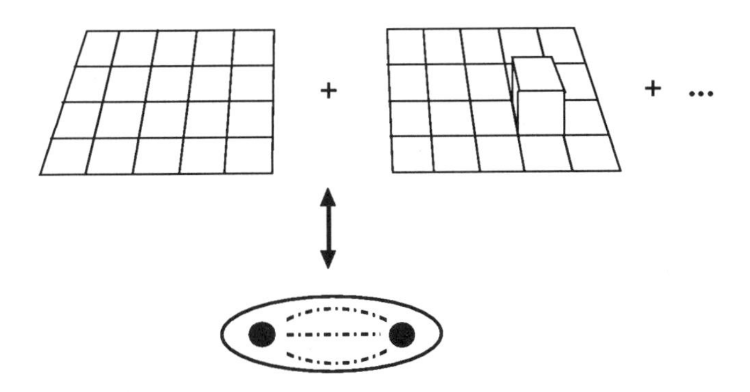

Figure 1: Strong coupling diagrams for a Wilson loop and the confining string

the strong coupling expansion leads to confinement of static quarks, and the above-mentioned surfaces are related to confining strings between colour sources [2, 3].

So there appears to be some relation between lattice gauge theory and strings. Can this relation be made more precise? In particular, is it possible to describe lattice gauge theory in terms of an effective string theory? Many attempts have been made in this direction, e.g. by Nielsen and Olesen, 't Hooft, Nambu, Gervais and Neveu, Polyakov, Migdal and others [4]. It appears to be difficult to obtain concrete results.

As mentioned above, strings appear in the strong coupling expansion. It is, however, difficult to reformulate lattice gauge on the basis of the strong coupling expansion as a theory of strings. The main problem comes from the nontrivial weights of the diagrams.

Another point, where strings appear in lattice gauge theory, is the $1/N$-expansion [5], but it is difficult to obtain a string formulation for finite N from that.

My impression is that the question of an effective string theory for gauge fields is still open. Let us therefore turn to simpler field theoretic models and look for strings in them. In particular, let us consider scalar fields. The simplest model with a scalar field is the Ising model. Its field $s(x)$ is associated with the points of a lattice and only assumes values $s(x) = \pm 1$, representing spins pointing up or down. The action is given by

$$S = -\kappa \sum_{<xy>} s(x)s(y)\,, \tag{9}$$

where $< xy >$ is a link between nearest neighbour points x and y on the lattice.

In the same universality class is ϕ^4-theory. In the phase with broken symmetry the action can be written as

$$S = \int d^d x \left\{ \frac{1}{2}(\partial_\mu \phi)^2 + \frac{g}{4!}(\phi^2 - v^2)^2 \right\}. \tag{10}$$

The minima of the double well potential are located at $\phi = \pm v$.

In the Ising model at large κ ("low temperatures"), as well as in ϕ^4-theory in the broken symmetric phase interfaces appear. They are $(d-1)$-dimensional surfaces separating regions with opposite values of the field. In the Ising model they are domain walls between regions with $s(x) = +1$ and $s(x) = -1$. For large enough κ fluctuations are small and these interfaces are rather well defined objects. On a finite rectangular lattice with appropriate boundary conditions nearly flat interfaces can be prepared. Similarly, in the ϕ^4-model interfaces separate regions with $\phi(x) \approx +v$ from those with $\phi(x) \approx -v$.

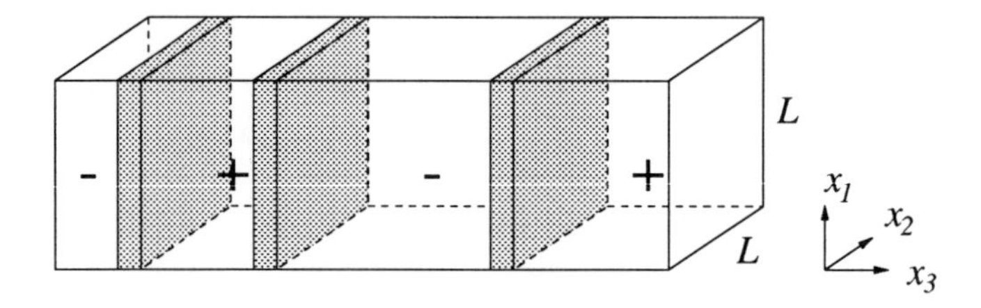

Figure 2: Interfaces in a finite Ising-like system

The interfaces of scalar field theory are string-like, i.e. two-dimensional, only for $d = 3$, whereas the strings of lattice gauge theory are always two-dimensional. In $d = 3$, where we have both types of "strings", there is an interesting additional relation between them. This is duality, which can be made quite explicit for models

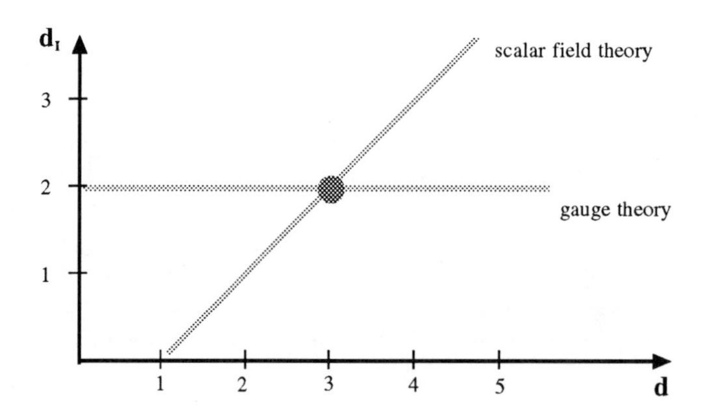

Figure 3: Dimensionality d_I of surfaces in d dimensions

with an abelian symmetry group. An abelian model, whose fields are p-forms and whose interaction terms are defined on (p + 1)-cells, is mapped by duality onto an equivalent model of $(d-\mathrm{p}-2)$-forms and interactions on $(d-\mathrm{p}-1)$-cells. For $d = 3$ duality maps the Ising model (p=0) at low temperatures onto the Z_2 gauge theory at high temperatures and vice versa. Therefore the Ising string in 3 dimensions is really the same as the Z_2 gauge theory string.

In the following we shall concentrate on the three-dimensional situation more closely.

2 Interfaces in $d = 3$

There are many systems of statistical mechanics in three dimensions for which interfaces play an interesting role. These include

- liquid gas coexistence
- binary liquid mixtures
- anisotropic ferromagnetism
- ferroelectrics
- superconductors
- crystal growth.

Near a critical point ($T \approx T_c$) of such a system one observes universal behaviour of certain quantities related to interfaces. Consider the interface tension τ or the reduced interface tension $\sigma = \tau/kT$, where k is Boltzmann's constant. It is positive for $T < T_c$, but vanishes according to

$$\sigma \sim \sigma_0 \left| \frac{T - T_c}{T_c} \right|^{\mu} \tag{11}$$

as the temperature T approaches T_c. The value of the critical exponent is approximately $\mu \approx 1.26$ and appears to be universal. On the other hand, the amplitude σ_0 is not universal.

The critical law for the correlation length ξ, which diverges at the critical point, is

$$\xi \sim \xi_0 \left| \frac{T - T_c}{T_c} \right|^{-\nu} . \tag{12}$$

Widom's scaling law [6] relates the indices μ and ν. In $d = 3$ it reads

$$\mu = 2\nu \,. \tag{13}$$

Consequently the product $\sigma\xi^2$ approaches a finite value at the critical point:

$$\sigma\xi^2 \longrightarrow \sigma_0 \xi_0^2 \doteq R_- \,. \tag{14}$$

The number R_- appears to be universal, too. It is a so-called universal amplitude product. In the past it has been studied experimentally for various systems, by means of Monte Carlo calculations, and by field theoretic methods. Other quantities, which have been investigated in connection with interfaces, are the interface width, the interface profile, the interface stiffness etc.. Closely related to R_- is

$$R_+ = \sigma_0 (\xi_0^+)^2 = R_- \left(\frac{\xi_0^+}{\xi_0^-} \right)^2 , \tag{15}$$

where ξ_0^+ is the correlation length amplitude of the high temperature phase (which is easier accessible experimentally), and ξ_0^- the one of the low temperature phase.

Why should one study R_- or R_+? Early results from the ϵ-expansion [7] and from Monte-Carlo calculations [8] were in strong conflict with experimental numbers $R_+ = 0.38(2)$ (for a brief summary of the history and relevant references see [9]). Therefore the question of universality for these interface-related quantities arose. Furthermore it became desirable to learn about the status of the theoretical predictions.

So, let us turn to the theoretical calculation of the interface tension.

3 Description of interfaces by field theory

The critical phenomena of systems in the universality class of the three-dimensional Ising model, like those mentioned in the previous section, can be calculated in the framework of massive ϕ^4-theory. The scalar field $\phi(x)$ represents the order parameter. For example in the case of a liquid mixture this would be the difference of densities of the two liquids. The action for the scalar field is given in Eq. (10). In order to study a planar interface we consider the theory in a rectangular box with quadratic cross-section L^2 in the $x_1 - x_2$-plane and antiperiodic boundary conditions in the x_3-direction. Alternatively, one might choose

$$\phi(x) \longrightarrow \begin{cases} +v, & \text{as } x_3 \to \infty \\ -v, & \text{as } x_3 \to -\infty . \end{cases} \tag{16}$$

A classical solution of the field equations is given by the kink

$$\phi_c = \sqrt{\frac{3m^2}{g}} \tanh \frac{m}{2}(x_3 - a) , \tag{17}$$

centered at $x_3 = a$, where $m^2 = gv^2/3$.

Its classical action is

$$S_c = 2 \frac{m^3}{g} L^2 . \tag{18}$$

Thus the saddle point approximation to the functional integral with the boundary conditions specified above,

$$Z_{+-} = \int_{\pm} D\phi \, e^{-S} , \tag{19}$$

is given by

$$Z_{+-} \approx e^{-S_c} . \tag{20}$$

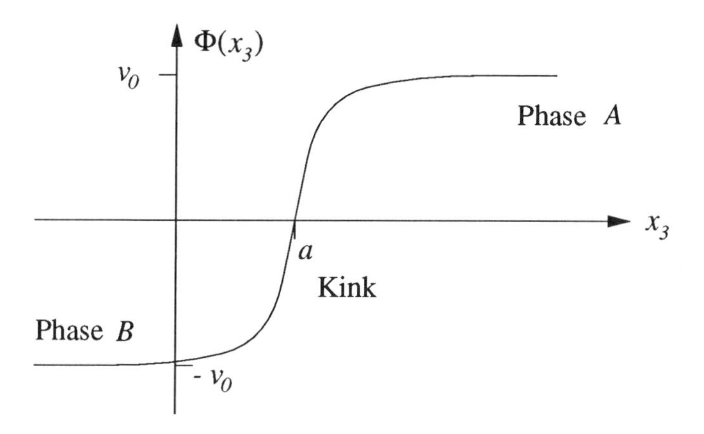

Figure 4: The kink as an interface between two phases

Because Z_{+-} is the partition function of a system with an interface, it should depend on L like

$$Z_{+-} \sim e^{-\sigma L^2}. \tag{21}$$

We can read off the interface tension in the saddle point approximation

$$\sigma \approx 2\frac{m^3}{g}, \tag{22}$$

and for the product of interest we obtain

$$\sigma\xi^2 = \frac{\sigma}{m^2} \approx 2\frac{m}{g}. \tag{23}$$

Introducing the dimensionless coupling constant

$$u \equiv \frac{g}{m}, \tag{24}$$

we write our tree level result as

$$\frac{\sigma}{m^2} \approx \frac{2}{u}. \tag{25}$$

Now let us take into account fluctuations

$$\phi(x) = \phi_c(x) + \eta(x) \tag{26}$$

around the classical solution. The action

$$S = S_c + \frac{1}{2}\int d^3x\, \eta(x) M \eta(x) + O(\eta^3) \tag{27}$$

contains the fluctuation operator

$$M = -\partial_\mu \partial^\mu + m^2 - \frac{3}{2} m^2 \cosh^{-2}(\frac{m}{2} x_3). \tag{28}$$

For the case of periodic boundary conditions, where an interface need not be present, the partition function is denoted Z_{++}, and M is replaced by the Helmholtz operator

$$M_0 = -\partial_\mu \partial^\mu + m^2 \,. \tag{29}$$

The relevant ratio of partition functions can then expressed as

$$\frac{Z_{+-}}{Z_{++}} = \sqrt{\frac{S_c}{2\pi}} \left(\frac{\det' M}{\det M_0} \right)^{-1/2} \times \exp\Big\{ - S_c + \frac{1}{2} \; \; + \frac{1}{8} \Big[\; - \; \Big] + \frac{1}{8} \Big[\; - \; \Big] + \frac{1}{12} \Big[\; - \; \Big] + O(g^2) \Big\}, \tag{30}$$

where the meaning of the graphs can be found in [10]. To evaluate this expression, first of all renormalization has to be carried out in the usual way. Moreover, the contribution of multi-kink configurations has been taken into account, but I refuse to reveal any details here.

Luckily, in the one-loop approximation the determinants can be calculated exactly [11] and one obtains

$$\frac{Z_{+-}}{Z_{++}} = C e^{-\sigma L^2} \,, \tag{31}$$

with

$$\frac{\sigma}{m^2} = \frac{2}{u_R} \left(1 + \sigma_1 \frac{u_R}{4\pi} + \ldots \right) \tag{32}$$

and exact expressions for the constants C and σ_1. The quantities m and u_R are the renormalized mass and coupling now.

Believe it or not, it has been possible to evaluate the two-loop contribution to the ratio σ/m^2, too [10]. Written in the form

$$\sigma_2 \left(\frac{u_R}{4\pi} \right)^2 \tag{33}$$

the coefficient $\sigma_2 = -0.0076(8)$ turns out to be rather small. In order to obtain the desired universal ratio we have to evaluate the function

$$\frac{\sigma}{m^2} \equiv R_-(u_R) \tag{34}$$

at the fixed point value $u_R = u_R^*$. The most recent value $u_R^* = 14.3(1)$ from Monte Carlo calculations [12] is consistent with an estimate of 14.2(2) from three-dimensional field theory [13]. At this value of the coupling the one-loop contribution is 24% and the two-loop contribution roughly 1% of the tree-level term. The final result from field theory for the amplitude product is

$$R_- = 0.1065(9) \,. \tag{35}$$

It compares well with the recent Monte Carlo result $R_- = 0.1040(8)$ by Hasenbusch and Pinn [14]. The corresponding numbers for R_+, using theoretical values for ξ_0^+/ξ_0^-, lie in the range from 0.40(1) to 0.42(1) and are compatible with the recent experimental result $R_+ = 0.41(4)$ [15], which is higher than the earlier average of 0.37(3).

4 Effective string description

Field theory describes fluctuating interfaces, as we have seen. The relevant partition function is analogous to a functional integral over fluctuating string world-sheets, Eq. (3). A proposal to describe the dynamics of fluctuating interfaces in terms of an effective string model is the "capillary wave model" or "drumhead model" [16]. The interface is considered to be a surface without overhangs, which can be described by a height function $x_3 = h(x_1, x_2)$. The action is, as in the Nambu-Goto case, given

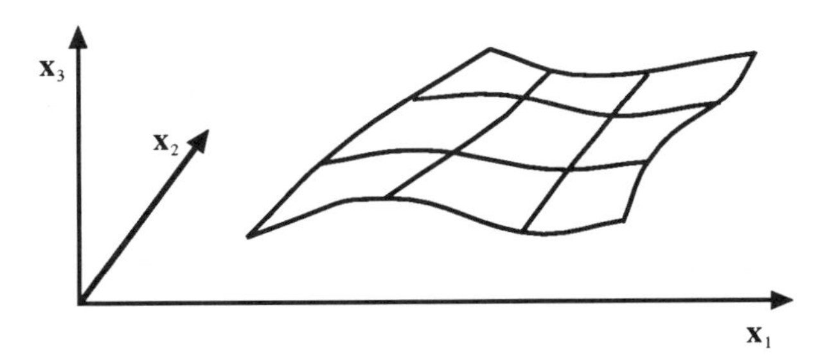

Figure 5: Smooth surface in the drumhead model

by the area:

$$S = \tilde{\sigma} \int_0^{L_1} dx_1 \int_0^{L_2} dx_2 \sqrt{1 + \left(\frac{\partial h}{\partial x_1}\right)^2 + \left(\frac{\partial h}{\partial x_2}\right)^2} . \tag{36}$$

Expanding in powers of h we get

$$\begin{aligned} S &= \tilde{\sigma} L_1 L_2 + \frac{\tilde{\sigma}}{2} \int dx_1 \int dx_2 \left[\left(\frac{\partial h}{\partial x_1}\right)^2 + \left(\frac{\partial h}{\partial x_2}\right)^2 \right] \\ &\quad - \frac{\tilde{\sigma}}{8} \int dx_1 \int dx_2 \left[\left(\frac{\partial h}{\partial x_1}\right)^2 + \left(\frac{\partial h}{\partial x_2}\right)^2 \right]^2 + \cdots \end{aligned} \tag{37}$$

The second term, the Gaussian action S_G, is quadratic in the field $h(x_1, x_2)$ and describes a massless scalar field in two dimensions. The expansion of the action above leads to an expansion of the partition function which is organized in powers of $1/\tilde{\sigma} L_1 L_2$:

$$Z_{+-} = e^{-\tilde{\sigma} L_1 L_2} \, Z_1 \cdot Z_2 \cdot \ldots \tag{38}$$

where the one-loop term

$$Z_1 = \int Dh \, e^{-S_G} = (\det \Delta_{(2)})^{-1/2} \tag{39}$$

can be expressed in terms of the determinant of the two-dimensional Laplace operator with appropriate boundary conditions. This is a well known object in 2d conformal invariant field theory with central charge $c = 1$ on a torus [17], and has been calculated explicitly:

$$Z_1 = \frac{1}{\sqrt{-i\tau}} \left| \frac{\eta(\tau)}{\eta(i)} \right|^{-2} \cdot \text{const.}, \tag{40}$$

where

$$\eta(\tau) = q^{1/24} \prod_{n=1}^{\infty} (1 - q^n), \qquad q = e^{2\pi i \tau} \tag{41}$$

is Dedekind's eta-function, and the parameter τ is given by the aspect ratio

$$\tau = i \frac{L_1}{L_2}. \tag{42}$$

This opens the possibility to test the capillary wave model by studying the dependence of Z_1 on L_1/L_2. Comparing two partition functions with equal area, $L^2 = L_1 L_2$, the leading term cancels out and one has

$$\frac{Z_{+-}(L_1, L_2)}{Z_{+-}(L, L)} = \sqrt{\frac{L_2}{L_1}} \left| \frac{\eta(iL_1/L_2)}{\eta(i)} \right|^{-2}. \tag{43}$$

A comparison of this formula with Monte Carlo results for the $d = 3$ Ising model shows very good agreement, supporting the capillary wave model [18].

The capillary wave model also predicts the roughening phenomenon [16]. The width w of an interface of size $L \times L$, given by

$$w^2 = \frac{1}{L^2} \int dx_1 dx_2 \, \langle (h(x_1, x_2) - \langle h \rangle)^2 \rangle, \tag{44}$$

can be calculated in the Gaussian capillary wave model [19, 20]. For large L it diverges like

$$w^2 = \frac{1}{2\pi\sigma} \log \frac{L}{R_0}, \tag{45}$$

with some cutoff length R_0. This behaviour indicates the dominance of longwave fluctuations of the interface.

The Gaussian approximation, considered so far, is not specific to the capillary wave model. In fact, it is an infrared fixed point for a whole class of effective models. In order to test the capillary wave model one should go beyond the Gaussian approximation. This has been done in [21] (see also [22]). They have evaluated the two-loop contribution Z_2 to the partition function and get

$$\begin{aligned} Z_2 &= 1 + \frac{\tilde{\sigma}}{8} \int dx_1 \int dx_2 \, \langle ((\nabla h)^2)^2 \rangle_G \\ &= 1 + \frac{1}{4\sigma L^2} \end{aligned} \tag{46}$$

for $L_1 = L_2 = L$. Included is a renormalization of the interface tension σ. This term has been nicely confirmed by Monte Carlo calculations [21].

At this point one might wonder whether it is possible to derive the string description, i.e. the capillary wave model, directly from ϕ^4-theory in three dimensions. In the Gaussian approximation this can indeed be done, see [23]. The fluctuation operator M, Eq. (28), decomposes into the two-dimensional Laplacean $\Delta_{(2)}$ and a Schrödinger operator Q: $M = -\Delta_{(2)} + Q$. The operator Q has two discrete eigenvalues, 0 and $3m^2/4$, and a continuous spectrum. The zero-mode is associated with translations of the interface along x_3. Consequently, the determinant of M contains $\det \Delta_{(2)}$ as a factor. But this is just the contribution of a two-dimensional free massless scalar field, and is identical to the Gaussian capillary wave model. The remaining factor belongs to massive modes on the interface and does not dominate the long-wavelength behaviour. It contributes to the renormalization of σ.

To derive the string model from ϕ^4 theory beyond the Gaussian approximation of interface fluctuations is more difficult. First of all, for a given field $\phi(x)$ the interface variables $h(x_1, x_2) = F[\phi(x)]$ have to be defined suitably, at least for fields not too far away from the kink solution ϕ_c. Formally one would then write

$$\int D[\phi(x)]\; e^{-S} \equiv \int D[h(\xi)]\; e^{-S_h} \tag{47}$$

with

$$e^{-S_h} = \int D[\phi(x)]\; e^{-S} \delta[h(\xi) - F[\phi(x)]\,. \tag{48}$$

This approach has been studied e.g. in [24, 25]. In the low temperature limit, $T \to 0$, the interface (or string) action S_h indeed approaches the Nambu-Goto action

$$S_h = \sigma \cdot \text{Area} + \text{corrections}\,. \tag{49}$$

This is, however, far away from the critical point.

5 Questions

Many questions concerning the relation of critical interfaces to strings are still open. Let us consider fluctuations. Near the critical point the fluctuations of the interface are strong. In the field theoretic approach this means that a typical field ϕ is by no means similar to the classical solution ϕ_c. Why then, does the semiclassical expansion work so well?

We can interpret this effect as a result of renormalization. Fluctuations on short scales produce ultraviolet divergencies, which lead to the renormalization of the mass m and the interface tension σ. In the renormalized propagator in a kink background the UV-fluctuations are summed up effectively. In terms of renormalized quantities we can thus expect a smoother behaviour. So it is the usual picture of renormalization, which is at work.

On the side of the string description the same question arises. Near the critical point the interface is far from smooth. Overhangs, bubbles and handles appear. Why does the capillary wave model work?

Figure 6: Overhangs, bubbles and handles

Attempts to answer this question have been made in [26]. The analogue of renormalization is claimed to be a "condensation of handles". Near each point of the interface thin tubes can be attached, which represent overhangs or handles. They yield a renormalization factor proportional to the area, which in turn can be absorbed into the renormalization of the tension σ. The average size of a handle is expected to be microscopic and independent of the large scale geometry of the interface, so that these condensed handles do not influence the long-wavelength behaviour.

Another question concerns the conformal anomaly. The bosonic string can be quantized consistently without Liouville modes only in 26 dimensions. Why does the effective string model work in 3 dimensions?

A more careful treatment of the transformation from field variables to string variables would take into account the arising Jacobian J:

$$e^{-S_h} = \int D[\phi(x)]\, e^{-S} J[h(\xi)]\, \delta[h(\xi) - F[\phi(x)]] \,. \tag{50}$$

A calculation of J in the framework of a four-dimensional abelian Higgs-model [27] gave

$$J = \text{const.} \exp\left\{ \int d^2\xi \frac{22}{96\pi} (\partial_a \log\sqrt{g})^2 + \ldots \right\} . \tag{51}$$

This term contributes to the Virasoro generators L_n, and their algebra reads

$$[L_n, L_m] = (n-m)L_{n+m} + \frac{d - 26 + 22}{12}(m^3 - m)\delta_{n+m,0} \,. \tag{52}$$

The anomaly term now vanishes in $d = 4$ dimensions, as desired. The string action is given by

$$S_h = \int d^2\xi \left\{ \sigma\sqrt{g} - \frac{11}{48\pi}(\partial_a \log\sqrt{g})^2 - \beta\sqrt{g}(\partial_a t_{\mu\nu})^2 + \ldots \right\} , \tag{53}$$

where $t_{\mu\nu}$ is the extrinsic curvature. The second piece, the Liouville term, has been first proposed in [28].

Because the Jacobian J is of a pure geometric nature, the same effect is expected to take place in three-dimensional ϕ^4-theory.

To summarize, there are interesting relations between string theory and fluctuating interfaces in critical statistical systems, and there are several open points, which deserve further study.

References

[1] A.M. Polyakov, *Phys. Lett.* **B 103** (1981) 207

[2] K.G. Wilson, *Phys. Rev.* **D 10** (1974) 2445

[3] J. Kogut, L. Susskind, *Phys. Rev.* **D 11** (1975) 395

[4] H.B. Nielsen, P. Olesen, *Nucl. Phys.* **B 61** (1973) 45
G. 't Hooft, *Nucl. Phys.* **B 72** (1974) 461
Y. Nambu, *Phys. Lett.* **B 80** (1979) 372
J.L. Gervais, A. Neveu, *Nucl. Phys.* **B 163** (1980) 189
A.M. Polyakov, *Nucl. Phys.* **B 164** (1980) 179
A.A. Migdal, *Nucl. Phys.* **B 189** (1981) 253

[5] B. de Wit, G. 't Hooft, *Phys. Lett.* **B 69** (1977) 61

[6] B. Widom, *J. Chem. Phys.* **43** (1965) 3892

[7] E. Brézin, S. Feng, *Phys. Rev.* **B 29** (1984) 472

[8] K. Binder, *Phys. Rev.* **A 25** (1982) 1699

[9] G. Münster, *Int. J. Mod. Phys.* **C 3** (1992) 879

[10] P. Hoppe, G. Münster, *Phys. Lett.* **A** (1998), in press

[11] G. Münster, *Nucl. Phys.* **B 340** (1990) 559

[12] M. Caselle, M. Hasenbusch, *J. Phys.* **A 30** (1997) 4963

[13] C. Gutsfeld, J. Küster, G. Münster, *Nucl. Phys.* **B 479** (1996) 654

[14] M. Hasenbusch, K. Pinn, *Physica* **A 245** (1997) 366

[15] T. Mainzer, D. Woermann, *Physica* **A 225** (1996) 312

[16] F.P. Buff, R.A. Lovett, F.H. Stillinger Jr., *Phys. Rev. Lett.* **15** (1965) 621

[17] C. Itzykson, J.-B. Zuber, *Nucl. Phys.* **B 275** (1986) 580

[18] M. Caselle, F. Gliozzi, S. Vinti, *Phys. Lett.* **B 302** (1993) 74

[19] M. Lüscher, G. Münster, P. Weisz, *Nucl. Phys.* **B 180** (1981) 1

[20] M. Hasenbusch, K. Pinn, *Physica* **A 192** (1993) 342

[21] M. Caselle et al., *Nucl. Phys.* **B 432** (1994) 590

[22] K. Dietz, T. Filk, *Phys. Rev.* **D 27** (1983) 2944

[23] P. Provero, S. Vinti, *Nucl. Phys.* **B 441** (1995) 562

[24] H.W. Diehl, D.M. Kroll, H. Wagner, *Z. Phys.* **B 36** (1980) 329

[25] R.K.P. Zia, *Nucl. Phys.* **B 251** (1985) 676

[26] M. Ademollo et al., *Nucl. Phys.* **B 94** (1975) 221
M. Caselle, F. Gliozzi, S. Vinti, *Nucl. Phys. (Proc. Suppl.)* **B 34** (1994) 726
V. Dotsenko et al., *Phys. Rev. Lett.* **71** (1993) 811

[27] E.T. Akhmedov, M.N. Chernodub, M.I. Polikarpov, M.A. Zubkov, *Phys. Rev.* **D 53** (1996) 2087

[28] J. Polchinski, A. Strominger, *Phys. Rev. Lett.* **67** (1991) 1681

Gaugino condensation in M-theory

Hans Peter Nilles

Physikalisches Institut, Universität Bonn,
Nussallee 12, D-53115 Bonn, Germany

Abstract: We investigate gaugino condensation in the framework of the strongly coupled heterotic $E_8 \times E_8$ string (M–theory). Supersymmetry is broken in a hidden sector and gravitational interactions induce soft breaking parameters in the observable sector. They are of order of the gravitino mass and lead to satisfactory phenomenological and cosmological consequences.

Recently there have been attempts to study string theories in the region of intermediate and strong coupling. The strongly coupled version of the $E_8 \times E_8$ theory is believed to be an orbifold of 11–dimensional M–theory, an interval in $d = 11$ with $E_8 \times E_8$ gauge fields restricted to the two $d = 10$ dimensional boundaries respectively [1]. Applied to the question of unification [2] the following picture emerges: the GUT–scale $M_{\rm GUT} = 3 \times 10^{16}$ GeV is identified with $1/R$ where $V = R^6$ is the volume of compactified six–dimensional space. $\alpha_{\rm GUT} = 1/25$ and the correct value of the $d = 4$ reduced Planck mass $M_P = 2.4 \times 10^{18}$ GeV can be obtained by choosing the length of the $d = 11$ interval appropriately. The fundamental mass scale of the $d = 11$ theory $M_{11} = \kappa^{-2/9}$ (with κ the $d = 11$ Einstein gravitational coupling) has to be chosen a factor 2 larger than $M_{\rm GUT}$ and at that scale $\alpha_{\rm string} = g^2_{\rm string}/4\pi$ is of order unity. This then represents a rather natural framework for the unification of coupling constants.

As in the weakly coupled theory, supersymmetry might be broken dynamically by gaugino condensation [3] in the hidden E_8 on one boundary of space–time [4, 5, 6]. Gravitational interactions will play the role of messengers to the observable sector at the opposite boundary.

In this talk we shall discuss this mechanism in detail and compute the predictions for the low energy effective theory. We find results very similar to the situation in the $d = 10$ weakly coupled case, with one notable exception: *gaugino masses are of comparable size to the gravitino mass, thus solving the problem of small gaugino masses that occurred in the weakly coupled case* [5]. This might have important phenomenological consequences [7].

The effective action of the strongly coupled $E_8 \times E_8$ – M–theory [1] is given by

$$L = \frac{1}{\kappa^2}\int d^{11}x\sqrt{g}\left[-\frac{1}{2}R - \frac{1}{2}\bar{\psi}_I\Gamma^{IJK}D_J\left(\frac{\Omega+\hat{\Omega}}{2}\right)\psi_K - \frac{1}{48}G_{IJKL}G^{IJKL} + \ldots\right] \quad (1)$$
$$+\frac{1}{2\pi(4\pi\kappa^2)^{2/3}}\int d^{10}x\sqrt{g}\left[-\frac{1}{4}F^a_{AB}F^{aAB} - \frac{1}{2}\bar{\chi}^a\Gamma^A D_A(\hat{\Omega})\chi^a + \ldots\right].$$

Compactifying to $d = 4$ we obtain [2, 8]

$$G_N = 8\pi\kappa_4^2 = \frac{\kappa^2}{8\pi^2 V\rho}, \qquad \alpha_{GUT} = \frac{(4\pi\kappa^2)^{2/3}}{V} \quad (2)$$

with $V = R^6$ and $\pi\rho = R_{11}$. Fitting G_N and $\alpha_{GUT} = 1/25$ then gives $R_{11}M_{11} \approx 12$ and $M_{11}R \approx 2.3$ ($M_{11} \approx 6 \times 10^{16}$ GeV). The rather large value of the $d = 4$ reduced Planck Mass $M_P = \kappa_4^{-1}$ is obtained as a result of the fact that R_{11} is large compared to R.

We now perform a compactification using the method of reduction and truncation [9]. For the metric we write [5, 6]

$$g_{MN} = \begin{pmatrix} e^{-\gamma}e^{-2\sigma}g_{\mu\nu} & & \\ & e^{\sigma}\delta_{mn} & \\ & & e^{2\gamma}e^{-2\sigma} \end{pmatrix} \quad (3)$$

with $M, N = 1\ldots 11$; $\mu,\nu = 1\ldots 4$; $m, n = 5\ldots 10$; $2R_{11} = 2\pi\rho = M_{11}^{-1}e^{\gamma}e^{-\sigma}$ and $V = e^{3\sigma}M_{11}^6$. At the classical level this leads to the Kähler potential

$$K = -\log(\mathcal{S}+\bar{\mathcal{S}}) - 3\log(\mathcal{T}+\bar{\mathcal{T}} - 2C_i\bar{C}_i) \quad (4)$$

with

$$\mathcal{S} = \frac{2}{(4\pi)^{2/3}}\left(e^{3\sigma} \pm i24\sqrt{2}D\right), \quad \mathcal{T} = \frac{\pi^2}{(4\pi)^{4/3}}\left(e^{\gamma} \pm i6\sqrt{2}C_{11}\right) \quad (5)$$

where D and C_{11} are the usual axion fields.

The imaginary part of $\mathcal{S}$ (Im$\mathcal{S}$) corresponds to the model independent axion, and the gauge kinetic function is $f = \mathcal{S}$. This is very similar to the weakly coupled case. Before drawing any conclusion from these formulae, however, we have to discuss a possible obstruction at the one loop level. It can be understood from the mechanism of anomaly cancellation [2]. For the 3–index tensor field H in $d = 10$ supergravity to be well defined one has to satisfy $dH = \mathrm{tr}F_1^2 + \mathrm{tr}F_2^2 - \mathrm{tr}R^2 = 0$ cohomologically. In the simplest case of the standard embedding one assumes $\mathrm{tr}F_1^2 = \mathrm{tr}R^2$ locally and the gauge group is broken to $E_6 \times E_8$. Since in the M–theory case the two different gauge groups live on the two different boundaries of space–time such a cancellation point by point is no longer possible. We expect nontrivial vacuum expectation values (vevs) of

$$(dG) \propto \sum_i \delta(x^{11} - x_i^{11})\left(\mathrm{tr}F_i^2 - \frac{1}{2}\mathrm{tr}R^2\right) \quad (6)$$

at least on one boundary (x_i^{11} is the position of i–th boundary). In the case of the standard embedding we would have $\mathrm{tr}F_1^2 - \frac{1}{2}\mathrm{tr}R^2 = \frac{1}{2}\mathrm{tr}R^2$ on one and $\mathrm{tr}F_2^2 - \frac{1}{2}\mathrm{tr}R^2 = -\frac{1}{2}\mathrm{tr}R^2$ on the other boundary. This might pose a severe problem since a nontrivial vev of G might be in conflict with supersymmetry ($G_{11ABC} = H_{ABC}$). The supersymmetry transformation law in $d = 11$ reads

$$\delta\psi_M = D_M\eta + \frac{\sqrt{2}}{288}G_{IJKL}\left(\Gamma_M^{IJKL} - 8\delta_M^I\Gamma^{JKL}\right)\eta + \ldots \tag{7}$$

Supersymmetry will be broken unless e.g. the derivative term $D_M\eta$ compensates the nontrivial vev of G. Witten has shown [2] that such a cancellation can occur and constructed the solution in the linearized approximation (linear in the expansion parameter $\kappa^{2/3}$) which corresponds to the large T–limit in the weakly coupled theory*. The supersymmetric solution leads to a nontrivial dependence [5, 6] of the σ and γ fields with respect to x^{11}:

$$\frac{\partial\gamma}{\partial x^{11}} = -\frac{\partial\sigma}{\partial x^{11}} = \frac{\sqrt{2}}{24}\frac{\int d^6x\sqrt{g}\omega^{AB}\omega^{CD}G_{ABCD}}{\int d^6x\sqrt{g}} \tag{8}$$

where the integrals are over the Calabi–Yau manifold and ω is the corresponding Kähler form. A definition of our $\mathcal{S}$ and $\mathcal{T}$ fields in the four-dimensional theory would then require an average over the 11–dimensional interval. We would therefore write

$$\mathcal{S} = \frac{2}{(4\pi)^{2/3}}\left(e^{3\bar{\sigma}} \pm i24\sqrt{2}\bar{D}\right), \quad \mathcal{T} = \frac{\pi^2}{(4\pi)^{4/3}}\left(e^{\bar{\gamma}} \pm i6\sqrt{2}\bar{C}_{11}\right) \tag{9}$$

where bars denote averaging over the 11th dimension. It might be of some interest to note that the combination $\mathcal{S}\mathcal{T}^3$ is independent of x^{11} even before this averaging procedure took place.

$\exp(3\sigma)$ represents the volume of the six–dimensional compact space in units of M_{11}^{-6}. The x^{11} dependence of σ then leads to the geometrical picture that the volume of this space varies with x^{11} and differs at the two boundaries. In the given approximation, this variation is linear, and for growing R_{11} the volume on the E_8 side becomes smaller and smaller. At a critical value of R_{11} the volume will thus vanish and this will provide us with an upper limit on R_{11}. For the phenomenological applications we then have to check whether our preferred choice of R_{11} (that fits the correct value of the $d = 4$ Planck mass) satisfies this bound. Although the coefficients are model dependent we find in general that the bound can be satisfied, but that R_{11} is quite close to its critical value. A choice of R_{11} much larger than $(5 \times 10^{15}\mathrm{GeV})^{-1}$ is therefore not permitted.

This variation of the volume is the analogue of the one loop correction of the gauge kinetic function in the weakly coupled case and has the same origin, namely a Green–Schwarz anomaly cancellation counterterm. In fact, also in the strongly coupled case we find corrections for the gauge coupling constants [5] at the E_6 and E_8 side. ¿From

*For a discussion beyond this approximation in the weakly coupled case see ref. [12].

(8) we see that gauge couplings will no longer be given by the (averaged) $\mathcal{S}$–field, but by that combination of the (averaged) $\mathcal{S}$ and $\mathcal{T}$ fields which corresponds to the $\mathcal{S}$–field before averaging at the given boundary: a trivial calculation yields

$$f_{6,8} = \mathcal{S} \pm \alpha\mathcal{T} \tag{10}$$

at the E_6 (E_8) side respectively[†]. The critical value of R_{11} will correspond to infinitely strong coupling at the E_8 side $\mathcal{S} - \alpha\mathcal{T} = 0$ (similar to the weakly coupled case). Since we are here close to criticality a correct phenomenological fit of $\alpha_{\rm GUT} = 1/25$ should include this correction $\alpha_{\rm GUT}^{-1} = \mathcal{S} + \alpha\mathcal{T}$ where $\mathcal{S}$ and $\alpha\mathcal{T}$ give comparable contributions. This is a difference to the weakly coupled case, where in $f = S + \epsilon T$ the latter contribution was small compared to S. Observe that this picture of a loop correction $\alpha\mathcal{T}$ to be comparable to the tree level result still makes sense in the perturbative expansion, since f does not receive further perturbative corrections beyond one loop [13, 14].

In a next step we are now ready to discuss the dynamical breakdown of supersymmetry via gaugino condensation in the strongly coupled M–theory picture. In analogy to the previous discussion we start investigating supersymmetry transformation laws for ψ_A and ψ_{11} in the higher–dimensional (now $d = 11$) field theory [4] where gaugino bilinears appear in the right hand side of both expressions. It can therefore be expected that gaugino condensation breaks supersymmetry. Still the details have to be worked out. In the $d = 10$ example, the gaugino condensate and the three–index tensor field H contributed to the scalar potential in a full square. This lead to a vanishing cosmological constant as well as the fact that $F_S = 0$ at the classical level. Hořava has observed [4] that a similar mechanism might be in operation in the $d = 11$ theory

$$\frac{1}{12\kappa^2}\int_{M^{11}} d^{11}x\sqrt{g}\left(G_{ABC11} - \frac{\sqrt{2}}{16\pi}\left(\frac{\kappa}{4\pi}\right)^{2/3}\delta(x^{11})\bar{\chi}^a\Gamma_{ABC}\chi^a\right)^2 . \tag{11}$$

After a careful calculation this leads to a vanishing variation $\delta\psi_A = 0$. In our model (based on reduction and truncation) we can now compute these quantities explicitly. We assume gaugino condensation to occur at the E_8 boundary $\langle\bar{\chi}^a\Gamma_{ijk}\chi^a\rangle = \Lambda^3\epsilon_{ijk}$ where $\Lambda < M_{\rm GUT}$ and ϵ_{ijk} is the covariantly constant holomorphic 3–form. This leads to a nontrivial vev of G_{11ABC} at this boundary and supersymmetry is broken. At that boundary we obtain $F_{\mathcal{S}} = 0$ and $F_{\mathcal{T}} \neq 0$ as expected from the fact that the component ψ_{11} of the 11–dimensional gravitino plays the role of the goldstino.

In the effective $d = 4$ theory we now have to average over the 11th dimension leading to

$$\langle F_{\mathcal{T}}\rangle \approx \frac{1}{2}\mathcal{T}\frac{\int dx^{11}\delta\psi_{11}}{\int dx^{11}} \tag{12}$$

as the source of SUSY breakdown. This will then allow us to compute the size of supersymmetry breakdown on the observable E_6 side. Gravitational interactions

[†]With the normalization of the $\mathcal{T}$ field as in (9), α is a quantity of order 1.

play the role of messengers that communicate between the two boundaries. This effect can be seen from (2): large R_{11} corresponds to large M_P and $\langle F_T \rangle$ gives the effective size of SUSY breaking on the E_6 side ($R_{11} \to \infty$ implies $M_P \to \infty$). The gravitino mass is given by

$$m_{3/2} = \frac{\langle F_{\mathcal{T}} \rangle}{\mathcal{T} + \bar{\mathcal{T}}} \approx \frac{\Lambda^3}{M_P^2} \tag{13}$$

as in the weakly coupled case, and we expect this to represent the scale of soft supersymmetry breaking parameters in the observable sector. These soft masses are determined by the coupling of the corresponding fields to the goldstino multiplet. As we have seen before, we cannot compute the scalar masses reliably in our approximation: $m_0 = 0$ because of the no–scale structure that appears as an artifact of our approximation. Fields of different modular weight will receive contributions to m_0 that are of order $m_{3/2}$. So far this is all similar to the weakly coupled case. An important difference appears, however, when we turn to the discussion of observable sector gaugino masses. In the weakly coupled case they were zero at tree level and appeared only because of the radiative corrections at one loop. As a result of this small correction, gaugino masses were expected to be much smaller than $m_{3/2}$. In the strongly coupled case we still have

$$m_{1/2} = \frac{\frac{\partial f_6}{\partial S} F_S + \frac{\partial f_6}{\partial \mathcal{T}} F_{\mathcal{T}}}{2\mathrm{Re} f_6} \tag{14}$$

and the 1–loop effect is encoded in the variation of the σ and γ fields from one boundary to the other. Here, however, the loop corrections are sizable compared to the classical result because of the fact that R_{11} is close to its critical value. As a result we expect observable gaugino masses of the order of the gravitino mass. The problem of the small gaugino masses does therefore not occur in this situation. Independent of the question whether F_S or $F_{\mathcal{T}}$ are the dominant sources of supersymmetry breakdown, the gauginos will be heavy of the order of the gravitino mass. The exact relation between the soft breaking parameters m_0 and $m_{1/2}$ will be a question of model building. If in some models $m_0 \ll m_{1/2}$ this might give a solution to the flavor problem. The no–scale structure found above might be a reason for such a suppression of m_0. As we have discussed above, this structure, however, is an artifact of our simplified approximation and does not survive in perturbation theory. At best it could be kept exact (but only for the fields with modular weight -1) in the $R_{11} \to \infty$ limit. The upper bound on R_{11} precludes such a situation. With observable gaugino masses of order $m_{3/2}$ we also see that $m_{3/2}$ cannot be arbitrarily large and should stay in the TeV – range.

In recent months several other groups have studied similar questions in detail [15, 16, 17]. The discussion was explicitly done at the classical level. Some conclusions different from ours (concerning large values of $m_{3/2}$ and/or R_{11}) can only be obtained in that approximation. The one loop corrections, however, require $R_{11} < R_{\mathrm{critical}}$ as well as $m_{3/2}$ in the TeV – range. Related work beyond the classical level has been reported in [18, 19].

Acknowledgements

I would like to thank J. Conrad, M. Olechowski and M. Yamaguchi for discussions and collaboration. This work was supported by the European Commission programs ERBFMRX–CT96–0045 and CT96–0090.

References

[1] P. Hořava and E. Witten, Nucl. Phys. **B460** (1996) 506; Nucl. Phys. **B475** (1996) 94.

[2] E. Witten, Nucl. Phys. **B471** (1996) 135.

[3] J.P. Derendinger, L.E. Ibáñez and H.P. Nilles, Phys. Lett. **B155** (1985) 65,Nucl. Phys. **B267** (1986) 365; M.Dine, R. Rohm, N. Seiberg and E. Witten, Phys. Lett. **B156** (1985) 55.

[4] P. Hořava, Phys. Rev. **D54** (1996) 7561.

[5] H. P. Nilles, M. Olechochowski and M. Yamaguchi, hep-th/9707143, Phys. Lett. **B415** (1997) 24

[6] H. P. Nilles, M. Olechochowski and M. Yamaguchi, hep-th/9801030

[7] For a review see: H. P. Nilles, Phys. Reports **110** (1984) 1

[8] J. Conrad, hep/th-9708031

[9] E. Witten, Phys. Lett. **B155** (1985) 151.

[10] S. Ferrara, L. Girardello and H.P. Nilles, Phys. Lett. **B125** (1983) 457.

[11] L.E. Ibáñez and H.P. Nilles, Phys. Lett. **B169** (1986) 354.

[12] H.P. Nilles and S. Stieberger, hep–th/9702110.

[13] M. Shifman and A. Vainshtein, Nucl. Phys. **B277** (1986) 456.

[14] H.P. Nilles, Phys. Lett. **B180** (1986) 240.

[15] I. Antoniadis and M. Quirós, Phys. Lett. **B392** (1997) 61; hep–th/9705037.

[16] T. Li, J.L. Lopez and D.V. Nanopoulos, hep–ph/9702237; hep–ph/9704247.

[17] E. Dudas and C. Grojean, hep–th/9704177.

[18] Z. Lalak and S. Thomas, hep-th/9707223

[19] A. Lukas, B. Ovrut and D. Waldram, hep-th/9710208

Renormalization of strongly coupled U(1) gauge theories

J. Jersák

Institute of Theoretical Physics E,
RWTH Aachen, D-52056 Aachen, Germany

Abstract: Recent numerical studies of the 4D pure compact U(1) lattice gauge theory, I have participated in, are reviewed. We look for a possibility to construct an interesting nonperturbatively renormalizable continuum theory at the phase transition between the confinement and Coulomb phases. First I describe the numerical evidence, obtained from calculation of bulk observables on spherical lattices, that the theory has a non-Gaussian fixed point. Further the gauge-ball spectrum in the confinement phase is presented and its universality confirmed. The unexpected result is that, in addition to massive states, the theory contains a very light, possibly massless scalar gauge ball. I also summarize results of studies of the compact U(1) lattice theory with fermion and scalar matter fields and point out that at strong coupling it represents a model of dynamical fermion mass generation.

1 Introduction

Quantum field theories used in the standard model and its supersymmetric extensions are either asymptotically free or so-called trivial theories. Both are defined in the vicinity of Gaussian fixed points and can be studied perturbatively. In various collaborations we address the question whether in four dimensions (4D) there exist quantum field theories which are not accessible to or anticipated by perturbation theory. I describe the results of our systematic numerical study of critical behavior in several compact U(1) gauge models on the lattice, both pure and with matter fields. Their common feature is confinement when the bare gauge coupling is strong. It has never been clear whether this is only a lattice artefact or whether a confining continuum U(1) gauge theory with confinement can be constructed. Though not yet conclusive, our results are promising and we hope to stimulate more theoretical attention.

2 Renormalization of lattice field theories

In numerical simulations of lattice field theories the concept of renormalizability requires the existence of critical behaviour somewhere in the bare coupling parameter space. Dimensionful observables, e.g. masses $m_s, m_1, m_2, \ldots$, n-point functions, etc, can be calculated in the lattice constant units a as functions of couplings in the vicinity of a critical point or manifold. The dimensionless correlation lengths $\xi_i = 1/am_i \to \infty$ when a critical point is approached.

Renormalizability further requires the existence of "lines of constant physics" in the bare coupling parameter space, along which the observables scale, i.e. their dimensionless ratios $r_1 = m_1/m_s, r_2 = m_2/m_s, \ldots$, stay (approximately, in practice) constant. If such a line hits the critical point, one can construct the continuum limit and obtain the values of dimensionful observables by fixing one mass scale, m_s, in physical units. Then $a = 1/\xi_s m_s \to 0$ and $m_i = r_i m_s$. These results are usually universal, i.e. independent of the detailed choice of the coupling space and lattice structure, being governed by a limited number of fixed points.

The lines of constant physics can approach a critical manifold but, before hitting it, leave the parameter space. Or they can enter phase transitions of weak first order. In both cases the correlation lengths grow, but stay finite. Some inherent cut-off $\Lambda = \xi_s^{max} m_s$ remains. If it is large with respect to the physical scale m_s, i.e. if $\xi_s^{max} >> 1$, the theory is renormalizable in a restricted sense. In this way e.g. the familiar trivial theories arise.

3 Pure gauge theory on spherical lattices

Recently we have reconsidered the oldest candidate for a non-Gaussian fixed point in the 4D lattice field theory, the phase transition between the confinement and the Coulomb phases in the pure compact U(1) gauge theory. We have used the extended Wilson action in order to enlarge the possibility of finding a second order phase transition,

$$S = -\sum_P w_P \left[\beta \cos(\Theta_P) + \gamma \cos(2\Theta_P)\right]. \tag{1}$$

Here $w_P = 1$ and $\Theta_P \in [0, 2\pi)$ is the plaquette angle, i.e. the argument of the product of U(1) link variables along a plaquette P. Taking $\Theta_P = a^2 g F_{\mu\nu}$, where a is the lattice spacing, and $\beta + 4\gamma = 1/g^2$, one obtains for weak coupling g the usual continuum action $S = \frac{1}{4}\int d^4x F_{\mu\nu}^2$.

Earlier investigations performed as usual on toroidal lattices suggested that the phase transition, which is clearly of first order at positive γ, might be of second order at negative γ. Though a weak two-state signal is present there [1], it might be a finite size effect. It has been found that it disappears on lattices with sphere-like topology [2, 3, 4, 5]. As this type of lattice does not change the universality class [6, 7], one can use it as well as the usual toroidal one.

It has turned out that on spherical lattices the transition has properties typical for a second-order transition. This concerns, in particular, the dependence on the size of the lattice. Use of modern finite size scaling (FSS) analysis techniques, and large computer resources, allowed to determine on finite lattices the critical properties of the phase transition. A more detailed account of our work, as well as relevant references, can be found in Refs. [3, 4]. Here I list only the most important findings.

The measurements have been performed at $\gamma = 0, -0.2, -0.5$ on lattices of the volumina up to about 20^4. The FSS behavior of the Fisher zero, specific heat, some cumulants and pseudocritical temperatures gave consistent results. The value of the correlation length critical exponent ν has been found in the range $\nu = 0.35 - 0.40$.

The most reliable measurement of ν has been provided by the FSS analysis of the Fisher zero, i.e. of the first zero z_0 of the partition function in the complex plane of the coupling β. The expected behavior with increasing volume V is $\mathrm{Im}\, z_0 \propto V^{-1/D\nu}$. The joint fit to the data at all three γ values gives $\nu = 0.365(8)$. The scaling behavior of various pseudocritical temperatures, which have been determined from several other observables, is consistent with this value (Fig. 3 in [4]).

The consistency of the data with the FSS theory suggests that the phase transition is of second order, as it implies growing correlation lengths. Unfortunately, it does not exclude that this growth stops on lattices substantially larger than those we have used. Very weak first order phase transition is still possible [8, 9], which would mean that the cut-off cannot be made really infinite. The scaling behaviour could be governed by a fixed point "behind" the phase transition.

In a very recent paper [10] Campos et al. suggest that this is what happens. Of course, also their data require an extrapolation to the infinite volume. For example latent heat, associated with the two-state signal present on toroidal lattices, decreases with volume. It can be extrapolated both to nonvanishing values and zero, depending on the ansatz. At some second-order transitions, two-state signals due to finite size effects are known to vanish extremely slowly with volume, mocking up a first order. It is most probably not possible to clarify the issue beyond any doubt by numerical methods. But the theory might be interesting even if the cut-off could not be made infinite but only much larger than the physical scale, as in physical applications we are dealing with effective theories anyhow.

4 Gauge-ball spectrum in the confinement phase

In any case the scaling behaviour of as many observables as possible should be determined. Therefore we have investigated at $\gamma = -0.2$ and -0.5 the spectrum of the theory in the confinement phase [11, 12]. It consists of massive "gauge balls", states analogous to glue balls in pure QCD. In particular, no massless photon is present in this phase. Also the string tension σ has been estimated.

The gauge-ball masses m_j in various channels j of the cubic group have been measured at both γ for different β. Then their scaling behaviour in the form

$$m_j = c_j(\beta_c^j - \beta)^{\nu_j} \tag{2}$$

(and similar for $\sqrt{\sigma}$) has been determined in each channel j individually. We found two groups of masses with strikingly different scaling behaviour. Within the whole β

range a large group of the gauge-ball masses scale with roughly the same exponents ν_j close to the non-Gaussian value 0.365(8) found in [3, 4, 5]. Though its accuracy is not yet satisfactory, the exponent of $\sqrt{\sigma}$ seems to be consistent with this value, too. However, in several channels getting contribution from the 0^{++} gauge ball the values of ν_j are approximately Gaussian, i.e. 1/2. This is shown in Fig. 3 of Ref. [13]. The values β_c^j are quite consistent with each other in all channels for each γ. Therefore, in the further analysis we have assumed the same value of β_c in all channels.

Assuming the same exponent ν_j for each group, denoted ν_{ng} and ν_{g} for the non-Gaussian and Gaussian group, respectively, a joint fit was performed with these two exponents, common β_c, and the individual amplitudes c_j as free parameters:

$$m_j = c_j \tau^{\nu_{\text{f}}}, \quad \text{f} = \text{ng}, \text{g} \tag{3}$$

The resulting values of the exponents at $\gamma = -0.2$ are

$$\begin{aligned} \nu_{\text{ng}} &= 0.367(14) \\ \nu_{\text{g}} &= 0.51(3). \end{aligned} \tag{4}$$

Results at $\gamma = -0.5$ are fully consistent with these values [12].

From these results we conclude that the system has two mass scales which we denote by m_{ng} and m_{g}. In Fig. 5 of Ref. [13] we show their behaviour. Except 0^{++}, the gauge-ball masses and presumably $\sqrt{\sigma}$ scale proportional to m_{ng}. The ratios $r_j = m_j/m_{\text{ng}}$ are thus constant. This holds for both γ values. Thus, within the limits of numerical determination, we have found two lines of constant physics. We have further found that r_j are independent of γ, which indicates universality. This suggests that the scaling behaviour of the pure U(1) gauge theory belongs at different γ to the same universality class governed by one non-Gaussian fixed point. The parameter γ is irrelevant in this class as long as it is kept negative.

Choosing m_{ng} for a physical scale, we can consider a continuum theory (or a theory with large cut-off) with the spectrum given by the values of r_j. They are given in [11]. As $m_{\text{g}}/m_{\text{ng}}$ approaches zero, this theory would contain massless (or very light) scalar, which could possibly decouple.

What kind of theory would it be? The procedure we have applied to the spectrum and σ can be extended to all observables and, in general, to n-point functions. If scaling in corresponding powers of m_{ng} is found, one can, in principle, construct the theory without having any continuum Lagrangian for it. Or it could be some non-polynomial Lagrangian in the fields corresponding to the states in the spectrum. As σ would possibly be finite and nonvanishing in the m_{ng}^2 units, it would be a confining theory. An interesting proposal has been made in Ref. [14] and reported at this workshop by N. Sasakura.

The pure U(1) lattice gauge theory with the Villain (periodic Gaussian) action belongs to the same universality class [5]. Rigorous dual relationships valid for that action imply that also the following 4D models are governed by the same non-Gaussian fixed point: the Coulomb gas of monopole loops [15], the noncompact U(1) Higgs model at large negative squared bare mass (frozen 4D superconductor) [16, 17], and an effective string theory equivalent to this Higgs model [18, 19].

5 Compact U(1) gauge theory with matter fields

These findings raise once again the question, whether in strongly interacting 4D gauge field theories further non-Gaussian fixed points exist, which might possibly be of interest for theories beyond the standard model. The pursuit of this question requires an introduction of matter fields. Therefore we have investigated some extensions of the pure compact U(1) gauge theory which might have interesting fixed points. We introduce fermion and scalar matter fields of unit charge, either each separately or both simultaneously. Because of space limits I give only a very brief overview of the results.

The compact fermionic QED has some analogous promising properties in the quenched approximation [20]. In compact scalar QED we confirm the Gaussian behavior at the endpoint of the Higgs phase transition line [21]. In a theory with both scalar and fermion matter fields ($\chi U\phi$ model), the chiral symmetry is broken, and the mass of unconfined composite fermions $F = \phi^{\dagger}\chi$ is generated dynamically [22]. It might be an explanation or alternative to the Higgs-Yukawa sector of the SM. In 2D [23] and 3D [24] it belongs very probably to the universality class of the 2D and 3D Gross-Neveu model, respectively, and is thus nonperturbatively renormalizable. In the 4D $\chi U\phi$ model we demonstrate the existence of a tricritical point where the scaling behaviour is distinctly difrerent from the four-fermion theory, enhancing the chances for renormalizability [22, 25, 21]. Here, apart from the massive fermion and Goldstone boson, the spectrum contains also a massive 0^{++} gauge ball. The particular role of this state both in pure U(1) and in the $\chi U\phi$ model is remarkable and needs a theoretical explanation.

I thank J. Cox, W. Franzki, C. B. Lang, T. Neuhaus, and P.W. Stephenson for collaboration on the topics I have described, and J. Ambjorn, I. Campos, and H. Pfeiffer for discussions. The computations have been performed on the computers of HLRZ Jülich, RWTH Aachen, and KFU Graz. Our work was supported by Deutsches BMBF and DFG.

References

[1] H. G. Evertz, J. Jersák, T. Neuhaus, and P. M. Zerwas, *Nucl. Phys.* **B251** **[FS13]** (1985) 279.

[2] C. B. Lang and T. Neuhaus, *Nucl. Phys.* **B431** (1994) 119.

[3] J. Jersák, C. B. Lang, and T. Neuhaus, *Phys. Rev. Lett.* **77** (1996) 1933.

[4] J. Jersák, C. B. Lang, and T. Neuhaus, *Phys. Rev.* **D54** (1996) 6909.

[5] C. B. Lang and P. Petreczky, *Phys. Lett.* **387B** (1996) 558.

[6] C. Hoebling, A. Jakovac, J. Jersák, C. B. Lang, and T. Neuhaus, *Nucl. Phys. B (Proc. Suppl.)* **47** (1996) 815.

[7] C. Hoebling and C. B. Lang, *Phys. Rev.* **B54** (1996) 3434.

[8] K. Decker, A. Hasenfratz, and P. Hasenfratz, *Nucl. Phys.* **B295 [FS21]** (1988) 21.

[9] A. Hasenfratz, *Phys. Lett.* **201B** (1988) 492.

[10] I. Campos, A. Cruz, and A. Tarancón, *First order signatures in 4D pure compact U(1) gauge theory with toroidal and spherical topologies*, hep-lat/9711045.

[11] J. Cox, W. Franzki, J. Jersák, C. B. Lang, T. Neuhaus, and P. W. Stephenson, *Nucl. Phys. B (Proc. Suppl.)* **53** (1997) 696.

[12] J. Cox et al., *Scaling of gauge balls and static potential in the confinement phase of the pure U(1) lattice gauge theory*, HLRZ1997_40, hep-lat/9709054.

[13] J. Cox, W. Franzki, J. Jersák, C. B. Lang, T. Neuhaus, and P. W. Stephenson, *Nucl. Phys.* **B499** (1997) 371.

[14] J. Ambjørn, D. Espriu, and N. Sasakura, *U(1) lattice gauge theory and $N = 2$ supersymmetric Yang-Mills theory*, hep-th/9707095.

[15] T. Banks, R. Myerson, and J. Kogut, *Nucl. Phys.* **B129** (1977) 493.

[16] M. E. Peskin, *Ann. Phys.* **113** (1978) 122.

[17] J. Fröhlich and P. A. Marchetti, *Europhys. Lett.* **2** (1986) 933.

[18] J. Polchinski and A. Strominger, *Phys. Rev. Lett.* **67** (1991) 1681.

[19] M. I. Polikarpov, U.-J. Wiese, and M. A. Zubkov, *Phys. Lett.* **309B** (1993) 133.

[20] J. Cox, W. Franzki, J. Jersák, C. B. Lang, and T. Neuhaus, *Strongly coupled compact lattice QED with staggered fermions*, PITHA 97/22, HLRZ 19/97, hep-lat/9705043.

[21] W. Franzki and J. Jersák, *Dynamical fermion mass generation at a tricritical point in strongly coupled U(1) lattice gauge theory*, PITHA 97/43, HLRZ1997_66, hep-lat/9711039.

[22] C. Frick and J. Jersák, *Phys. Rev.* **D52** (1995) 340.

[23] W. Franzki, J. Jersák, and R. Welters, *Phys. Rev.* **D54** (1996) 7741.

[24] I. M. Barbour, W. Franzki, and N. Psycharis, *3d* $\chi U\phi$, contribution to the conference Lattice '97.

[25] W. Franzki and J. Jersák, *Strongly coupled U(1) lattice gauge theory as a microscopic model of Yukawa theory*, PITHA 97/42, HLRZ1997_65, hep-lat/9711038.

SYM on the lattice

István Montvay

Deutsches Elektronen Synchrotron DESY,
Notkestr. 85, D-22603 Hamburg, Germany

Abstract: Non-perturbative predictions and numerical simulations in supersymmetric Yang-Mills (SYM) theories are reviewed.

1 Introduction

The investigation of non-perturbative properties of supersymmetric (SUSY) gauge theories is presently a rather theoretical subject: it is not known whether (broken) supersymmetry is realized in nature or not. Moreover, the simple ("minimal") supersymmetric extensions of the standard model do not involve strong interactions near or above the scale of supersymmetry breaking, hence the knowledge of the non-perturbative supersymmetric dynamics is not directly required. In spite of this there is a continuing interest in studying strongly interacting quantum field theories with supersymmetry. (For an early review of the subject see ref. [1].)

The motivation to investigate non-perturbative features of supersymmetric gauge theories is partly coming from the desire to understand relativistic quantum field theories better in general: the supersymmetric points in the parameter space of all quantum field theories are very special since they correspond to situations of a high degree of symmetry. As recent work of Seiberg and Witten [2] and other related papers showed, there is a possibility to approach non-perturbative questions in four dimensional quantum field theories by starting from exact solutions in some highly symmetric points and treat the symmetry breaking as a small perturbation. Beyond this, the knowledge of non-perturbative dynamics in supersymmetric quantum field theories can also be helpful in understanding the greatest puzzle of the standard model, with or without supersymmetric extensions, namely the existence of a large number of seemingly free parameters. As we know from QCD, strong interactions in non-abelian gauge theories are capable to reproduce from a small number of input parameters a large number of dynamically generated parameters for quantities characterizing bound states. This is a possible solution also for the parameters of the standard model if new strong interactions are active beyond the electroweak symmetry breaking scale.

1.1 $N = 1$ SYM

The simplest supersymmetric gauge theory is the supersymmetric extension of Yang-Mills theory. The action of Yang-Mills theory with $N = 1$ supersymmetry is conventionally given as

$$\int d^4x\, d^2\theta\, \mathrm{Tr}(W^\alpha W_\alpha)$$
$$= \int d^4x\, \mathrm{Tr}\left\{ -\frac{1}{2}F_{\mu\nu}F^{\mu\nu} + \frac{i}{2}F_{\mu\nu}\tilde{F}^{\mu\nu} - i\lambda\sigma^\mu(D_\mu\bar{\lambda}) + i(D_\mu\bar{\lambda})\bar{\sigma}^\mu\lambda + D^2 \right\} , \quad (1)$$

where the first line is written in terms of the spinorial field strength superfield $W(x, \theta, \bar{\theta})_\alpha$ which depends on the four-coordinate x and the anticommuting Weyl-spinor variables $\theta_\alpha, \bar{\theta}_{\dot{\alpha}}$ $(\alpha, \dot{\alpha} = 1, 2)$. After performing the Grassmannian integration on θ, one obtains the second form in terms of the component fields. The field strength tensor $F_{\mu\nu}$ and its dual are defined, as usual, as

$$F_{\mu\nu}(x) \equiv -igF^r_{\mu\nu}(x)T_r \ , \qquad \tilde{F}_{\mu\nu} \equiv \frac{1}{2}\epsilon_{\mu\nu\rho\sigma}F^{\rho\sigma} \ . \quad (2)$$

$\lambda, \bar{\lambda}$ represent a Majorana fermion field in the adjoint representation and D is an auxiliary field.

The action in (1) includes a Θ-term, therefore it is natural to introduce the complex coupling

$$\tau \equiv \frac{\Theta}{2\pi} + \frac{4\pi i}{g^2} \quad (3)$$

and then, with arbitrary Θ, the $N = 1$ SYM action becomes:

$$\frac{1}{4\pi}\Im\left\{\tau \int d^4x\, d^2\theta\, \mathrm{Tr}(W^\alpha W_\alpha)\right\}$$
$$= \frac{1}{g^2}\int d^4x\, \mathrm{Tr}\left[-\frac{1}{2}F_{\mu\nu}F^{\mu\nu} - i\lambda\sigma^\mu(D_\mu\bar{\lambda}) + i(D_\mu\bar{\lambda})\bar{\sigma}^\mu\lambda + D^2\right]$$
$$+\frac{\Theta}{16\pi^2}\int d^4x\, \mathrm{Tr}\left[F_{\mu\nu}\tilde{F}^{\mu\nu}\right] \ . \quad (4)$$

The remarkable feature of the action in (4) is that, after performing the trivial Gaussian integration over the auxiliary field D, it is nothing else than an ordinary Yang-Mills action with a massless Majorana fermion in the adjoint representation. This shows that this theory is "automatically" supersymmetric. Introducing a non-zero gaugino mass $m_{\tilde{g}}$ breaks supersymmetry "softly". Such a mass term is:

$$m_{\tilde{g}}(\lambda^\alpha\lambda_\alpha + \bar{\lambda}^{\dot{\alpha}}\bar{\lambda}_{\dot{\alpha}}) = m_{\tilde{g}}(\bar{\Psi}\Psi) \ . \quad (5)$$

Here in the first form the Majorana-Weyl components $\lambda, \bar{\lambda}$ are used, in the second form the Dirac-Majorana field Ψ.

The Yang-Mills theory of a Majorana fermion in the adjoint representation is, in a general sense, similar to QCD: besides the special Majorana-feature the only difference is that the fermion is in the adjoint representation and not in the fundamental one. As in QCD, a central feature of low-energy dynamics is the realization of the global chiral symmetry. As there is only a single Majorana adjoint "flavour", the global chiral symmetry of $N = 1$ SYM is $U(1)_\lambda$, which coincides with the so called *R-symmetry* generated by the transformations

$$\theta'_\alpha = e^{i\varphi}\theta_\alpha \ , \qquad \bar{\theta}'_{\dot{\alpha}} = e^{-i\varphi}\bar{\theta}_{\dot{\alpha}} \ . \tag{6}$$

This is equivalent to

$$\lambda'_\alpha = e^{i\varphi}\lambda_\alpha \ , \ \bar{\lambda}'_{\dot{\alpha}} = e^{-i\varphi}\bar{\lambda}_{\dot{\alpha}} \ , \ \Psi' = e^{-i\varphi\gamma_5}\Psi \ . \tag{7}$$

The $U(1)_\lambda$-symmetry is anomalous: for the corresponding axial current $J_\mu \equiv \bar{\Psi}\gamma_\mu\gamma_5\Psi$, in case of $SU(N_c)$ gauge group with coupling g, we have

$$\partial^\mu J_\mu = \frac{N_c g^2}{32\pi^2}\epsilon^{\mu\nu\rho\sigma}F^r_{\mu\nu}F^r_{\rho\sigma} \ . \tag{8}$$

However, the anomaly leaves a Z_{2N_c} subgroup of $U(1)_\lambda$ unbroken. This can be seen, for instance, by noting that the transformations

$$\Psi \to e^{-i\varphi\gamma_5}\Psi \ , \qquad \bar{\Psi} \to \bar{\Psi}e^{-i\varphi\gamma_5} \tag{9}$$

are equivalent to

$$m_{\tilde{g}} \to m_{\tilde{g}}e^{-2i\varphi\gamma_5} \ , \ \Theta_{SYM} \to \Theta_{SYM} - 2N_c\varphi \ , \tag{10}$$

where Θ_{SYM} is the Θ-parameter of gauge dynamics. Since Θ_{SYM} is periodic with period 2π, for $m_{\tilde{g}} = 0$ the $U(1)_\lambda$ symmetry is unbroken if

$$\varphi = \varphi_k \equiv \frac{k\pi}{N_c} \ , \qquad (k = 0, 1, \ldots, 2N_c - 1) \ . \tag{11}$$

For this statement it is essential that the topological charge is integer.

The discrete global chiral symmetry Z_{2N_c} is expected to be spontaneously broken by the non-zero *gaugino condensate* $\langle\lambda\lambda\rangle \neq 0$ to Z_2 defined by $\{\varphi_0, \varphi_{N_c}\}$ (note that $\lambda \to -\lambda$ is a rotation). The consequence of this spontaneous chiral symmetry breaking pattern is the existence of a first order phase transition at zero gaugino mass $m_{\tilde{g}} = 0$. For instance, in case of $N_c = 2$ there exist two degenerate ground states with opposite signs of the gaugino condensate. The symmetry breaking is linear in $m_{\tilde{g}}$, therefore the two ground states are exchanged at $m_{\tilde{g}} = 0$ and there is a first order phase transition.

The non-perturbative features of the SYM theory can be investigated in a lattice formulation. As always, the lattice action is not unique (see section 3.1). A possible formulation was given by Curci and Veneziano [3] based on the well known lattice formulation of QCD introduced by Wilson. In the lattice action the *bare gauge coupling* (of $SU(N_c)$) is convetionally represented by $\beta \equiv 2N_c/g^2$ and the *bare gaugino mass* by the *hopping parameter* K. In the plane of (β, K) there is a *critical line* corresponding to zero gaugino mass and the expected phase structure is the one shown in figure 1.

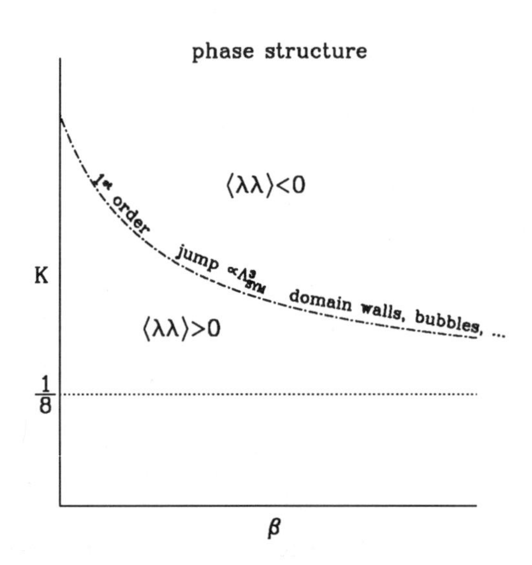

Figure 1: Expected phase structure of SU(2) Yang-Mills theory with a Majorana fermion in adjoint representation in the (β, K)-plane. The dashed-dotted line is a first order phase transition at zero gluino mass, where supersymmetry is expected.

1.2 $N = 2$ SYM

The SYM theory with $N = 2$ *extended supersymmetry* is a highly constrained theory which has, however, more structure than the relatively simple $N = 1$ case discussed above. In particular, besides the $N = 1$ "vector superfield" containing the gauge boson and gaugino (A_μ, λ), it also involves an $N = 1$ "chiral superfield" (ϕ, λ') in the adjoint representation which consists of the complex scalar $\phi \equiv A + iB$ and the Majorana fermion λ'. The Majorana pair (λ, λ') can be combined to a Dirac-fermion ψ and then the vector-like (non-chiral) nature of this theory can be made explicit.

The Euclidean action of $N = 2$ SYM theory in component notation, for simplicity in case of an SU(2) gauge group, is the following:

$$S_{SYM}^{N=2} = \int d^4x \left\{ \frac{1}{4} F_{\mu\nu}^r(x) F_{\mu\nu}^r(x) + \frac{1}{2}(D_\mu A^r(x))(D_\mu A^r(x)) + \frac{1}{2}(D_\mu B^r(x))(D_\mu B^r(x)) \right.$$
$$+ \overline{\psi}^r(x)\gamma_\mu D_\mu \psi^r(x) + ig\epsilon_{rst}\overline{\psi}^r(x)[A^s(x) + i\gamma_5 B^s(x)]\psi^t(x)$$
$$\left. + \frac{g^2}{2}[A^r(x)A^r(x)B^s(x)B^s(x) - A^r(x)B^r(x)A^s(x)B^s(x)] \right\} . \qquad (12)$$

This is a massless adjoint Higgs-Yukawa model with special Yukawa- and quartic couplings given in terms of the gauge coupling.

In $N = 2$ SUSY possible couplings are so strongly constrained by the symmetry that only gauge couplings are allowed. Another important feature is that the symmetry also implies that the matter field content is always vector-like. Therefore

$N = 2$ SUSY theories are always non-chiral and hence well suited for lattice sudies. A lattice formulation of $N = 2$ SYM based on the Wilsonian formulation of QCD has been investigated in [4].

The main new feature of $N = 2$ SYM compared to $N = 1$ SYM is that it also contains scalar fields, hence there is the possibility of Higgs mechanism. Let us here only consider the simplest case of an $SU(2)$ gauge group. In the Higgs phase the vacuum expectation value of the scalar field is non-zero. In terms of the real components $\phi(x) \equiv A(x) + iB(x)$ we have

$$\langle A^r(x) \rangle \neq 0 \,, \qquad \langle B^r(x) \rangle \neq 0 \,. \tag{13}$$

This implies the Higgs mechanism breaking of $SU(2) \to U(1)$, similarly to the Georgi-Glashow model. Due to the Higgs mechanism the "charged" gauge bosons become heavy. The low-energy effective theory is $N = 2$ SYM with $U(1)$ gauge group.

Seiber and Witten proved [2] that extended SUSY and asymptotic freedom can be exploited to determine exactly the low-energy effective action in the Higgs phase, if the vacuum expectation values are large. At strong couplings there are two singularities of the effective action corresponding to light monopoles and dyons, respectively.

The expectation value of the complex scalar field $\langle \phi \rangle$ parametrizes the *moduli space* of zero-energy degenerate vacua. The degeneracy is a consequence of $N = 2$ supersymmetry. This phenomenon is usually referred to as the existence of *flat directions*: the potential identically vanishes for $B^r = cA^r$. In the present case the moduli space is a non-compact manifold with two parameters. The presence of non-compact flat directions requires the breaking of supersymmetry for the definition of the path integral over the scalar fields: otherwise the path integral would be divergent. It is also plausible that soft breaking with mass terms is not enough in the Higgs phase, where the mass-squared terms in the potential are negative. Therefore small hard breaking by dimensionless couplings is also required.

These arguments are quite general. In the special case of $N = 2$ SYM the general renormalizable scalar potential with the given set of scalar fields is

$$V(A, B) \equiv \frac{1}{2} m_A^2 A^r A^r + \frac{1}{2} m_B^2 B^r B^r$$

$$+\lambda_A (A^r A^r)^2 + \lambda_B (B^r B^r)^2 + \lambda_{[AB]} A^r A^r B^s B^s - \lambda_{(AB)} (A^r B^r)^2 \,. \tag{14}$$

N=2 supersymmetry is at the point of parameter space where

$$m_A = m_B = \lambda_A = \lambda_B = 0 \,, \quad 2\lambda_{[AB]} = 2\lambda_{(AB)} = g^2 \,. \tag{15}$$

In order that the path integral over the scalar fields be convergent, the quartic couplings have to fulfil the following conditions:

$$\lambda_A > 0 \qquad \text{AND} \qquad \lambda_B > 0 \qquad \text{AND}$$

$$\left\{ \lambda_{[AB]} \geq \max(0, \lambda_{(AB)}) \qquad \text{OR} \qquad 4\lambda_A \lambda_B > \max[\lambda_{[AB]}^2, (\lambda_{[AB]} - \lambda_{(AB)})^2] \right\} . \tag{16}$$

This is in conflict with the supersymmetry conditions.

The consequence of the conflict between supersymmetry and the convergence of the path integral over the scalar fields is that in a path integral formulation of the quantized theory the supersymmetry has to be broken. On the lattice this means that supersymmetry is broken as long as the lattice spacing is non-zero and can only be restored in the continuum limit $a \to 0$.

The tuning to the supersymmetric point in $N = 2$ SYM can be studied, for instance, in lattice perturbation theory [4]. It can be shown that the compact flat direction is reproduced for $a \to 0$ on a specific phase transition where three different kinds of Higgs-phases meet. The emergence of the non-compact flat direction is a result of cancelling of quantum correction contributions from scalars and fermions. This is similar to the situation which occurs if the so called vacuum stability boundary in Higgs-Yukawa models reaches zero fermion mass. (For a lattice investigation of the vacuum stability bound see, for instance, ref. [5].)

2 Non-perturbative predictions for $N = 1$ SYM

In analogy with QCD, one expects that the spectrum of the SYM model consists of colourless bound states formed out of the fundamental excitations, namely gluons and gluinos. (In this context we shall use the name "gluino" instead of the more general term "gaugino".) In the supersymmetric point at zero gluino mass these bound states should be organized in supersymmetry multiplets, according to the representations of the SUSY extension of the Poincaré algebra. For the description of lowest energy bound states one can use an effective field theory in terms of suitably chosen colourless composite operators.

For $N = 1$ SYM the effective action was constructed by Veneziano and Yankielowicz (VY) [6]. The composite operator appearing in the VY effective action is a chiral supermultiplet S containing as component fields the expressions for the anomalies [7]:

$$S \equiv A(y) + \sqrt{2}\theta\Psi(y) + \theta^2 F(y) \; , \tag{17}$$

where, for instance, the scalar component is proportional to the gluino bilinear

$$A \propto \lambda^\alpha \lambda_\alpha \; . \tag{18}$$

The other components contain gluino-gluino and gluino-gluon combinations. Therefore, as far as a constituent picture is applicable to the bound states formed by strong interactions, the particle content of the lowest supersymmetry multiplet is: a pseudoscalar gluino-gluino bound state, a Majorana spinor gluon-gluino bound state and a scalar gluino-gluino bound state. In terms of S the VY effective action has the form

$$S_{VY} = \frac{1}{\alpha} \int d^4x \, d^2\theta \, d^2\bar{\theta} \, (S^\dagger S)^{1/3} + \gamma \, [\int d^4x \, d^2\theta \, (S \log \frac{S}{\Lambda} - S) + \text{h.c.}] \; . \tag{19}$$

Here α and γ are positive constants and Λ is the usual mass parameter for the asymptotically free coupling defined at scale μ:

$$\Lambda \equiv \mu e^{-1/2\beta_0 g(\mu)^2} , \qquad \beta_0 = \frac{3N_c}{16\pi^2} . \tag{20}$$

As usual, β_0 denotes the first coefficient of the β-function and we consider here the gauge group $SU(N_c)$.

The effective action in (19) incorporates the breaking of the discrete Z_{2N_c} chiral symmetry by the *gluino condensate*

$$\langle\lambda\lambda\rangle = C\Lambda^3 e^{2\pi i k/N_c} . \tag{21}$$

The phase factor depending on the integer k refers to the different ground states defined in (11). The proportionality factor C depends, of course, on the renormalization scheme belonging to Λ. Instanton calculations and other reasonings imply that we have $C = 32\pi^2$ in the dimensional reduction scheme $\Lambda = \Lambda_{DR}$ [8].

The main assumption needed to derive the VY effective action (20) is the choice of the chiral superfield S as the dominant degree of freedom of low energy dynamics. Making a more general ansatz also containing gluon-gluon composites leads to a generalization and to two mixed supermultiplets in the low energy spectrum [9]. Even if S is accepted as the dominant variable, one can argue about the existence of a chirally symmetric ground state, in addition to the N_c ground states with broken chiral symmetry given by the integer $k < N_c$ in (21) [10].

An interesting question is how the spectrum of glueballs, gluinoballs and gluino-glueballs is influenced by the soft supersymmetry breaking due to a non-zero gluino mass $m_{\tilde{g}} \neq 0$. For small $m_{\tilde{g}}$ it is possible to derive the coefficients of the terms linear in $m_{\tilde{g}}$ in the mass formulas [11]. (Note, however, that the two papers in this reference arrive to different results.)

A general consequence of the chiral symmetry breaking is the existence of a first order phase transition at $m_{\tilde{g}} = 0$. At this point the different ground states in (21) are degenerate and a coexistence of the corresponding phases is possible. In a mixed phase situation, as usual at first order phase transitions, the different phases are separated by "bubble wall" interfaces. The *interface tension* of the walls can be exactly derived from the central extension of the $N = 1$ SUSY algebra [12]. The result is that the energy density of the interface wall is related to the jump of the gluino condensate by

$$\epsilon = \frac{N_c}{8\pi^2} \left| \langle\lambda\lambda\rangle_1 - \langle\lambda\lambda\rangle_2 \right| . \tag{22}$$

Combining this with eq. (21) implies that the dimensionless ratio $\epsilon/|\langle\lambda\lambda\rangle|$ is predicted independently of the renormalization scheme.

In order to compare the predictions (21) and (22) to the results of lattice Monte Carlo simulations it is convenient to switch to the lattice Λ-parameter Λ_{LAT}. First one can use [8]

$$\Lambda_{DR}/\Lambda_{\overline{MS}} = \exp\{-1/18\} \tag{23}$$

and then for the Curci-Veneziano lattice action [13]

$$\Lambda_{\overline{MS}}/\Lambda_{LAT} = \exp\left\{-\frac{1}{\beta_0}\left[\frac{1}{16N_c} - N_c P + \frac{N_c n_a}{2} P_3\right]\right\} ,$$

$$\beta_0 = \frac{N_c}{48\pi^2}(11 - 2n_a) , \qquad P = 0.0849780(1) , \qquad P_3 = 0.0066960(1) . \tag{24}$$

Here n_a is the number of Majorana fermions in the adjoint representation, that is for SYM we set $n_a = 1$.

For transforming (21) and (22) to lattice units we need, in fact, the value of $a\Lambda_{LAT}$ at the particular values of interest of the lattice bare parameters β, K. Before performing the lattice simulations this is, of course, not known. An order of magnitude estimate can be obtained from pure gauge theory (at $K = 0$) by noting that both for $N_c = 2$ and $N_c = 3$ we have for the lowest gluball mass M [14]

$$a\Lambda_{LAT} \simeq \frac{aM}{200} . \tag{25}$$

Assuming this approximate relation also at the critical line for zero gluino mass, we can use eqs. (21)-(25) for estimating orders of magnitudes. In the region where $aM = \mathcal{O}(1)$ we obtain

$$a^3|\langle\lambda\lambda\rangle_1 - \langle\lambda\lambda\rangle_2| = \mathcal{O}(1) , \qquad a^3\epsilon = \mathcal{O}(10^{-1}) . \tag{26}$$

As these numbers show, the predicted first order phase transition is, in fact, strong enough for a relatively easy observation in lattice simulations.

3 Numerical Monte Carlo simulations

The lattice Monte Carlo simulations of quantum field theories are performed in Euclidean space-time. For SYM, and more generally for a Yang-Mills theory of Majorana fermions in the adjoint representation ("gaugino" or in the context of strong interactions "gluino") with arbitrary mass we need first of all the definition of Majorana fermions in Euclidean space-time.

In the literature one may sometimes find the statement that there are no Euclidean Majorana spinors (see, for instance, [15]). This is only true as long as one is concentrating on the hermiticity properties of fields, as in Minkowski space. The definition required for an Euclidean path integral can be based on the appropriate analytic continuation of expectation values [16]. The essential point is that for Majorana fermions the Grassmann variables Ψ and $\overline{\Psi}$ are not independent, as is the case for Dirac fermions, but are related by

$$\overline{\Psi} = \Psi^T C , \tag{27}$$

with C the charge conjugation Dirac matrix. In fact, starting from an Euclidean Dirac fermion field represented by the pair $\psi, \overline{\psi}$ one can define two Majorana fermion fields satisfying (27) by

$$\Psi^{(1)} \equiv \frac{1}{\sqrt{2}}(\psi + C\overline{\psi}^T) , \qquad \Psi^{(2)} \equiv \frac{i}{\sqrt{2}}(-\psi + C\overline{\psi}^T) . \tag{28}$$

Using also the inverse relations

$$\psi = \frac{1}{\sqrt{2}}(\Psi^{(1)} + i\Psi^{(2)}) , \qquad \psi_c \equiv C\overline{\psi}^T = \frac{1}{\sqrt{2}}(\Psi^{(1)} - i\Psi^{(2)}) , \tag{29}$$

one can easily relate expectation values of Majorana and Dirac fermion fields [17].

3.1 Lattice actions and algorithms

Following Curci and Veneziano [3], we can take for the fermionic part of the SYM action the well known Wilson formulation. If the Grassmanian fermion fields in the adjoint representation are denoted by ψ_x^r and $\overline{\psi}_x^r$, with r being the adjoint representation index ($r = 1, .., N_c^2 - 1$ for SU(N_c)), then the fermionic part of the lattice action is:

$$S_f = \sum_x \left\{ \overline{\psi}_x^r \psi_x^r - K \sum_{\mu=1}^{4} \left[\overline{\psi}_{x+\hat{\mu}}^r V_{rs,x\mu}(1+\gamma_\mu)\psi_x^s + \overline{\psi}_x^r V_{rs,x\mu}^T (1-\gamma_\mu)\psi_{x+\hat{\mu}}^s \right] \right\} . \tag{30}$$

Here K is the hopping parameter, the irrelevant Wilson parameter removing the fermion doublers in the continuum limit is fixed to $r = 1$, and the matrix for the gauge-field link in the adjoint representation is defined as

$$V_{rs,x\mu} \equiv V_{rs,x\mu}[U] \equiv 2\mathrm{Tr}(U_{x\mu}^\dagger T_r U_{x\mu} T_s) = V_{rs,x\mu}^* = V_{rs,x\mu}^{-1T} . \tag{31}$$

The generators $T_r \equiv \frac{1}{2}\lambda_r$ satisfy the usual normalization $\mathrm{Tr}\,(\lambda_r \lambda_s) = \frac{1}{2}$. In case of SU(2) ($N_c = 2$) we have $T_r \equiv \frac{1}{2}\tau_r$ with the isospin Pauli-matrices τ_r. The normalization of the fermion fields in (30) is the usual one for numerical simulations. The full lattice action is the sum of the pure gauge part and fermionic part:

$$S = S_g + S_f . \tag{32}$$

Here the standard Wilson action for the SU(N_c) gauge field S_g is a sum over the plaquettes

$$S_g = \beta \sum_{pl} \left(1 - \frac{1}{N_c} \mathrm{Re}\,\mathrm{Tr}\, U_{pl} \right) , \tag{33}$$

with the bare gauge coupling given by $\beta \equiv 2N_c/g^2$.

Using the relations in (29) one can decompose S_f as a sum over the two Majorana components:

$$S_f = \sum_{xu,yv} \overline{\psi}_y^v Q_{yv,xu} \psi_x^u = \frac{1}{2} \sum_{j=1}^{2} \sum_{xu,yv} \overline{\Psi}_y^{(j)v} Q_{yv,xu} \Psi_x^{(j)u} , \tag{34}$$

where the *fermion matrix* Q is defined in (30). Using this, the fermionic path integral for Dirac fermions can be written as

$$\int [d\overline{\psi} d\psi] e^{-S_f} = \int [d\overline{\psi} d\psi] e^{-\overline{\psi} Q \psi} = \det Q = \prod_{j=1}^{2} \int [d\Psi^{(j)}] e^{-\frac{1}{2}\overline{\Psi}^{(j)} Q \Psi^{(j)}} . \tag{35}$$

For Majorana fields the path integral involves only $[d\Psi^{(j)}]$, either with $j = 1$ or $j = 2$. For $\Psi \equiv \Psi^{(1)}$ or $\Psi \equiv \Psi^{(2)}$ we have

$$\int [d\Psi] e^{-\frac{1}{2}\overline{\Psi} Q \Psi} = \pm\sqrt{\det Q} . \tag{36}$$

In order to define the sign in (36) one has to consider the *Pfaffian* of the antisymmetric matrix

$$M \equiv CQ . \tag{37}$$

This can be defined for a general complex antisymmetric matrix $M_{\alpha\beta} = -M_{\beta\alpha}$ with an even number of dimensions $(1 \leq \alpha, \beta \leq 2N)$ by a Grassmann integral as

$$\mathrm{Pf}(M) \equiv \int [d\phi] e^{-\frac{1}{2}\phi_\alpha M_{\alpha\beta} \phi_\beta} = \frac{1}{N! 2^N} \epsilon_{\alpha_1 \beta_1 \dots \alpha_N \beta_N} M_{\alpha_1 \beta_1} \dots M_{\alpha_N \beta_N} . \tag{38}$$

Here, of course, $[d\phi] \equiv d\phi_{2N} \dots d\phi_1$, and ϵ is the totally antisymmetric unit tensor. One can easily show that

$$[\mathrm{Pf}(M)]^2 = \det M . \tag{39}$$

If M is taken from (37) one also has $\det M = \det Q$.

The relations in (36) or (39) show that, in order to represent a Majorana fermion, in the path integral over the gauge field the square root of the fermion determinant (or the Pfaffian of the matrix in (37)) has to be taken. In this sense a Majorana fermion corresponds to a flavour number $\frac{1}{2}$. Concerning the sign of the square root, in numerical simulations it is easier to take always the absolute value. This presumably does not have an influence in the continuum limit because in the continuum the (real) eigenvalues of the Dirac matrix come in pairs and the square root is always positive (see, for instance, [18]).

In general, the lattice action describing a given "target" continuum quantum field theory is not unique. Besides the Curci-Veneziano action discussed up to now, another possibility is based on five-dimensional domain walls [19, 20, 21]. In this approach one knows the value of the bare fermion mass where the supersymmetric continuum limit is best approached and one has advantages from the point of view of the speed of symmetry restoration. The price one has to pay is the proliferation of (auxiliary) fermion flavours. Another proposal for reaching supersymmetric quantum field theories is to try direct dimensional reduction on the lattice [22].

In order to perform Monte Carlo simulations with effective flavour number $\frac{1}{2}$ corresponding to Majorana fermions in the lattice formulation of Curci and Veneziano, one can either use the *multi-bosonic technique* [23, 17] or apply the *hybrid classical dynamics algorithm* [24]. Exploratory studies have been started recently. (For a recent review and status report see [25].)

The first step in numerical simulations is to consider the *quenched approximation*, which neglects the dynamical effects of gluinos [26, 27]. Since quenching breaks supersymmetry explicitly, this mainly serves as a testing ground for mass measurements and helps to localize the physically interesting bare parameter range.

A first large scale numerical simulation of SU(2) SYM with dynamical gluinos has been started recently by the DESY-Münster collaboration [28] using the supercomputers at HLRZ, Jülich and DESY, Zeuthen. The main goals of this collaboration are: to find the first order phase transition at zero gluino mass and to determine the masses of the lowest bound states formed out of gluons and gluinos in the interesting range of the gluino mass.

Acknowledgements

I thank Peter Weisz for correspondence and for communicating his result on the ratio of Λ-parameters in (24).

References

[1] D. Amati, K. Konishi, Y. Meurice, G.C. Rossi and G. Veneziano, *Phys. Rep.* **162** (1988) 169

[2] N. Seiberg and E. Witten, *Nucl. Phys.* **B426** (1994) 19; ERRATUM *ibid.* **B430** (1994) 485

[3] G. Curci and G. Veneziano, *Nucl. Phys.* **B292** (1987) 555

[4] I. Montvay, *Phys. Lett.* **B344** (1995) 176 and *Nucl. Phys.* **B445** (1995) 399

[5] L. Lin, I. Montvay, G. Münster, M. Plagge, H. Wittig, *Phys. Lett.* **B317** (1993) 143

[6] G. Veneziano, S. Yankielowicz, *Phys. Lett.* **B113** (1982) 231

[7] S. Ferrara, B. Zumino, *Nucl. Phys.* **B87** (1975) 207

[8] D. Finnell, P. Pouliot, *Nucl. Phys.* **B453** (1995) 225.

[9] G.R. Farrar, G. Gabadadze, M. Schwetz, hep-th/9711166

[10] A. Kovner, M. Shifman, *Phys. Rev.* **D56** (1997) 2396

[11] A. Masiero, G. Veneziano, *Nucl. Phys.* **B249** (1985) 593;
N. Evans, S.D.H. Hsu, M. Schwetz, hep-th/9707260.

[12] A. Kovner, M. Shifman, A. Smilga, *Phys. Rev.* **D56** (1997) 7978

[13] A. Hasenfratz, P. Hasenfratz, *Phys. Lett.* **93B** (1980) 165;
P. Weisz, *Phys. Lett.* **100B** (1981) 331 and private communication

[14] C. Michael, M. Teper, *Phys. Lett.* **B199** (1987) 95;
H. Chen, J. Sexton, A. Vaccarino, D. Weingarten, *Nucl. Phys. Proc. Suppl.* **34** (1994) 357;
M.J. Teper, hep-lat/9711011

[15] H. Leutwyler, A. Smilga, *Phys. Rev.* **D46** (1992) 5607

[16] H. Nicolai, *Nucl. Phys.* **B140** (1978) 294;
P. van Nieuwenhuizen, A. Waldron, *Phys. Lett.* **B389** (1996) 29

[17] I. Montvay, *Nucl. Phys.* **B466** (1996) 259

[18] S.D.H. Hsu, hep-th/9704149

[19] R. Narayanan, H. Neuberger, *Nucl. Phys.* **B443** (1995) 305;
P. Huet, R. Narayanan, H. Neuberger, *Phys. Lett.* **B380** (1996) 291;
H. Neuberger, hep-lat/9710089

[20] J. Nishimura, *Phys. Lett.* **B406** (1997) 215 and hep-lat/9709112;
T. Hotta, T. Izubuchi, J. Nishimura, hep-lat/9709075 and hep-lat/9712009

[21] S. Aoki, K. Nagai, S.V. Zenkin, *Nucl. Phys.* **B508** (1997) 715 and hep-lat/9709058.

[22] N. Maru, J. Nishimura, hep-th/9705152

[23] M. Lüscher, *Nucl. Phys.* **B418** (1994) 637

[24] A. Donini, M. Guagnelli, *Phys. Lett.* **B383** (1996) 301

[25] I. Montvay, hep-lat/9709080, to appear in the *Proceedings of the Lattice '97 Conference in Edinburgh.*

[26] G. Koutsoumbas, I. Montvay, *Phys. Lett.* **B398** (1997) 130

[27] A. Donini, M. Guagnelli, P. Hernandez, A. Vladikas, hep-lat/9708006 and hep-lat/9710065.

[28] G. Koutsoumbas, I. Montvay, A. Pap, K. Spanderen, D. Talkenberger, J. Westphalen, hep-lat/9709091, to appear in the *Proceedings of the Lattice '97 Conference in Edinburgh.*

The Status of String Theory

John H. Schwarz

California Institute of Technology, Pasadena, CA 91125 USA

Abstract: There have been many remarkable developments in our understanding of superstring theory in the past few years, a period that has been described as "the second superstring revolution." Several of them are discussed here. The presentation is intended primarily for the benefit of nonexperts.

1 Introduction

This manuscript presents a brief overview of some of the advances in understanding superstring theory that have been achieved in the last few years. It is aimed at physicists who are not experts in string theory, but who are interested in hearing about recent developments. Where possible, the references cite review papers rather than original sources.

It is now clear that what had been regarded as five distinct superstring theories in ten dimensions are better viewed as five special points in the moduli space of consistent vacua of a single theory. Morover, another special limit corresponds to a vacuum with Lorentz invariance in eleven dimensions. Some of the evidence that supports this picture is reviewed in Sect. 2. The "second superstring revolution" is characterized by the discovery of various non-perturbative properties of superstring theory. An important aspect of this is the occurrence of p-dimensional excitations, called p-branes. Their properties are under good mathematical control when they preserve some of the underlying supersymmetry. The maximally supersymmetric p-branes that occur in 10 or 11 dimensions are surveyed in Sect. 3. For detailed reviews of the material in Sect. 2 and Sect. 3, see [1] – [6].

Sect. 4 describes how suitably constructed brane configurations can be used to derive, and make more geometrical, some of the non-perturbative properties of supersymmetric gauge theories that have emerged in recent years. Sect. 5 presents evidence for the existence of new non-gravitational quantum theories in six dimensions. In particular, there are pairs of theories with (2,0) and (1,1) supersymmetry that are related by T duality. Finally, in Sect. 6, the Matrix Theory proposal, which is a candidate for a non-perturbative description of M theory in a certain class of backgrounds, is sketched. This subject has been reviewed recently in [7, 8]. This is a rapidly developing subject, which appears likely to be a major focus of research in the next couple of years.

There have been other interesting developments in the past few years, which are omitted from this survey. The most remarkable, perhaps, is the application of D-brane technology to the study of black hole physics. This has led to a microscopic explanation of the origin of black hole thermodynamics in wide classes of examples. (For reviews see [9] and [10].) Other omitted topics include applications to particle physics phenomenology and to cosmology. For two other surveys of recent developments in string theory, see [11] and [12].

2 M Theory

A schematic representation of the relationship between the five superstring vacua in 10d and the 11d vacuum, characterized by 11d supergravity at low energy, is given in Fig. 1. The idea is that there is some large moduli space of consistent vacua of a single underlying theory – denoted by M here. The six limiting points, represented as circles, are special in the sense that they are the ones with (super) Poincaré invariance in ten or eleven dimensions. The letters on the edges refer to the type of transformation relating a pair of limiting points. The numbers 16 or 32 refer to the number of unbroken supersymmetries. In 10d the minimal spinor is Majorana–Weyl and has 16 real components, so the conserved supercharges correspond to just one MW spinor in three cases (type I, HE, and HO). Type II superstrings have two MW supercharges, with opposite chirality in the IIA case and the same chirality in the IIB case. In 11d the minimal spinor is Majorana with 32 real components.

The 11d vacuum, including 11d supergravity, is characterized by a single scale – the 11d Planck scale m_p. It is proportional to $G_N^{-1/9}$, where G_N is the 11d Newton constant. The connection to type IIA theory is obtained by taking one of the ten spatial dimensions to be a circle (S^1 in the diagram) of radius R. Type IIA string theory in 10d has a dimensionless coupling constant g_s, which is given by the vev of e^{ϕ}, where ϕ is the dilaton field – a massless scalar field belonging to the IIA supergravity multiplet. In addition, the IIA theory has a mass scale, m_s, whose square gives the tension of the fundamental IIA string. The relationship between the parameters of the 11d and IIA descriptions is given by

$$m_s^2 = R m_p^3 \tag{1}$$

$$g_s = R m_s. \tag{2}$$

Numerical factors (such as 2π) are not important for present purposes and have been dropped. The significance of these equations will emerge later. However, one point can be made immediately. The conventional perturbative analysis of the IIA theory is an expansion in powers of g_s with m_s fixed. The second relation implies that this is an expansion about $R = 0$, which accounts for the fact that the 11d interpretation was not evident in studies of perturbative string theory. The radius R is a modulus – the vev of a massless scalar field with a flat potential. One gets from the IIA point

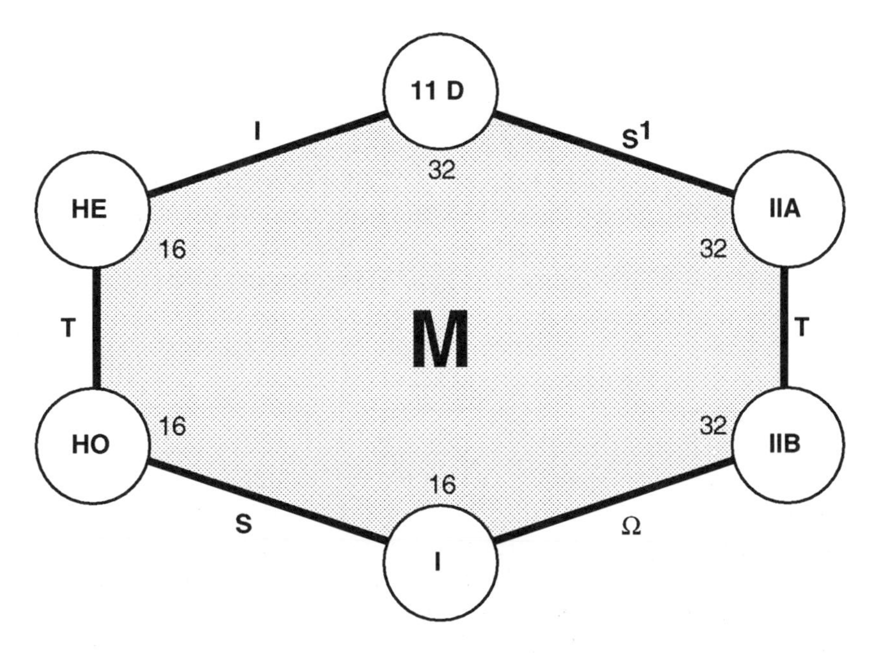

Figure 1: The M theory moduli space.

to the 11d point by continuing this vev from zero to infinity. This is the meaning of the edge of Fig. 1 labeled S^1.

The relationship between the $E_8 \times E_8$ heterotic string vacuum (denoted HE) and 11d is very similar. The difference is that the compact spatial dimension is a line interval (denoted I in the diagram) instead of a circle. The same relations in eqs. (1) and (2) apply in this case. This compactification leads to an 11d space-time that is a slab with two parallel 10d faces. One set of E_8 gauge fields is confined to each face, whereas the gravitational fields reside in the bulk. There is a nice generalization of the 10d anomaly cancellation mechanism to this 11d setting [13]. It only works for E_8 gauge groups.

The two edges of Fig. 1 labeled T connect vacua related by T duality. For example, if the IIA theory is compactified on (another) circle of radius R_A leaving nine noncompact dimensions, this is equivalent to compactifying the IIB theory on a circle of radius

$$R_B = (m_s^2 R_A)^{-1}. \tag{3}$$

Thus, continuing the modulus R_A from infinity to zero (or R_B from zero to infinity) gives an interpolation between the IIA and IIB theories. The T duality relating the two heterotic theories (HE and HO) is essentially the same, though there are additional technical details in this case.

The edge connecting the HO vacuum and the type I vacuum is labeled by S in the diagram, since these two vacua are related by S duality. Specifically, denoting the two string coupling constants by $g_s^{(HO)}$ and $g_s^{(I)}$, the relation is

$$g_s^{(I)} g_s^{(HO)} = 1. \tag{4}$$

In other words, the two dilatons satisfy $\phi^{(I)} + \phi^{(HO)} = 0$, and the edge connecting the HO and I points in Fig. 1 represents a continuation from weak coupling ($\phi = -\infty$) to strong coupling ($\phi = +\infty$). It has been known for a long time that the two vacua have the same gauge symmetry ($SO(32)$) and the same supersymmetry, but it was unclear how they could be equivalent because type I strings and heterotic strings are very different. The explanation is that heterotic strings appear as nonperturbative excitations in the type I description. The converse is not quite true, because type I strings disintegrate at strong coupling.

The final link, labeled Ω in Fig. 1, connects the type IIB and type I vacua. Ω represents an "orientifold projection," which involves modding out by a particular Z_2 discrete symmetry. Starting from the IIB picture, the Z_2 in question is an orientation reversal of the IIB string ($\sigma \to -\sigma$) [14, 15]. This results in unoriented closed strings ("untwisted sector") and unoriented open strings carrying $SO(32)$ gauge symmetry ("twisted sector"). In the modern viewpoint, the open strings can be regarded as ending on 32 superimposed D9-branes. We will say more about D-branes later.

3 p-branes

Supersymmetry algebras with central charges admit "short representations", the existence of which is crucial for testing conjectured non-perturbative properties of theories that previously were only defined perturbatively. Schematically, when a state carries a central charge Q, the supersymmetry algebra implies that its mass is bounded below ($M \geq |Q|$). Moreover, when the state is "BPS saturated," *i.e.*, $M = |Q|$, the representation theory changes, and a state can belong to a short representation of the algebra. This phenomenon is already familiar for the case of Poincaré symmetry in 4d, which allows a massless photon to have just two helicity states (a short representation), whereas a massive vector boson must have three helicity states.

This BPS saturation property arises not only for point particles, characterized by a mass M, but for extended objects with p spatial dimensions, called p-branes. In this case the central charge is a rank p tensor. At first sight, this might seem to be in conflict with the Coleman–Mandula theorem, which forbids finite tensorial central charges. However, the p-branes carry a finite charge per unit volume, so that the total charge is infinite for a BPS p-brane that is an infinite hyperplane, and there is no contradiction. The BPS saturation condition in this case implies that the tension (or mass per unit volume) of the p-brane equals the charge density. Another way of viewing BPS p-branes is as solitons that preserve some of the supersymmetry of the underlying theory.

The theories in question (I will focus on the ones with 32 supercharges) are approximated at low energy by supergravity theories that contain various antisymmetric tensor gauge fields. They are conveniently represented by differential forms

$$A_n \equiv A_{\mu_1\mu_2\ldots\mu_n} dx^{\mu_1} \wedge dx^{\mu_2} \wedge \ldots \wedge dx^{\mu_n}. \tag{5}$$

In this notation, the corresponding gauge-invariant field strength is given by an $(n+1)$-form $F_{n+1} = dA_n$ plus possible additional terms. A type II or 11d supergravity theory with such a gauge field has two kinds of BPS p-brane solutions, which preserve one-half of the supersymmetry. One, which can be called "electric," has $p = n - 1$. The other, called "magnetic," has $p = D-n-3$, where D is the space-time dimension (ten or eleven for the cases considered here).

A hyperplane with p spatial dimensions in a space-time with $D - 1$ spatial dimensions can be surrounded by a sphere S^{D-p-2}. If A is a $(p + 1)$-form potential for which a p-brane is the source, the electric charge Q_E of the p-brane is given by a straightforward generalization of Gauss's law:

$$Q_E \sim \int_{S^{D-p-2}} *F, \tag{6}$$

where S^{D-p-2} is a sphere surrounding the p-brane and $*F$ is the Hodge dual of the $(p+2)$-form field strength F. Similarly, a dual $(D-p-4)$-brane has magnetic charge given by

$$Q_M \sim \int_{S^{p+2}} F. \tag{7}$$

The Dirac quantization condition, for electric and magnetic 0-branes in $D = 4$, has a straightforward generalization to a p-brane and a dual $(D - p - 4)$-brane in D dimensions

$$\frac{1}{2\pi} Q_E Q_M \in \mathbf{Z}. \tag{8}$$

An approximate description of the classical dynamics of a "thin" p-brane is given by a generalized Nambu–Goto formula

$$S_p = T_p \int \left(\sqrt{-\det G_{\mu\nu}} + \ldots\right) d^{p+1}\sigma, \tag{9}$$

where

$$G_{\mu\nu} = g_{MN}(X)\partial_\mu X^M \partial_\nu X^N. \tag{10}$$

Here $G_{\mu\nu}(\sigma)$ is a metric on the $(p+1)$-dimension world-volume of the p-brane obtained as a pullback of the D-dimensional space-time metric $g_{MN}(X)$. The functions $X^M(\sigma)$ describe the embedding of the p-brane in space-time. The coefficient T_p is the p-brane tension – its universal mass per unit volume. Note that (for $\hbar = c = 1$) $T_p \sim (\text{mass})^{p+1}$. This integral is just the volume of the embedded p-brane, generalizing the invariant length of the world-line of a point particle or the area of the world-sheet of a string. The dots represent terms involving other world-volume degrees of freedom required by supersymmetry.

Superstring theories in 10d have three distinct classes of p-branes. These are distinguished by how the tension T_p depends on the string coupling constant g_s.

A "fundamental" p-brane has $T_p \sim (m_s)^{p+1}$, with no dependence on g_s. Such p-branes only occur for $p = 1$ – the fundamental strings. Since these are the only objects that survive at $g_s = 0$, they are the only ones that can be used as the fundamental degrees of freedom in a perturbative description. A second class of p-branes, called "solitonic," have $T_p \sim (m_s)^{p+1}/g_s^2$. These only occur for $p = 5$, the five-branes that are the magnetic duals of the fundamental strings. This dependence on the coupling constant is familiar from field theory. A good example is the mass of an 't Hooft–Polyakov monopole in gauge theory. The third class of p-branes, called "Dirichlet" (or Dp-branes), have $T_p \sim (m_s)^{p+1}/g_s$. This behavior, intermediate between "fundamental" and "solitonic," was not previously known in field theory. In 10d type II theories D-branes occur for all $p \leq 9$ – even values in the IIA case and odd ones in the IIB case. They are all interrelated by T dualities; moreover, the magnetic dual of a Dp-brane is a Dp'-brane with $p' = 6 - p$. D-branes are very important, and so we will have more to say about them later.

Eleven-dimensional supergravity contains a three-form potential. Therefore, the 11d vacuum admits two basic kinds of p-branes – the M2-brane (also known as the supermembrane) and the M5-brane. These are EM duals of one another. Since the only parameter of the 11d vacuum is the Planck mass m_p, their tensions are necessarily $T_{M2} = (m_p)^3$ and $T_{M5} = (m_p)^6$, up to numerical coefficients.

We can use the relation between the 11d theory compactified on a circle of radius R and the IIA theory in 10d to deduce the tensions of certain IIA p-branes. Starting with the M2-brane we can either allow one of its dimensions to wrap the circular dimension, leaving a string in the remaining dimensions, or we can simply embed it in the non-compact dimensions, where it is then still viewed as a 2-brane. In the latter case, the tension remains m_p^3. Using eqs. (1) and (2), we can recast this as $T = (m_s)^3/g_s$, which we recognize as the tension of the D2-brane of IIA theory. On the other hand, the wrapped M2-brane leaves a string with tension $T = m_p^3 R = m_s^2$. Thus we see that eq. (1) reflects the fact that a fundamental IIA string is actually a wrapped M2 brane. Starting with the M5-brane, we can carry out analogous calculations. If it is not wrapped we obtain a IIA 5-brane with tension $T = m_p^6 = m_s^6/g_s^2$, which is the correct relation for the solitonic 5-brane (usually called the NS5-brane). If it is wrapped on the circle, one is left with a IIA 4-brane with tension $T = m_p^6 R = m_s^5/g_s$. This has the correct tension to be identified as a D4-brane. In other words, the D4-brane is actually a wrapped M5-brane.

There are a couple basic facts about D-branes in type II superstring theories that should be pointed out. First of all, they can be understood in the weak coupling limit (which makes them heavy) as surfaces on which fundamental type II strings can end. This is where the Dirichlet boundary conditions come in. This has a number of implications. One is that the dynamics of D-branes at weak coupling can be deduced from that of fundamental strings using perturbative methods. Another is that since a type II string carries a conserved charge that couples to a two-form potential, the end of a string must carry a point charge, which gives rise to electric flux of a Maxwell field. This implies that the world-volume theory of a D-brane contains a $U(1)$ gauge field. In fact, for strong fields that vary slowly it is actually a non-linear theory of

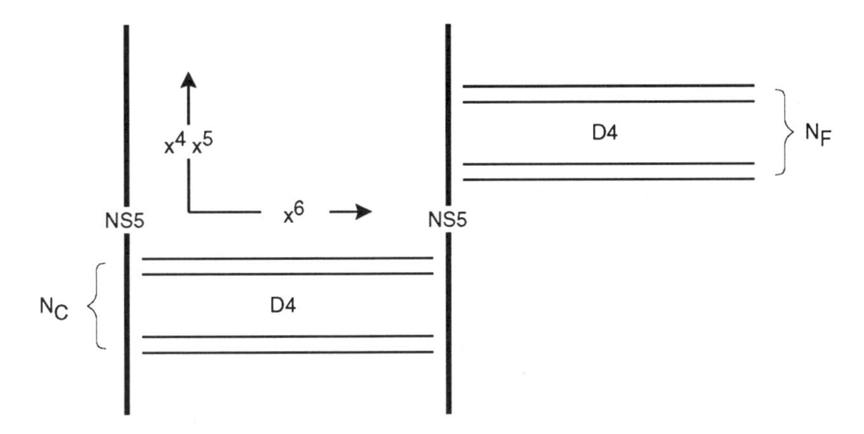

Figure 2: Brane configuration for an $N = 2$ 4d gauge theory.

the Born–Infeld type. The $U(1)$ gauge field can be regarded as arising as the lowest excitation of an open string with both ends attached to the D-brane.

Consider now k parallel Dp-branes, which are $(p+1)$-dimensional hyperplanes in $\mathbf{R}^{10}$. In this case, open strings can end on two different branes. The lowest mode of a string connecting the ith and jth D-brane is a gauge field that carries i and j type electric charges at its two ends. Altogether one has a $U(k)$ gauge theory in $p+1$ dimensions. Classically, this can be constructed as the dimensional reduction of $U(k)$ super Yang–Mills theory in 10d. The separations of the D-branes are given by the vevs of scalar fields, which break the gauge group to a subgroup. For $p \leq 3$, these gauge theories have a straightforward quantum interpretation, but for $p > 3$ the gauge theories are non-renormalizable. I will return to this issue in Sect. 5.

4 Brane-Configuration Constructions of SUSY Gauge Theories

In the last section we saw that a collection of k parallel D-branes gives a supersymmetric $U(k)$ gauge theory. The unbroken supersymmetry in this case is maximal (16 conserved supercharges). In this section we describe more complicated brane configurations, which break additional supersymmetries, and give susy gauge theories in 4d with a richer structure. This is an active subject, which can be approached in several different ways. Here we will settle for two examples in one particular approach. (For a different approach see [16].)

The first example [17] is a configuration of NS5-branes and D4-branes in type IIA theory depicted in Fig. 2. This configuration gives rise to an $SU(N_C)$ gauge theory in 4d with $N = 2$ supersymmetry (8 conserved supercharges). To explain why, one must first describe the geometry. All of the branes are embedded in 10d so as to completely fill the dimensions that will be identified as the 4d space-time with coordinates x^0, x^1, x^2, x^3. In addition, the NS5-branes also fill the x^4 and x^5

dimensions, which are represented by the vertical direction in the figure, and they have fixed values of x^6, x^7, x^8, x^9. The D4-branes, on the other hand, have a specified extension in the x^6 direction, depicted horizontally in the figure, and they have fixed values of x^4, x^5, x^7, x^8, x^9. The idea is that the gauge theory lives on the N_C D4-branes, which are suspended between the NS5-branes. The x^6 extension becomes negligible for energies $E \ll 1/L$, where L is the separation between the NS5-branes. In this limit the 5d theory on the D4-branes is effectively four dimensional. In addition there are N_F semi-infinite D4-branes, which result in N_F hypermultiplet flavors in the fundamental representation of the gauge group. These states arise as the lowest modes of open strings connecting the two types of D4-branes. The presence of the NS5-branes is responsible for breaking the supersymmetry from $N = 4$ to $N = 2$.

This picture is valid at weak coupling, because the gauge coupling constant g_{YM} is given by $g_{YM}^2 = g_s/(Lm_s)$, and the IIA picture is valid for small g_s. Substituting eq. (2), we see that $g_{YM}^2 = R/L$, where R is the radius of a circular eleventh dimension. So far, the description of the geometry omits consideration of this eleventh dimension, but by taking it into account we can see what happens to the gauge theory when g_{YM}^2 is not small and quantum effects become important. The key step is to recall that a D4-brane is actually an M5-brane wrapped around the circular eleventh dimension. Thus, reinterpreted as a brane configuration embedded in 11d, the entire brane configuration corresponds to a single smooth M5-brane. The junctions are now smoothed out in a way that can be made quite explicit. The correct configuration is one that is a stable static solution of the M5-brane equation of motion, which degenerates to the IIA configuration we have described in the limit $R \to 0$. There is a simple method, based on complex analysis, for finding such solutions. If space is described as a complex manifold, with a specific choice of complex structure, then the brane configuration is a stable static solution if its spatial dimensions are embedded holomorphically. In the example at hand, the relevant dimensions are two dimensions of the M5-brane, which are embedded in the four dimensions denoted x^4, x^5, x^6, x^{10}, where x^{10} is the circular eleventh dimension. A complex structure is specified by choosing as holomorphic coordinates $v = x^4 + ix^5$ and $t = \exp[(x^6 + ix^{10})/R]$, which is single-valued. Then a holomorphically embedded submanifold is specified by a holomorphic equation of the form $F(t, v) = 0$. The appropriate choice of F is a polynomial in t and v with coefficients that correspond in a simple way to the positions of the NS5-branes and D4-branes. (For further details see Ref. [17].) This 2d surface is precisely the Seiberg–Witten Riemann surface (or "curve") that characterizes the exact non-perturbative low-energy effective action of the gauge theory. When first discovered, this curve was introduced as an auxiliary mathematical construct with no evident geometric significance. We now see that the Seiberg–Witten solution is given by an M5-brane with four of its six dimensions giving the space time and the other two giving the Seiberg–Witten curve! This simple picture makes the exact non-perturbative low energy physics of a wide class of $N = 2$ gauge theories almost trivial to work out.

Let me briefly mention how the brane configuration described above can be modified to describe certain $N = 1$ susy gauge theories. One way to achieve this is to rotate one of the two NS5-branes so that it fills the dimensions x^8, x^9 and has fixed x^5, x^6 coordinates. When this is done the N_C D4-branes running between the NS5-branes are forced to be coincident. The rotation breaks the supersymmetry to $N = 1$. One of the remarkable discoveries of Seiberg is that an $N = 1$ susy gauge theory with gauge group $SU(N_C)$ and $N_F \geq N_C$ flavors is equivalent in the infrared to an $SU(N_F - N_C)$ gauge theory with a certain matter content. This duality can be realized geometrically in the brane configuration picture by smoothly deforming the picture so as to move one NS 5-brane to the other side of the other one [18, 19]. Such a move certainly changes the exact quantum vacuum described by the configuration. However, the parameters involved are irrelevant in the infrared limit, so one achieves a simple understanding of Seiberg duality.

5 New Non-gravitational 6d Quantum Theories

We have seen that it is interesting and worthwhile to consider the world volume theory of a collection of coincident or nearly coincident branes. For such a theory to be regarded in isolation in a consistent way, it is necessary to define a limit in which the brane degrees of freedom decouple from those of the surrounding spacetime "bulk." Such a limit was implicitly involved in the discussion of the preceding section. (This involves some subtleties, which we did not address.) In this section we wish to consider the 6d world-volume theory that lives on a set of (nearly) coincident 5-branes. If one can define a limit in which the degrees of freedom of the world-volume theory decouple from those of the bulk, but still remain self-interacting, then we will have defined a consistent non-trivial 6d quantum theory [20]. (The only assumption that underlies this is that M theory/superstring theory is a well-defined quantum theory.) The 6d quantum theories that are obtained this way do not contain gravity. The existence of consistent quantum theories without gravity in dimensions greater than four came as quite a surprise to many people.

As a first example consider k parallel M5-branes embedded in flat 11d spacetime. This neglects their effect on the geometry, which is consistent in the limit that will be considered. The only parameters are the 11d Planck mass m_p and the brane separations L_{ij}. In 11d an M2-brane is allowed to terminate on an M5-brane. Therefore, a pair of M5 branes can have an M2-brane connect them. When the separation L_{ij} becomes small, this M2-brane is well approximated by a string of tension $T_{ij} = L_{ij} m_p^3$. The limit that gives decoupling of the bulk degrees of freedom is $m_p \to \infty$. By letting the separations approach zero at the same time, this limit can be carried out holding the string tensions T_{ij} fixed. In the limit one obtains a chiral 6d quantum theory with $(2, 0)$ supersymmetry containing k massless tensor supermultiplets and a spectrum of strings with tensions T_{ij}. There are five massless scalars associated to each brane (parametrizing their transverse excitations). They are coordinates for the moduli space of the resulting theory, which is $(\mathbf{R}^5)^k / S_k$. The permutation group S_k is due to quantum statistics for identical branes. String tensions depend on position in moduli space, and specific ones approach zero at its singularities.

A closely related construction is to consider k parallel NS 5-branes in the IIA theory. The difference in this case is that one of the transverse directions (parametrized by one of the five scalars) is the circular eleventh dimension. In carrying out the decoupling limit one can send the radius R to zero at the same time, holding the fundamental type IIA string tension $T = m_s^2 = m_p^3 R$ fixed. The resulting decoupled 6d theory contains this string in addition to the ones described above. It becomes bound to the NS5-branes in the limit, as the amplitude to come free vanishes in the limit $g_s \to 0$. The resulting theory has the moduli space $(\mathbf{R}^4 \times S^1)^k / S_k$. This theory contains fundamental strings and has a chiral extended supersymmetry, features that are analogous to type IIB superstring theory in 10d. However, it is actually a class of non-gravitational theories (labeled by k) in 6d. Because of the analogy some authors refer to this class of theories as *iib* string theories. Six-dimensional non-gravitational analogs of type IIA string theory, denoted *iia* string theories, are obtained by means of a similar decoupling limit applied to a set of parallel NS5-branes in IIB theory. These *iia* and *iib* string theories are related by T duality. Explicitly, compactifying one spatial dimension on a circle of radius R_a or R_b, the theories (with given k) become equivalent for the identification $m_s^2 R_a R_b = 1$. This feature is directly inherited from the corresponding property of the IIA and IIB theories.

There are various generalizations of these theories that will not be described here. There are also 6d non-gravitational counterparts of the two 10d heterotic theories. These have chiral $(1, 0)$ supersymmetry. In the notation of Fig. 1, they could be referred to as *he* and *ho* theories. They, too, are related by T duality. Although the constructions make us confident about the existence and certain general properties of these theories, they are not very well understood. The 10d string theories have been studied for many years, whereas these 6d string theories are only beginning to be analyzed. Like their 10d counterparts, the fact that they have T dualities implies that they are not conventional quantum field theories.

6 The Matrix Theory Proposal

The discovery of string dualities and the connection to 11d has taught us a great deal about non-perturbative properties of superstring theories, but it does not constitute a complete non-perturbative formulation of the theory. In October 1996, Banks, Fischler, Shenker, and Susskind made a specific conjecture for a complete nonperturbative definition of the theory in eleven uncompactified dimensions called 'Matrix Theory' [21, 8]. In this approach, as we will see, other compactification geometries require additional inputs. It is far from obvious that the BFSS proposal is well-defined and consistent with everything we already know. However, it seems to me that there is enough that is right about it to warrant the intense scrutiny that it has received and is continuing to receive. At the time of this writing, the subject is in a state of turmoil. On the one hand, there is a new claim that the BFSS prescription (as well as a variant due to Susskind [22]) can be derived from previous knowledge [23]. On the other, some people [24, 25, 26] are (cautiously) claiming to have found specific settings in which it gives wrong answers! In the following, we do not comment further upon these claims. Instead, we describe the basic ideas of Matrix Theory, as well as some of its successes and limitations.

One of the p-branes that has not been discussed yet is the D0-brane of type IIA theory in 10d. Being a D-brane, its mass is $M = m_s/g_s$. Using eq. (2), one sees that $M = 1/R$, which means that it can be understood as the first Kaluza–Klein excitation of the 11d supergravity multiplet on the circular eleventh dimension. In fact, this is a good way of understanding (and remembering) eq. (2). Like all the type II D-branes it is a BPS state that preserves half of the supersymmetry, so one has good mathematical control. From the 11d viewpoint it can be viewed as a wave going around the eleventh dimension with a single quantum of momentum. Higher Kaluza–Klein excitations with $M = N/R$ are also BPS states. From the IIA viewpoint these are bound states of N D0-branes with zero binding energy. The existence of a bound state at a threshold is a very subtle dynamical question, which must be true in this case. This has in fact been proved for $N = 2$ in ref. [27] and for all prime values of N in [28].

By the prescription given in Sect. 4, the dynamics of N D0-branes is described by the dimensional reduction of $U(N)$ super Yang–Mills theory in 10d to one time dimension only. When this is done, the spatial coordinates of the N D0-branes are represented by $N \times N$ Hermitian matrices! This theory has higher order corrections, in general. However, one can speculate that these effects are suppressed by viewing the N D0-branes in the infinite momentum frame (IMF). This entails letting $p_{11} = N/R$ approach infinity at the same time as $R \to \infty$. The techniques involved here are reminiscent of those developed in connection with the parton model of hadrons in the late 1960's. The BFSS conjecture is that this IMF frame $N \to \infty$ limit of the D0-brane system constitutes an exact non-perturbative description of the 11d quantum theory. The $N \to \infty$ limit is awkward, to say the least, for testing this conjecture. A stronger version of the conjecture, due to Susskind, is applicable to finite N. It asserts that the IMF D0-brane system, with fixed N, provides an exact non-perturbative description of the 11d theory compactified on a light-like circle with N units of (null) momentum along the circle.

One of the first issues to be addressed was how this conjecture should be generalized when additional dimensions are compact, specifically if they form an n-torus T^n. The reason this is a non-trivial problem is that open strings connecting pairs of D0-branes can lie along many topologically distinct geodesics. It turns out that all these modes can be taken into account very elegantly by replacing the 1-dimensional quantum theory of the D0-branes by an $(n+1)$-dimensional quantum theory, where the n spatial dimensions lie on the dual torus $\tilde{T}^n$. The extra dimensions precisely account for all the possible stretched open strings. This picture had some immediate successes. For example, it nicely accounted for all the duality symmetries for various values of n. However, $(n+1)$-dimensional super Yang–Mills theory is non-renormalizable for $n > 3$, so this description of the theory is certainly incomplete in those cases. The new theories described in Sect. 5 provide natural candidates when $n = 4$ or 5, but when $n > 5$ there are no theories of this type, and so we seem to be stuck.

In conclusion, Matrix Theory is a very interesting proposal for defining M theory non-perturbatively. Whether it is correct, or needs to be modified, is very much up in the air at the present time. However, even if it is right, it is unclear how to define vacua with more than five compact dimensions. This fact is very intriguing, since this is precisely what is required to describe the world that we observe.

References

[1] C. Vafa, "Lectures on Strings and Dualities," hep-th/9702201.

[2] P.K. Townsend, "Four Lectures on M Theory," hep-th/9612121.

[3] M.J. Duff, "Supermembranes," p. 219 in *Fields, Strings, and Duality* (TASI 96), eds. C. Efthimiou and B. Greene, World Scientific 1997, hep-th/9611203.

[4] P. Aspinwall, "K3 Surfaces and String Duality," p. 421 in *Fields, Strings, and Duality* (TASI 96), eds. C. Efthimiou and B. Greene, World Scientific 1997, hep-th/9611137.

[5] J. Polchinski, "Lectures on D-Branes," p. 293 in *Fields, Strings, and Duality* (TASI 96), eds. C. Efthimiou and B. Greene, World Scientific 1997, hep-th/9611050.

[6] J.H. Schwarz, "Lectures on Superstring and M Theory Dualities," p. 359 in *Fields, Strings, and Duality* (TASI 96), eds. C. Efthimiou and B. Greene, World Scientific 1997, hep-th/9607201.

[7] A. Bilal, "M(atrix) Theory: a Pedagogical Introduction," hep-th/9710136.

[8] T. Banks, "Matrix Theory," hep-th/9710231.

[9] J. Maldacena, "Black Holes in String Theory," hep-th/9607235; "Black Holes and D-Branes," hep-th/9705078.

[10] D. Youm, "Black Holes and Solitons in String Theory," hep-th/9710046.

[11] S. Mukhi, "Recent Developments in String Theory: A Brief Review for Particle Physicists," hep-ph/9710470.

[12] W. Lerche, "Recent Developments in String Theory," hep-th/9710246.

[13] P. Hořava and E. Witten, "Heterotic and Type I String Dynamics from Eleven Dimensions," *Nucl. Phys.* **B460** (1996) 506, hep-th/9510209.

[14] A. Sagnotti, "Open Strings and Their Symmetry Groups," in: "Non-Perturbative Quantum Field Theory," eds: G. Mack et al (Pergamon Press, 1988)

[15] P. Hořava, "Strings on World-Sheet Orbifolds," *Nucl. Phys.* **B327** (1989) 461.

[16] W. Lerche, "Introduction to Seiberg–Witten Theory and Its Stringy Origin," *Nucl. Phys. B (Proc. Suppl.)* **55B** (1997) 83, hep-th/9611190.

[17] E. Witten, "Solutions of Four-Dimensional Field Theories Via M Theory," *Nucl. Phys.* **B500** (1997) 3, hep-th/9703166.

[18] S. Elitzur, A. Giveon, and D. Kutasov, "Branes and $N = 1$ Duality in String Theory," *Phys. Lett.* **B400** (1997) 269, hep-th/9702014;
S. Elitzur, A. Giveon, D. Kutasov, and E. Rabinovici, "Brane Dynamics and $N = 1$ Supersymmetric Gauge Theory," hep-th/9704104.

[19] H. Ooguri and C. Vafa, "Geometry of $N = 1$ Dualities in Four Dimensions," *Nucl. Phys.* **B500** (1997) 62, hep-th/9702180.

[20] N. Seiberg, "New Theories in Six Dimensions and Matrix Description of M-Theory on T^5 and T^5/Z_2," *Phys. Lett.* **B408** (1997) 111, hep-th/9705221.

[21] T. Banks, W. Fischler, S, Shenker and L. Susskind, *Phys.Rev.* **D55** (1997) 112, hep-th/9610043.

[22] L. Susskind, "Another Conjecture about M(atrix) Theory," hep-th/9704080.

[23] N. Seiberg, "Why is the Matrix Model Correct?" hep-th/9710009.

[24] M. Dine and A. Rajaraman, "Multigraviton Scattering in the Matrix Model," hep-th/9710174.

[25] M.R. Douglas and H. Ooguri, "Why Matrix Theory is Hard," hep-th/9710178.

[26] K. Becker, M. Becker, E. Keski-Vakkuri, and P. Kraus, private communications.

[27] S. Sethi and M. Stern, "D-Brane Bound States Redux," hep-th/9705046.

[28] M. Porrati and A. Rozenberg, "Bound States at Threshold in Supersymmetric Quantum Mechanics," hep-th/9708119.

The "Ether–world" and Elementary Particles

F. Jegerlehner

DESY–Zeuthen
Platanenallee 6, D–15738 Zeuthen, Germany

Abstract:

We discuss a scenario of "the path to physics at the Planck scale" where todays theory of the interactions of elementary particles, the so called Standard Model (SM), emerges as a low energy effective theory describing the long distance properties of a sub–observable medium existing at the Planck scale, which we call "ether". Properties of the ether can only be observable to the extent that they are relevant to characterize the universality class of the totality of systems which exhibit identical low energy behavior. In such a picture the SM must be embedded into a "Gaussian extended SM" (GESM), a quantum field theory (QFT) which includes the SM but is extended in such a way that it exhibits a quasi infrared (IR) stable fixed point in all its couplings. Some phenomenological consequences of such a scenario are discussed.

1 The Path to Physics at the Planck Scale

The current development in the theory of elementary particles is largely triggered by the attempt to unify gravity with the SM interactions at the Planck scale $\Lambda_P \sim 10^{19}$ GeV. A high degree of symmetry is required in order to cure the problems with ultraviolet divergences. The well known symmetry pattern is: **M–THEORY** $\sim$ **STRINGS** and all that $\leftarrow$ **SUGRA** $\leftarrow$ **SUSY** $\leftarrow$ **SM**, with arrows pointing towards more symmetry, provided we neglect symmetry breakings by masses and other soft breaking terms.

Experience from condensed matter physics and a number of known facts suggest that a completely different picture could be behind what we observe as elementary particle interactions at low energies. It might well be that many known features and symmetries we observe result as a consequence of "blindness for details" at long distances of some unknown kind of medium which exhibits as a fundamental cutoff the Planck length. The symmetry pattern thus could look like: **ETHER** $\sim$ **Planck medium** $\rightarrow$ **QFT** $\simeq$ **GESM** $\rightarrow$ **SM**. Unlike in renormalized QFT, here the relationship between bare and renormalized parameters obtains a physical meaning. Such ideas are quite old ([1, 2, 3, 4] and many others) and in some aspects are now commonly accepted among particle physicists. Physics at the Planck scale cannot be described by local quantum field theory. The curvature of space-time is relevant and special relativity is modified by gravitational effects. One expects a world which exhibits an intrinsic cutoff corresponding to the fundamental length $a_P \simeq 10^{-33}$ cm. But not only Poincaré invariance may break down, also the laws of quantum mechanics need not hold any longer at Λ_P . The "microscopic" theory at distances a_P is unknown, but we know it belongs to the "universality class" of possible theories which exhibit as a universal low energy effective asymptote the known electroweak and strong interactions as well as classical gravity. Long distance universality is a well known phenomenon from condensed matter physics, where we know that a ferromagnet, a liquid-gas system and a superconductor may exhibit identical long range properties (phase diagram, critical exponents etc.). Our hypothesis could be that there exist some kind of a "Planck solid", for example. We should mention right here that there is a principal difference between a normal solid and a "Planck solid"; of the latter we only can observe its long range properties, the critical or quasi–critical behavior, its true short range properties will never be observable since we will never be able to built a "Planck microscope" which would allow us to perform experiments at Λ_P . Also the observations which tell us about the properties of the early universe will never suffice to pin down in detail the structure at distances a_P. A possible "ether-theory" can be only a theory of universality classes, dealing with the totality of possible systems which exhibit identical critical behavior. It is a non–trivial task to specify possible candidate models belonging to the universality class which manifests itself as the SM at low energies. Here at best, we may illustrate some points, which allow us to make plausible the viability of such a picture. The approach discussed here is based on the experimentally well established physics and theory of critical phenomena which teaches us the emergence of local Euclidean renormalizable QFT as a low energy effective structure (see e.g.[2, 5]).

2 Low energy effective theories

The typical example we have in mind is the Landau-Ginsburg theory as an effective "macroscopic" description of a real superconductor. For a given microscopic system we may envisage the construction of the low energy effective theory by means of a renormalization semi–group transformation á la Kadanoff (block spin picture) Wilson (cutoff renormalization group), which is a very general physical concept in statistical physics not restricted to QFT. Let us assume that a possible object in the universality class of interest is described by a classical statistical system with fluctuation variables $S_\alpha = S_{\alpha 0} + S_{\alpha 1}$, where the S_i $(i = 0, 1)$ in momentum space have support $S_0 :\ 0 \leq p \leq \Lambda/2$ and $S_1 :\ \Lambda/2 \leq p \leq \Lambda$, respectively, and eliminating the short distance fluctuations in the partition function yields

$$Z = \int \prod dS_\alpha e^{-H(S_\alpha)|_g} = \int \prod dS_{\alpha 0} \prod dS_{\alpha 1} e^{-H(S_{\alpha 0}+S_{\alpha 1})|_g} = \int \prod dS_{\alpha 0} e^{-H'(S_{\alpha 0})|_{g'=Rg}}$$

where g' are the effective couplings of the effective theory and the effective theory is suitable for calculating properties of the original system at $p < \Lambda/2$. This process of lowering the cutoff by a factor of two may be iterated in order to find the long range (low energy) asymptote we are looking for.

Multi-pole forces arise in a natural way in such a scenario, for which it is important to assume "space-time" to have some arbitrary dimension $d \geq 4$ (see below). Since we assume all kind of fields and excitations living at the Planck scale, there exists a long ranged potential behaving as $\Phi \sim -\frac{1}{r^{d-2}}$ for $(r \to \infty)$. For weaker decay the thermodynamic limit would not exist. Stronger decay leads to sub-leading terms at long distances. A multi-pole expansion leads to moments of the form

$$\partial_i \Phi = (d-2)\,\frac{x_i}{r^d}\;,\quad \partial_j \partial_i \Phi = (d-2)\left\{\frac{\delta_{ij}}{r^d} - d\frac{x_i x_j}{r^{d+2}}\right\}\;,\quad \cdots$$

which naturally mediate interactions between fluctuation variables q, A_i, Q_{ij}, $\cdots$ characterized by "energy" forms:

$$H = q_1 q_2 \Phi\;,\quad -q_1 A_{2i} \partial_i \Phi\;,\quad q_1 Q_{2ij} \partial_j \partial_i \Phi\;,\quad A_{1i} A_{2j} \partial_j \partial_i \Phi\;,\quad \cdots .$$

"Charge neutrality" for large distances requires $q = 0$. Dipole–dipole interaction, for example,

$$H = -\sum K_{x-y,ik} A^i_x A^k_y$$

thus have a kernel

$$\tilde{K}_{ik}(q) = m^2 \left(d\frac{q_i q_k}{q^2} + \delta_{ik}\right) + c\,\left(-d q_i q_k + q^2 \delta_{ik}\right) + O(q^4)$$

and the propagators shows a from known from the massive gauge-boson in the 't Hooft gauge

$$\tilde{G}_{q,ij} = \left(\tilde{K}_q\right)^{-1}_{ij} = \left(\delta_{ij} - \frac{g + bq^2}{g + bq^2 + m^2 + q^2}\,\frac{q_i q_j}{q^2}\right)\frac{1}{m^2 + q^2}\;,$$

which demonstrates that spin 1 gauge bosons enter in a natural way. Note that modes are observable only if they *propagate*, which implies that the leading low q–terms $\bar{\psi}\gamma^\mu \partial_\mu \psi$, $\partial_\mu \phi \partial^\mu \phi$, $F_{\mu\nu} F^{\nu\mu}$, $\cdots$ determine the normalization (wave function renormalization) and this fixes the rules for dimensional counting. Also note that by "renormalization" of the Bose fields and the couplings one can always arrange the q^2–term in the bilinear part to be Euclidean invariant (Liu-Stanley theorem). A similar statement should hold for fermions. A detailed investigation of possible low energy structures is very elaborate. Here we mention a simplified Ansatz (assumed to), which was discussed as a way to derive non–Abelian gauge–theories from "tree unitarity" requirements [6]. For simplicity we assume an Euclidean invariant action at the Planck scale. Consider only three types of particle species: scalars ϕ_a, fermions ψ_α and vector-bosons $W_{i\mu}$ with covariant propagators. Since here we do not refer to tree unitarity but to low energy expansion (IR power–counting) we need consider only terms which are not manifestly irrelevant

$$\begin{aligned}\mathcal{L} &= \bar{\psi}_\alpha \left\{ L^i_{\alpha\beta} P_- + R^i_{\alpha\beta} P_+ \right\} \gamma^\mu \psi_\beta W_{\mu i} + \tfrac{1}{2} D_{ijk} W^k_\mu \left(W^j_\alpha \partial_\mu W^{\alpha i} - W^i_\alpha \partial_\mu W^{\alpha j} \right) \\ &+ \bar{\psi}_\alpha \left\{ C^{+b}_{\alpha\beta} P_+ + C^{-b}_{\alpha\beta} P_- \right\} \psi_\beta \phi^b + \tfrac{1}{2} K^b_{ij} W^i_\mu W^j_\mu \phi_b \\ &+ \tfrac{1}{2} T^i_{ba} W^i_\mu \left(\phi_a \partial_\mu \phi_b - \phi_b \partial_\mu \phi_a \right) + \tfrac{1}{4} M^{ij}_{ab} W_{\mu i} W_{\mu j} \phi_a \phi_b\end{aligned}$$

with arbitrary interaction matrices of the fields. The extraction of the leading low energy asymptote is equivalent to the requirement of renormalizability of S–matrix elements, and this has been shown to necessarily be a non–Abelian gauge theory which must have undergone a Higgs mechanism if the gauge bosons are not strictly massless. Since terms of order $O(E/\Lambda_P)$ are automatically suppressed in the low energy regime only a renormalizable effective field theory can survive as a tail, the possible renormalizable theories on the other hand are known and are easy to classify. Thus gauge symmetries and in particular the non–Abelian ones appear as a conspiracy of different modes "self–arranged" in such a way the $O((E/\Lambda_P)^n)$–terms $(n > 0)$ are absent. Also anomaly–cancelation and the related quark–lepton duality (family structure) are easily understood and natural in such a context. To an accuracy of $E/\Lambda_P = 10^{-3}$ we are thus dealing with a renormalizable local QFT of the "spontaneously broken gauge theory" (SBGT) type at a scale 10^{16} GeV. As we shall argue below this cannot be just the SM.

The fact that there are only a few possible forms for the low energy effective theories is particularly attractive and tells us that symmetries and particular mathematical properties may be interpreted to emerge as low energy patterns.

The low energy expansion in terms of field monomials makes sense only in the vicinity of a second (or higher) order phase transition point where the system exhibits long range correlations and is described in the long range limit by an effective conformal quantum field theory characterized by an infrared stable fixed point. It is well known that for dimensions $d > 4$ such effective theories turn out to be Gaussian (free field) theories. Non–trivial theories with a stable ground state are possible only for dimensions $d \leq 4$. At the boarder case $d = 4$ non-trivial long range interactions set in and we expect effective couplings to be weak and therefore perturbation theory to work[2]. The quasi–triviality but non-triviality in this scenario is due to the fact that a huge but finite cutoff, namely Λ_P, exist in the underlying physics. Such a scenario explains why elementary particle interactions, up to scales explored so far, are described by renormalizable quantum field theory and why we can do perturbation theory. Note that space-time "compactifies itself" by the decoupling of the $n = d - 4$ extra dimensions.

In $d = 4$ space-time dimensions there exist infinitely many infrared "irrelevant" (non–renormalizable) operators of dimension > 4 (scaling like $(E/\Lambda_P)^n$ with $n \geq 1$); but there exist only relatively few infrared "marginal" (strictly renormalizable) dimension 4 operators (scaling like $\ln^n(E/\Lambda_P)$ with $n \geq 1$) and even fewer infrared "relevant" (super–renormalizable) operators of dimension < 4 (scaling like $(\Lambda_P/E)^n$ with $n \geq 1$). The dimension ≤ 4 operators characterize a renormalizable QFT. The relevant operators must be tuned for criticality in order that the low energy expansion makes sense. This fine tuning is of course the main obstacle for such a vision to be convincing. Unlike in a condensed matter physics laboratory we cannot tune by hand the temperature and the external fields for criticality. However, since we have good reasons to assume that there exist a dense variety of fluctuations and modes it is conceivable that at long distances we just see those modes which "conspire" precisely in such a way as to allow for long distance fluctuations. This conspiracy is nothing but the "symmetry patterns" which emerge at long distances. Other existing modes just are frozen at short distances and are not observable. The "critical dimension" $d = 4$ is crucial for the scenario to work because weakly interacting large scale fluctuations are expected to govern the quasi critical region. As we mentioned above, gauge symmetries are particularly easy to understand within this context. The gauge groups expected in such a scenario of course are the ones which follow by conspiracy of "particles" in singlets, doublets, triplets, etc. exactly as we observe them in the real world. Thus, while a $U(1) \otimes SU(2) \otimes SU(3) \otimes \cdots$ pattern looks to emerge in a natural way, we never would expect higher dimensional multiplets to show up if the particular symmetry would not be there already a the Planck scale. Also the repetition of fermion family patterns, known in the SM, looks to be a rather natural possibility in our approach.

3 Natural properties at low energies

Above we outlined that known empirical facts about structural properties of elementary particle theory find a natural explanation in low energy effective theories. Usually quantum mechanics and special relativity, four dimensionality and renormalizability are independent inputs. Detailed investigations confirm that all these properties may be understood as consequences of the existence of an "ether" in the appropriate universality class. At long distances we observe: 1.) Local quantum field theory; note that the equivalence of Euclidean QFT and Minkowski QFT is a general property of any renormalizable QFT (Osterwalder–Schrader theorem). The analyticity properties allow for the necessary Wick rotation. In its Minkowski version QFT incorporates quantum mechanics and special relativity, which thus show up as low energy structures more or less automatically[5]. 2.) Space–time dimension $d = 4 = 3 + 1$. 3.) Interactions are renormalizable and thus described by a Lagrangian which includes low dimensional monomials of fields only. 4.) Weak coupling and perturbative nature of elementary particle interactions; this is natural only if we require the low energy effective QFT in $d = 4$ to exhibit a trivial (Gaussian) IR fixed point, which is the natural candidate for the low energy effective theory to stabilize. Since in addition we have to require the SM to part of it we call it "Gaussian extended SM" (GESM). It is weakly interacting at low energies due to the existence of the large but finite cutoff. 5.) Local gauge symmetries with small gauge groups; they provide the dynamical principle which fixes the interactions of the SM and of Einstein's theory of gravity (equivalence principles). 6.) The existence of a large finite physical cutoff implies that the relationship between bare and renormalized quantities are physical.

Basis of all this is the equivalence of statistical mechanics near criticality and quantum field theory. There are only a few low range theories possible (conformal quantum field theories characterized by a few properties like global symmetries, dimension etc.) Also the equivalence of the path integral quantization and the canonical quantization is more than an accident within this context.

Further consequences are briefly discussed in the following: i) Since we need a quasi Gaussian IR fixed point, asymptotic freedom as seen in the SM must be lost at higher energies; this requires $N \geq 9$ families to exist. The asymptotic freedom of QCD and in the $SU(2)_L$ coupling are a consequence of the decoupling of the heavier fermions. ii) Since the relationship between bare and renormalized parameters must be physical, positivity of counter terms etc. must be required, which has direct consequences for vacuum stability and positivity of both the bare and the renormalized Higgs potential, for example. iii) QFT properties are expected to be violated once $E/\Lambda_P,\ (E/\Lambda_P)^2, \cdots$ terms come into play; at $E \simeq 10^{16}$ GeV one might expect 0.1% effects. This might be important to remember in the attempts to solve the puzzle of baryogenesis, for example.

The simplest GESM may be obtained by adding more (heavier) fermion families to the SM. For a SM with N families the one–loop counter terms for the $U(1)_Y$, $SU(2)_L$ and $SU(3)_c$ couplings read $\delta g/g = c_g \ln(\Lambda^2/\mu^2)$ with $c_{g'} = \frac{g'^2}{24\pi^2}(5/3N+1/8)$, $c_g = \frac{g^2}{24\pi^2}(N-11/2+1/8)$ and $c_{g_s} = \frac{g_s^2}{16\pi^2}(4/3N-11)$, respectively. The corresponding β–functions must all be positive (IR fixed point condition) in the weak coupling limit. For the Abelian coupling we have $c_{g'} > 0$ in any case. For the non-Abelian couplings $c_g > 0$ provided $N \geq 6$ and $c_{g_s} > 0$ provided $N \geq 9$, i.e. they must be matter dominated. Note that for the unbroken $U(1)_{\rm em}$ $c_e = \sin^2\Theta_W c_g + \cos^2\Theta_W c_{g'} = \frac{e^2}{24\pi^2}\,(8/3\ N - 11/2 + 1/4) > 0$ for $N \geq 2$. In our scheme there is a prediction for $\tan^2\Theta_{W\rm eff} = (g'^2/g^2)_{\rm eff} = \frac{g'^2(1+2\delta g'/g')}{g^2(1+2\delta g/g)} \sim \frac{24N-129}{40N+3}$. It is positive only provided $N \geq 6$ and we obtain $\sin^2\Theta_W$=0.05814,0.16321,0.19333,0.23356 and 0.37500 for N=6,8,9,11 and ∞. Note that more realistic estimates must include appropriate threshold/decoupling effects. Of course the existence of additional fermion families is possible, although additional light neutrinos are excluded as we know.

We finally have to worry about the quadratic divergences i.e. the tuning of the relevant parameters for criticality. In the SM, utilizing dimensional regularization, the quadratic divergences show up as poles at $d = 2$ and this solely concerns the Higgs mass counter term [7] $\delta m_H^2 = \frac{1}{16\pi^2 v^2}\{A_0(m_H)\ 3m_H^2 + A_0(M_Z)\ (m_H^2 + 6M_Z^2) + A_0(M_W)\ (2m_H^2 + 12M_W^2) + \sum_{f_s} A_0(m_f)\ (-8m_f^2) + \cdots\}$ where $A_0(m) = \Lambda^2(m^2/\mu^2)^{(d/2-1)}\,(4\pi)^{-d/2}\,\Gamma(1-d/2)$ and thus $\delta m_H^2 \sim 6(\Lambda/v)^2(m_H^2 + M_Z^2 + 2M_W^2 - 4m_f^2)$ and the IR fixed point condition requires $m_H \simeq (4(m_t^2+m_b^2)-M_Z^2-2M_W^2)^{1/2} \sim$ 318 GeV[8]. This lowest non–trivial order consideration seems to predict a reasonable value of the Higgs mass. Since we are in a perturbative regime higher order perturbative corrections modify the precise value of the prediction but they cannot affect the existence of a solution which is not ruled out by experiment. Actually, current precision measurements very strongly suggest that the Higgs coupling is fairly weak (SM fits favor $m_H < 420$ GeV at 95 % C.L. and thus $\lambda = m_H^2/(2v^2) < 1.5$

which leads to an expansion parameter about $\alpha_\lambda = \lambda^2/(4\pi) \sim 0.18$). In fact the symmetry which constrains the scalar mass here is dilatation invariance. Of course, like in SUSY theories, the cancelations of the contributions is only possible between fermionic and bosonic degrees of freedom. This prediction should not be taken too serious. First of all it does not include the effects from the extra heavy fermion families which must exist in this scheme. More serious is the expectation that such a result cannot be universal, it is expected to depend on the actual structure of the "bare" theory, which is unknown. On the other hand renormalizable SBGT is in effect up to energies of about 10^{16} GeV and below that standard gauge invariance and RG arguments apply.

At this point we should remember 't Hooft's naturalness argument: "Small" masses are natural only if setting them to zero increases the symmetry of the system[4]. Indeed a light particle spectrum (IR relevant terms) must be the result of a "conspiracy" i.e. modes conspire to form approximately multiplets of some symmetry which protects the masses from large renormalizations: light fermions require approximate chiral symmetry, light vector bosons require approximate local gauge symmetry, light scalars require approximate super symmetry.

The view developed in the previous sections has to be worked out in more details in many respects. There are many open problems, for example, concerning the origin of fermions. I think this is a promising framework which should be considered seriously. One big advantage is that it has non–trivial phenomenological consequences which are testable in the not too far future.

References

[1] L. D. Landau, in "Niels Bohr and the Development of Modern Physics", ed. W. Pauli, Mc Graw-Hill, New York, 1955, p.52

[2] K. Wilson, *Phys. Rev.* **B4** (1971) 2174, ibid. 3184
K. Wilson, M.E. Fisher, *Phys. Rev. Lett.* **28** (1972) 240;
K. Wilson, J. Kogut, *Phys. Rept.* **12** (1974) p. 75

[3] F. Jegerlehner, *Helv. Phys. Acta* **51** (1978) 783

[4] G. 't Hooft, in "Recent Developments in Gauge Theories", G. 't Hooft et al. (eds.), Plenum Press, New York, 1980, p. 135

[5] F. Jegerlehner, in "Trends in Elementary Particle Theory", eds. H. Rollnik, K. Dietz, Springer, Berlin, 1975, p.114, J. Kogut, *Rev. Mod. Phys.***51** (1979) 659

[6] J.M. Cornwall, D.N. Levin, G. Tiktopoulos, *Phys. Rev.* **D10** (1974) 1145.

[7] J. Fleischer, F. Jegerlehner, *Phys. Rev.* **D23** (1981) 2001

[8] M. Veltman, *Acta Phys. Polonica* **B12** (1981) 437

$N = 1$ Finite Unified Theories Predictions and Dualities

George Zoupanos

Physics Department, National Technical University,
Zografou, GR-15780 Athens, Greece.
and
Institut f. Physik, Humboldt-Universität,
D10115 Berlin, Germany

Abstract: $N = 1$, all-loop Finite Unified Theories (FUTs) are very interesting not only since they realize an old theoretical dream, but also due the remarkable predictive power of particular models as well as for providing candidates that might shed light in non-perturbative Physics. Here we discuss (a) the recent developments concerning the soft supersymmetry breaking (SSB) sector of these theories and the resulting predictions in very interesting realistic models, and (b) the results of a recent search for duals of $N = 1$, all-loop FUTs.

1 Introduction

Finiteness is an outstanding issue in the various theoretical endeavors that attempt to achieve a deeper understanding of Nature. Most theorists, quite independently from the school of thinking they belong to, tend to believe that the divergencies of ordinary field theory are not of fundamental nature, but rather they are signaling the existence of New Physics at higher scales. Therefore, it is very natural to believe that the final theory, which hopefully unifies all interactions, should be completely finite. Given the searches of strings, non-commutative geometry and q-groups that aim, among others, to achieve finiteness, it is worth stressing that *finiteness does not require necessarily gravity.* The latter statement is well established in gauge theories with extended supersymmetry. For instance, it is well known that all $N = 4$ and some $N = 2$ supersymmetric gauge theories are free from ultraviolet (UV) divergencies at all-orders in perturbation theory (PT). Of particular interest is the existence of N = 1 supersymmetric gauge theories [2, 5], which are finite at all-orders in PT and the construction in that framework of a realistic finite $SU(5)$ GUT, which succesfully has predicted, among others, the top quark mass [1].

In the following we shall first report on the recent progress that has been done, and which increases appreciably the predictive power of the models that can be constructed and provides us with new predictions for the Higgs particles masses and the s-spectrum, which were not available in the previous attempts.

The discussion about finite $N = 1$ gauge theories till recently was limited to perturbative aspects of these theories while non-perturbative problems, like their bound state spectrum, were left open. The knowledge of the perturbative properties were sufficient for the study of GUTs. However, the recent progress that has been done might permit also studies at strong couplings. The basic idea that could allow us to address non-perturbative problems in these theories is the hope that there exist some kind of electric-magnetic, strong-weak coupling duality, like the one that is believed to be exhibited in $N = 4$ supersymmetric gauge theories. Moreover, the low energy effective action of $N = 2$ supersymmetric gauge theories can be even solved using duality and holomorphy [8]. Finally, $N = 1$ supersymmetric gauge theories, which can describe realistically the so far observed world, exhibit a weaker version of duality symmetry [9], namely theories differing in the UV can describe the same Physics in the infrared (IR).

In the following we shall also report on a systematic search for duals of $N = 1$, all-loop finite gauge theories [7]. This search covers most of the chiral finite gauge theories including the realistic $SU(5)$ model mentioned earlier and several vector-like models. Surpisingly, we found that the duals of the $N = 1$, all-loop finite gauge theories are usually not finite, with the exception of an $SO(10)$ theory which emerges from this search as a candidate to exhibit also S-duality.

2 Finiteness and Reduction of Couplings in $N = 1$ SUSY Gauge Theories

Consider a chiral, anomaly free, $N = 1$ globally supersymmetric gauge theory based on a group G with gauge coupling constant g. The superpotential of the theory is given by

$$W = \frac{1}{2} m^{ij} \Phi_i \Phi_j + \frac{1}{6} C^{ijk} \Phi_i \Phi_j \Phi_k \ , \tag{1}$$

where m^{ij} and C^{ijk} are gauge invariant tensors and the matter field Φ_i transforms according to the irreducible representation R_i of the gauge group G.

The one-loop β-function of the gauge coupling g is given by

$$\beta_g^{(1)} = \frac{dg}{dt} = \frac{g^3}{16\pi^2} [\sum_i l(R_i) - 3\, C_2(G)] \ , \tag{2}$$

where $l(R_i)$ is the Dynkin index of R_i and $C_2(G)$ is the quadratic Casimir of the adjoint representation of the gauge group G. The β-functions of C^{ijk}, by virtue of the non-renormalization theorem, are related to the anomalous dimension matrix γ_i^j

of the matter fields Φ_i as:

$$\beta_C^{ijk} = \frac{d}{dt} C^{ijk} = C^{ijp} \sum_{n=1} \frac{1}{(16\pi^2)^n} \gamma_p^{k(n)} + (k \leftrightarrow i) + (k \leftrightarrow j) \,. \tag{3}$$

At one-loop level γ_i^j is

$$\gamma_i^{j(1)} = \frac{1}{2} C_{ipq} C^{jpq} - 2\, g^2\, C_2(R_i) \delta_i^j \,, \tag{4}$$

where $C_2(R_i)$ is the quadratic Casimir of the representation R_i, and $C^{ijk} = C^*_{ijk}$.

As one can see from Eqs. (2) and (4) all the one-loop β-functions of the theory vanish if $\beta_g^{(1)}$ and $\gamma_i^{j(1)}$ vanish, i.e.

$$\sum_i \ell(R_i) = 3C_2(G) \,, \qquad \frac{1}{2} C_{ipq} C^{jpq} = 2\delta_i^j g^2 C_2(R_i) \,. \tag{5}$$

A very interesting result is that the conditions (5) are necessary and sufficient for finiteness at the two-loop level.

A natural question to ask is what happens at higher loop orders. The finiteness conditions (5) impose relations between gauge and Yukawa couplings. We would like to guarantee that such relations leading to a reduction of the couplings hold at any renormalization point. The necessary, but also sufficient, condition for this to happen is to require that such relations are solutions to the reduction equations (REs)

$$\beta_g \frac{dC^{ijk}}{dg} = \beta^{ijk} \tag{6}$$

and hold at all-orders. Remarkably the existence of all-order power series solutions to (6) can be decided at the one-loop level [4].

A very interesting theorem [2] guarantees the vanishing of the β-functions to all-orders in perturbation theory, if we demand reduction of couplings, and that all one-loop anomalous dimensions of the matter fields in the completely and uniquely reduced theory vanish identically.

3 All-Loop Finite Unified Theories

A predictive Finite Unified $SU(5)$ model which is finite to all-orders, in addition to the requirements mentioned already, should also have the following properties:

1. One-loop anomalous dimensions are diagonal, i.e., $\gamma_i^{(1)\,j} \propto \delta_i^j$.
2. Three fermion generations, $\overline{\mathbf{5}}_i$ $(i = 1, 2, 3)$, obviously should not couple to $\mathbf{24}$. This can be achieved for instance by imposing $B - L$ conservation.
3. The two Higgs doublets of the MSSM should mostly be made out of a pair of Higgs quintet and anti-quintet, which couple to the third generation.

In the following we discuss two versions of the all-order finite model.

A: The model of ref. [1].
B: A slight variation of the model **A**.

The superpotential which describe the two models takes the form [1, 6]

$$W = \sum_{i=1}^{3}\left[\,\frac{1}{2}g_i^u\,\mathbf{10}_i\mathbf{10}_i H_i + g_i^d\,\mathbf{10}_i\overline{\mathbf{5}}_i\,\overline{H}_i\,\right] + g_{23}^u\,\mathbf{10}_2\mathbf{10}_3 H_4 \quad (7)$$
$$+g_{23}^d\,\mathbf{10}_2\overline{\mathbf{5}}_3\,\overline{H}_4 + g_{32}^d\,\mathbf{10}_3\overline{\mathbf{5}}_2\,\overline{H}_4 + \sum_{a=1}^{4} g_a^f\,H_a\,\mathbf{24}\,\overline{H}_a + \frac{g^\lambda}{3}\,(\mathbf{24})^3\,,$$

where H_a and $\overline{H}_a$ $(a = 1,\ldots,4)$ stand for the Higgs quintets and anti-quintets.

The non-degenerate and isolated solutions to $\gamma_i^{(1)} = 0$ for the models $\{\mathbf{A}\,,\,\mathbf{B}\}$ are:

$$(g_1^u)^2 = \{\frac{8}{5},\frac{8}{5}\}g^2\,,\ (g_1^d)^2 = \{\frac{6}{5},\frac{6}{5}\}g^2\,,\ (g_2^u)^2 = (g_3^u)^2 = \{\frac{8}{5},\frac{4}{5}\}g^2\,, \quad (8)$$
$$(g_2^d)^2 = (g_3^d)^2 = \{\frac{6}{5},\frac{3}{5}\}g^2\,,\ (g_{23}^u)^2 = \{0,\frac{4}{5}\}g^2\,,\ (g_{23}^d)^2 = (g_{32}^d)^2 = \{0,\frac{3}{5}\}g^2\,,$$
$$(g^\lambda)^2 = \frac{15}{7}g^2\,,\ (g_2^f)^2 = (g_3^f)^2 = \{0,\frac{1}{2}\}g^2\,,\ (g_1^f)^2 = 0\,,\ (g_4^f)^2 = \{1,0\}g^2\,.$$

According to the theorem of ref. [2] these models are finite to all-orders. After the reduction of couplings the symmetry of W is enhanced [6].

The main difference of the models **A** and **B** is that three pairs of Higgs quintets and anti-quintets couple to the **24** for **B** so that it is not necessary to mix them with H_4 and $\overline{H}_4$ in order to achieve the triplet-doublet splitting after the symmetry breaking of $SU(5)$.

4 Supersymmetry Breaking and Predictions of Low Energy Parameters

The above models are completed as realistic theories by adding SSB terms as follows:

$$-\mathcal{L}_{\rm SB} = \frac{1}{6}\,h^{ijk}\,\phi_i\phi_j\phi_k + \frac{1}{2}\,b^{ij}\,\phi_i\phi_j + \frac{1}{2}\,(m^2)^j_i\,\phi^{*\,i}\phi_j + \frac{1}{2}\,M\,\lambda\lambda + \text{h.c.}\,, \quad (9)$$

where the ϕ_i are the scalar parts of the chiral superfields Φ_i , λ are the gauginos and M their unified mass.

Concerning the supersymmetry breaking sector of the theory it has been recently shown that the requirement of two-loop finiteness of SSB in a finite gauge theory leads to

a. the relation $h^{ijk} = -MY^{ijk}$ and,

b. the soft scalar-mass sum rule $(m_i^2 + m_j^2 + m_k^2)/MM^\dagger = 1 + g^2/16\pi^2\Delta^{(1)}$ for i,j,k with $C^{ijk} \neq 0$, where $\Delta^{(1)}$ is the two-loop correction, which vanishes for the universal choice [3], but also in the considered models without universal masses.

Since the gauge symmetry is spontaneously broken below $M_{\rm GUT}$, the finiteness conditions do not restrict the renormalization property at low energies, and all it remains are boundary conditions on the gauge and Yukawa couplings (8), the $h = -MC$ relation and the soft scalar-mass sum rule at $M_{\rm GUT}$. We have examined the evolution of these parameters according to their renormalization group equations at two-loop for dimensionless parameters and at one-loop for dimensional ones with these boundary conditions. Below $M_{\rm GUT}$ their evolution is assumed to be governed by the MSSM. It is further assumed a unique supersymmetry breaking scale M_s so that below M_s the SM is the correct effective theory.

The predictions for the top quark mass M_t are ~ 183 and ~ 174 GeV in models **A** and **B** respectively. Comparing these predictions with the most recent experimental value $M_t = (175.6 \pm 5.5)$ GeV, and recalling that the theoretical values for M_t may suffer from a correction of less than $\sim 4\%$ [5], we see that they are consistent with the experimental data.

Using the sum rule we can now determine the spectrum of realistic models in terms of just a few parameters. In addition to the successful prediction of the top quark mass the characteristic features of the spectrum are that 1) the lightest Higgs mass is predicted ~ 120 GeV and 2) the s-spectrum starts above 200 GeV. Therefore, the next important test of these Finite Unified Theories will be given with the measurement of the Higgs mass, for which the models show an appreciable stability in their prediction.

5 Dualities of Finite Gauge Theories

In ref. [7] we have done an extensive search for dual gauge theories of all-loop finite, $N = 1$ supersymmetric gauge theories. It is shown how to find explicitly the dual gauge theories of almost all chiral, $N = 1$, all-loop finite gauge theories, while several models have been discussed in detail, including the realistic finite $SU(5)$ unified theory of ref. [1]. As we have seen one- and two-loop finiteness of a gauge theory is guaranteed by, first, choosing the particle content such that the one-loop gauge β-function vanishes, and subsequently, by adding a superpotential such that all one-loop matter field anomalous dimensions are zero. Furthermore, the all-loop finiteness requires that the relations among gauge and Yukawa couplings, obtained by demanding the vanishing of the one-loop anomalous dimensions, should be unique solutions of the reduction equations, which in turn guarantees that they can be uniquely determined to all-orders in PT. Therefore, using established methods for searching for duals à la Seiberg, we have examined almost all known $N = 1$ supersymmeric gauge theories (with the exception of E_6 models and $SO(10)$ containing anti-spinors) and vanishing one-loop β-functions. These theories have, first, been promoted to all-loop finite ones, by adding appropriate superpotential and then by meeting the requirements of all-loop finiteness. In addition, certain vector-like, all-loop finite, $N = 1$ gauge theories and their duals have been discussed in the standard field theory framework but also by using the derivation of gauge theories from branes. However, the brane picture still encounters several difficulties in the corresponding hunting for finite gauge theories.

The result of our search is that the duals à la Seiberg of all-loop finite gauge theories are asymptotically free. In certain cases looking to the IR limit of the theories, where both theories of the dual pair are describing the same physics, we found that the spontaneously broken theories seem superficially asymptotically non-free (as the $U(1)$ of the Standard Model embedded in a GUT).

There is only one model which is one-loop finite in the IR. The model is based on the gauge group $SO(10)$ and has matter content consisting of 8 vector and 8 spinor superfields. The dual of this theory is based on the gauge group $SU(17) \times SP(14)$. The first factor is asymptotically free, while the second is one-loop finite. However, in the IR, after spontaneous symmetry breaking of the $SU(17)$ down to $SU(9)$, both gauge factors are one-loop finite. We found evidence that the IR dual is all-loop finite and therefore the dual pair is a candidate to exhibit $N = 1$ S-duality.

The dual of the realistic finite unified theory [1] based on $SU(5)$ has been determined and discussed in some detail. However, the resulting dual theory is rather complicated and it does not give, so far, any hint for any useful use of it. On the other hand, we should note that the dual was constructed by using the deconfinement method [10], which does not lead to unique results. Therefore, we should not exclude the possibility that more interesting duals can be constructed.

6 Conclusions

The search for realistic Finite Unified theories started a few years ago [1, 5] with the successful prediction of the top quark mass, and it has now been complemented with a new important ingredient concerning the finiteness of the SSB sector of the theory. Specifically, a sum rule for the soft scalar masses has been obtained which quarantees the finiteness of the SSB parameters up to two-loops [6], avoiding at the same time serious phenomenological problems related to the previously known "universal" solution. It is found that this sum rule coincides with that of a certain class of string models in which the massive string modes are organized into $N = 4$ supermultiplets.

Motivated from the recent developments on dualities of gauge theories with extended supersymmetry we have been searching for candidates with S-duality among the $N = 1$ supersymmetric gauge theories, which have richer dynamics and are much more promissing in describing the real world, as the one discussed above. The strategy was to look for duals à la Seiberg of all-loop, $N = 1$ FUTs which are FUTs too and therefore exhibiting S-duality. From our search, one chiral, $N = 1$, all-loop FUT has been singled out giving promises of having S-duality as the $N = 4$ gauge theories.

7 Acknowledgements

It is a pleasure to thank the Organizing Committee for the warm hospitality. Work supported partly by the projects FMBI-CT96-1212 and ERBFMRXCT960090, and PENED95/1170;1981.

References

[1] D. Kapetanakis, M. Mondragón and G. Zoupanos, *Zeit. f. Phys.* **C60** (1993) 181; M. Mondragón and G. Zoupanos, *Nucl. Phys.* **B** (Proc. Suppl) **37C** (1995) 98.

[2] C. Lucchesi, O. Piguet and K. Sibold, *Helv. Phys. Acta* **61** (1988) 321; *Phys. Lett.* **B201** (1988) 241.

[3] I. Jack and D.R.T. Jones, *Phys. Lett.* **B333** (1994) 372.

[4] W. Zimmermann, *Com. Math. Phys.* **97** (1985) 211; R. Oehme and W. Zimmermann *Com. Math. Phys.* **97** (1985) 569.

[5] For an extended discussion and a complete list of references see: J. Kubo, M. Mondragón and G. Zoupanos, *Acta Phys. Polon.* **B27** (1997) 3911.

[6] T. Kobayashi, J. Kubo, M. Mondragón and G. Zoupanos, *Constraints on Finite Soft SUSY Breaking Terms*, hep-ph/9707425, to be published in *Nucl. Phys.* **B**.

[7] A. Karch, D. Lüst, G. Zoupanos, *Dualities in all-order finite $N = 1$ gauge theories*, hep-th/9711157.

[8] N. Seiberg and E. Witten, *Nucl. Phys.* **B426**, (1994) 19, hep-th/9407087; *Nucl. Phys.* **B431** (1994) 484, hep-th/9408099.

[9] N. Seiberg, *Nucl. Phys.* **B435** (1995) 129, hep-th/9411149.

[10] M. Berkooz, *Nucl. Phys.* **B452** (1995) 513, hep-th/9505067.

Status of the electroweak phase transitions

Z. Fodor

KEK Theory Division
Tsukuba, 1-1 Oho, Ibaraki, 305, JAPAN

Abstract: The observed baryon asymmetry of the universe has finally been determined at the finite temperature electroweak phase transition. In order to understand the baryon asymmetry a quantitative description of the electroweak phase transition is needed. In this talk some features of this phase transition are summarized. Particular interest is paid to the analytical and lattice estimates on the sphaleron transition and to the nature of the electroweak phase transition in the standard model and in the minimal supersymmetric standard model. On the one hand due to the large top-quark mass no SM-like Higgs boson can give strong enough phase transition. Thus, in the standard model the electroweak baryogenesis scenario can be excluded. On the other hand there is a small region in the parameter space of the minimal supersymmetric standard model, which might explain the observed baryon asymmetry. Since this small region is predicted perturbatively a lattice confirmation is needed.

1 Introduction

According to the standard picture of the cosmological phase transitions at high temperatures (e.g. in the early universe) the electroweak gauge symmetry has been restored. As the universe expands and supercools there is a phase transition between the high temperature "symmetric" and low temperature "broken" phases. The characteristics of this phase transition (critical temperature: T_c, surface tension: σ, etc.) are clearly of interest. There are evidences that the world is made of exclusively matter, at least on the $10^{13} M_{\odot}$ scale. The quantitative measure of the baryon asymmetry is the dimensionless ratio of the baryon number density to the entropy density

$$\Delta_B = B/s \approx 5 \cdot 10^{-11}. \tag{1}$$

Since no known mechanism can separate on the above huge scales we have to understand the origin of this asymmetry. There are basically two possibilities. The baryon asymmetry might have been an initial condition of our universe or it could have been generated in the early universe. The second possibility (baryogenesis) is clearly more attractive.

In order produce the observed baryon asymmetry three conditions must be satisfied.
a. baryon number violating processes
b. C and CP violation
c. departure from the thermal equilibrium (the basic question is the strength of the phase transition).
It is easy to understand the role of these conditions.
a. In the lack of baryon number violation the baryoniC charge of the universe were constant. Assuming zero baryon number as an initial condition would result in no baryon asymmetry today.
b. If C or CP were conserved, then the rate of processes producing baryons (a) would be the same as the rate of processes producing antibaryons. No net baryon number could have been generated.
c. In thermal equilibrium the universe stationary, it has no time dependence. If the initial condition were zero baryon number, it would remain zero forever.

An attractive proposal is to understand the observed baryon asymmetry as produced at the finite temperature electroweak phase transition [1]. Section 2 will discuss the first (a) Section 3 the last (c) condition in this scenario. Section 4 contains the conclusion.

2 Sphaleron rate at finite temperature

The basic source of the baryon number violating processes in the standard model is the electroweak anomaly for baryon number, which demands that

$$\Delta B \sim g^2 \int dt d^3x \mathrm{tr} F\tilde{F}, \tag{2}$$

where g is the weak coupling and F is the field strength of the weak gauge field. This anomaly equation relates the produced baryon number to topological transitions of the weak gauge fields. The vacua with different topological numbers (N_{CS}) are separated by a potential barrier.

Note, that the above anomaly equation holds not only for baryons but for leptons, too. Thus, baryon and lepton numbers are violated simultaneously ($B+L$ violation).

At zero temperature the transition between the vacua is a tunneling event with an unobservably small probability: $\approx 10^{-170}$. The transition at high temperatures, but below T_c is a thermal jump. The system jumps up to the top of the barrier (a saddle point: sphaleron) and rolls down to the neighbouring vacuum. In the meantime baryon and lepton numbers are violated. The jump to the barrier, therefore the transition rate is suppressed by the Boltzmann factor

$$\Gamma \propto \exp\left[-E_{sph}/T\right], \tag{3}$$

where the sphaleron energy is proportional to the mass of the W-boson $E_{sph} \propto m_W/\alpha_W$ [2].

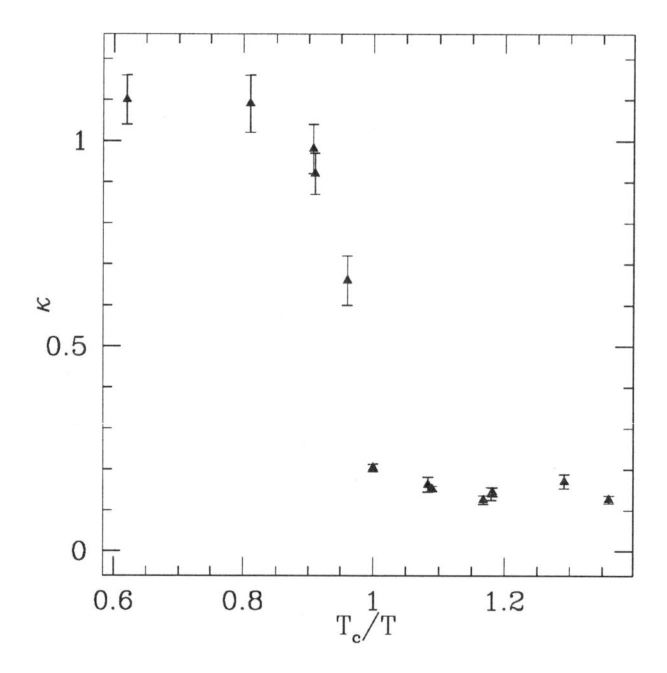

Figure 1: Transition rate normalized by T^4 as a function of the temperature.

Rapid baryon violation processes in the broken phase can wash out the asymmetry generated before. Therefore a "minimal" condition for a successful baryogenesis is that these processes should be slower than the expansion rate of the universe. Comparing (3) with the Hubble constant the minimal condition can be written as $\varphi_b/T_b > 1$, where φ_b is the value of the Higgs field at the transition temperature T_b.

Above T_c the exponential suppression of the transition rate is absent. Naive power counting suggest

$$\Gamma = \kappa \cdot (\alpha_w T)^4, \tag{4}$$

where the constant κ is of order 1.

Recently it has been argued [3] that the assumptions used to derive 4 are not valid. The real dependence on α_w is a different one

$$\Gamma = \kappa' \cdot \alpha_w^5 T^4. \tag{5}$$

The origin of this damping is the fact that nearly static magnetic fields can be absorbed by the system, which results in a loss of energy for the magnetic fields of interest. In the symmetric phase the baryon number violating processes are much faster than the expansion rate of the universe.

In recent years there has been considerable activity in order to determine the sphaleron transition rate at both sides of the phase transition. No successful numerical method is known for a full Minkowskian theory; however, important results have

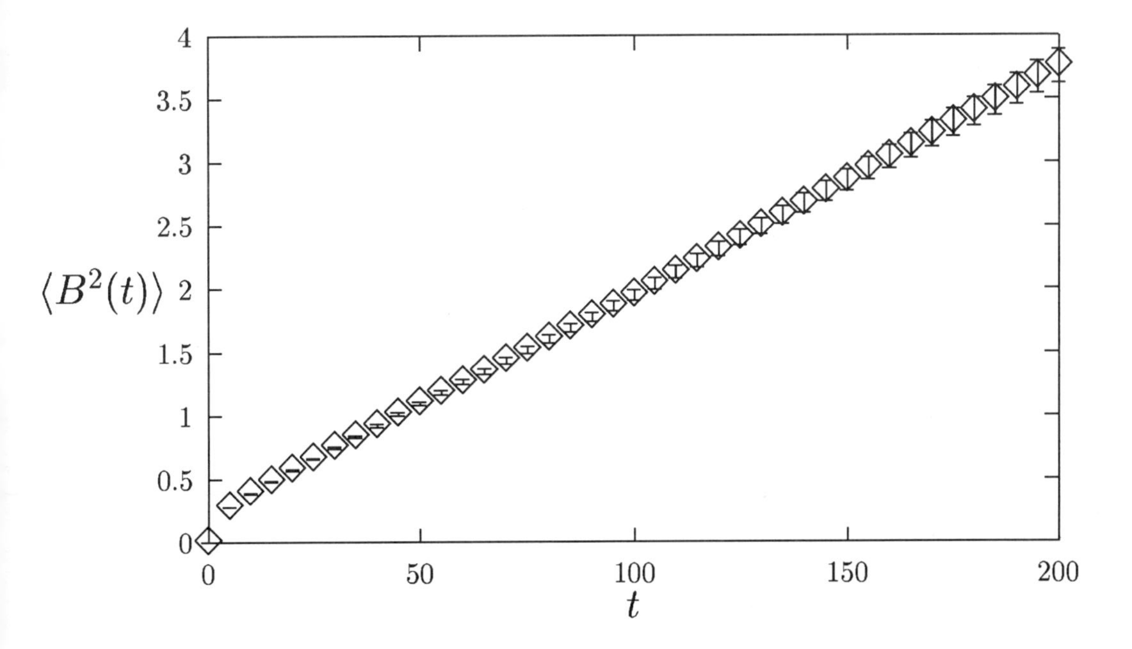

Figure 2: Diffusion of N_{CS}.

been obtained by real time simulations in the classical approximation for the finite temperature theory.

The used procedure contains several steps. The classical theory is formulated in terms of fields and their conjugate momenta on a spatial lattice. The phase-space of the system is then sampled with the statistical weight $\exp(-H/T)$. Starting with some initial configuration from the thermal sample the classical canonical equations give the real-time evolution of the fields. As a function of time one can determine different observables and their averages (such as changes of the topological charge). There are fewer physical degrees of freedom in the theory than the number of the phase space variables used to formulate it. It is a highly non-trivial task to find effective thermalization algorithms for constrained systems. At present there are two good solutions to this problem [4].

In the broken phase numerical simulations [5] indicated an extreme difference between the lattice results and (3) The observed sphaleron transition rate was a bit smaller than in the symmetric phase; however, as it can be seen on Fig. 1 no Boltzmann suppression has been observed. Going down into the broken phase the difference between the analytical and lattice results is several orders of magnitude. It is argued [3] that the reason is the problematic definition of the topological charge on a lattice, which give a systematic error in the rate.

A large scale numerical study for the sphaleron rate in the symmetric phase [6]

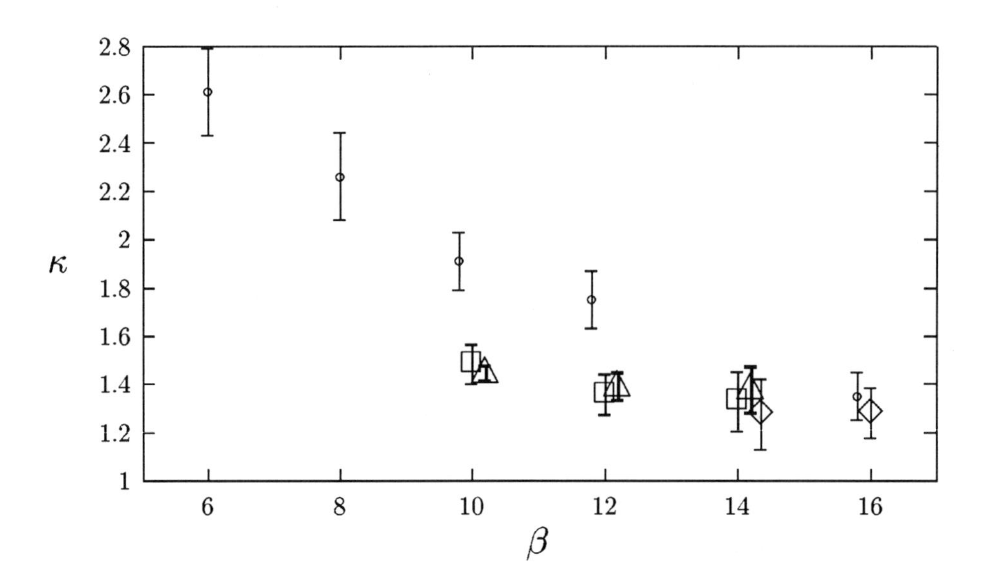

Figure 3: The dimensionless coefficient κ as a function of the inverse lattice spacing.

favoured the naive law

$$\Gamma = \kappa(\alpha_w T)^4. \tag{6}$$

The classical motion of the Chern-Simons number consists of two pieces. The thermal fluctuation, which linearly diverges with the lattice cutoff, and a random walk between different vacua (see Fig. 2).

$$< B^2(t) >= c + \Gamma V t. \tag{7}$$

A detailed finite size and finite scaling analysis suggested that the above naive formula is correct with a coefficient of $\kappa = 1.09 \pm 0.05$.

According to [3] the observed rate might be only a lattice artifact. The authors reanalysed their data with an improved technique (cooling of lattice field configurations to determine the time evolution of the topological charge) [7].

They introduced, along with the real time t, a cooling time τ. The dynamical variables are functions of them: $p_i(t,\tau), q_i(t,\tau)$. The t evolution is given by the standard equation of motions, while for the evolution in the τ direction an overdamped motion is used

$$\partial_\tau q_i(t,\tau) = -\partial_{q,i} V[q(t,\tau)]. \tag{8}$$

It is easy to see that in the vicinity of a static solution cooling leads to an exponential decay of stable eigenmodes and exponential growth of unstable ones. Moreover, the rate of decay (growth) is exponentially rapid for high frequency mode. The new results has shown some deviation from the older one; however, the $\Gamma = \kappa' \alpha_w^5 T^4$ law of the symmetric phase can not be confirmed.

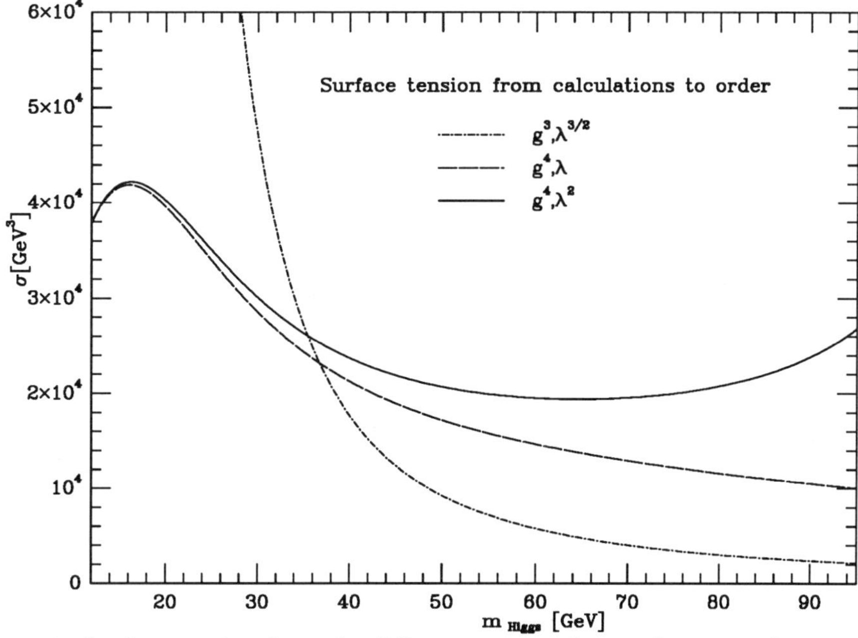

Figure 4: Surface tension from the different potentials as a function of m_H.

Fig. 3 shows the result of this analysis (open squares and triangles) and that of [8] (open dots). According to [3] the rate should have a $1/\beta$ behaviour. Clearly, larger measurement samples and careful systematic error analysis is needed to resolve the issue of the lattice spacing dependence of the rate.

A definitely positive outcome of the cooling technique is the new simulation in the broken phase. With the cooling technique no transition between the different vacua has been observed. This result resolves the discrepancy between [5] and eq. (3); however, no definite answer to the numerical value of the transition rate could be given. (Recently the multicanonical technique has been successfully applied to determine the nonperturbative transition rate in the broken phase [9].)

3 The nature of the phase transition

Perturbative studies show that in the realistic Higgs mass range ($m_H > 70\ GeV$) the perturbative approach breaks down [11], it predicts $\mathcal{O}(100\%)$ corrections [10] for the relevant quantities (e.g. interface tension, latent heat or correction to the course of the phase transition).

A nice illustration (c.f. Fig. 4) for that is the interface tension as a function of the Higgs-boson mass in different orders of the perturbation theory. As it can be seen in the physically allowed region (according to the LEP experiments the lower bound for the standard model Higgs-boson mass is relatively large: $m_H > 75$ GeV)

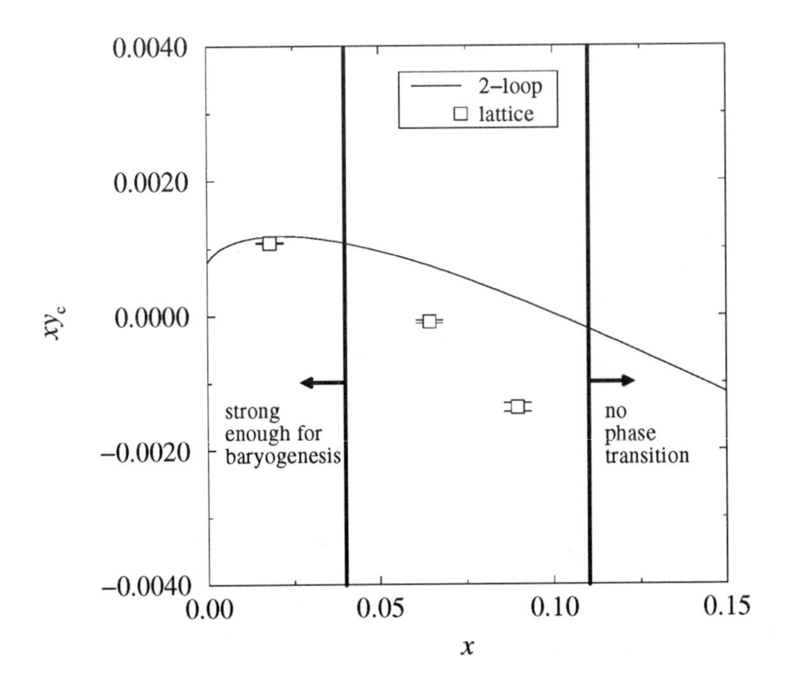

Figure 5: Phase diagram of the three-dimensional SU(2)-Higgs model.

the perturbation theory breaks down, its predictions can not be believed.

A popular way to study this basically non-perturbative problem is first to perform a dimensional reduction in perturbation theory. One starts with the original theory (e.g. Standard Model) and integrates out the heavy degrees of freedom perturbatively. The obtained theory is a three-dimensional bosonic one. The temperature dependent parameters of this theory are determined by the matching conditions between the full theory and the reduced one.

The theory has a dimensionful coupling, which fixes the overall scale. The two dimensionless quantities are x and y. One of them is x which determines the properties of the phase transition (connected to the zero temperature features of the original four-dimensional theory)

$$x \sim \frac{1}{8}\frac{m_H^2}{m_W^2} + c_2\frac{m_{top}^4}{m_W^4}. \tag{9}$$

The other one is y which is used to tune the system in order to find the phase transition point (connected to the temperature of the original four-dimensional theory)

$$y \sim c_1 \frac{T - T_c^{(pert)}}{T_c^{(pert)}}. \tag{10}$$

Here c_1 and c_2 are some constants.

Static thermodynamical properties, mass spectrum and other related features have been studied both analytically and numerically for these three-dimensional models (a recent summary is [12] and see references therein).

According to these results the electroweak phase transition is of first order for small Higgs-boson masses; however, it turns out to be an analytic cross-over above $m_H \approx 67$ GeV (critical Higgs-boson mass value for the SU(2)-Higgs model) [13]. Due to the large mass of the top-quark the phase transition can not fulfill the φ_b/T_b condition for any choice of the Higgs-boson mass.

Fig. 5 shows the phase diagram of the three-dimensional theory. The solid line corresponds to the two-loop perturbation theory and the open symbols represent the result of the lattice simulations. For x values smaller than ~ 0.03 the phase transition is strong enough to satisfy the minimal condition $\varphi_b/T_b > 1$. The phase transition ends around $x \sim 0.1$ (we will discuss the determination of the endpoint). For large enough x values no phase transition can be observed, only a rapid cross-over occurs.

The determination of the endpoint of the finite temperature EWPT, thus a characteristic feature of the phase diagram, can be done by the use of the Lee-Yang zeros of the partition function $\mathcal{Z}$ [13]. One analytically continues $\mathcal{Z}$ to complex values of the couplings by reweighting the available data. Denoting κ_0 the lowest zero of $\mathcal{Z}$, i.e. the zero closest to the real axis, one expects in the vicinity of the endpoint the scaling law $\text{Im}(\kappa_0) = d_1(x)V^{-\nu} + d_2(x)$. In order to pin down the endpoint one looks for a x value for which d_2 vanishes. Again, the change in x can been done by reweighting (or by direct simulations at different x couplings).

Using this Lee-Yang technique the Bielefeld group obtained $x_{end} = 0.0951(16)$, whereas the result of the Leipzig group is $x_{end} = 0.1020(20)$ (the two groups used different lattice spacings in their simulations).

Another possibility in order to understand the non-perturbative features of the electroweak phase transition is to study the full four-dimensional theory by lattice simulations. Since fermions always have nonzero Matsubara frequencies, the perturbative treatment of these, at high temperatures very massive, modes could be satisfactory. Thus, the starting point of the lattice analyses is the $SU(2)$-Higgs model, which contains the essential features of the standard model of electroweak interactions.

In the last three years our group (DESY-Electroweak collaboration) presented a series of papers (see e.g. [14]) in order to clarify the details of the phase transition on four-dimensional lattices. Our work has been done on computers at HLRZ Jülich (CRAY-T90) and DESY-Ifh, Zeuthen (APE-Quadrics). The simulations have been performed for four set of parameters ($m_W = 80$ GeV): $m_H \approx 18$ GeV, 35 GeV, 49 GeV and 80 GeV.

For the first three masses the usual lattice formulation can be applied, thus identical lattice spacings in the spacial and temporal directions. In these cases a fairly good agreement is found between the lattice results and perturbation theory [16].

As an illustration (see Fig. 6) T_c/m_H is presented for $m_H \approx 35$ GeV. For this quantity very precise results exist and an extrapolation to zero lattice spacing is also possible. We have used $L_t = 2, 3, 4$ and 5 to extrapolate to the continuum limit. As it can be seen there is approximately three standard deviation discrepancy between

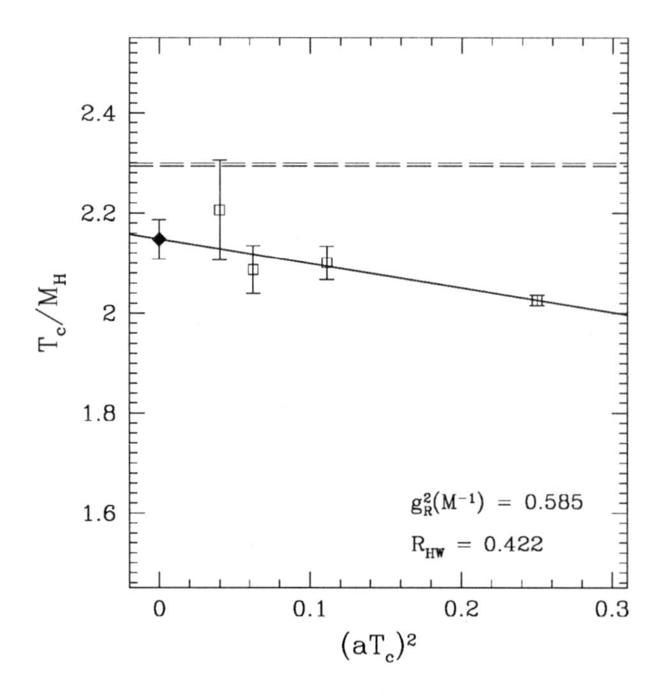

Figure 6: Critical temperature as a function of the lattice spacing.

the lattice and perturbative result. Note, that the result of the three-dimensional technique predicts a T_c/m_H much closer to the perturbative result. The lattice artifacts of the four-dimensional approach are expected to be proportional to a^2, whereas in the three-dimensional method they are proportional to a.

The $m_H \approx 80$ GeV case is much more difficult. The phase transition gets weaker, the lowest excitations have masses small compared to the temperature, T. From this feature one expects that a finite temperature simulation on isotropic lattice would need several hundred lattice points in the spatial directions even for $L_t = 2$ temporal extension.

In order to solve this problem we have used the simple idea that finite temperature field theory can be conveniently studied on asymmetric lattices, i.e. lattices with different spacings in temporal and spacial directions [15]. The resulting action contains anisotropies in the couplings. This action has been studied in perturbation theory and on the lattice.

One wants to tune the bare parameters in a way that the one-loop renormalized masses are finite in the continuum limit (however, their values in lattice units vanish $a_s M_{ren} = 0$ for $a_s \to 0$). At the same time the vacuum expectation value of the scalar field will be also zero in lattice units ($a_s v = 0$ for $a \to 0$) , i.e. we are at the phase transition point between the spontaneously broken Higgs phase and the SU(2) symmetric phase. The condition is fulfilled by an appropriate choice of the parameters (e.g. critical hopping parameter). The ratios of the other couplings are still free parameters and can be fixed by two additional conditions. We demand rotational (Lorenz) invariance for the scalar and vector propagators on the one-loop level. This ensures that the propagators with one-loop corrections have the same form in the z and t directions.

Clearly, arbitrary couplings for different directions would not lead to such rotationally invariant two-point functions. Technically the corrections to the anisotropies in the kinetic parts of the tree level propagators should be cancelled by the kinetic parts of the self-energies. This requires the knowledge of the wave function correction term in our theory. We have carried out both the perturbative and the lattice determinations of anisotropies. They are in complete agreement.

The values of the thermodynamical quantities (σ/T_c^3 and $\Delta\epsilon/T_c^4$) for $m_H \approx$ 80 GeV are substantially smaller than their perturbative values. They are even consistent with a no first order phase transition scenario on the approximately 1-σ level. These results can be interpreted as a sign for the endpoint for the finite temperature electroweak phase transition.

Similarly to the standard model case perturbative two-loop results exist for the minimal supersymmetric standard model phase transition [17]. An interesting feature of the result that it opens the baryogenesis window for light stop. Setting sun QCD diagrams (stop-gluon) can give large logarithmic contributions which increase the strength of the phase transition. Typically, setting sun diagrams give contribution proportional to $\varphi^2 \log \varphi$.

In the case of stop-gluon graph the prefactor is proportional to the strong coupling, which resulted in an enhancement effect on φ_b/T_b. In order not to have infrared divergencies for the stop sector the authors restrict themself to $m_U^2 > 0$ sector (negative m_U^2 are related to the existence of colour breaking minimum [18]). Fig. 7 shows that the ratio of the jump in the Higgs field divided by the temperature can be larger than one for low values of $\tan\beta$ (solid lines) and large enough m_A. Very low $\tan\beta$ corresponds to a light neutral Higgs (lines of constant Higgs masses are dashed). The LEP limit tells us that this mass must be larger than $\approx$ 70 GeV. Using this experimental bound the window for baryogenesis can be summarized as: $m_H < 85$ GeV; $2.25 < \tan\beta < 3.60$ and $m_A > 120$ GeV.

The allowed parameter space is very constrained. In particular $m_H < 85$ GeV will be soon tested at LEP. The authors claim that the large correction compared to the one-loop results (approximately 100%) is not a serious problem here and it does not mean the breakdown of the perturbation theory. The reason for that is the fact that the QCD corrections are leading order on the two-loop level; thus the large correction is not a symptom of the perturbative expansion, it is merely the appearance of a new interaction term.

Nevertheless, the setting-sun diagrams have bad infrared behaviour in the standard model. Direct lattice Monte-Carlo approach is probably needed in order to answer the question in the MSSM case, too.

In order to answer the question for the MSSM the three-dimensional approach (dimensional reduction plus lattice Monte-Carlo simulations in three-dimensions) is used, too (see e.g. [12]). The results are qualitatively the same as summarized above.

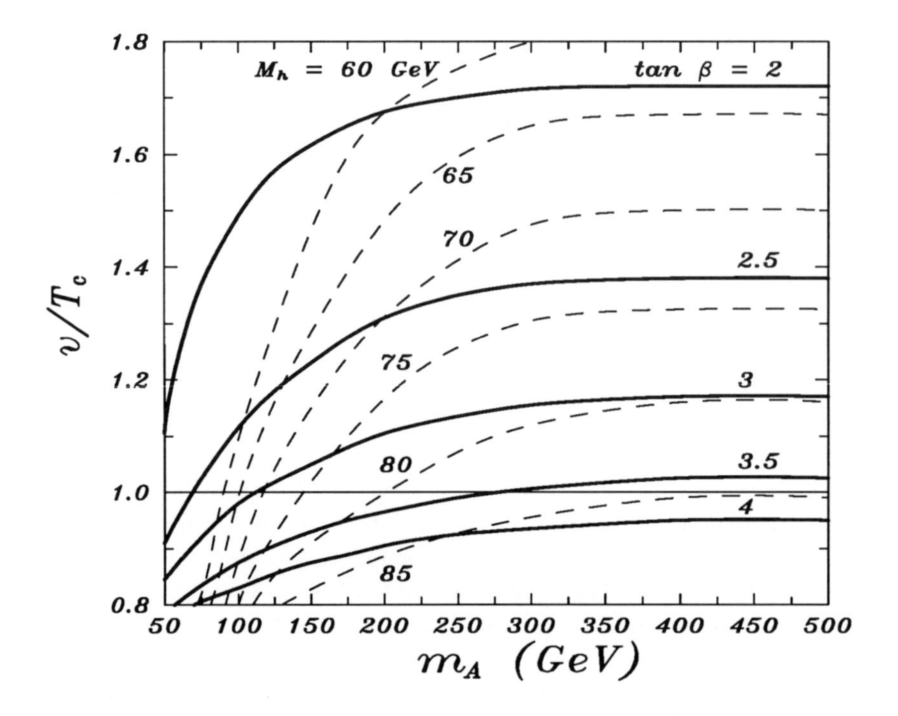

Figure 7: The normalized jump of the Higgs field in the MSSM.

However, these analyses are less reliable precisely in the small m_U^2 region, which is the interesting one for electroweak baryogenesis.

4 Conclusions

Analytical and real time numerical result for sphaleron processes are still not in complete agreement. Above T_c new analytical estimates suggest a $\Gamma \propto \alpha_w^5 T^4$ behaviour for the transition rate, which is not seen unambigously in numerical studies. The present numerical techniques are not sensitive enough to determine the sphaleron transition rate below T_c. Perturbative studies show, that the electroweak phase transition can not be described above $m_H = 70$ GeV by perturbative methods. Lattice results exclude successful baryogenesis in the minimal standard model. The present bounds on the Higgs-boson mass most probably results in a cross-over (no phase transition scenario). The minimal supersymmetric standard model still has some constrained parameter region, which perturbatively predicts a phase transition, strong enough for baryogenesis.

Acknowledgements

The author acknowledges the organizers for the kind invitation, and the partial support of Hung. Sci. Foundation Grants T016240/T022929 and FKFP-0128/1997.

References

[1] V. A. Kuzmin, V. A. Rubakov and M. E. Shaposhnikov, Phys. Lett. B155 (1985) 36.

[2] P. Arnold, L. McLerran, Phys. Rev. D36 (1987) 581.

[3] P. Arnold, D. Son, L. Yaffe, Phys. Rev. D55 (1997) 6264.

[4] A. Krasnitz, Nucl. Phys. B455 (1995) 320; G.D. Moore Nucl. Phys. B480 (1996) 657.

[5] J. Smit, W.H. Tang, Nucl. Phys. B482 (1996) 265.

[6] J. Ambjorn, A. Krasnitz, Phys. Lett. B362 (1995) 97.

[7] J. Ambjorn, A. Krasnitz, Nucl. Phys. B506 (1997) 387.

[8] . G.D. Moore, N. Turok, hep-ph/9703266.

[9] . G.D. Moore, hep-ph/9801204.

[10] Z. Fodor, A. Hebecker, Nucl. Phys. B432 (1994) 127.

[11] W. Buchmüller, Z. Fodor, T. Helbig and D. Walliser, Ann. Phys. (NY) 234 (1994) 260.

[12] M. Laine, hep-ph/9707415.

[13] F. Karsch, T. Neuhaus, A. Patkós, J. Rank, Nucl. Phys. Proc.Suppl. 53 (1997) 623; M. Gurtler, E.M. Ilgenfritz, A. Schiller, Phys. Rev. D56 (1997) 3888

[14] Z. Fodor et al. Phys. Lett. B334 (1994) 405; Nucl. Phys. B439 (1995) 147; Csikor et al. Nucl. Phys. B474 (1996) 421.

[15] F. Csikor, Z. Fodor, Phys. Lett. B380 (1996) 113.

[16] W. Buchmüller, Z. Fodor and A. Hebecker, Nucl. Phys. B447 (1995) 317.

[17] B. de Carlos, J.R. Espinosa, Phys. Lett. B407 (1997) 12.

[18] M. Carena, M. Quiros and C.E.M. Wagner, Phys. Lett. B380 (1996) 81.

Two-dimensional quantum geometry

J. Ambjørn , J. L. Nielsen and J. Rolf

Niels Bohr Institute
Blegdamsvej 17, DK-2100 Ø, Copenhagen
Denmark

Abstract:

We discuss various aspects of the fractal structure of two-dimensional quantum gravity coupled to conformal field theory. In particular we ague that the intrinsic Hausdorff dimension of pure two-dimensional quantum gravity is four, while the so-called spectral dimension is two. We also point out that a naive use of the Regge formulation of quantum gravity leads to erroneous results.

1 Introduction

In classical gravity one probes the structure of space-time in two ways. One can put out rods and clocks and in this way directly determine the metric. An alternative and less direct way is to study the propagation of test particles. When we turn to a theory of quantum gravity both methods encounter conceptional problems since we are instructed to perform the quantum average over all geometries. At present we have no genuine quantum gravity theory in four-dimensional space-time and maybe the final theory will be such that we will understand how to probe the structure of the full quantum space-time. However, it might be useful to try to address the same questions in two-dimensional quantum gravity where a theory does in fact exist. Such a theory contains no field-theoretical degrees of freedom (no graviton), only quantum mechanical degrees of freedom (i.e. a *finite* number of degrees of freedom). Nevertheless many of the conceptional questions which cause problems in four dimensions are already present in two dimensions.How do we define observables which are invariant under reparameterization, how do we define propagation of particles and how do we at all determine simple properties of space-time when there is no fixed reference geometry? In a certain way the quantum nature of these questions is as pronounced in 2d as in 4d, despite the fact that the action is trivial and non-dynamical in 2d. The reason for this is precisely the triviality of the action. It implies that all 2d geometries have to be included with the same weight in the quantum average. The concept of some fixed background geometry around which one can study small perturbations is absent.

In the next section we describe shortly the quantum gravity equivalence to the classical rod-clock measurement of geometry, and in the section to follow some details are provided in a discussion of the propagation of test particles in quantum space-time. The final section contains a discussion of the results, the possibilities for generalizations to higher dimensions and the subtle question of a correct choice of regularization.

2 The functional integral over 2d geometries

The partition function for two-dimensional (Euclidean) geometries is

$$Z(\Lambda) = \int \mathcal{D}[g]\, \mathrm{e}^{-\Lambda V_g}, \qquad V_g \equiv \int_{\mathcal{M}} d^2\xi \sqrt{g(\xi)}. \tag{1}$$

It is sometimes convenient to consider the partition function where the volume V of space-time is kept fixed. We define it by

$$Z(V) = \int \mathcal{D}[g]\, \delta(V - V_g), \tag{2}$$

such that

$$Z(\Lambda) = \int_0^\infty dV\, \mathrm{e}^{-V\Lambda} Z(V). \tag{3}$$

Here we will discuss the set of reparameterization invariant observables denoted Hartle-Hawkings wave-functionals and the two-point functions. The Hartle-Hawking wave-functional is defined by

$$W(L;\Lambda)) = \int_L \mathcal{D}[g]\, \mathrm{e}^{-S(g;\Lambda)} \tag{4}$$

where L symbolizes the *boundary* of the manifold $\mathcal{M}$. In dimensions higher than two one should specify (the equivalence class of) the metric on the boundary and the functional integration is over all equivalence classes of metrics having this boundary metric. In two dimensions the equivalence class of the boundary metric is uniquely fixed by its length and we take L to be the length of the boundary. It is often convenient to consider boundaries with variable length L by introducing a *boundary cosmological term* in the action:

$$S(g;\Lambda,\Lambda_b) = \Lambda \int_{\mathcal{M}} \mathrm{d}^2\xi \sqrt{g(\xi)} + \Lambda_B \int_{\partial\mathcal{M}} \mathrm{d}s, \tag{5}$$

where $\mathrm{d}s$ is the invariant line element corresponding to the boundary metric induced by g and Λ_B is called the boundary cosmological constant. We can then define

$$W(\Lambda_B,\Lambda) = \int \mathcal{D}[g]\, \mathrm{e}^{-S(g;\Lambda,\Lambda_B)}. \tag{6}$$

The wave-functions $W(L;\Lambda)$ and $W(\Lambda_B,\Lambda)$ are related by a Laplace transformation in the boundary length:

$$W(\Lambda_B,\Lambda)=\int_0^\infty \mathrm{d}L\,\mathrm{e}^{-\Lambda_B L}\,W(L;\Lambda). \tag{7}$$

The two-point function is defined by

$$G(R;\Lambda)=\int \mathcal{D}[g]\,\mathrm{e}^{-S(g,\Lambda)}\iint \mathrm{d}^2\xi\sqrt{g(\xi)}\,\mathrm{d}^2\eta\sqrt{g(\eta)}\,\delta(D_g(\xi,\eta)-R), \tag{8}$$

where $D_g(\xi,\eta)$ denotes the geodesic distance between ξ and η in the given metric g. Again, it is sometimes convenient to consider a situation where the space-time volume V is fixed. This function, $G(R;V)$ will be related to (8) by a Laplace transformation, as above for the partition function Z:

$$G(R;\Lambda)=\int_0^\infty \mathrm{d}V\,\mathrm{e}^{-V\Lambda}\,G(R;V). \tag{9}$$

It is seen that $G(R;\Lambda)$ and $G(R;V)$ has the interpretation of partition functions for universes with two marked points separated a given geodesic distance R. If we denote the average volume of a spherical shell of geodesic radius R in the class of metrics with space-time volume V by $S_V(R)$, we have by definition

$$S_V(R)=\frac{G(R;V)}{VZ(V)}. \tag{10}$$

It is seen that $S_V(R)$ serves as the quantum equivalent to the classical spherical shell of radius R in a fixed geometry. In this sense it is the quantum version of a "rod and clock" measurement of the geometry. In the same spirit one can define an intrinsic fractal dimension, d_H, of the ensemble of metrics by

$$\lim_{R\to 0} S_V(R)\sim R^{d_H-1}(1+O(R)). \tag{11}$$

Alternatively, one could use the following d_H:

$$\langle V\rangle_R\sim-\frac{\partial\log G(R;\Lambda)}{\partial\Lambda}\sim R^{d_H} \tag{12}$$

for a suitable range of R related to the value of Λ. As will be clear, the two definitions agree in the case of pure gravity. Eq. (11) can be viewed as a "local" definition of d_H, while eq. (12) is "global" definition. Since the two definitions result in the same d_H two-dimensional gravity has a genuine fractal dimension over all scales.

Eq. (2) shows that the calculation of $Z(V)$ is basically a counting problem: each geometry, characterized by the equivalence class of metrics $[g]$, appears with the same weight. The same is true for the other observables defined above. One way of performing the summation is to introduce a suitable regularization of the set of geometries by means of a cut-off, to perform the summation with this cut-off and then remove the cut-off, like in the case of geometric paths considered above.

2.1 The Regularization

In the calculations referred to below we will use the regularization of the integral over geometries denoted *dynamical triangulations.* In 2d gravity and string theory it was introduced in [1, 2, 3]. The natural building blocks will be equilateral triangles with side lengths ε, but there will be no integration over positions in some target space[1]. We can glue the triangles together to form a triangulation of a two-dimensional manifold $\mathcal{M}$ with a given topology. If we view the triangles as flat in the interior, we have in addition a unique piecewise linear metric assigned to the manifold, such that the volume of each triangle is $dA_\varepsilon = \sqrt{3}\varepsilon^2/4$ and the total volume of a triangulation T consisting of N_T triangles will be $N_T dA_\varepsilon$, i.e. we can view the triangulation as associated with a Riemannian manifold $(\mathcal{M}, g)$. ¿From this point of view a summation over triangulations of the kind mentioned above will form a grid in the class of Riemannian geometries associated with a given manifold $\mathcal{M}$. The hope is that the grid is sufficient dense and uniform to be able the describe correctly the functional integral over all Riemannian geometries when $\varepsilon \to 0$.

We will show that it is the case by explicit calculations, where some of the results can be compared with the corresponding continuum expressions. They will agree. But the surprising situation in two-dimensional quantum gravity is that the analytical power of the regularized theory seems to exceed that of the formal continuum manipulations. Usually the situation is the opposite: regularized theories are either used in a perturbative context to remove infinities order by order, or introduced in a non-perturbative setting in order make possible numerical simulations. Here we will derive analytic (continuum) expressions with an ease which can presently not be matched by formal continuum manipulations.

Let us end this subsection by emphasizing some general problems of regularization. A regularization as Dynamical Triangulation can be viewed as being explicit invariant under reparameterization in the sense that we can view each triangulation as representing a *geometry*, i.e. the equivalence class of metrics related by reparameterization. However, we have presently no real control of how dense or how uniform a grid this class of geometries form in the class of all geometries. One reason is that we have not yet a mathematical definition of the measure on the space of geometries. However, for a number of observables one can use different regularizations and compare the results. Using continuum methods, like the conformal bootstrap approach of KPZ and DDK one gets agreement with the results obtained by the use of dynamical triangulations. One can thus have a fair amount of confidence in these results (and the healthiness) even if a rigorous mathematical proof is absent.

However, not all regularization methods work. One standard way of discretizing classical gravity is the Regge prescription, which in some ways is similar to the dynamical triangulation approach in the sense that both approaches use piecewise linear manifolds. In The Regge approach one fixes the connectivity of a given trian-

[1] We could introduce such embedding in R^d, but in that case we would not consider two-dimensional gravity but rather bosonic string theory, where the embedded surface was the world sheet of the string [2, 4].

gulation and the dynamical freedom is obtained by varying the length of the links. This is contrary to the philosophy used in the dynamical triangulation approach where the connectivity of the triangulation is allowed to change, but the length of the links are kept fixed. While the Regge approach works perfectly in the context of classical gravity, it has not yet been so useful in the context of quantum gravity. One reason is that many link-length assignments might correspond to the same geometry, as is easily seen by considering the case of the triangulation in flat space. Clearly there are many different link assignments which correspond to flat space. Once we have the vertices corresponding to flat space, one can move the vertices around in flat space without changing the flat geometry. One need a Fadeev-Popov determinant to single out the physical degrees of freedom. Nobody has calculated it explicitly as a function of the link-length. People have tried to ignore this problem. However, as was shown in [7] this leads to a partition function which in general will be ill defined and it will always give rise to "spikes", i.e. singular geometries where the results of calculations of two-point functions etc. do not agree with the ones reported below and thus not with the standard Liouville theory approach to two-dimensional quantum gravity. The conclusion is that one at least has to include some kind of Fadeev-Popov determinant if the Regge approach shall be useful in quantum gravity. Nobody knows presently how to do this.

The choice of regularization is thus somewhat non-trivial, but we are in the lucky situation that dynamical triangulations seem to provide a correct prescription in two-dimensional quantum gravity.

2.2 The Hartle-Hawking Wave-Functional and the two-point function

As already remarked the calculations of the Hartle-Hawking wave-functionals and the two-point functions reduce to counting problems. These counting problems now become mathematically well-defined when formulated in the framework of dynamical triangulations. We refer to [12] for detailed accounts and a discussion of how to take the continuum limit and discuss here only the continuum results obtained after taking the *scaling limit* of the discretized results.

For the Hartle-Hawking wave-functional the result is [5, 6]

$$W(\Lambda_B, \Lambda) \sim (\Lambda_B - \frac{1}{2}\sqrt{\Lambda})\sqrt{\Lambda_B + \sqrt{\Lambda}}. \tag{13}$$

By an inverse Laplace transformation one obtains

$$W(L, \Lambda) = L^{-5/2}(1 + L\sqrt{\Lambda})\, e^{-L\sqrt{\Lambda}}. \tag{14}$$

It is seen how universes with boundaries longer than the natural length scale $\sqrt{\Lambda}$ are exponentially suppressed. While this result is natural and not surprising, it is more difficult to understand the divergence for small L. As we shall see shortly, it is related to the fact that the average geometry is genuinely fractal at scales less that $\sqrt{\Lambda}$.

The counting related to the Hartle-Hawking wave-functionals can be refined to cover a situation where two marked triangles in a triangulation is separated a geodesic distance r, where we by a geodesic distance r on the triangulation mean that the shortest path between the two marked triangles via neighboring triangles is r triangles. Again we have to refer to [12, 8] for details. ¿From the discretized expression one can again take the continuum limit and obtains for the two-point function $G(R;\Lambda)$ where two marked points are separated a geodesic distance R:

$$G(R;\Lambda) = \Lambda^{3/4} \frac{\cosh\left[R\sqrt[4]{\Lambda}\right]}{\sinh^3\left[R\sqrt[4]{\Lambda}\right]}. \tag{15}$$

Using this expression it follows that

$$\langle V \rangle_R \sim -\frac{\partial \log G(R;\Lambda)}{\partial \Lambda} \sim R^4(1 + O(R^4/V)), \tag{16}$$

indicating that $d_H = 4$ rather than 2, as one naively would have expected. Observe also that $G(R;\Lambda)$ falls off exponentially for $R \gg \sqrt[4]{\Lambda}$. It has been shown that this is a consequence of the same kind of sub-additivity arguments which lead to the exponential fall off or the massive particle propagator [8]. Note finally the *anomalous dimension* of R relatively to Λ. This is a necessary consequence of the fact that $d_H \neq 2$.

The d_H above is the "globally defined" Hausdorff dimension in the sense discussed below (12) as is clear from (15). We can determine the "local" d_H, defined by eq. (11), by performing the inverse Laplace transformation of $G(R;\Lambda)$ to obtain $G(R;V)$. The average volume $S_V(R)$ of a spherical shell of geodesic radius R in the ensemble of universes with space-time volume V can then calculated from (10). One obtains

$$S_V(R) = R^3 F(R/V^{\frac{1}{4}}), \qquad F(0) > 0, \tag{17}$$

where $F(x)$ can be expressed in terms of certain generalized hyper-geometric functions [9]. Eq. (17) shows that also the "local" $d_H = 4$.

2.3 Summary

It has been shown how it is possible to calculate the functional integral over two-dimensional geometries, in close analogy to the functional integral over random paths. One of the most fundamental results from the latter theory is that the generic random path between two points in R^d, separated a geodesic distance R, is *not* proportional to R but to R^2. This famous result has a direct translation to the theory of random two-dimensional geometries: the generic volume of a closed universe of radius R is *not* proportional to R^2 but to R^4.

3 The spectral dimension of space-time

Above we described the quantum version of the classical "rod and clock" description of the geometry. Now we turn to the test particle aspect. Consider the propagation of test particles in a fixed geometry. By observation we can determine the propagator. By an inverse Laplace transformation we can find the associated heat kernel (see below). The heat kernel allows an asymptotic expansion where the various terms have a direct geometry interpretation allowing us to calculate various contractions of the Riemann curvature and other geometric quantities. One can wonder how much is left of this information in a genuine theory of quantum gravity where we integrate over all geometries. Such a study leads natural to the introduction of the spectral dimension as a measure of the structure of the quantum space-time.

The most intuitive definition of the spectral dimension is based on the diffusion equation on a (compact) manifold with metric g_{ab}. Let Δ_g denote the Laplace-Beltrami operator corresponding to g_{ab}. The probability distribution $K(\xi,\xi';T)$ of diffusion is related to the massless scalar propagator $(-\Delta_g)^{-1}$ by [2]

$$\left\langle \xi' \middle| (-\Delta_g)^{-1} \middle| \xi \right\rangle' = \int_0^\infty dT\, K'(\xi,\xi';T). \tag{18}$$

In particular, the average return probability distribution at time T has the following small T behavior:

$$RP'_g(T) \equiv \frac{1}{V_g}\int \mathrm{d}^d\xi \sqrt{g}\, K'(\xi,\xi;T) \sim \frac{1}{T^{d/2}}(1+O(T)), \tag{19}$$

where V_g denotes the volume of the compact manifold with metric g. The important point, in relation to quantum gravity, is that $RP'_g(T)$ is invariant under reparameterization. Thus the quantum average over geometries can be defined:

$$RP'_V(T) \equiv \frac{1}{Z_V}\int \mathcal{D}[g]_V\, \mathrm{e}^{-S_{\text{eff}}([g])} RP'_g(T), \tag{20}$$

where Z_V denotes the partition function of quantum gravity for fixed space-time volume V (see (25) for more details about Z_V and $\mathcal{D}[g]_V$), and $S_{\text{eff}}([g])$ denotes the effective action of quantum gravity after the integration over possible matter fields. The *spectral* dimension d_s in quantum gravity is now defined by the small T behavior of the functional average $RP'_V(T)$

$$RP'_V(T) \sim \frac{1}{T^{d_s/2}}(1+O(T)). \tag{21}$$

The $O(T)$ term in (19) has a well known asymptotic expansion in powers of T, where the coefficient of T^r is an integral over certain powers and certain contractions of the curvature tensor. This asymptotic expansion breaks down when $T \sim V^{2/d}$ at which point the exponential decay in T of the heat kernel K takes over. If we average over all geometries as in (20), it is natural to expect that the only invariant left will be the volume V which is kept fixed. Thus we expect that we can write

$$RP'_V(T) = \frac{1}{T^{d_s/2}} F\Big(\frac{T}{V^{2/d_s}}\Big), \tag{22}$$

[2] Since we consider compact manifolds the Laplace-Beltrami operator Δ_g has zero modes. Eq. (18) should be understood with these zero modes projected out. This is indicated with a prime.

where $F(0) > 0$ and $F(x)$ falls off exponentially fast for $x \to \infty$.

For fixed manifold of dimension d and a given smooth geometry $[g]$ we have $d = d_s$ by definition. The functional average can *a priori* change this, i.e. the dimension of T can become anomalous. A well known example of similar nature can be found for the ordinary free particle. In the path integral representation of the free particle, any smooth path of course has fractal dimension equal to one. Nevertheless the short distance properties of the free particle reflects that the generic path contributing to the path integral has fractal dimension (the *extrinsic* Hausdorff dimension – in the target space R^D) $D_H = 2$ with probability one. In the same way the functional integral over geometries might change d_s from the "naive" value d. In two-dimensional quantum gravity it is known, as mentioned above, that the intrinsic Hausdorff dimension *is* different from $d = 2$ and the generic geometry is in this sense fractal, with probability one. When one considers diffusion on fixed fractal structures (often embedded in R^D) it is well known that d_s can be different from both D and the fractal dimension d_h of the structure. If δ denotes the so-called anomalous gap exponent, defined by the relation between the diffusion time T and the average spread of diffusion on the fractal structure, but measured in R^D:

$$\left\langle r^2(T) \right\rangle \sim T^{2/\delta}, \tag{23}$$

then the relation between the fractal dimension (*intrinsic* Hausdorff dimension) d_h of the structure, the spectral dimension of the diffusion and the gap exponent is:

$$d_s = \frac{2d_h}{\delta}. \tag{24}$$

If δ is not anomalous, i.e. $\delta = 2$ as for diffusion on a smooth geometry, we have $d_s = d_h$, which is the analogue of $d_s = d$ for fixed smooth geometries. However, in general $\delta \neq 2$ (for a review of diffusion on fractal structure, see e.g. [10]).

3.1 Spectral dimension for 2d quantum gravity

We will derive a simple relation between the spectral dimension and the extrinsic Hausdorff dimension for dynamical self-similar systems like two-dimensional quantum gravity, branched polymers etc. Since the extrinsic Hausdorff dimension is known for these systems it will allow a determination of the spectral dimension. As a typical example of such models (and maybe the most interesting), we consider two-dimensional quantum gravity. The partition function for two-dimensional quantum gravity coupled to D Gaussian fields X_μ is given by

$$Z_V = \int \mathcal{D}[g]_V \mathcal{D}[X_\mu]_{\mathrm{cm}}\, \mathrm{e}^{-\int \mathrm{d}^2\xi \sqrt{g}\, g^{ab} \partial_a X_\mu \partial_b X_\mu}, \tag{25}$$

where $\int \mathcal{D}[g]_V$ denotes the integration over *geometries*, i.e. equivalence classes of metrics on the two-dimensional manifold of fixed space-time volume V of the manifold, while $\mathcal{D}[X_\mu]_{\mathrm{cm}}$ denotes the functional integration over the D Gaussian fields X_μ, but with the center of mass fixed (to zero). The extrinsic Hausdorff dimension D_H is usually defined as

$$\left\langle X^2 \right\rangle_V \sim V^{2/D_H} \quad \text{for} \quad V \to \infty, \tag{26}$$

where

$$\left\langle X^2 \right\rangle_V \equiv \frac{1}{Z_V} \int \mathcal{D}[g]_V \mathcal{D}[X_\mu]_{\text{cm}} \, \mathrm{e}^{-\int d^2\xi \sqrt{g}\, g^{ab} \partial_a X_\mu \partial_b X_\mu} \, \frac{1}{DV} \int d^2\xi \sqrt{g}\, X_\mu^2(\xi). \tag{27}$$

The Gaussian action in X implies that:

$$\begin{aligned}
\left\langle X^2 \right\rangle_V &= \frac{1}{DVZ_V} \frac{\partial}{\partial\omega} \int \mathcal{D}[g]_V \mathcal{D}[X_\mu]_{\text{cm}} \, \mathrm{e}^{-\int d^2\xi \sqrt{g}\, g^{ab}\partial_a X_\mu \partial_b X_\mu + \omega \int d^2\xi \sqrt{g} X_\mu^2(\xi)} \bigg|_{\omega=0} \\
&= \frac{1}{DVZ_V} \frac{\partial}{\partial\omega} \int \mathcal{D}[g]_V \Big(\det{}'(-\Delta_g - \omega)\Big)^{-D/2} \bigg|_{\omega=0} \\
&= \frac{1}{2VZ_V} \int \mathcal{D}[g]_V \Big(\det{}'(-\Delta_g)\Big)^{-D/2} \, \mathrm{Tr}\,' \left[\frac{1}{-\Delta_g}\right] \\
&= \frac{1}{2V} \left\langle \mathrm{Tr}\,' \left[\frac{1}{-\Delta_g}\right] \right\rangle_V
\end{aligned} \tag{28}$$

where the primes on the determinants and traces again mean that zero modes are excluded. Formula (28) is used to define $\langle X^2 \rangle_V$ when D is non-integer.

Using (18) and (22) we get

$$\left\langle X^2 \right\rangle_V = \frac{1}{2} \int_0^\infty dT \, \frac{1}{T^{d_s/2}} F\Big(\frac{T}{V^{2/d_s}}\Big) \sim V^{2/d_s - 1}, \tag{29}$$

for V going to infinity. From (26) we now conclude that

$$\frac{1}{d_s} = \frac{1}{D_H} + \frac{1}{2}. \tag{30}$$

Several remarks are in order. Strictly speaking the above derivation assumes that $d_s < 2$. If $d_s > 2$ we have to introduce a small-T cut-off ε in (29). In this case it is convenient to consider instead $\langle (X^2)^n \rangle_V \sim V^{2n/D_H}$ where $n = [d_s/2] + 1$ for non-integer d_s. It is then easy to show that the leading large V behavior on the right hand side of the equation corresponding to (29) will be $V^{2n/d_s - 1}$ and we get

$$\frac{1}{d_s} = \frac{1}{D_H} + \frac{1}{2n} \quad \text{for} \quad 2n - 2 < d_s < 2n \ (\ n = 1, 2, \dots \ ). \tag{31}$$

In the study of diffusion on fixed fractal structures one usually encounters $d_s < 2$. In the following we will *assume* $d_s \leq 2$ due to the following reasoning. We expect that $d_s \to 2$ for $D \to -\infty$ since large negative D implies that a saddle point calculation of (27) around a fixed geometry should be reliable. A strictly fixed geometry implies $d_s = 2$ and $D_H = \infty$ (in agreement with (30)). Also the saddle point calculation results in $D_H = \infty$ and should be valid in a neighborhood of $D = -\infty$. Hence $d_s = 2$

in a neighborhood of $D = -\infty$. If doing anything, one would expect fluctuating geometries to decrease D_H since there are many generate geometries where $D_H < \infty$, e.g. the branched polymer-like geometries to be discussed later. Thus it is reasonable to assume $d_s \leq 2$.

Once this assumption is made, it follows immediately that $d_s = 2$ for all $D \leq 1$ since it is known from Liouville theory that $D_H = \infty$. Let us just recall the argument[3] [14, 15]. Define the two-point function in random surface theory by:

$$G(p) = \left\langle \int\!\!\int \mathrm{d}^2\xi_1 \sqrt{g(\xi_1)}\, \mathrm{d}^2\xi_2 \sqrt{g(\xi_2)}\, \mathrm{e}^{ip(X(\xi_1)-X(\xi_2))} \right\rangle_V \tag{32}$$

If we use the following definition of $\langle X^2 \rangle_V$ (which is equivalent to (27) for large V)

$$\left\langle X^2 \right\rangle_V = \frac{1}{DV^2} \left\langle \int\!\!\int \mathrm{d}^2\xi_1 \sqrt{g(\xi_1)}\, \mathrm{d}^2\xi_2 \sqrt{g(\xi_2)} \big(X(\xi_1) - X(\xi_2)\big)^2 \right\rangle_V, \tag{33}$$

it follows that

$$\left\langle X^2 \right\rangle_V = -\frac{1}{DV^2} \frac{\partial^2}{\partial p^2} G(p)\Big|_{p=0}. \tag{34}$$

Since it is known that $G(p)$ behaves as $V^{2-\Delta_0(p)}$ in flat space with $\Delta_0(p) \propto p^2$, the KPZ formula allows us to calculate $\Delta(p)$ after coupling to gravity:

$$\Delta(p) = \frac{\sqrt{1 - D + 24\Delta_0(p)} - \sqrt{1-D}}{\sqrt{25-D} - \sqrt{1-D}}. \tag{35}$$

It follows that for $D < 1$ we have

$$\left\langle X^2 \right\rangle_V \sim \log V, \tag{36}$$

while for $D = 1$

$$\left\langle X^2 \right\rangle_V \sim \log^2 V. \tag{37}$$

In both cases $D_H = \infty$ and thus $d_s = 2$ from (30).

4 Discussion

We have discussed various aspects of two-dimensional Euclidean quantum gravity which we believe are of general interest also i higher dimensions. One caveat in such a discussion is the use of Euclidean signature. Some of the "observables" are indeed not very natural from the point of view of a Lorentzian signature of space-time. Contrary to the situation in ordinary field theory where it is well established how to rotate from Euclidean to Lorentzian space-time, we do not know how to perform this

[3]The treatment in [14] is based on a more general scaling assumption than is need in two-dimensional quantum gravity and this leaves open the possibility of a scaling different from the one given here. In [15] the treatment was narrowed down to the one presented here.

rotation even in two-dimensional quantum gravity. However, again it is tempting to study the question in more detail precisely in two dimensions since we can treat two-dimensional quantum gravity with (almost) standard field theoretical tools. It might give us valuable information about the general principles underlying possible rotations also in higher dimensional quantum gravity.

In principle the many of above definitions, given in the context of two-dimensional quantum gravity, apply for any theory of quantum gravity (see for instance [11, 12, 13] for definitions in the case of four dimensional simplicial quantum gravity) and we believe they are of such general nature that they might be valid (with suitable modifications) also in a full four-dimensional theory of quantum gravity when it is finally formulated.

References

[1] F. David, *Nucl. Phys.* **B257** (1985) 45; *Nucl. Phys.* **B257** (1985) 543;

[2] J. Ambjørn, B. Durhuus and J. Fröhlich, *Nucl. Phys.* **B257** (1985) 433;

[3] V.A. Kazakov, I.K. Kostov and A.A. Migdal, *Phys. Lett.* **157B** (1985) 295.

[4] J. Ambjørn, B. Durhuus J. Fröhlich and P. Orland, *Nucl. Phys.* **B270** (1986) 457;

[5] F. David, *Mod. Phys. Lett.* **A5** (1990) 1019.

[6] J. Ambjørn, J. Jurkiewicz and Y. M. Makeenko, *Phys. Lett.* **B251** (1990) 517.

[7] J. Ambjørn, J.L. Nielsen, J. Rolf and G. Savvidy, *Class. Quant. Grav.* **14** (1997) 3225.

[8] J. Ambjørn and Y. Watabiki, *Nucl. Phys.* **B445** (1995) 129.

[9] J. Ambjørn, J. Jurkiewicz and Y. Watabiki, *Nucl. Phys.* **B454** (1995) 313.

[10] S. Havlin and D. Ben-Avraham, *Adv. Phys.* **36**, (1987) 695.

[11] J. Ambjørn and J. Jurkiewicz, *Phys. Lett.* **B278** (1992) 42; *Nucl. Phys.* **B451** (1995) 643, hep-th/9503006.

[12] J. Ambjørn, B. Durhuus and T. Jonsson, *Quantum Geometry*, Cambridge Monographs on Mathematical Physics, Cambridge 1997.

[13] J. Ambjørn, M. Carfora and A. Marzuoli, *The Geometry of Dynamical Triangulations*, Lectures Notes in Physics, m50, Springer 1997.

[14] J. Distler, Z. Hlousek, H. Kawai, *Int. J. Mod. Phys.* **A5** (1990) 391.

[15] H. Kawai, *Nucl. Phys. Proc. Suppl.* **26** (1992) 93.

Supermembranes and Super Matrix Theory

Bernard de Wit

Institute for Theoretical Physics, Utrecht University
Princetonplein 5, 3508 TA Utrecht, The Netherlands

Abstract: We review recent developments in the theory of supermembranes and their relation to matrix models.

1 Supersymmetric quantum mechanics

Consider the class of supersymmetric Hamiltonians of the form

$$H = \frac{1}{g}\mathrm{Tr}\Big[\tfrac{1}{2}\mathbf{P}^2 - \tfrac{1}{4}[X^a, X^b]^2 + \tfrac{1}{2}g\,\theta^{\mathrm{T}}\gamma_a[X^a,\theta]\Big]\,, \tag{1}$$

depending on a number of d-dimensional coordinates $\mathbf{X} = (X^1,\ldots,X^d)$, corresponding momenta $\mathbf{P}$, as well as real spinorial anticommuting coordinates θ_α, all taking values in the matrix representation of some Lie algebra. The phase space is restricted to the subspace invariant under the corresponding (compact) Lie group and is therefore subject to Gauss-type constraints. The above Hamiltonians arise in the zero-volume limit of supersymmetric Yang-Mills theories, which explains the presence of these constraints.

The theories based on (1) were proposed long ago as extended models of supersymmetric quantum mechanics with more than four supersymmetries [1]. The spatial dimension d and the corresponding spinor dimension are severely restricted. The models exist for $d = 2, 3, 5$, or 9 dimensions; the (real) spinor dimension equals $2, 4, 8$, or 16, respectively. Naturally this is also the number of independent supercharges. In what follows we restrict ourselves to the highest-dimensional case, where the model contains 16 supercharges. However, additional charges can be obtained by splitting off an abelian factor of the gauge group (we will mainly consider the gauge group $\mathrm{U}(N)$),

$$Q^+ = \mathrm{Tr}\Big[(P^a\gamma_a + \tfrac{1}{2}i[X^a,X^b]\gamma_{ab})\,\theta\Big]\,, \qquad Q^- = g\,\mathrm{Tr}\,[\,\theta\,]\,. \tag{2}$$

The Q^+ generate the familiar supersymmetry algebra (in the group-invariant subspace),

$$\{Q^+_\alpha, Q^+_\beta\} \approx H\,\delta_{\alpha\beta}\,. \tag{3}$$

A central theme of this lecture is that the supermembrane in the light-cone formulation is described by a quantum-mechanical model of the type above with an infinite-dimensional gauge group corresponding to the area-preserving diffeomorphisms of the membrane spacesheet [2]; the coupling constant g is then equal to the total light-cone momentum $(P_-)_0$, which in a flat target space equals P_0^+. In 11 spacetime dimensions the supermembrane is subject to 32 supercharges. The 16 charges Q^- given in (2) are then associated with the center-of-mass superalgebra. The connection with the supermembrane shows that the manifest SO(9) symmetry, which from the viewpoint of the supermembrane is simply the exact transverse rotational invariance of the lightcone formulation, extends to the 11-dimensional Lorentz group in the limit of an appropriate infinite-dimensional gauge group [3, 4].

Classical zero-energy configurations require all commutators $[X^a, X^b]$ to vanish. Dividing out the gauge group implies that zero-energy configurations are thus parametrized by $\mathbf{R}^{9N}/S_N$. The zero-energy valleys in the potential extend all the way to infinity where they become increasingly narrow. Their existence raises questions about the nature of the spectrum of the Hamiltonian (1). In the bosonic versions of these models the wave function cannot freely extend to infinity, because at large distances it becomes more and more squeezed in the valley. By the uncertainty principle, this gives rise to kinetic-energy contributions which increase monotonically along the valley. Another way to see this effect is by noting that oscillations perpendicular to the valleys give rise to a zero-point energy, which induces an effective potential barrier that confines the wave function. This confinement causes the spectrum to be discrete. However, for the supersymmetric models defined by (1) the situation is different. Supersymmetry can cause a cancelation of the transverse zero-point energy. Then the wave function is no longer confined, indicating that the supersymmetric models have a continuous spectrum. The latter was rigourously proven for the gauge group SU(N) [5].

For the supermembrane, the classical zero-mass configurations correspond to zero-area stringlike configurations of arbitrary length. As the supermembrane mass is described by a Hamiltonian of the type (1), the mass spectrum of the supermembrane is continuous for the same reasons as given above. For a supermembrane moving in a target space with compact dimensions, winding may raise the mass of the membrane state. This is so because winding in more than one direction gives rise to a nonzero central charge in the supersymmetry algebra, which sets a lower limit on the membrane mass. This fact should not be interpreted as an indication that the spectrum becomes discrete. The possible continuity of the spectrum hinges on the two features mentioned above. First the system should possess continuous valleys of classically degenerate states. Qualitatively one recognizes immediately that this feature is not directly affected by winding. A classical membrane with winding can still have stringlike configurations of arbitrary length, without increasing its area. Hence the classical instability persists. The second feature is supersymmetry. Without winding it is clear that the valley configurations are supersymmetric, so that one concludes that the spectrum is continuous. With winding the latter aspect is more subtle. However, we note that, when the winding density is concentrated in one part of the spacesheet, then valleys can emerge elsewhere corresponding to stringlike configurations with supersymmetry. Hence, as a space-sheet local field theory, supersymmetry can be broken in one region where the winding is concentrated and unbroken in another. In the latter region stringlike configurations can form, which, at least semiclassically, will not be suppressed by quantum corrections [6]. However, in this case we can only describe the generic features of the spectrum. Our arguments do not preclude the existence of mass gaps.

Finally, whether or not the Hamiltonian (1) allows normalizable or localizable zero-energy states, superimposed on the continuous spectrum, is a subtle question. Early discussion on the existence of such zero-energy states can be found in [2, 7]; more recent discussions can be found in [8, 9]. According to [9] such states do indeed exist in $d = 9$. There is an important difference between states whose energy is exactly equal to zero and states of positive energy. The supersymmetry algebra implies that zero-energy states are annihilated by the supercharges. Hence, they are supersinglets. The positive-energy states, on the other hand, must constitute full supermultiplets. So they are multiplets consisting of multiples of $1+1$, $2+2$, $8+8$, or $128+128$ bosonic + fermionic states, corresponding to $d = 2, 3, 5$ or 9, respectively.

To prove or disprove the existence of discrete states with winding is even more difficult. While the contribution of the bosonic part of the Hamiltonian increases by concentrating the winding density on part of the spacesheet, the matrix elements in the fermionic directions will also grow large, making it difficult to estimate the eigenvalues. At this moment the only rigorous result is the BPS bound that follows from the supersymmetry algebra. Obviously, the state of lowest mass for given winding numbers is always a BPS state, which is invariant under some residual supersymmetry. The counting of states proceeds in a way that is rather similar to the case of no winding.

2 Supermembranes

Fundamental supermembranes can be described in terms of actions of the Green-Schwarz type, possibly in a nontrivial but restricted (super)spacetime background [10]. Such actions exist for supersymmetric p-branes, where $p = 0, 1, \ldots$ defines the spatial dimension of the brane. Thus for $p = 0$ we have a superparticle, for $p = 1$ a superstring, for $p = 2$ a supermembrane, and so on. The dimension of spacetime in which the superbrane can live is very restricted. These restrictions arise from the fact that the action contains a Wess-Zumino-Witten term, whose supersymmetry depends sensitively on the spacetime dimension. If the coefficient of this term takes a particular value then the action possesses an additional fermionic gauge symmetry, the so-called κ-symmetry. This symmetry is necessary to ensure the matching of (physical) bosonic and fermionic degrees of freedom. In the following we restrict ourselves to supermembranes (i.e., $p = 2$) in 11 dimensions.

The supermembrane action [10] is written in terms of superspace embedding coordinates $Z^M(\zeta) = (X^\mu(\zeta), \theta(\zeta))$, which are functions of the three world-volume coordinates ζ^i ($i = 0, 1, 2$). It takes the following form,

$$S[Z(\zeta)] = \int \mathrm{d}^3\zeta \left[-\sqrt{-g(Z(\zeta))} - \tfrac{1}{6}\varepsilon^{ijk}\,\Pi_i^A\Pi_j^B\Pi_k^C\, B_{CBA}(Z(\zeta))\right], \tag{4}$$

where $\Pi_i^A = \partial Z^M/\partial\zeta^i\, E_M^A$ and the induced metric equals $g_{ij} = \Pi_i^r\Pi_j^s\,\eta_{rs}$, with η_{rs} the constant Lorentz-invariant metric. Flat superspace is characterized by

$$\begin{aligned}
E_\mu{}^r &= \delta_\mu{}^r\,, & E_\mu{}^a &= 0\,,\\
E_\alpha{}^a &= \delta_\alpha{}^a\,, & E_\alpha{}^r &= -(\bar\theta\Gamma^r)_\alpha\,,\\
B_{\mu\nu\alpha} &= (\bar\theta\Gamma_{\mu\nu})_\alpha\,, & B_{\mu\alpha\beta} &= (\bar\theta\Gamma_{\mu\nu})_{(\alpha}\,(\bar\theta\Gamma^\nu)_{\beta)}\,,\\
B_{\alpha\beta\gamma} &= (\bar\theta\Gamma_{\mu\nu})_{(\alpha}\,(\bar\theta\Gamma^\mu)_\beta\,(\bar\theta\Gamma^\nu)_{\gamma)}\,, & B_{\mu\nu\rho} &= 0\,.
\end{aligned} \tag{5}$$

The gamma matrices are denoted by Γ^r; gamma matrices with more than one index denote antisymmetrized products of gamma matrices with unit weight. In flat superspace the supermembrane Lagrangian, written in components, reads (in the notation and conventions of [2]),

$$\mathcal{L} = -\sqrt{-g(X,\theta)} - \varepsilon^{ijk}\,\bar\theta\Gamma_{\mu\nu}\partial_k\theta\left[\tfrac{1}{2}\,\partial_i X^\mu(\partial_j X^\nu + \bar\theta\Gamma^\nu\partial_j\theta) + \tfrac{1}{6}\,\bar\theta\Gamma^\mu\partial_i\theta\,\bar\theta\Gamma^\nu\partial_j\theta\right], \tag{6}$$

The target space can have compact dimensions which permit winding membrane states [6]. In flat superspace the induced metric,

$$g_{ij} = (\partial_i X^\mu + \bar\theta\Gamma^\mu\partial_i\theta)(\partial_j X^\nu + \bar\theta\Gamma^\nu\partial_j\theta)\,\eta_{\mu\nu}\,, \tag{7}$$

is supersymmetric. Therefore the first term in (6) is trivially invariant under spacetime supersymmetry. In 4, 5, 7, or 11 spacetime dimensions the second term proportional to ε^{ijk} is also supersymmetric (up to a total divergence) and the full action is invariant under κ-symmetry.

In the case of the open supermembrane, κ-symmetry imposes boundary conditions on the fields [11]. They must ensure that the following integral over the boundary of the membrane world volume vanishes,

$$\begin{aligned}
\int_{\partial M} \Big[&\tfrac{1}{2}\mathrm{d}X^\mu\wedge(\mathrm{d}X^\nu + \bar\theta\Gamma^\nu\mathrm{d}\theta)\,\bar\theta\Gamma_{\mu\nu}\delta_\kappa\theta + \tfrac{1}{6}\bar\theta\Gamma^\mu\mathrm{d}\theta\wedge\bar\theta\Gamma^\nu\mathrm{d}\theta\,\bar\theta\Gamma_{\mu\nu}\delta_\kappa\theta\\
&+\tfrac{1}{2}(\mathrm{d}X^\mu - \tfrac{1}{3}\bar\theta\Gamma^\mu\mathrm{d}\theta)\wedge\bar\theta\Gamma_{\mu\nu}\mathrm{d}\theta\,\bar\theta\Gamma^\nu\delta_\kappa\theta\Big] = 0\,.
\end{aligned} \tag{8}$$

This can be achieved by having a "membrane D-p-brane" at the boundary with $p = 1, 5$, or 9, which is defined in terms of $(p+1)$ Neumann and $(10-p)$ Dirichlet boundary conditions for the X^μ, together with corresponding boundary conditions on the fermionic coordinates. More explicitly, we define projection operators

$$\mathcal{P}_\pm = \tfrac{1}{2}\Big(\mathbf{1} \pm \Gamma^{p+1}\,\Gamma^{p+2}\cdots\Gamma^{10}\Big)\,, \tag{9}$$

and impose the Dirichlet boundary conditions

$$\begin{aligned}
\partial_\parallel X^M| &= 0\,, \qquad M = p+1,\ldots,10\,,\\
\mathcal{P}_-\theta| &= 0\,,
\end{aligned} \tag{10}$$

where $\partial_\perp$ and $\partial_\|$ define the world-volume derivatives perpendicular or tangential to the surface swept out by the membrane boundary in the target space. Note that the fermionic boundary condition implies that $\mathcal{P}_-\partial_\|\theta = 0$. Furthermore, it implies that spacetime supersymmetry is reduced to only 16 supercharges associated with spinor parameters $\mathcal{P}_+\epsilon$, which is *chiral* with respect to the $(p+1)$-dimensional world volume of the D-p-brane at the boundary. With respect to this reduced supersymmetry, the superspace coordinates decompose into two parts, one corresponding to $(X^M, \mathcal{P}_-\theta)$ and the other corresponding to $(X^m, \mathcal{P}_+\theta)$ where $m = 0, 1, \ldots, p$. While for the five-brane these superspaces exhibit a somewhat balanced decomposition in terms of an equal number of bosonic and fermionic coordinates, the situation for $p = 1, 9$ shows heterotic features in that one space has an excess of fermionic and the other an excess of bosonic coordinates. Moreover, we note that supersymmetry may be further broken, e.g. by choosing different Dirichlet conditions on nonconnected segments of the supermembrane boundary.

The Dirichlet boundary conditions can be supplemented by the following Neumann boundary conditions,

$$\begin{aligned} \partial_\perp X^m| &= 0 \qquad m = 0, 1, \ldots, p\,, \\ \mathcal{P}_+\partial_\perp\theta| &= 0\,. \end{aligned} \tag{11}$$

These do not lead to a further breakdown of the rigid spacetime symmetries.

We now continue and follow the light-cone quantization described in [2]. The supermembrane Hamiltonian takes the form

$$H = \frac{1}{P_0^+} \int \mathrm{d}^2\sigma \sqrt{w} \left[\frac{P^a P_a}{2\,w} + \tfrac{1}{4}\{X^a, X^b\}^2 - P_0^+ \bar\theta \gamma_- \gamma_a \{X^a, \theta\} \right]. \tag{12}$$

Here the integral runs over the spatial components of the world volume denoted by σ^1 and σ^2, while $P^a(\sigma)$ $(a = 2, \ldots, 9)$ are the momenta conjugate to the transverse coordinates X^a. In this gauge the light-cone coordinate $X^+ = (X^1 + X^0)/\sqrt{2}$ is linearly related to the world-volume time denoted by τ. The momentum P_- is time independent and proportional to the center-of-mass value $P_0^+ = (P_-)_0$ times some density $\sqrt{w(\sigma)}$ of the spacesheet, whose spacesheet integral is normalized to unity. The center-of-mass momentum P_0^- is equal to minus the Hamiltonian (12) subject to the gauge condition $\gamma_+\theta = 0$. And finally we made use of the Poisson bracket $\{A, B\}$ defined by

$$\{A(\sigma), B(\sigma)\} = \frac{1}{\sqrt{w(\sigma)}}\, \varepsilon^{rs}\, \partial_r A(\sigma)\, \partial_s B(\sigma). \tag{13}$$

Note that the coordinate $X^- = (X^1 - X^0)/\sqrt{2}$ itself does not appear in the Hamiltonian (12). It is defined via

$$P_0^+\, \partial_r X^- = -\frac{\mathbf{P}\cdot\partial_r\mathbf{X}}{\sqrt{w}} - P_0^+\, \bar\theta\gamma_-\partial_r\theta\,, \tag{14}$$

and implies a number of constraints that will be important in the following. Obviously, the right-hand side of (14) must be closed; without winding in X^-, it must be exact.

The equivalence of the large-N limit of SU(N) quantum mechanics with the closed supermembrane model is based on the residual invariance of the supermembrane action in the light-cone gauge. This invariance corresponds to the area-preserving diffeomorphisms of the membrane surface. These are defined by transformations of the worldsheet coordinates

$$\sigma^r \to \sigma^r + \xi^r(\sigma)\,, \tag{15}$$

with

$$\partial_r(\sqrt{w(\sigma)}\,\xi^r(\sigma)\,) = 0. \tag{16}$$

It is convenient to rewrite this condition in terms of dual spacesheet vectors by

$$\sqrt{w(\sigma)}\,\xi^r(\sigma) = \varepsilon^{rs}\,F_s(\sigma)\,. \tag{17}$$

In the language of differential forms the condition (16) may then be simply recast as $\mathrm{d}F = 0$. The trivial solutions are the exact forms $F = \mathrm{d}\xi$, or in components,

$$F_s = \partial_s\xi(\sigma)\,, \tag{18}$$

for any globally defined function $\xi(\sigma)$. The nontrivial solutions are the closed forms which are not exact. On a Riemann surface of genus g there are precisely $2g$ linearly independent non-exact closed forms, whose integrals along the homology cycles are normalized to unity[1]. In components we write

$$F_s = \phi_{(\lambda)\,s}\,, \qquad \lambda = 1,\ldots,2g\,. \tag{19}$$

The commutator of two infinitesimal area-preserving diffeomorphisms is determined by the product rule

$$\xi_r^{(3)} = \partial_r\left(\frac{\epsilon^{st}}{\sqrt{w}}\xi_s^{(2)}\xi_t^{(1)}\right)\,, \tag{20}$$

where both $\xi_r^{(1,2)}$ are closed vectors. Because $\xi_r^{(3)}$ is exact, the exact vectors thus generate an invariant subgroup of the area-preserving diffeomorphisms. As we shall discuss in the next section this subgroup can be approximated by SU(N) in the large-N limit, at least for closed membranes. For open membranes the boundary conditions on the fields (10) lead to a smaller group, such as SO(N).

The presence of the closed but non-exact forms is crucial for the winding of the embedding coordinates. More precisely, while the momenta $\mathbf{P}(\sigma)$ and the fermionic coordinates $\theta(\sigma)$ remain single valued on the spacesheet, the embedding coordinates, written as one-forms with components $\partial_r\mathbf{X}(\sigma)$ and $\partial_r X^-(\sigma)$, are decomposed into

[1] In the mathematical literature the globally defined exact forms are called "hamiltonian vector fields", whereas the closed but not exact forms which are not globally defined go under the name "locally hamiltonian vector fields".

closed one-forms. Their non-exact contributions are multiplied by an integer times the length of the compact direction. The constraint alluded to above amounts to the condition that the right-hand side of (14) is closed.

Under the full group of area-preserving diffeomorphisms the fields X^a, X^- and θ transform according to

$$\delta X^a = \frac{\varepsilon^{rs}}{\sqrt{w}}\,\xi_r\,\partial_s X^a\,, \quad \delta X^- = \frac{\varepsilon^{rs}}{\sqrt{w}}\,\xi_r\,\partial_s X^-\,, \quad \delta\theta^a = \frac{\varepsilon^{rs}}{\sqrt{w}}\,\xi_r\,\partial_s\theta\,, \tag{21}$$

where the time-dependent reparametrization ξ_r consists of closed exact and non-exact parts. Accordingly there is a gauge field ω_r, which is therefore closed as well and transforming as

$$\delta\omega_r = \partial_0\xi_r + \partial_r\Big(\frac{\varepsilon^{st}}{\sqrt{w}}\,\xi_s\,\omega_t\Big)\,. \tag{22}$$

Corresponding covariant derivatives are

$$D_0X^a = \partial_0X^a - \frac{\varepsilon^{rs}}{\sqrt{w}}\,\omega_r\,\partial_sX^a\,, \qquad D_0\theta = \partial_0\theta - \frac{\varepsilon^{rs}}{\sqrt{w}}\,\omega_r\,\partial_s\theta\,, \tag{23}$$

and likewise for D_0X^-.

The action corresponding to the following Lagrangian density is then gauge invariant under the transformations (21) and (22),

$$\begin{aligned}\mathcal{L} \;=\; & P_0^+\sqrt{w}\Big[\tfrac{1}{2}\,(D_0\mathbf{X})^2 + \bar\theta\,\gamma_-\,D_0\theta - \tfrac{1}{4}\,(P_0^+)^{-2}\,\{X^a,X^b\}^2 \\ & +(P_0^+)^{-1}\,\bar\theta\,\gamma_-\,\gamma_a\,\{X^a,\theta\} + D_0X^-\Big]\,,\end{aligned} \tag{24}$$

where we draw attention to the last term proportional to X^-, which can be dropped in the absence of winding. Moreover, we note that for open supermembranes, (24) is invariant under the transformations (21) and (22) only if $\xi_{||} = 0$ holds on the boundary. This condition defines a subgroup of the group of area-preserving transformations, which is consistent with the Dirichlet conditions (10). Observe that here $\partial_{||}$ and $\partial_\perp$ refer to the *spacesheet* derivatives tangential and perpendicular to the membrane boundary[2].

The action corresponding to (24) is also invariant under the supersymmetry transformations

$$\begin{aligned}\delta X^a &= -2\,\bar\epsilon\,\gamma^a\,\theta\,, \\ \delta\theta &= \tfrac{1}{2}\gamma_+\,(D_0X^a\,\gamma_a + \gamma_-)\,\epsilon + \tfrac{1}{4}(P_0^+)^{-1}\,\{X^a,X^b\}\,\gamma_+\,\gamma_{ab}\,\epsilon, \\ \delta\omega_r &= -2\,(P_0^+)^{-1}\,\bar\epsilon\,\partial_r\theta\,.\end{aligned} \tag{25}$$

The supersymmetry variation of X^- is not relevant and may be set to zero. For the open case one finds that the boundary conditions $\omega_{||} = 0$ and $\epsilon = \mathcal{P}_+\,\epsilon$ must

[2]Consistency of the Neumann boundary conditions (11) with the area-preserving diffeomorphisms (21) further imposes $\partial_\perp\xi^{||} = 0$ on the boundary, where indices are raised according to (17).

be fulfilled in order for (25) to be a symmetry of the action. In that case the theory takes the form of a gauge theory coupled to matter. The pure gauge theory is associated with the Dirichlet and the matter with the Neumann (bosonic and fermionic) coordinates.

In the case of a 'membrane D-9-brane' one now sees that the degrees of freedom on the 'end-of-the world' 9-brane precisely match those of 10-dimensional heterotic strings. *On* the boundary we are left with eight propagating bosons X^m (with $m = 2, \ldots, 9$), as X^{10} is constant on the boundary due to (10), paired with the 8-dimensional chiral spinors θ (subject to $\gamma_+\theta = \mathcal{P}_-\theta = 0$), i.e., the scenario of Hořava-Witten [12].

The full equivalence with the membrane Hamiltonian is now established by choosing the $\omega_r = 0$ gauge and passing to the Hamiltonian formalism. The field equations for ω_r then lead to the membrane constraint (14) (up to exact contributions), partially defining X^-. Moreover the Hamiltonian corresponding to the gauge theory Lagrangian of (24) is nothing but the light-cone supermembrane Hamiltonian (12). Observe that in the above gauge theoretical construction the space-sheet metric w_{rs} enters only through its density $\sqrt{w}$ and hence vanishing or singular metric components do not pose problems.

We are now in a position to study the full 11-dimensional supersymmetry algebra of the winding supermembrane. For this we decompose the supersymmetry charge Q associated with the transformations (25), into two 16-component spinors,

$$Q = Q^+ + Q^-, \quad \text{where} \quad Q^\pm = \tfrac{1}{2}\gamma_\pm\gamma_\mp Q\,, \tag{26}$$

to obtain

$$\begin{aligned} Q^+ &= \int \mathrm{d}^2\sigma \left(2\,P^a\gamma_a + \sqrt{w}\,\{X^a, X^b\}\,\gamma_{ab}\right)\theta\,,\\ Q^- &= 2\,P_0^+ \int \mathrm{d}^2\sigma\,\sqrt{w}\,\gamma_-\,\theta. \end{aligned} \tag{27}$$

In the presence of winding the supersymmetry algebra takes the form [6]

$$\begin{aligned} (\,Q^+_\alpha, \bar{Q}^+_\beta\,)_{\rm DB} &= 2\,(\gamma_+)_{\alpha\beta}\,H - 2\,(\gamma_a\gamma_+)_{\alpha\beta}\int \mathrm{d}^2\sigma\,\sqrt{w}\,\{X^a, X^-\}\,,\\ (\,Q^+_\alpha, \bar{Q}^-_\beta\,)_{\rm DB} &= -(\gamma_a\gamma_+\gamma_-)_{\alpha\beta}\,P_0^a - \tfrac{1}{2}\,(\gamma_{ab}\gamma_+\gamma_-)_{\alpha\beta}\int \mathrm{d}^2\sigma\,\sqrt{w}\,\{X^a, X^b\}\,,\\ (\,Q^-_\alpha, \bar{Q}^-_\beta\,)_{\rm DB} &= -2\,(\gamma_-)_{\alpha\beta}\,P_0^+\,, \end{aligned} \tag{28}$$

where use has been made of the Dirac brackets of the phase-space variables and the defining equation (14) for $\partial_r X^-$.

The new feature of this supersymmetry algebra is the emergence of the central charges in the first two anticommutators, which are generated through the winding contributions. They represent topological quantities obtained by integrating the winding densities

$$z^a(\sigma) = \varepsilon^{rs}\,\partial_r X^a\,\partial_s X^- \tag{29}$$

and

$$z^{ab}(\sigma) = \varepsilon^{rs}\,\partial_r X^a\,\partial_s X^b \tag{30}$$

over the space-sheet. It is gratifying to observe the manifest Lorentz invariance of (28). Here we should point out that, in adopting the light-cone gauge, we assumed that there was no winding for the coordinate X^+. In [13] the corresponding algebra for the matrix regularization was studied. The result coincides with ours in the large-N limit, in which an additional longitudinal five-brane charge vanishes, provided that one identifies the longitudinal two-brane charge with the central charge in the first line of (28). This identification requires the definition of X^- in the matrix regularization, a topic that we return to in the next section. The form of the algebra is another indication of the consistency of the supermembrane-supergravity system.

Until now we discussed the general case of a flat target space with possible winding states. To make the identification with the matrix models more explicit, let is ignore the winding and split off the center-of-mass (CM) variables. First of all, the constant P_0^+ represents the membrane CM momentum in the direction associated with the coordinate X^-,

$$P_0^+ = \int \mathrm{d}^2\sigma\, P^+ . \tag{31}$$

The other CM coordinates and momenta are

$$\mathbf{P}_0 = \int \mathrm{d}^2\sigma\, \mathbf{P}\,, \qquad \mathbf{X}_0 = \int \mathrm{d}^2\sigma \sqrt{w(\sigma)}\,\mathbf{X}(\sigma)\,, \qquad \theta_0 = \int \mathrm{d}^2\sigma \sqrt{w(\sigma)}\,\theta(\sigma)\,. \tag{32}$$

In the light-cone gauge we are left with the transverse coordinates $\mathbf{X}$ and corresponding momenta $\mathbf{P}$, which transform as vectors under the SO(9) group of transverse rotations. Only sixteen fermionic components θ remain, which transform as SO(9) spinors. Furthermore we have the CM momentum P_0^+ and the CM coordinate X_0^- (the remaining modes in X^- are dependent), while the CM momentum P_0^- is equal to minus the supermembrane Hamiltonian and takes the following form

$$H = \frac{\mathbf{P}_0^2}{2P_0^+} + \frac{\mathcal{M}^2}{2P_0^+}\,. \tag{33}$$

Here $\mathcal{M}$ is the supermembrane mass operator, which does *not* depend on any of the CM coordinates or momenta. The explicit expression for $\mathcal{M}^2$ is

$$\mathcal{M}^2 = \int \mathrm{d}^2\sigma\, \sqrt{w(\sigma)}\Big[\frac{[\mathbf{P}^2(\sigma)]'}{w(\sigma)} + \tfrac{1}{2}\big(\{X^a, X^b\}\big)^2 - 2P_0^+\,\bar\theta\gamma_-\gamma_a\{X^a,\theta\}\Big]\,, \tag{34}$$

where $[\mathbf{P}^2]'$ indicates that the contribution of the CM momentum $\mathbf{P}_0$ is suppressed.

The structure of the Hamiltonian (33) shows that the wave functions for the supermembrane now factorize into a wave function pertaining to the CM modes and a wave function of the supersymmetric quantum-mechanical system that describes the other modes. For the latter the mass operator plays the role of the Hamiltonian. When the mass operator vanishes on the state, then the 32 supercharges act exclusively on the CM coordinates and generate a massless supermultiplet of eleven-dimensional supersymmetry. In case there is no other degeneracy beyond that caused by supersymmetry, the resulting supermultiplet is the one of supergravity, describing the graviton, the antisymmetric tensor and the gravitino. In terms of the $SO(9)$ helicity representations, it consists of $\mathbf{44} \oplus \mathbf{84}$ bosonic and $\mathbf{128}$ fermionic states. For an explicit construction of these states, see [14]. When the mass operator does not vanish on the states, we are dealing with huge supermultiplets consisting of multiples of $2^{15} + 2^{15}$ states.

3 The matrix approximation

The expressions for the Hamiltonian (12), the supercharges (27) and the constraints associated with (14) are clearly in direct correspondence with the Hamiltonian, supersymmetry charges and the Gauss constraints for the matrix models introduced in section 1. This correspondence between de supermembrane and supersymmetric quantum mechanics becomes exact after one replaces P_0^+ by the coupling constant g and rewrites the spinor coordinates in terms of a real SO(9) spinor basis. In order to make the relation more explicit one may expand functions on the spacesheet in a complete set of functions Y_A with $A = 0, 1, 2, \ldots, \infty$. It is convenient to choose $Y_0 = 1$. Furthermore we choose a basis of the closed one-forms, consisting of the exact ones, $\partial_r Y_A$, and a set of closed nonexact forms denoted by $\phi_{(\lambda)r}$. Completeness implies the following decompositions,

$$\begin{aligned} \{Y_A, Y_B\} &= f_{AB}{}^C\, Y_C\,, \\ \frac{\varepsilon^{rs}}{\sqrt{w}}\, \phi_{(\lambda)r}\, \partial_s Y_A &= f_{\lambda A}{}^B\, Y_B\,, \\ \frac{\varepsilon^{rs}}{\sqrt{w}}\, \phi_{(\lambda)r}\, \phi_{(\lambda')s} &= f_{\lambda\lambda'}{}^A\, Y_A\,, \end{aligned} \tag{35}$$

so that the constants $f^{AB}{}_C$, $f_{\lambda A}{}^B$ and $f_{\lambda\lambda'}{}^A$ represent the structure constants of the infinite-dimensional group of area-preserving diffeomorphisms. Lowering of indices can be done with the help of the invariant metric

$$\eta_{AB} = \int \mathrm{d}^2\sigma\, \sqrt{w(\sigma)}\, Y_A(\sigma)\, Y_B(\sigma)\,. \tag{36}$$

There is no need to introduce a metric for the λ indices. Observe that we have $\eta_{00} = 1$. Furthermore it is convenient to choose the functions Y_A with $A \geq 1$ such that $\eta_{0A} = 0$. Completeness implies

$$\eta^{AB}\, Y_A(\sigma)\, Y_B(\rho) = \frac{1}{\sqrt{w(\sigma)}}\, \delta^{(2)}(\sigma, \rho)\,. \tag{37}$$

After lowering of upper indices, the structure constants are defined as follows [3, 6],

$$f_{ABC} = \int \mathrm{d}^2\sigma\, \varepsilon^{rs}\, \partial_r Y_A(\sigma)\, \partial_s Y_B(\sigma)\, Y_C(\sigma)\,,$$

$$
\begin{aligned}
f_{\lambda BC} &= \int \mathrm{d}^2\sigma\, \varepsilon^{rs}\, \phi_{(\lambda)\,r}(\sigma)\, \partial_s Y_B(\sigma)\, Y_C(\sigma)\,, \\
f_{\lambda\lambda' C} &= \int \mathrm{d}^2\sigma\, \varepsilon^{rs}\, \phi_{(\lambda)\,r}(\sigma)\, \phi_{(\lambda')\,s}(\sigma)\, Y_C(\sigma)\,.
\end{aligned} \tag{38}
$$

Note that we have $f_{AB0} = f_{\lambda B0} = 0$.

Using the above basis one may write down the following mode expansions for the phase-space variables of the supermembrane,

$$
\begin{aligned}
\partial_r \mathbf{X}(\sigma) &= \sum_\lambda \mathbf{X}^\lambda\, \phi_{(\lambda)\,r}(\sigma) + \sum_A \mathbf{X}^A\, \partial_r Y_A(\sigma)\,, \\
\mathbf{P}(\sigma) &= \sum_A \sqrt{w(\sigma)}\, \mathbf{P}^A\, Y_A(\sigma)\,, \\
\theta(\sigma) &= \sum_A \theta^A\, Y_A(\sigma)\,,
\end{aligned} \tag{39}
$$

introducing winding modes for the transverse coordinates $\mathbf{X}$. A similar expansion exists for X^-.

Other tensors are needed, for instance, to write down the Lorentz algebra generators [3]. An obvious tensor is given by

$$
d_{ABC} = \int \mathrm{d}^2\sigma\, \sqrt{w(\sigma)}\, Y_A(\sigma)\, Y_B(\sigma)\, Y_C(\sigma)\,, \tag{40}
$$

which is symmetric in all three indices and satisfies $d_{AB0} = \eta_{AB}$. Another tensor, whose definition is more subtle, arises when expressing X^- in terms of the other coordinates and momenta. We recall that X^- is restricted by (14), which implies the following Gauss-type constraint,

$$
\varphi^A = f_{BC}{}^A \Big[\mathbf{P}^B \cdot \mathbf{X}^C + P_0^+\, \bar\theta^B \gamma_- \theta^C\Big] + f_{B\lambda}{}^A\, \mathbf{P}^B \cdot \mathbf{X}^\lambda \approx 0\,. \tag{41}
$$

The coordinate X^- receives contributions proportional to $Y_A(\sigma)$, which can be parametrized by ($A \neq 0$)

$$
X_A^- \approx \frac{1}{2P_0^+}\, c^A{}_{BC} \Big[\mathbf{P}^B \cdot \mathbf{X}^C + P_0^+\, \bar\theta^B \gamma_- \theta^C\Big] + \frac{1}{2P_0^+}\, c^A{}_{B\lambda}\, \mathbf{P}^B \cdot \mathbf{X}^\lambda\,. \tag{42}
$$

In addition X^- has CM and winding modes. Observe that the tensors $c^A{}_{BC}$ and $c^A{}_{B\lambda}$ are somewhat ambiguous, as (42) is only defined up to the constraints (41). The symmetric component of $c^A{}_{BC}$ is, however, fixed and given by $c^A{}_{BC} + c^A{}_{CB} = -2d_{ABC}$. Note that $c^A{}_{B0} = 0$. There are many other identities between the various tensors, such as [3],

$$
\begin{aligned}
& f_{[AB}{}^E\, f_{C]E}{}^D = d_{(AB}{}^E\, f_{C)E}{}^D = d_{ABC}\, f_{[DE}{}^B\, f_{FG]}{}^C = \\
& c_{DE}{}^{[A}\, f^{BC]E} = d_{EA[B} d_{C]D}{}^E = 0\,.
\end{aligned} \tag{43}
$$

If we replace the group of the area-preserving diffeomorphisms by a finite group, then (34) defines the Hamiltonian of a supersymmetric quantum-mechanical system based on a finite number of degrees of freedom [15]. In the limit to the infinite-dimensional group we thus recover the supermembrane. This observation enables one to regularize the supermembrane in a supersymmetric way by considering a limiting procedure based on a sequence of groups whose limit yields the area-preserving diffeomorphisms. For closed membranes of certain topology it is known how to approximate a (sub)group of the area-preserving diffeomorphisms as a particular $N \to \infty$ limit of SU(N). To be precise, it can be shown that the structure constants of SU(N) tend to those of the diffeomorphism subgroup associated with the hamiltonian vectors, up to corrections of order $1/N^2$. While some of the identities (43) remain valid at finite N, others receive corrections of order $1/N^2$. Furthermore, the tensors $c^A{}_{BC}$ and $c^A{}_{B\lambda}$ are intrinsically undefined at finite N. Therefore, the expression for X^- is ambiguous for the matrix model and Lorentz invariance holds only in the large-N limit [3, 4].

The nature of the large-N limit itself is subtle and depends on the membrane topology. As long as N is finite, no distinction can be made with regard to the topology. In some sense, all topologies are thus included at the level of finite N. However, the diffeomorphisms associated with the harmonic vectors are problematic, because they cannot be incorporated for finite N, at least not at the level of the Lie algebra. This was shown in [3], where it was established that the finite-N approximation to the structure constants $f_{\lambda BC}$ violates the Jacobi identities for a toroidal membrane. Therefore it seems impossible to present a matrix model regularization of the supermembrane with winding contributions. There exists a standard prescription for dealing with matrix models with winding [16], however, which is therefore conceptually different. The consequences of this difference are not well understood. The prescription amounts to adopting the gauge group $[\mathrm{U}(N)]^M$, for winding in one dimension, which in the limit $M \to \infty$ leads to supersymmetric Yang-Mills theories in $1+1$ dimensions [16]. Hence, in this way it is possible to extract extra dimensions from a suitably chosen infinite-dimensional gauge group. Obviously this approach can be generalized to a hypertorus.

Finally we add that the matrix regularization works also for the case of open supermembranes. In that case one deals with certain subgroups of SU(N). We refer to [11] for further details.

4 Membranes and matrix models in curved space

So far we considered a supermembrane moving in a flat target superspace. To that order we substituted the flat superspace expressions (5) into the supermembrane action (4). However, these expression can in principle be evaluated for nontrivial backgrounds, such as those induced by a nontrivial target-space metric, a target-space tensor field and a target-space gravitino field, corresponding to the fields of (on-shell) 11-dimensional supergravity. This background can in principle be incorporated into superspace by a procedure known as 'gauge completion' [17]. For 11-dimensional supergravity, the first steps of this procedure have been carried out long ago [18], but unfortunately only to first order in fermionic coordinates θ.

For brevity of the presentation, let us just confine ourselves to the purely bosonic case and present the light-cone formulation of the membrane in a background consisting of the metric $G_{\mu\nu}$ and the tensor gauge field $C_{\mu\nu\rho}$ [19]. The Lagrangian density for the bosonic membrane follows directly from (4),

$$\mathcal{L} = -\sqrt{-g} + \tfrac{1}{6}\varepsilon^{ijk}\partial_i X^\mu\, \partial_j X^\nu\, \partial_k X^\rho\, C_{\mu\nu\rho}\,, \tag{44}$$

where $g_{ij} = \partial_i X^\mu\, \partial_j X^\nu\, \eta_{\mu\nu}$. For the light-cone formulation, the coordinates are treated in the usual fashion in terms of light-cone coordinates $X^\pm$ and transverse coordinates $\mathbf{X}$. Furthermore we use the diffeomorphisms in the target space to bring the metric in a convenient form [20],

$$G_{--} = G_{a-} = 0\,. \tag{45}$$

Following the same steps as for the membrane in flat space, discussed in section 2, one again derives a Hamiltonian formulation. Interestingly enough, the constraint takes the same form as (14). Of course, the definition of the momenta in terms of the coordinates and their derivatives does involve the background fields, but at the end all explicit dependence on the background cancels out.

The Hamiltonian now follows straightforwardly. After additional gauge choices,

$$C_{+-a} = 0\,, \quad C_{-ab} = 0\,, \quad G_{+-} = 1\,, \tag{46}$$

it takes the form

$$\begin{aligned} H &= \int \mathrm{d}^2\sigma \Big\{ \frac{1}{P_-}\Big[\tfrac{1}{2}(P_a - C_a - P_-\, G_{a+})^2 + \tfrac{1}{4}(\varepsilon^{rs}\,\partial_r X^a\,\partial_s X^b)^2\Big] \\ &\qquad -\tfrac{1}{2}P_-\, G_{++} - \tfrac{1}{2}\varepsilon^{rs}\,\partial_r X^a\,\partial_s X^b\, C_{+ab}\Big\}\,, \end{aligned} \tag{47}$$

We want to avoid explicit time dependence of the background fields, so we assume the metric and the tensor field to be independent of X^+. If we assume, in addition, that they are independent of X^-, it turns out that P_- becomes τ-independent. This allows us to set $P_-(\sigma) = (P_-)_0\,\sqrt{w(\sigma)}$, exactly as in flat space. With these restrictions, it is possible to write down a gauge theory of area-preserving diffeomorphisms for the membrane in the presence of background fields. Its Lagrangian density equals

$$\begin{aligned} w^{-1/2}\,\mathcal{L} &= \tfrac{1}{2}(D_0 X^a)^2 + D_0 X^a\left(\tfrac{1}{2}C_{abc}\{X^b, X^c\} + G_{a+}\right) \\ &\qquad -\tfrac{1}{4}\{X^a, X^b\}^2 + \tfrac{1}{2}G_{++} + \tfrac{1}{2}C_{+ab}\{X^a, X^b\}\,, \end{aligned} \tag{48}$$

where we used the metric G_{ab} to contract transverse indices; the Poisson bracket and the covariant derivatives were already introduced in section 2. For convenience we have set $(P_-)_0 = 1$.

The action corresponding to (48) is manifestly invariant under area-preserving diffeomorphisms in the presence of the background fields. It is now straightforward to write it in terms of a matrix model, by truncating the mode expansion for coordinates and momenta as explained in the previous section. Matrix models in curved space have been discussed before [21]; for more recent papers dealing with matrix models in the presence of certain backgrounds, see [22]. A more explicit derivation of the results of this section and their supersymmetric extension will appear in a forthcoming publication [23].

5 The continuous supermembrane mass spectrum

The continuous mass spectrum of the supermembrane forms an obstacle in interpreting the membrane states as elementary particles, in analogy to what is done in string theory. Instead the continuity of the spectrum should be viewed as a result of the fact that supermembrane states do not really exist as asymptotic states. The membrane collapses into stringlike configurations and is to be interpreted as a multimembrane state. Obviously such states exhibit a continuous mass spectrum. As we alluded to earlier, there is evidence that massless ground states exist, probably associated with the states of 11-dimensional supergravity [9]. In the winding sector there may exist massive BPS states, which are the lowest-mass states for given winding number. Whether additional non-BPS bound states exist is not known. It could be that beyond the massless and BPS winding states, there is nothing than a continuum of multimembrane states.

Qualitatively, the situation is the same for the matrix models (1) based on a finite number of degrees of freedom. Among the zero-energy states there are those where the matrices take a block-diagonal form, which can be regarded as a direct product of states belonging to lower-rank matrix models [24]. The fact that the moduli space of ground states, whose nature is protected by supersymmetry at the quantum-mechanical level, is isomorphic to $\mathbf{R}^{9N}/S_N$, is already indicative of a corresponding description in terms of an N-particle Fock space. The finite-N matrix models have an independent interpretation in string theory. Strings can end on certain defects by means of Dirichlet boundary conditions. These defects are called D-branes (for further references, see [25]). They can have a p-dimensional spatial extension and carry Ramond-Ramond charges [26]. D-Branes play an important role in the non-perturbative behaviour of string theory. The models of section 1 are relevant for D0-branes (Dirichlet particles), but we note in passing that there are similar models relevant for higher-dimensional D-branes, which emerge in the zero-volume limit of supersymmetric gauge theories coupled to matter.

The effective short-distance description for D-branes can be derived from simple arguments [27]. As the strings must be attached to the p-dimensional branes, we are dealing with open strings whose endpoints are attached to a p-dimensional subspace. At short distances, the interactions caused by these open strings are determined by the massless states of the open string, which constitute the ten-dimensional Yang-Mills supermultiplet, propagating in a reduced $(p+1)$-dimensional spacetime. Because the endpoints of open strings carry Chan-Paton factors the effective short-distance behaviour of N D-branes can be described in terms of a $\mathrm{U}(N)$ ten-dimensional supersymmetric gauge theory reduced to the $(p+1)$-dimensional world volume of the D-brane. The $\mathrm{U}(1)$ subgroup is associated with the center-of-mass motion of the N D-branes.

In the type-IIA superstring one has Dirichlet particles moving in a 9-dimensional space. As the world volume of the particles is one-dimensional ($p = 0$), the short-distance interactions between these particle is thus described by the model of section 1 with gauge group $\mathrm{U}(N)$ and $d = 9$. The continuous spectrum without gap is natural here, as it is known that, for static D-branes, the Ramond-Ramond repulsion cancels against the gravitational and dilaton atraction, a similar phenomenon as for BPS monopoles. With this gauge group the coordinates can be described in terms of $N \times N$ hermitean matrices. The valley configurations correspond to the situation where all these matrices can be diagonalized simultanously. The eigenvalues then define the positions of N D-particles in the 9-dimensional space. As soon as one or several of these particles coincide then the $[\mathrm{U}(1)]^N$ symmetry that is left invariant in the valley, will be enhanced to a nonabelian subgroup of $\mathrm{U}(N)$. Clearly there are more degrees of freedom than those corresponding to the D-particles, which are associated with the strings stretching between the D-particles. As we alluded to above the model naturally incorporates configurations corresponding to widely separated clusters of D-particles, each of which can be described by a supersymmetric quantum-mechanics model based on the product of a number of $\mathrm{U}(k)$ subgroups forming a maximal commuting subgroup of $\mathrm{U}(N)$. When all the D-particles move further apart this corresponds to configurations deeper and deeper into the potential valleys. These D-particles thus define an independent perspective on the models introduced in section 1, which can be used to study their dynamics. We refer to [28] for work along these lines.

The study of D-branes was further motivated by a conjecture according to which the degrees of freedom of M-theory are fully captured by the $\mathrm{U}(N)$ super-matrix models in the $N \to \infty$ limit [24]. The elusive M-theory is defined as the strong-coupling limit of type-IIA string theory and is supposed to capture all the relevant degrees of freedom of all known string theories, both at the perturbative and the nonperturbative level [29, 30]. In this description the various string-string dualities are fully incorporated. At large distances M-theory is described by 11-dimensional supergravity. A direct relation between supermembranes and type-IIA string theory was emphasized in [29], based on the relation between extremal black holes in 10-dimensional supergravity [31] and the Kaluza-Klein states of 11-dimensional supergravity in an S^1 compactification. In this compactification the Kaluza-Klein photon coincides with the Ramond-Ramond vector field of type-IIA string theory. Therefore Kaluza-Klein states are BPS states whose Ramond-Ramond charge is proportional to their mass. Hence they have the same characteristics as the Dirichlet particles. On the other hand, the effective interaction between infinitely many Dirichlet particles leads to a theory that is identical to that of an elementary supermembrane. There are alternative compactifications of M-theory which make contact with other string theories. Supermembranes have been used to provide evidence for the duality of M-theory on $\mathbf{R}^{10} \times S_1/\mathbf{Z}_2$ and 10-dimensional $E_8 \times E_8$ heterotic strings [12]. Finally the so-called double-dimensional reduction of membranes leads to fundamental string states [32].

Acknowledgements

Most of the work reported here was carried out in collaboration with K. Peeters and J.C. Plefka. I thank the organizers of this symposium for providing a stimulating atmosphere.

References

[1] M. Claudson and M.B. Halpern, *Nucl. Phys.* **B250** (1985) 689; R. Flume, *Ann. Phys.* **164** (1985) 189; M. Baake, P. Reinicke, and V. Rittenberg, *J. Math. Phys.* **26** (1985) 1070.

[2] B. de Wit, J. Hoppe and H. Nicolai, *Nucl. Phys.* **B305** [FS23] (1988) 545.

[3] B. de Wit, U. Marquard and H. Nicolai, *Commun. Math. Phys.* **128** (1990) 39.

[4] K. Ezawa, Y. Matsue and K. Murakami, *Lorentz symmetry of supermembrane in light cone gauge formulation*, `hep-th/9705005`;
S. Melosch, Diplomarbeit, Univ. Hamburg (unpublished).

[5] B. de Wit, M. Lüscher and H. Nicolai, *Nucl. Phys.* **B320** (1989) 135.

[6] B. de Wit, K. Peeters and J.C. Plefka, *Phys. Lett.* **B409** (1997) 117, `hep-th/9705225`; *Nucl. Phys. B (Proc. Suppl.)* **62A-C** (1998) 405, `hep-th/9707261`.

[7] B. de Wit and H. Nicolai, in proc. Trieste Conference on Supermembranes and Physics in $2+1$ dimensions, p. 196, eds. M.J. Duff, C.N. Pope and E. Sezgin (World Scient., 1990);
B. de Wit, *Nucl. Phys. B (Proc. Suppl.)* **56B** (1997) 76, `hep-th/9701169`.

[8] J. Fröhlich and J. Hoppe, *On zero mass ground states in supermembrane matrix models*, `hep-th/9701119`.

[9] S. Sethi and M. Stern, *D-Brane bound states redux*, `hep-th/9705046`;
M. Porrati and A. Rosenberg, *Bound states at threshold in supersymmetric quantum mechanics*, `hep-th/9708119`;

[10] E. Bergshoeff, E. Sezgin and P.K. Townsend, *Phys. Lett.* **189B** (1987) 75; *Ann. Phys.* **185** (1988) 330.

[11] A. Strominger, *Phys. Lett.* **B383** (1996) 44, `hep-th/9512059`;
K. Becker and M. Becker, *Nucl. Phys.* **B472** (1996) 221, `hep-th/9602071`;
M. Li, *Phys. Lett.* **B397** (1997) 37, `hep-th/9612144`;
N. Kim, and S.-J. Rey, *Nucl. Phys.* **B504** (1997) 189, `hep-th/9701139`;
Ph. Brax and J. Mourad, *Phys. Lett.* **B408** (1997) 142, `hep-th/9704165`; *Open supermembranes coupled to M Theory fivebranes*, `hep-th/9707246`;
K. Ezawa, Y. Matsuo and K. Murakami, *Matrix regularization of open supermembrane: Towards M Theory five-brane via open supermembrane*, `hep-th/9707200`;
B. de Wit, K. Peeters and J.C. Plefka, *Open and closed supermembranes with winding*, proc. Strings '97, *Nucl. Phys. B (Proc. Suppl.)*, to appear, `hep-th/9710215`.

[12] P. Hořava and E. Witten, *Nucl. Phys.* **B460** (1996) 506, hep-th/9510209; *Nucl. Phys.* **B475** (96) 94, hep-th/9603142.

[13] T. Banks, N. Seiberg and S.H. Shenker, *Nucl. Phys.* **B490** (1997) 91, hep-th/9612157.

[14] J. Plefka and A. Waldron, *Nucl. Phys.* **B512** (1998) 460, hep-th/9710104; *Asymptotic supergraviton states in Matrix Theory,* these proceedings, hep-th/9801093.

[15] J. Goldstone, unpublished; J. Hoppe, in proc. Int. Workshop on Constraint's Theory and Relativistic Dynamics, eds. G. Longhi and L. Lusanna (World Scient., 1987).

[16] W. Taylor, *Phys. Lett.* **B394** (1997) 283, hep-th/9611042;
O. Ganor, S. Rangoolam and W. Taylor, *Nucl. Phys.* **B492** (1997) 191, hep-th/9611202.

[17] L. Brink, M. Gell-Mann, P. Ramond and J.H. Schwarz, *Phys. Lett.* **74B** (1978) 336;
P. van Nieuwenhuizen and S. Ferrara, *Ann. Phys.* **127** (1980) 274.

[18] E. Cremmer and S. Ferrara, *Phys. Lett.* **91B** (1980) 61.

[19] B. de Wit, K. Peeters and J.C. Plefka, *Supermembranes and supermatrix models,* València workshop *Beyond the Standard Model; from Theory to Experiment,* October 13–17, 1997, to appear, hep-th/9712082.

[20] M. Goroff and J.H. Schwarz, *Phys. Lett.* **127B** (1983) 61.

[21] M.R. Douglas, H. Ooguri and S.H. Shenker, *Phys. Lett.* **B402** (1997) 36, hep-th/9702203;
M.R. Douglas, *D-branes in curved space,* hep-th/9703056;
M.R. Douglas, A. Kato and H. Ooguri, *D-brane actions on Kähler manifolds,* hep-th/9708012;
M.R. Douglas and H. Ooguri, *Why matrix theory is hard,* hep-th/9710178.

[22] A. Connes, M.R. Douglas and A. Schwarz, *Noncommutative geometry and matrix theory: compactification on tori,* hep-th/9711162;
M.R. Douglas and C. Hull, *D-branes and the Noncommutative Torus,* hep-th/9711165;
N.A. Obers, B. Pioline and E. Rabinovici, *M-Theory and U-duality on T^d with gauge backgrounds,* hep-th/9712084.

[23] B. de Wit, K. Peeters and J.C. Plefka, in preparation.

[24] T. Banks, W. Fischler, S.H. Shenker and L. Susskind, *Phys. Rev.* **D55** (1997) 5112, hep-th/9610043.

[25] J. Polchinski, S. Chaudhuri and C.V. Johnson, *Notes on D-branes*, hep-th/9602052;
J. Polchinski, *TASI Lectures on D-branes*, hep-th/9611050;
C. Bachas, in *Gauge Theories, Applied Supersymmetry and Quantum Gravity II*, p. 3, eds. A. Sevrin, K.S. Stelle, K. Thielemans, A. Van Proeyen, (Imperial College Press, 1997), hep-th/9701019.

[26] J. Polchinski, *Phys. Rev. Lett.* **75** (1995) 4724, hep-th/9510017.

[27] E. Witten, *Nucl.* Phys. **B460** (1996) 335, hep-th/9510135.

[28] U. Danielson, G. Ferretti and B. Sundborg, *Int. J. Mod. Phys.* **A11** (1996) 5463, hep-th/9603081;
D. Kabat and P. Pouliot, *Phys. Rev. Lett.* **77** (1996) 1004, hep-th/9603127;
M.R. Douglas, D. Kabat, P. Pouliot and S.H. Shenker, *Nucl. Phys.* **B458** (1997) 85, hep-th/9608024;
J. Polchinski and P. Pouliot, *Phys. Rev.* **D56** (1997) 6601, hep-th/9704029;
O. Aharony and M. Berkooz, *Nucl. Phys.* **B491** (1997) 184, hep-th/9611215;
G. Lifschytz and S.D. Mathur, *Nucl. Phys.* **B507** (1997) 621, hep-th/9612087;
D. Berenstein and R. Corrado, *Phys. Lett.* **B406** (1997) 37, hep-th/9702108;
V. Balasubramanian and F. Larsen, *Nucl. Phys.* **B506** (1997) 61, hep-th/9703039;
K. Becker and M. Becker, *Nucl. Phys.* **B506** (1997) 48, hep-th/9705091;
K. Becker and M. Becker, J. Polchinski and A. Tseytlin, *Phys. Rev.* **D65** (1997) 3174, hep-th/9706072.

[29] P.K. Townsend, *Phys. Lett.* **B350** (1995) 184, hep-th/9501068, *Phys. Lett.* **B373** (1996) 68, hep-th/9512062; *Four Lectures on M-theory*, lectures given at the 1996 ICTP Summer School in High Energy Physics and Cosmology, Trieste, hep-th/9612121.

[30] E. Witten, *Nucl. Phys.* **B443** (1995) 85, hep-th/9503124.

[31] G. Horowitz and A. Strominger, *Nucl. Phys.* **B360** (1991) 197.

[32] M.J. Duff, P.S. Howe, T. Inami and K.S. Stelle, *Phys. Lett.* **191B** (1987) 70.

Discretized random geometries; an approach to the non-perturbative quantization of gravity

Bengt Petersson

Fakultät für Physik, Universität Bielefeld,
D-33501 Bielefeld

1 Introduction

The statistical mechanics of random manifolds provides a possible framework for a non-perturbative construction of quantum theory of gravity (there exist several excellent reviews, e.g. [1], [2]). In this construction the standard recipes of lattice field theory are adopted, with a notable extension: the lattice itself becomes a dynamical object instead of being an inert scaffolding. For fixed topology, the summation over geometries involved in the partition function, approximating Feynman's path integral, is best implemented using the method of dynamical triangulations. The successes of this approach are particularly spectacular in two dimensions. The results obtained with models exactly solvable in the continuum formalism have been reproduced. Furthermore, completely new results, hardly attainable with another approach have been obtained. I refer to the talk of Jan Ambjorn at this conference for more details.

The method of dynamical triangulations, also called simplicial quantum gravity (SQG) can be generalized to higher dimensions, making possible a non-perturbative approach to Euclidean four dimensional quantum gravity. In this context, the discrete counterpart of the Einstein-Hilbert action arises in a most natural way, as I will explain below. The construction of a viable theory in four dimensions, if there is one, is more difficult than in two dimensions. In particular there exists up to now no useful analytic approach, like the matrix model in 2d. The theory in the continuum is perturbatively non-renormalizable, and the action is unbounded in the Euclidean region. It may be that a non-perturbative approach may cure these difficulties.

As discussed some time ago by Weinberg [3], one may use a $2 + \epsilon$ expansion, which indicates a non-trivial ultraviolet fixed point. This approach has been further developed by Kawai and collaborators [4].

In the SQG approach, the first task is to establish the phase structure. Then one may search for non-trivial fixed points. Furthermore one should establish the universality classes. In particular one would like to know the influence of matter fields. In two dimensions, as discussed above, this is to a large extent resolved, at least for scalar fields. In four dimensions it is still very much work in progress.

In this talk I will give an overview of what is known about the phase structure up to now. I will first discuss the two dimensional case. In particular I will introduce some analytic arguments of a mean field type. Furthermore I will report on some recent arguments that there actually is a change of regime for central charge $c > 1$. Generalizing this approach to four dimensions, I will in particular discuss the influence of (gauge) matter fields, and present some quite intriguing new results on this sector of the model.

2 2d simplicial quantum gravity coupled to matter fields

In two dimensions it has been shown that the SQG is equivalent to a matrix model. The $1/N$ expansion of the matrix model corresponds to an expansion in topology. Furthermore there are strong indications that the model is equivalent to Liouville theory for $c \leq 1$. It is defined by the ensemble of two dimensional manifolds built from equilateral triangles with fixed and equal side lengths glued together along their sides. It is assumed, and corroborated by the results, that summing over such triangulations one approximates correctly the integral over all metrics. A triangulation has N_0 vertices, N_1 links and N_2 triangles. Due to the manifold condition and the condition of fixed topology, which we will adopt here, they are not all independent:

$$\begin{aligned} 2N_1 &= 3N_2 \\ N_2 - N_1 + N_0 &= \chi \end{aligned} \tag{1}$$

where χ is the Euler characteristic of the surface. When the metric is defined in the above mentioned way, the curvature is concentrated at the vertices. Let q_i be the coordination number of a vertex i, i.e. the number of links meeting at the vertex. Then, the local volume and the local curvature are given by

$$\begin{aligned} \frac{1}{3}q_i &\sim \sqrt{g}d^2\xi \\ R_i = 2\pi\frac{(6-q_i)}{q_i} &\sim R(\xi) \end{aligned} \tag{2}$$

We then have

$$\begin{aligned} \frac{1}{2}\sum_{i=1}^{N_0} = N_2 &\sim \int \sqrt{g}d^2\xi = V \\ \frac{1}{3}\sum_{i=1}^{N_0} = 4\pi\chi &\sim \int R\sqrt{g}d^2\xi = 4\pi\xi \end{aligned} \tag{3}$$

The partition function of SQG coupled to scalar fields is defined by

$$Z(\lambda, \alpha, c) = \sum_{\{\tau\}} \frac{1}{C(\tau)} e^{-\lambda N_2} \int \pi_i (d^c x q_i^\alpha) e^{-\sum_{<i,j>} (x_i - x_j)^2} \tag{4}$$

where τ denotes a specific triangulation and where x_i is a c component vector at the vertex i. One may also understand this as a discretization of the Polyakov string. It is known that the number of triangulations grows exponentially for fixed topology. We will not discuss here the attempts to sum over topologies. In Figure 1 is shown the phase diagram in the two parameters α and c.

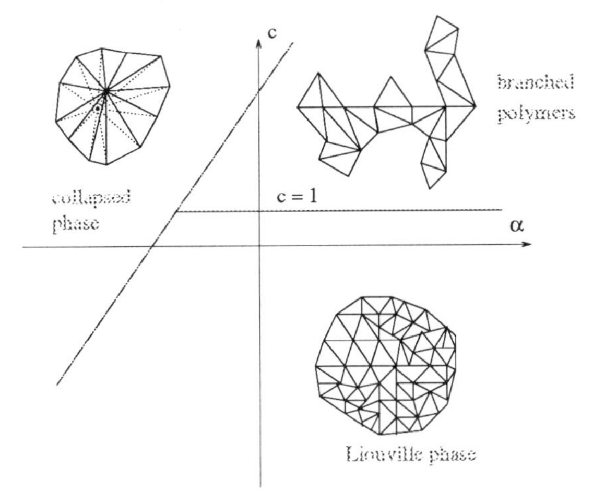

Figure 1

The phases can be characterized by the two critical exponents. One is the string susceptibility exponent γ_s, defined through

$$Z(\lambda, ...) = \sum_{N_2} e^{-\lambda N_2} Z_{N_2}$$
$$Z_{N_2} = e^{\lambda_c N_2} N_2^{\gamma_s - 3} (1 + 0(1/N_2)) \tag{5}$$

This can be seen to be related to the fractal structure, e.g by the probability to have a baby universe of size N connected to the rest of the surface by a minimal neck. In fact

$$P(N) \propto N^{\gamma_s - 2} (N_2 - N)^{\gamma_s - 2} \tag{6}$$

Another critical exponent is the internal Hausdorff dimension d_h, which measures the number of simplices n at a distance r from a fixed simplex through $n \propto r^{d_h - 1}$. In the branched polymer phase, $\gamma_s = 1/2$ and $d_h = 2$. In the Liouville phase γ_s is negative. For pure gravity, $d_h = 4$. For more discussion on the critical indices, I refer to the talk of Jan Ambjorn at this conference.

In the phase diagram, one may expect α to be irrelevant, because it corresponds to introducing higher powers of the curvature. There is, however, strong evidence for a transition to a singular crumpled phase. This can be understood in a simple mean-field model, which we introduced [5]. It is also known in a slightly different version as the Backgammon model. Suppose one has M boxes (which will correspond to the vertices) and N balls (corresponding to the triangles). One puts q_i balls in the i:th box with the probability $1/q_i^\beta$, with the obvious constraint that the total number of balls is N. The model is defined by the partition function

$$Z(\rho,\beta) = \sum_{\{q_i\}} \prod_{i=1}^{M} \left(\frac{1}{q_i^\beta}\right) \delta\left(\frac{1}{N}\sum_i^M q_i - \rho\right) \tag{7}$$

Then let $N \to \infty$ keeping $\rho = M/N$ fixed. There are in fact two phases, one where the balls are essentially equidistributed over the boxes but for $\beta > \beta_{crit}$ or $\rho > \rho_{crit}$, there is a condensed phase where essentially all balls are in one box, while the others have the minimum allowed number. In the triangulation, this means that there will be a singular vertex. Note that no geometry went into this argument. In fact in spherical topology, one will have two singular vertices.

The phase transition at $c = 1$ is more difficult to understand by simple arguments, because it is a change in the geometry. One suggestion, stemming from the continuum Liouville theory, is that the surfaces will become unstable to the formation of spikes [6]. The free energy of a spike has been calculated to be

$$F_{\text{spike}} = \frac{1-c}{12} \ln 1/a \tag{8}$$

where a is a short distance cut off. Obviously the formation of spikes is favoured for $c > 1$ and this may lead to a branched polymer phase. A more formal argument has been given by David [7], who considers an approximate renormalization group treatment of the matrix model. This leads naturally to a new coupling ρ to minimal necks. In the plane of the couplings (λ, ρ) there are two fixed points: one is the gravity fixed point, and the other the branched polymer fixed point. The first one disappears into the complex plane for $c > 1$. In previous numerical simulations (with $\rho = 0$) a transition at $c = 1$ was not seen, one rather saw a very slow change with c into the branched polymer phase, which was only apparent for $c > 5$. Together with Gudmar Thorleifsson I have recently made extensive numerical simulations including the coupling ρ [8]. In fact we confirm the suggestion of David, and so one of the last unsolved problems concerning the phase diagram in 2d has been solved.

3 4d simplicial quantum gravity, and the coupling to gauge fields

The formal definition of Euclidean quantum gravity is through the partition function

$$Z = \int \frac{Dg_{\mu\nu}}{\text{Diff}} e^{-S[g_{\mu\nu}]} \tag{9}$$

where S is the Einstein-Hilbert action

$$S\left[g_{\mu\nu}\right] = \int d^4\xi\sqrt{g(\xi)}\left\{\lambda - \frac{1}{16\pi G}R(\xi)\right\} + B.T. \tag{10}$$

Four dimensional SQG is defined in analogy with the two dimensional case discussed above. Therefore I will leave out most of the details, which are easy exercises for the reader. The four manifold is built by gluing together four-simplices. The number of (sub)simplices is denoted by N_0, N_1, N_2, N_3, N_4. Due to the manifold conditions, and for fixed topology, only two of these are independent, say N_4 and N_2. They correspond essentially to the total volume and the total curvature.

The discretized action becomes

$$S[T] = \kappa_4 N_4[T] - \kappa_2 N_2[T] \tag{11}$$

The discretized partition function becomes

$$Z(\kappa_2, \kappa_4) = \sum_{\{\tau\}} \frac{1}{C(\tau)} e^{\kappa_2 N_2 - \kappa_4 N_4} = \sum_{N_4} e^{-\kappa_4 N_4} e^{F(\kappa_2, N_4)} \tag{12}$$

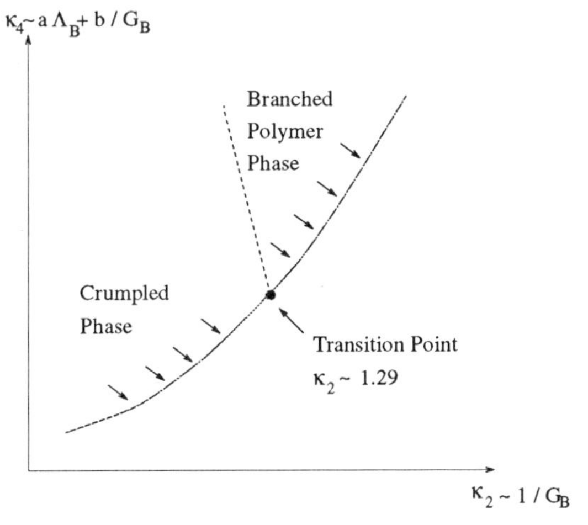

Figure 2

It is not proven, but strongly suggested by extensive numerical calculations, that the free energy density, i.e. $F(\kappa_2, N_4)/N_4$, has a finite limit when $N_4 \to \infty$. The phase structure in the (κ_2, κ_4) plane is shown in Figure 2. There is a critical line. Along this line, for large κ_2, i.e. small bare Newtons constant, there is a branched polymer phase. For small κ_2 there is a strongly crumpled phase with singular vertices, i.e. with a connectivity that grows linearly with the volume. None of these phases look like an extended universe. One had long hoped that the transition between them would be of second order and define a non-trivial fixed point. However, we have shown that the transition is first order, with a clear double peak in the free energy distribution [9]. It has later been shown that this double peak persists with growing volume, and that the distance between the peaks is constant.

Applying the balls-in-boxes model, described above, to this problem, where one now has to consider a exponentially weighted varying number of boxes, also predicts a first order phase transition [10]. However, the transition in the four dimensional SQG is much weaker. The gap in the mean curvature is much smaller than the total change in the curvature from the collapsed to the branched polymer phase.

It therefore seems that it would be possible, by a suitable modification of the action, to find a second order transition. It may also be that one may find a new phase, more like the Liouville phase, which could describe extended structures. Inspired by the arguments in two dimension, Mottola and collaborators have assumed that the conformal factor is describing the long distance physics of quantum gravity also in four dimension, and have written down an effective action, a generalization of Liouville theory [11]. However, this is only a suggestion, because one does not know how to integrate over the transverse (graviton) degrees of freedom. It was used to make a similar argument about the free energy of spikes, as the one described above for two dimensions [12]. In fact, it turns out that using the effective action of Mottola et al, introducing matter fields should suppress spikes. Vector fields should be most efficient.

Although these arguments are not very firmly founded, we found it interesting to investigate the effects of introducing vector (gauge) fields, although earlier investigations including matter fields had not shown any effects on the geometry. It is straightforward to introduce such gauge fields, represented by non-compact fields $A(l_{ab})$ living on the links l_{ab} of the manifold, and $A(l_{ab}) = -A(l_{ba})$. We add to the action in eq.(14) a matter part,

$$S_M = \sum_{t_{abc}} o(t_{abc}) \left[A(l_{ab}) + A(l_{bc}) + A(l_{ca})\right]^2 , \tag{13}$$

The symbol o denotes the order of the triangle, i.e. to how many four simplices it belongs. It is related to the local volume. One then also integrates over the real variables $A(l_{ab})$.

We have performed a strong coupling expansion, the first time it has been employed in 4d simplicial gravity, and also extensive numerical simulations. It turns out that introducing one vector field will not change the situation very much, although the first order gap becomes even smaller, about half of its value for pure gravity. Introducing three vector fields, however, radically changes the behaviour. The branched polymer phase disappears, and is possibly replaced by a quite new phase. In this phase we find γ negative, and the Hausdorff dimension near four.[13] These results are certainly very encouraging, although a lot of further work is needed to elucidate the properties of four dimensional simplicial gravity and its possible continuum limits.

4 Acknowledgements

I am very grateful to P. Bialas, S. Bilke, Z, Burda, A. Krzywicki, J. Tabaczek and G. Thorleifsson for inspiring collaborations. Furthermore I thank the Center for

Computational Physics, University of Tsukuba, where this contribution was written, for there kind hospitality.

References

[1] F. David, *Simplicial Quantum Gravity and Random Lattices*, Proc. Les Houches Summer School, Session LVII (1992).

[2] J. Ambjørn, *Quantization of Geometry*, Proc. Les Houches Summer School, Session LXII (1994), (hep-th/9411179).

[3] S. Weinberg, in General relativity, an Einstein-centenary survey, eds. S. W. Hawking and W. Israel, Cambridge University Press (1979)

[4] H. Kawai, Y. Kitazawa and M. Ninomiya, Nucl. Phys. **B467** (1996) 313, and references therein

[5] P. Bialas, Z. Burda, B. Petersson and J. Tabaczek, Nucl. Phys. **B495** (1997) 463.
P. Bialas and Z. Burda, Phys. Lett. **B384** (1996) 75.
P. Bialas, Z. Burda and D. Johnston, Nucl. Phys. **B493** (1997) 505.

[6] M.E. Cates, Europhys. Lett. **8** (1988) 719; A. Krzywicki, Phys. Rev. **D41** (1990) 3086; F. David, Nucl. Phys. **B368** (1992) 671.

[7] F. David, Nucl. Phys. **B487** (1997) 633

[8] B. Petersson and G. Thorleifsson, Nucl. Phys. Proc. Suppl. in press (hep-lat/9709072)

[9] P. Bialas, Z. Burda, A. Krzywicki and B. Petersson, Nucl. Phys. **B472** (1996) 293.

[10] P. Bialas and Z. Burda, (hep-lat/9707028);

[11] I. Antoniadis and E. Mottola, Phys. Rev. **D45** (1992) 2013; I. Antoniadis, P.O. Mazur and E. Mottola, Nucl. Phys. **B388** (1992) 627.

[12] J. Jurkiewicz and A. Krzywicki, Phys. Lett. **B392** (1997) 291. I. Antoniadis, P.O. Mazur and E. Mottola, Phys. Lett. **B394** (1997) 49.

[13] S. Bilke, Z. Burda, A. Krzywicki, B. Petersson, J. Tabaczek and G. Thorleifsson, (hep-lat/9710077)

On M-Theory

H. Nicolai

Max–Planck–Institut für Gravitationsphysik,
Schlaatzweg 1, D-14473 Potsdam, Germany

Abstract: This contribution gives a personal view on recent attempts to find a unified framework for non-perturbative string theories, with special emphasis on the hidden symmetries of supergravity and their possible role in this endeavor. A reformulation of $d = 11$ supergravity with enlarged tangent space symmetry $\mathrm{SO}(1,2) \times \mathrm{SO}(16)$ is discussed from this perspective, as well as an ansatz to construct yet further versions with $\mathrm{SO}(1,1) \times \mathrm{SO}(16)^\infty$ and possibly even $\mathrm{SO}(1,1)_+ \times \mathrm{ISO}(16)^\infty$ tangent space symmetry. It is suggested that upon "third quantization", dimensionally reduced maximal supergravity may have an equally important role to play in this unification as the dimensionally reduced maximally supersymmetric $SU(\infty)$ Yang Mills theory.

1 Introduction

Many theorists now believe that there is a unified framework for all string theories, which also accomodates $d = 11$ supergravity [1]. Much of the evidence for this elusive theory, called "M-Theory" [2], is based on recent work on duality symmetries in string theory which suggests that all string theories are connected through a web of non-perturbative dualities [3]. Although it is unknown what M-theory really is, we can probably assert with some confidence (i) that it will be a pregeometrical theory, in which space-time as we know it will emerge as a secondary concept (which also means that it makes little sense to claim that the theory "lives" in either ten or eleven dimensions), and (ii) that it should possess a huge symmetry involving new and unexplored types of Lie algebras (such as hyperbolic Kac Moody algebras), and perhaps other exotic structures such as quantum groups. In particular, the theory should be background independent and should be logically deducible from a vast generalization of the principles underlying general relativity.

According to a widely acclaimed recent proposal [4] M-Theory "is" the $N \to \infty$ limit of the maximally supersymmetric quantum mechanical $SU(N)$ matrix model [5] (see [6] for recent reviews, points of view and comprehensive lists of references). This model had already appeared in an earlier investigation of the $d = 11$ supermembrane

[7] in a flat background in the light cone gauge [8]. Crucial steps in the developments leading up to this proposal were the discovery of Dirichlet p-branes and their role in the description of non-perturbative string states [9], and the realization that the dynamics of an ensemble of such objects is described by dimensionally reduced supersymmetric Yang Mills theories [10]. Although there are a host of unsolved problems in matrix theory, two central ones can perhaps be singled out: one is the question whether the matrix model admits massless normalizable states for any N (see [11] for recent work in this direction); the other is related to the still unproven existence of the $N \to \infty$ limit. This would have to be a weak limit in the sense of quantum field theory, requiring the existence of a universal function $g = g(N)$ (the coupling constant of the $SU(N)$ matrix model) such that the limit $N \to \infty$ exists for all correlators. The existence of this limit would be equivalent to the renormalizability of the supermembrane [8]. However, even if these problems can be solved eventually, important questions remain with regard to the assertions made above: while matrix theory is pregeometrical in the sense that the target space coordinates are replaced by matrices, thus implying a kind of non-commutative geometry, the hidden exceptional symmetries of dimensionally reduced supergravities discovered long ago [12, 13] are hard to come by (see [14] and references therein).

In the first part of this contribution, I will report on work [15], which was motivated by recent advances in string theory as well as the possible existence of an Ashtekar-type canonical formulation of $d = 11$ supergravity. Although at first sight our results, which build on earlier work of [16, 17], may seem to be of little import for the issues raised above, I will argue that they could actually be relevant, assuming (as we do) that the success of the search for M-Theory will crucially depend on the identification of its underlying symmetries, and that the hidden exceptional symmetries of maximal supergravity theories may provide important clues as to where we should be looking. Namely, as shown in [16, 17], the local symmetries of the dimensionally reduced theories can be partially "lifted" to eleven dimensions, indicating that these symmetries may have a role to play also in a wider context than that of dimensionally reduced supergravity. The existence of alternative versions of $d = 11$ supergravity, which, though equivalent on-shell to the original version of [1], differ from it off-shell, suggests the existence of a novel kind of "exceptional geometry" for $d = 11$ supergravity and the bigger theory containing it. This new geometry would be intimately tied to the special properties of the exceptional groups, and would be characterized by relations such as (3)–(5) below, which have no analog in ordinary Riemannian geometry. The hope is, of course, that one may in this way gain valuable insights into what the (surely exceptional) geometry of M-Theory might look like, and that our construction may provide a simplified model for it. After all, we do not even know what the basic physical concepts and mathematical "objects" (matrices, BRST string functionals, spin networks,...?) of such a theory should be, especially if it is to be a truly pregeometrical theory of quantum gravity.

The second part of this paper discusses the infinite dimensional symmetries of $d = 2$ supergravities [13, 18, 19, 20, 21, 22, 23] and an ansatz that would incorporate them into the construction of [15, 16, 17]. The point of view adopted here is that the fundamental object of M-Theory could well be a kind of "Unendlichbein" belonging to an infinite dimensional coset space (cf. (27) below), which would generalize the space $GL(4, \mathbf{R})/\mathrm{SO}(1,3)$ of general relativity. This bein would be acted upon from the right side by a huge extension of the Lorentz group, containing not only space-time, but also internal symmetries, and perhaps even local supersymmetries. For the left action, one would have to appeal to some kind of generalized covariance principle. An intriguing, but also puzzling, feature of the alternative formulations of $d = 11$ supergravity is the apparent loss of manifest general covariance, as well as the precise significance of the global E_{11-d} symmetries of the dimensionally reduced theories. This could mean that in the final formulation, general covariance will have to be replaced by something else.

The approach taken here is thus different from and arguably even more speculative than current ideas based on matrix theory, exploiting the observation that instead of dimensionally reducing the maximally extended *rigidly* supersymmetric theory to one dimension, one might equally well contemplate reducing the maximally extended *locally* supersymmetric theory to one (light-like $\equiv$ null) dimension. While matrix theory acquires an infinite number of degrees of freedom only in the $N \to \infty$ limit, the chirally reduced supergravity would have an infinite number from the outset, being one half of a field theory in two dimensions. The basic idea is then that upon quantization the latter might undergo a similarly far-reaching metamorphosis as the quantum mechanical matrix model, its physical states being transmuted into "target space" degrees of freedom as in string theory [24]. This proposal would amount to a third quantization of maximal ($N = 16$) supergravity in two dimensions, where by "third quantization" I mean that the quantum treatment should take into account the gravitational degrees of freedom on the worldsheet, i.e. its (super)moduli for arbitrary genus. The model can be viewed as a very special example of $d = 2$ quantum cosmology; with the appropriate vertex operator insertions the resulting multiply connected $d = 2$ "universes" can be alternatively interpreted as multistring scattering diagrams [25]. One attractive feature of this proposal is that it might naturally bring in E_{10} as a kind of non-perturbative spectrum generating (rigid) symmetry acting on the third quantized Hilbert space, which would mix the worldsheet moduli with the propagating degrees of freedom. A drawback is that these theories are even harder to quantize than the matrix model (see, however, [26] and references therein).

2 $\mathrm{SO}(1,2) \times \mathrm{SO}(16)$ invariant supergravity in eleven dimensions

In [16, 17], new versions of $d = 11$ supergravity [1] with local $\mathrm{SO}(1,3) \times \mathrm{SU}(8)$ and $\mathrm{SO}(1,2) \times \mathrm{SO}(16)$ tangent space symmetries, respectively, have been constructed. [15] develops these results further (for the $\mathrm{SO}(1,2) \times \mathrm{SO}(16)$ invariant version of [17]), and also discusses a hamiltonian formulation in terms of the new variables. In both versions the supersymmetry variations acquire a polynomial form from which the corresponding formulas for the maximal supergravities in four and three dimensions can be read off directly and without the need for complicated duality redefinitions. This reformulation can thus be regarded as a step towards the complete fusion of the bosonic degrees of freedom of $d = 11$ supergravity (i.e. the elfbein E_M^A and the antisymmetric tensor A_{MNP}) in a way which is in harmony with the hidden symmetries of the dimensionally reduced theories.

For lack of space, and to exhibit the salient features as clearly as possible I will restrict the discussion to the bosonic sector. To derive the $SO(1,2) \times SO(16)$ invariant version of [17, 15] from the original formulation of $d = 11$ supergravity, one first breaks the original tangent space symmetry SO(1,10) to its subgroup $SO(1,2) \times SO(8)$ through a partial choice of gauge for the elfbein, and subsequently enlarges it again to $SO(1,2) \times SO(16)$ by introducing new gauge degrees of freedom. The symmetry enhancement of the transverse (helicity) group $SO(9) \subset SO(1,10)$ to SO(16) requires suitable redefinitions of the bosonic and fermionic fields, or, more succinctly, their combination into tensors w.r.t. the new tangent space symmetry. The construction thus requires a 3+8 split of the $d = 11$ coordinates and indices, implying a similar split for all tensors of the theory. It is important, however, that the dependence on all eleven coordinates is retained throughout.

The elfbein and the three-index photon are thus combined into new objects covariant w.r.t. to the new tangent space symmetry. In the special Lorentz gauge preserving $SO(1,2) \times SO(8)$ the elfbein takes the form

$$E_M^A = \begin{pmatrix} \Delta^{-1} e_\mu{}^a & B_\mu{}^m e_m{}^a \\ 0 & e_m{}^a \end{pmatrix} \tag{1}$$

where curved $d = 11$ indices are decomposed as $M = (\mu, m)$ with $\mu = 0,1,2$ and $m = 3, ..., 10$ (with a similar decomposition of the flat indices), and $\Delta := \det e_m{}^a$. In this gauge, the elfbein contains the (Weyl rescaled) dreibein and the Kaluza Klein vector $B_\mu{}^m$ both of which will be kept in the new formulation. By contrast, the internal achtbein is replaced by a rectangular 248-bein (e_{IJ}^m, e_A^m) containing the remaining "matter-like" degrees of freedom, where $([IJ], A)$ label the 248-dimensional adjoint representation of E_8 in the SO(16) decomposition. This 248-bein, which in the reduction to three dimensions contains all the propagating bosonic matter degrees of freedom of $d = 3, N = 16$ supergravity, is defined in a special SO(16) gauge by

$$(e_{IJ}^m, e_A^m) := \begin{cases} \Delta^{-1} e_a{}^m \Gamma^a_{\alpha\dot\beta} & \text{if } [IJ] \text{ or } A = (\alpha\dot\beta) \\ 0 & \text{otherwise} \end{cases} \tag{2}$$

where the SO(16) indices IJ or A are decomposed w.r.t. the diagonal subgroup $SO(8) \equiv (SO(8) \times SO(8))_{diag}$ of SO(16) (see [17] for details). Being the inverse densitized internal achtbein contracted with an SO(8) Γ-matrix, this object is very much analogous to the inverse densitized triad in the framework of Ashtekar's reformulation of Einstein's theory [27]. Note that, due to its rectangularity, there does not exist an inverse for the 248-bein (nor is one needed for the supersymmetry variations and the equations of motion!). In addition we need the composite fields (Q_μ^{IJ}, P_μ^A)

and (Q_m^{IJ}, P_m^A), which together make up an E_8 connection in eleven dimensions and whose explicit expressions in terms of the $d = 11$ coefficients of anholonomity and the four-index field strength F_{MNPQ} can be found in [17].

The new geometry is encoded into algebraic constraints between the vielbein components, which are without analog in ordinary Riemannian geometry because they rely in an essential way on special properties of the exceptional group E_8. We have

$$e_A^m e_A^n - \tfrac{1}{2} e_{IJ}^m e_{IJ}^n = 0 \tag{3}$$

and

$$\Gamma_{AB}^{IJ}\left(e_B^m e_{IJ}^n - e_B^n e_{IJ}^m\right) = 0 \qquad \Gamma_{AB}^{IJ} e_A^m e_B^n + 4 e_{K[I}^m e_{J]K}^n = 0 \tag{4}$$

where $\Gamma_{A\dot{A}}^I$ are the standard SO(16) Γ-matrices and $\Gamma_{AB}^{IJ} \equiv (\Gamma^{[I}\Gamma^{J]})_{AB}$, etc.; the minus sign in (3) reflects the fact that we are dealing with the maximally non-compact form $E_{8(+8)}$. While the SO(16) covariance of these equations is manifest, it turns out, remarkably, that they are also covariant under E_8. Obviously, (3) and (4) correspond to the singlet and the adjoint representations of E_8. More complicated are the following relations transforming in the **3875** representation of E_8

$$\begin{aligned}
e_{IK}^{(m} e_{JK}^{n)} - \tfrac{1}{16}\delta_{IJ} e_{KL}^m e_{KL}^n &= 0 \\
\Gamma_{\dot{A}B}^K e_B^{(m} e_{IK}^{n)} - \tfrac{1}{14}\Gamma_{\dot{A}B}^{IKL} e_B^{(m} e_{KL}^{n)} &= 0 \\
e_{[IJ}^{(m} e_{KL]}^{n)} + \tfrac{1}{24} e_A^m \Gamma_{AB}^{IJKL} e_B^n &= 0
\end{aligned} \tag{5}$$

Yet another set of relations involves the **27000** representation of E_8 [15].

The 248-bein and the new connection fields are subject to a "vielbein postulate" similar to the usual vielbein postulate stating the covariant constancy of the vielbein w.r.t. to generally covariant and Lorentz covariant derivative:

$$\begin{aligned}
(\partial_\mu - B_\mu{}^n\partial_n)e_{IJ}^m + \partial_n B_\mu{}^n e_{IJ}^m + \partial_n B_\mu{}^m e_{IJ}^n + 2\,Q_\mu{}^K{}_{[I} e_{J]K}^m + P_\mu^A \Gamma_{AB}^{IJ} e_m^B &= 0 \\
(\partial_\mu - B_\mu{}^n\partial_n)e_A^m + \partial_n B_\mu{}^m e_A^n + \partial_n B_\mu{}^n e_A^m + \tfrac{1}{4} Q_\mu^{IJ}\Gamma_{AB}^{IJ} e_B^m - \tfrac{1}{2}\Gamma_{AB}^{IJ} P_\mu^B e_{IJ}^m &= 0 \\
\partial_m e_{IJ}^n + 2\,Q_m{}^K{}_{[I} e_{J]K}^n + P_m^A \Gamma_{AB}^{IJ} e_B^n &= 0 \\
\partial_m e_A^n + \tfrac{1}{4} Q_m^{IJ}\Gamma_{AB}^{IJ} e_B^n - \tfrac{1}{2}\Gamma_{AB}^{IJ} P_m^B e_{IJ}^n &= 0
\end{aligned} \tag{6}$$

Like (3)–(5), these relations are E_8 covariant. It must be stressed, however, that the full theory of course does not respect E_8 invariance. A puzzling feature of (6) is that the covariantization w.r.t. an affine connection is "missing" in these equations, even though the theory is still invariant under $d = 11$ coordinate transformations. One can now show that the supersymmetry variations of $d = 11$ supergravity can be entirely expressed in terms of these new variables (and their fermionic partners).

The reduction of $d = 11$ supergravity to three dimensions yields $d = 3, N = 16$ supergravity [28], and is accomplished rather easily, since no duality redefinitions are needed any more, unlike in [12]. The propagating bosonic degrees of freedom in three dimensions are all scalar, and combine into a matrix $\mathcal{V}(x)$, which is an element of a non-compact $E_{8(+8)}/\mathrm{SO}(16)$ coset space, and whose dynamics is governed by a non-linear σ-model coupled to $d=3$ gravity. The identification of the 248-bein with the σ-model field $\mathcal{V} \in E_8$ is given by

$$e^m_{IJ} = \tfrac{1}{60}\mathrm{Tr}\,(Z^m \mathcal{V} X^{IJ} \mathcal{V}^{-1}) \qquad e^m_A = \tfrac{1}{60}\mathrm{Tr}\,(Z^m \mathcal{V} Y^A \mathcal{V}^{-1}) \tag{7}$$

where X^{IJ} and Y^A are the compact and non-compact generators of E_8, respectively, and where the Z^m for $m = 3, ..., 10$ are eight non-compact commuting generators obeying $\mathrm{Tr}(Z^m Z^n) = 0$ for all m and n (the existence of eight such generators is a consequence of the fact that the coset space $E_{8(+8)}/\mathrm{SO}(16)$ has real rank 8 and therefore admits an eight-dimensional maximal flat and totally geodesic submanifold [29]). This reduction provides a "model" for the exceptional geometry, where the relations (3)–(6) can be tested by means of completeness relations for the E_8 Lie algebra generators in the adjoint representation. Of course, this is not much of a test since all dependence on the internal coordinates is dropped in (7), and the terms involving $B_\mu{}^m$ disappear altogether. It would be desirable to find other "models" with non-trivial dependence on the internal coordinates. The only example of this type so far is provided by the S^7 truncation of $d=11$ supergravity for the $\mathrm{SO}(1,3)\times\mathrm{SU}(8)$ invariant version of $d=11$ supergravity [30].

3 More Symmetries

The emergence of hidden symmetries of the exceptional type in extended supergravities [12] was a remarkable and, at the time, quite unexpected discovery. It took some effort to show that the general pattern continues when one descends to $d=2$ and that the hidden symmetries become infinite dimensional [13, 18, 19, 20, 21, 22, 23], generalizing the Geroch group of general relativity [31]. As we will see, even the coset structure remains, although the mathematical objects one deals with become a lot more delicate. The fact that the construction described above works with a 4+7 and 3+8 split of the indices suggests that we should be able to go even further and to construct versions of $d=11$ supergravity with infinite dimensional tangent space symmetries, which would be based on a 2+9 or even a 1+10 split of the indices. This would also be desirable in view of the fact that the new versions are "simple" only in their internal sectors. The general strategy is thus to further enlarge the internal sector by absorbing more and more degrees of freedom into it, such that in the final step corresponding to a 1+10 split, only an einbein is left in the low dimensional sector. Although the actual elaboration of these ideas has to be left to future work, I will try to give at least a flavor of some anticipated key features.

3.1 Reduction to two dimensions

Let us first recall some facts about dimensional reduction of maximal supergravity to two dimensions. Following the empirical rules of dimensional reduction one is led to predict $E_9 = E_8^{(1)}$ as a symmetry for the dimensional reduction of $d = 11$ supergravity to two dimensions [13]. This expectation is borne out by the existence of a linear system for maximal $N = 16$ supergravity in two dimensions [24, 32] (see [33, 19] for the bosonic theory). The linear system requires the introduction of an extra "spectral" parameter t, and the extension of the σ-model matrix $\mathcal{V}(x)$ to a matrix $\widehat{\mathcal{V}}(x;t)$ depending on this extra parameter t, as is generally the case for integrable systems in two dimensions. An unusual feature is that, due to the presence of gravitational degrees of freedom, this parameter becomes coordinate dependent, i.e. we have $t = t(x;w)$, where w is an integration constant, sometimes referred to as the "constant spectral parameter" whereas t itself is called the "variable spectral parameter".

Here, we are mainly concerned with the symmetry aspects of this system, and with what they can teach us about the $d = 11$ theory itself. The coset structure of the higher dimensional theories has a natural continuation in two dimensions, with the only difference that the symmetry groups are infinite dimensional. This property is manifest from the transformation properties of the linear system matrix $\widehat{\mathcal{V}}$, with a global affine symmetry acting from the left, and a local symmetry corresponding to some "maximal compact" subgroup acting from the right:

$$\widehat{\mathcal{V}}(x;t) \longrightarrow g(w)\widehat{\mathcal{V}}(x;t)h(x;t) \tag{8}$$

Here $g(w) \in E_9$ with affine parameter w, and the subgroup to which $h(x;t)$ belongs is characterized as follows [18, 19]. Let τ be the involution characterizing the coset space $E_{8(+8)}/\mathrm{SO}(16)$: then $h(t) \in \mathrm{SO}(16)^\infty_\varepsilon$ is defined to consist of all τ^∞ invariant elements of E_9, where the extended involution τ^∞ is defined by $\tau^\infty(h(t)) := \tau h(\varepsilon t^{-1})$, with $\varepsilon = +1$ (or -1) for a Lorentzian (Euclidean) worldsheet. For $\varepsilon = 1$, which is the case we are mainly interested in, we will write $\mathrm{SO}(16)^\infty \equiv \mathrm{SO}(16)^\infty_\varepsilon$. We also note that $\mathrm{SO}(16)^\infty_\varepsilon$ is different from the affine extension of SO(16) for either choice of sign.

What has been achieved by the coset space description is the following: by representing the "moduli space of solutions" $\mathcal{M}$ (of the bosonic equations of motion of $d = 11$ supergravity with nine commuting space-like Killing vectors) as

$$\mathcal{M} = \frac{\text{solutions of field equations}}{\text{diffeomorphisms}} = \frac{E_9}{\mathrm{SO}(16)^\infty} \tag{9}$$

we have managed to endow this space, which a priori is very complicated, with a group theoretic structure, that makes it much easier to handle. In particular, the integrability of the system is directly linked to the fact that $\mathcal{M}$ possesses an infinite dimensional "isometry group" E_9. The introduction of infinitely many gauge degrees of freedom embodied in the subgroup $\mathrm{SO}(16)^\infty$ linearizes and localizes the action of this isometry group on the space of solutions. Of course, in making such statements, one should keep in mind that a mathematically rigorous construction of such spaces is a thorny problem. This is likewise true for the infinite dimensional groups[1] and their

[1]For instance, the Geroch group can be defined rigorously to consist of all maps from the

associated Lie algebras; the latter being infinite dimensional vector spaces, there are myriad ways of equiping them with a topology. We here take the liberty of ignoring these subleties, not least because these spaces ultimately will have to be "quantized" anyway.

There is a second way of defining the Lie algebra of $SO(16)^{\infty}_{\varepsilon}$ which relies on the Chevalley-Serre presentation. Given a finite dimensional non-compact Lie group G with maximal compact subgroup H, a necessary condition for this prescription to work is that $\dim H = \frac{1}{2}(\dim G - \operatorname{rank} G)$, and we will subsequently extend this prescription to the infinite Lie group. Let us first recall that any (finite or infinite dimensional) Kac Moody algebra is recursively defined in terms of multiple commutators of the Chevalley generators subject to certain relations [34]. More specifically, given a Cartan matrix A_{ij} and the associated Dynkin diagram, one starts from a set of $sl(2, \mathbf{R})$ generators $\{e_i, f_i, h_i\}$, one for each node of the Dynkin diagram, which in addition to the standard $sl(2, \mathbf{R})$ commutation relations

$$[h_i, h_j] = 0 \qquad [e_i, f_j] = \delta_{ij} h_j$$

$$[h_i, e_j] = A_{ij} e_j \qquad [h_i, f_j] = -A_{ij} f_j \tag{10}$$

are subject to the multilinear Serre relations

$$[e_i, [e_i, ...[e_i, e_j]...]] = 0 \qquad [f_i, [f_i, ...[f_i, f_j]...]] = 0 \tag{11}$$

where the commutators are $(1 - A_{ij})$-fold ones. The Lie algebra is then by definition the linear span of all multiple commutators which do not vanish by virtue of these relations.

To define the subalgebra $SO(16)^{\infty}_{\varepsilon}$, we first recall that the Chevalley involution θ is defined by

$$\theta(e_i) = -f_i \qquad \theta(f_i) = -e_i \qquad \theta(h_i) = -h_i \tag{12}$$

This involution, like the ones to be introduced below, leaves invariant the defining relations (10) and (11) of the Kac Moody algebra, and extends to the whole Lie algebra via the formula $\theta([x, y]) = [\theta(x), \theta(y)]$. It is not difficult to see that, for E_8 (and also for $sl(n, \mathbf{R})$), we have $\tau = \theta$, and the maximal compact subalgebras defined above correspond to the subalgebras generated by the multiple commutators of the θ invariant elements $(e_i - f_i)$ in both cases. The trick is now to carry over this definition to the affine extension, whose associated Cartan matrix has a zero eigenvalue. To do this, however, we need a slight generalization of the above definition; for this purpose, we consider involutions ω that can be represented as products of the form

$$\omega = \theta \cdot s \tag{13}$$

complex w plane to $SL(2, \mathbf{R})$ with meromorphic entries. With this definition, one obtains all multisoliton solutions of Einstein's equations, and on this solution space the group acts transitively by construction. Whether this is the right choice or not is then a matter of physics, not mathematics.

where the involution s acts as

$$s(e_i) = s_i e_i \qquad s(f_i) = s_i f_i \qquad s(h_i) = h_i \tag{14}$$

with $s_i = \pm 1$. It is important that different choices of s_i do not necessarily lead to inequivalent involutions (the general problem of classifying the involutive automorphisms of infinite dimensional Kac Moody algebras has so far not been completely solved, see e.g. [35][2]). In particular for E_9, which is obtained from E_8 by adjoining another set $\{e_0, f_0, h_0\}$ of Chevalley generators, we take $s_i = 1$ for all $i \geq 1$, whereas $s_0 = \varepsilon$, with ε as before, i.e. $\varepsilon = +1$ (or -1) for Lorentzian (Euclidean) worldsheet. Thus, on the extended Chevalley generators,

$$\omega(e_0) = -\varepsilon f_0 \qquad \omega(f_0) = -\varepsilon e_0 \qquad \omega(h_0) = -h_0 \tag{15}$$

With this choice, the involution ω coincides with the involutions defined before for the respective choices of ε, i.e. $\omega = \tau^\infty$, and therefore the invariant subgroups are the same, too. For $\varepsilon = 1$, the involution ω defines an infinite dimensional "maximal compact" subalgebra consisting of all the negative norm elements w.r.t. to the standard bilinear form

$$\langle e_i | f_j \rangle = \delta_{ij} \qquad \langle h_i | h_j \rangle = A_{ij} \tag{16}$$

(the norm of any given multiple commutator can be determined recursively from the fundamental relation $\langle [x,y] | z \rangle = \langle x | [y,z] \rangle$). The notion of "compactness" here is thus algebraic, not topological: the subgroup $SO(16)^\infty$ will not be compact in the topological sense (recall the well known example of the unit ball in an infinite dimensional Hilbert space, which is bounded but not compact in the norm topology). On the other hand, for $\varepsilon = -1$, the group $SO(16)^\infty_\varepsilon$ is not even compact in the algebraic sense, as $e_0 + f_0$ has positive norm. However, this is in accord with the expectation that $SO(16)^\infty_\varepsilon$ should contain the (non-compact) group SO(1,8) rather than SO(9) if one of the compactified dimensions is time-like.

3.2 2 + 9 split

Let us now consider the extension of the results described in section 2 to the situation corresponding to a 2+9 split of the indices. Elevating the local symmetries of $N = 16$ supergravity from two to eleven dimensions would require the existence of yet another extension of the theory, for which the Lorentz group SO(1,10) is replaced by $SO(1,1) \times SO(16)^\infty$; the subgroup $SO(16)^\infty$ can be interpreted as an extension of the transverse group SO(9) in eleven dimensions. Taking the hints from (1), we would now decompose the elfbein into a zweibein and nine Kaluza Klein vectors $B_\mu{}^m$ (with $m = 2, ..., 10$). The remaining internal neunbein would have to be replaced by an "Unendlichbein" $(e^m_{IJ}(x;t), e^m_A(x;t))$, depending on a spectral parameter t, necessary to parametrize the infinite dimensional extension of the symmetry group. However,

[2] I am very grateful to C. Daboul for helpful discussions on this topic.

in eleven dimensions, there is no anolog of the dualization mechanism, which would ensure that despite the existence of infinitely many dual potentials, there are only finitely many physical degrees of freedom. This indicates that if the construction works it will take us beyond $d = 11$ supergravity.

Some constraints on the geometry can be deduced from the requirement that in the dimensional reduction to $d = 2$, there should exist a formula analogous to (7), but with $\mathcal{V}$ replaced by the linear system matrix $\widehat{\mathcal{V}}$, or possibly even the enlarged linear system of [22]. Evidently, we would need a ninth nilpotent generator to complement the Z^m's of (7); an obvious candidate is the central charge generator c, since it obeys $\langle c|c\rangle = \langle c|Z^m\rangle = 0$ for all $m = 3, ..., 10$. The parameter t, introduced somewhat ad hoc for the parametrization of the unendlichbein, must obviously coincide with the spectral parameter of the $d = 2$ theory, and the generalized "unendlichbein postulate" should evidently reduce to the linear system of $d = 2$ supergravity in this reduction. To write it down, we need to generalize the connection coefficients appearing in the linear system. The latter are given by

$$\mathcal{Q}_\mu^{IJ} = Q_\mu^{IJ} + \dots \qquad \mathcal{P}_\mu^A = \frac{1+t^2}{1-t^2} P_\mu^A + \frac{2t}{1-t^2}\varepsilon_{\mu\nu} P^{\nu A} + \dots \tag{17}$$

with Q_μ^{IJ} and P_μ^A as before; the dots indicate t dependent fermionic contributions which we omit. A very important difference with section 2, where the tangent space symmetry was still finite dimensional, is that the Lie algebra of $SO(16)^\infty$ also involves the P's, and not only the Q's. More specifically, from the t dependence of the dimensionally reduced connections in (17) we infer that the connections $(\mathcal{Q}_M^{IJ}(x;t), \mathcal{P}_M^A(x;t))$ constitute an $SO(16)^\infty$ (and not an E_9) gauge connection. This means that the covariantizations in the generalized vielbein postulate are now in precise correpondence with the local symmetries, in contrast with the relations (6) which look E_8 covariant, whereas the full theory is invariant only under local SO(16).

To write down an ansatz, we put

$$\mathcal{D}_\mu := \partial_\mu - B_\mu{}^n \partial_n + \dots \tag{18}$$

where the dots stand for terms involving derivatives of the Kaluza Klein vector fields. Then the generalization of (6) should read

$$\begin{aligned}
\mathcal{D}_\mu e^m_{IJ}(t) + 2\,\mathcal{Q}_\mu{}^K{}_{[I}(t) e^m_{J]K}(t) + \mathcal{P}^A_\mu(t)\Gamma^{IJ}_{AB} e^m_B(t) &= 0 \\
\mathcal{D}_\mu e^m_A(t) + \tfrac{1}{4}\mathcal{Q}^{IJ}_\mu(t)\Gamma^{IJ}_{AB} e^m_B(t) - \tfrac{1}{2}\Gamma^{IJ}_{AB}\mathcal{P}^B_\mu(t) e^m_{IJ}(t) &= 0 \\
\partial_m e^n_{IJ}(t) + 2\,\mathcal{Q}_m{}^K{}_{[I}(t) e^n_{J]K}(t) + \mathcal{P}^A_m(t)\Gamma^{IJ}_{AB} e^n_B(t) &= 0 \\
\partial_m e^n_A(t) + \tfrac{1}{4}\mathcal{Q}^{IJ}_m(t)\Gamma^{IJ}_{AB} e^n_B(t) - \tfrac{1}{2}\Gamma^{IJ}_{AB}\mathcal{P}^B_m(t) e^n_{IJ}(t) &= 0
\end{aligned} \tag{19}$$

Of course, the challenge is now to find explicit expressions for the internal components $\mathcal{Q}^{IJ}_m(x;t)$ and $\mathcal{P}^A_m(x;t)$, such that (19) can be interpreted as a $d = 11$ generalization of the linear system of dimensionally reduced supergravity. Another obvious question concerns the fermionic partners of the unendlichbein: in two dimensions, the linear system matrix contains all degrees of freedom, including the fermionic ones, and the local $N = 16$ supersymmetry can be bosonized into a local $SO(16)^\infty$ gauge transformation [32]. Could this mean that there is a kind of bosonization in eleven dimensions or M-Theory? This idea may not be as outlandish as it sounds because a truly pregeometrical theory might be subject to a kind of "pre-statistics", such that the distinction between bosons and fermions arises only through a process of spontaneous symmetry breaking.

4 Yet more symmetries?

In 1982, B. Julia conjectured that the dimensional reduction of maximal supergravity to one dimension should be invariant under a further extension of the E-series, namely (a non-compact form of) the hyperbolic Kac Moody algebra E_{10} obtained by adjoining another set $\{e_{-1}, f_{-1}, h_{-1}\}$ of Chevalley generators to those of E_9 [36][3]. As shown in [38], the last step of the reduction requires a null reduction if the affine symmetry of the $d = 2$ theory is not to be lost. The reason is that the infinite dimensional affine symmetries of the $d = 2$ theories always involve dualizations of the type

$$\partial_\mu \varphi = \varepsilon_{\mu\nu} \partial^\nu \tilde{\varphi} \tag{20}$$

(in actual fact, there are more scalar fields, and the duality relation becomes non-linear, which is why one ends up with infinitely many dual potentials for each scalar degree of freedom). Dimensional reduction w.r.t. to a Killing vector ξ^μ amounts to imposing the condition $\xi^\mu \partial_\mu \equiv 0$ on *all* fields, including dual potentials. Hence,

$$\xi^\mu \partial_\mu \varphi = 0 \quad , \quad \xi^\mu \partial_\mu \tilde{\varphi} \equiv \eta^\mu \partial_\mu \varphi = 0 \tag{21}$$

where $\eta^\mu \equiv \varepsilon^{\mu\nu} \xi_\nu$. If ξ^μ and η^μ are linearly independent, this constraint would force all fields to be constant, which is clearly too strong a requirement. Hence we must demand that ξ^μ and η^μ are collinear, which implies

$$\xi^\mu \xi_\mu = 0, \tag{22}$$

i.e. the Killing vector must be null. Starting from this observation, it was shown in [38] that the Matzner Misner $sl(2, \mathbf{R})$ symmetry of pure gravity can be formally extended to an $sl(3, \mathbf{R})$ algebra in the reduction of the vierbein from four to one dimensions. Combining this $sl(3, \mathbf{R})$ with the Ehlers $sl(2, \mathbf{R})$ of ordinary gravity, or with the E_8 symmetry of maximal supergravity in three dimensions, one is led to the hyperbolic algebra $\mathcal{F}_3$ [39] for ordinary gravity, and to E_{10} for maximal supergravity. The transformations realizing the action of the Chevalley generators on the vierbein components can be worked out explicitly, and the Serre relations can be formally verified [38] (for E_{10}, this was shown more recently in [40]).

[3]The existence of a maximal dimension for supergravity [37] would thus be correlated with the existence of a "maximally extended" hyperbolic Kac Moody algebra, which might thus explain the occurrence of maximum spin 2 for massless gauge particles in nature.

There is thus some evidence for the emergence of hyperbolic Kac Moody algebras in the reduction to one null dimension, but the difficult open question that remains is what the configuration space is on which this huge symmetry acts. This space is expected to be much bigger than the coset space (9). Now, already for the $d = 2$ reduction there are extra degrees of freedom that must be taken into account in addition to the propagating degrees of freedom. Namely, the full moduli space involving all bosonic degrees of freedom should also include the moduli of the zweibein, which are not contained in (9). For each point on the worldsheet, the zweibein is an element of the coset space $\mathrm{GL}(2,\mathbf{R})/\mathrm{SO}(1,1)$; although it has no local degrees of freedom any more, it still contains the global information about the conformal structure of the world sheet Σ. Consequently, we should consider the Teichmüller space

$$\mathcal{T} = \frac{\{e_\mu{}^\alpha(x) \mid x \in \Sigma\}}{\mathrm{SO}(1,1) \times \mathrm{Weyl}(\Sigma) \times \mathrm{Diff}_0(\Sigma)} \tag{23}$$

as part of the configuration space of the theory (see [41] for a detailed description of $\mathcal{T}$). In fact, we should even allow for arbitrary genus of the worldsheet, and replace $\mathcal{T}$ by the "universal Teichmüller space" $\tilde{\mathcal{T}}$. This infinite dimensional space can be viewed as the configuration space space of non-perturbative string theory [42]. For the models under consideration here, however, even $\tilde{\mathcal{T}}$ is not big enough, as we must also take into account the dilaton ρ and the non-propagating Kaluza Klein vector fields in two dimensions. For the former, a coset space description was proposed in [22]. On the other hand, the Kaluza Klein vectors and the cosmological constant they could generate in two dimensions have been largely ignored in the literature. Even if one sets their field strengths equal to zero (there are arguments that the Geroch group, and hence infinite duality symmetries, are incompatible with a nonzero cosmological constant in two dimensions), there still remain topological degrees of freedom for higher genus world sheets.

The existence of inequivalent conformal structures is evidently important for the null reductions, as the former are in one-to-one correspondence with the latter. Put differently, the inequivalent null reductions are precisely parametrized by the space (23). The extended symmetries should thus not only act on one special null reduction (set of plane wave solutions of Einstein's equations), but relate different reductions. Indeed, it was argued in [40] that, for a toroidal worldsheet, the new $sl(2,\mathbf{R})$ transformations associated with the over-extended Chevalley generators change the conformal structure, but only for non-vanishing holonomies of the Kaluza Klein vector fields on the worldsheet. This indicates that the non-trivial realization of the hyperbolic symmetry requires the consideration of non-trivial worldsheet topologies. The dimensionally reduced theory thereby retains a memory of its two-dimensional ancestor. It is therefore remarkable that, at least for isomonodromic solutions of Einstein's theory, the $d = 2$ theory exhibits a factorization of the equations of motion akin to, but more subtle than the holomorphic factorization of conformal field theories [43]. In other words, there may be a way to think of the $d = 2$ theory as being composed of two chiral halves just as for the closed string. Consequently, a truncation to one null dimension may not be necessary after all if the theory factorizes all by itself.

In summary, what we are after here is a group theoretic unification of all these moduli spaces that would be analogous to (9) above, and fuse the matter and the topological degrees of freedom. No such description seems to be available for (23) (or $\tilde{\mathcal{T}}$), and it is conceivable that only the total moduli space $\widetilde{\mathcal{M}}$ containing both $\mathcal{M}$ and $\tilde{\mathcal{T}}$ as well as the dilaton and the Kaluza Klein, and perhaps even the fermionic, degrees of freedom is amenable to such an interpretation. Extrapolating the previous results, we are thus led to consider coset spaces E_{10}/H with $\mathrm{SO}(16)^\infty \subset H \subset E_{10}$. As before, the introduction of the infinitely many spurious degrees of freedom associated with the gauge group H would be necessary in order to "linearize" the action of E_{10}.

What are the choices for H? One possibility would be to follow the procedure of the foregoing section, and to define $H = \mathrm{SO}(16)^{\infty\infty} \subset E_{10}$ in analogy with $\mathrm{SO}(16)^\infty \subset E_9$ by taking its associated Lie algebra to be the linear span of all ω invariant combinations of E_{10} Lie algebra elements. To extend the affine involution to the full hyperbolic algebra, we would again invoke (13), setting $\varepsilon = +1$ in (15) (since we now assume the worldsheet to be Lorentzian), which leaves us with the two choices $s_{-1} = \pm 1$. For $s_{-1} = +1$ we would get the "maximal compact" subalgebra of E_{10}, corresponding to the compactification of ten spacelike dimensions. A subtlety here is that a definition in terms of the standard bilinear form is no longer possible, unlike for affine and finite algebras, as this would now also include part of the Cartan subalgebra of E_{10}: due to the existence of a negative eigenvalue of the E_{10} Cartan matrix, there exists a negative norm element $\sum_i n_i h_i$ of the Cartan subalgebra, which would have to be excluded from the definition of H (cf. the footnote on p. 438 of [22]). The alternative choice $s_{-1} = -1$ would correspond to reduction on a 9+1 torus.

However, for the null reduction advocated here, physical reasoning motivates us to propose yet another choice for H. Namely, in this case, H should contain the group $\mathrm{ISO}(9) \subset \mathrm{SO}(1,10)$ leaving invariant a null vector in eleven dimensions [44]. To identify the relevant parabolic subgroup of E_{10}, which we denote by $\mathrm{ISO}(16)^\infty$, we recall [38] that the over-extended Chevalley generators correspond to the matrices

$$e_{-1} = \frac{1}{\sqrt{2}} \begin{pmatrix} 0 & 0 & 1 \\ 0 & 0 & 1 \\ 0 & 0 & 0 \end{pmatrix} \qquad f_{-1} = \frac{1}{\sqrt{2}} \begin{pmatrix} 0 & 0 & 0 \\ 0 & 0 & 0 \\ 1 & 1 & 0 \end{pmatrix} \qquad h_{-1} = \frac{1}{2} \begin{pmatrix} 1 & 1 & 0 \\ 1 & 1 & 0 \\ 0 & 0 & -2 \end{pmatrix} \tag{24}$$

in a notation where we only write out the components acting on the $0, 1, 2$ components of the elfbein, with all other entries vanishing. Evidently, we have $h_{-1} = d - c_-$ with

$$d = \begin{pmatrix} 0 & 0 & 0 \\ 0 & 0 & 0 \\ 0 & 0 & -1 \end{pmatrix} \qquad c_- = -\frac{1}{2} \begin{pmatrix} 1 & 1 & 0 \\ 1 & 1 & 0 \\ 0 & 0 & 0 \end{pmatrix} \tag{25}$$

where d is the scaling operator on the dilaton ρ , and c_- is the central charge, alias the "level counting operator" of E_{10}, obeying $[c_-, e_{-1}] = -e_{-1}$ and $[c_-, f_{-1}] = +f_{-1}$

(and having vanishing commutators with all other Chevalley generators). Writing

$$c_{\pm} := -\frac{1}{2}\begin{pmatrix} 1 & 0 & 0 \\ 0 & 1 & 0 \\ 0 & 0 & 0 \end{pmatrix} \pm \frac{1}{2}\begin{pmatrix} 0 & 1 & 0 \\ 1 & 0 & 0 \\ 0 & 0 & 0 \end{pmatrix} \tag{26}$$

we see that the first matrix on the right scales the conformal factor, generating Weyl transformations (called Weyl(Σ) in (23)) on the zweibein, while the second generates the local SO(1,1) Lorentz transformations. In a lightcone basis, these symmetries factorize on the zweibein, which decomposes into two chiral einbeine. Consequently, Weyl transformations and local SO(1,1) can be combined into two groups $\mathrm{SO}(1,1)_{\pm}$ with respective generators $c_{\pm}$, and which act separately on the chiral einbeine. One of these, $\mathrm{SO}(1,1)_{-}$ (generated by c_{-}), becomes part of E_{10}. The other, $\mathrm{SO}(1,1)_{+}$, acts on the residual einbein and can be used to eliminate it by gauging it to one. Since $c_{\pm}$ acts in the same way on the conformal factor, we also recover the result of [18].

We wish to include both ISO(9) and $\mathrm{SO}(1,1)_{-}$ into the enlarged local symmetry $H = \mathrm{ISO}(16)^{\infty}$, and thereby unify the longitudinal symmetries with the "transversal" group $\mathrm{SO}(16)^{\infty}$ discussed before. Accordingly, we define $\mathrm{ISO}(16)^{\infty}$ to be the algebra generated by the $\mathrm{SO}(16)^{\infty}$ Lie algebra together with c_{-} and e_{-1}, as well as all their nonvanishing multiple commutators. The "classical" configuration space of M-Theory should then be identified with the coset space

$$\widetilde{\mathcal{M}} = \frac{E_{10}}{\mathrm{ISO}(16)^{\infty}} \tag{27}$$

Of course, we will have to worry about the fate of these symmetries in the quantum theory. Indeed, some quantum version of the symmetry groups appearing in (27) must be realized on the Hilbert space of third quantized $N = 16$ supergravity, such that E_{10} becomes a kind of spectrum generating (rigid) symmetry on the physical states, while the gauge group $\mathrm{ISO}(16)^{\infty}$ gives rise to the constraints defining them. Because "third quantization" here is analogous to the transition from first quantized string theory to string field theory, the latter would have to be interpreted as multi-string states in some sense (cf. [45] for earlier suggestions in this direction; note also that the coset space (27) is essentially generated by half of E_{10}, so there would be no "anti-string states"). According to [3], the continuous duality symmetries are broken to certain discrete subgroups over the integers in the quantum theory. Consequently, the quantum configuration space would be the left coset

$$\widetilde{\mathcal{F}} = E_{10}(\mathbf{Z})\backslash\widetilde{\mathcal{M}}$$

and the relevant partition functions would have to be new kinds of modular forms defined on $\widetilde{\mathcal{F}}$. However, despite recent advances [46, 47], the precise significance of the (discrete) "string Geroch group" remains a mystery, and it is far from obvious how to extend the known results and conjectures for finite dimensional duality symmetries to the infinite dimensional case (these statements apply even more to possible discrete hyperbolic extensions; see, however, [40, 48]). Moreover, recent work [49] confirms the possible relevance of quantum groups in this context (in the form of "Yangian doubles").

Returning to our opening theme, more should be said about the 1+10 split, which would lift up the $SO(1,1)_+ \times ISO(16)^\infty$ symmetry, and the "bein" which would realize the exceptional geometry alluded to in the introduction, and on which $ISO(16)^\infty$ would act as a generalized tangent space symmetry. However, as long as the 2+9 split has not been shown to work, and a manageable realization is not known for either E_{10} or $ISO(16)^\infty$, we must leave the elaboration of these ideas to the future. It could well prove worth the effort.

Acknowledgments: The results described in section 2 are based on work done in collaboration with S. Melosch. I would also like to thank C. Daboul, R.W. Gebert, H. Samtleben and P. Slodowy for stimulating discussions and comments.

References

[1] E. Cremmer, B. Julia, J. Scherk, *Phys. Lett.* **76B** (1978) 409

[2] E. Witten, *Nucl. Phys.* **B443** (1995) 85;
P.K. Townsend, *Phys. Lett.* **B350** (1995) 184; hep-th/9612121

[3] A. Font, L. Ibáñez, D. Lüst, F. Quevedo, *Phys. Lett.* **B249** (1990) 35;
S.J. Rey, *Phys. Rev.* **D43** (1991) 526;
A. Sen, *Phys. Lett.* **B303** (1993) 22; *Int. J. Mod. Phys.* **A9** (1994) 3707;
J. Schwarz, A. Sen, *Nucl. Phys.* **B411** (1994) 35;
M.J. Duff, R. Khuri, *Nucl. Phys.* **B411** (1994) 473;
A. Giveon, M. Porrati, E. Rabinovici, *Phys. Rep.* **244** (1994) 77;
C. M. Hull, P.K. Townsend, *Nucl. Phys.* **B438** (1995) 109;
E. Witten, *Nucl. Phys.* **B443** (1995) 85;
S. Kachru, C. Vafa, *Nucl. Phys.* **B450** (1995) 69;
J.H. Schwarz, *Phys. Lett.* **B360** (1995) 13; *Phys. Lett.* **B367** (1996) 97;
M.J. Duff, *Int. J. Mod. Phys.* **A11** (1996) 5623;
P. Horava, E. Witten, *Nucl. Phys.* **B460** (1996) 506

[4] T. Banks, W. Fischler, S.H. Shenker, L. Susskind, *Phys. Rev.* **D55** (1997) 5112

[5] M. Claudson, M. Halpern, *Nucl. Phys.* **B250** (1985) 689;
R. Flume, *Ann. Phys.* **164** (1985) 189;
M. Baake, P. Reinicke, V. Rittenberg, *J. Math. Phys.* **26** (1985) 107

[6] B. de Wit, hep-th/9701169;
T. Banks, hep-th/9710231;
L. Bigatti and L. Susskind, hep-th/9712072

[7] E. Bergshoeff, E. Sezgin, P.K. Townsend, *Phys. Lett.* **189B** (1987) 75; *Ann. of Phys.* **185** (1988) 330

[8] B. de Wit, J. Hoppe, H. Nicolai, *Nucl. Phys.* **B305** (1988) 545

[9] J. Polchinski, *Phys. Rev. Lett.* **75** (1995) 4724

[10] E. Witten, *Nucl. Phys.* **B460** (1996) 335;
J. Polchinski, hep-th/9611050

[11] J. Fröhlich, J. Hoppe, hep-th/9701119;
P. Yi, hep-th/9704098;
S. Sethi, M. Stern, hep-th/9705046;
M. Porrati, A. Rozenberg, hep-th/9708119;
J. Hoppe, hep-th/9709132;
M.B. Green, M. Gutperle, hep-th/9711107;
M.B. Halpern, C. Schwartz, hep-th/9712133

[12] E. Cremmer, B. Julia, *Phys. Lett.* **80B** (1978) 48; *Nucl. Phys.* **B159** (1979) 141

[13] B. Julia, in *Superspace and Supergravity*, eds. S.W. Hawking and M. Rocek (Cambridge University Press, 1981)

[14] S. Elitzur, A. Giveon, D. Kutasov, E. Rabinovici, hep-th/9707217

[15] S. Melosch, H. Nicolai, hep-th/9709227;
S. Melosch, PhD Thesis, in preparation

[16] B. de Wit, H. Nicolai, *Phys. Lett.* **155B** (1985) 47; *Nucl. Phys.* **B274** (1986) 363

[17] H. Nicolai, *Phys.Lett.* **187B** (1987) 363

[18] B. Julia, in *Unified Theories and Beyond*, Proc. 5th Johns Hopkins Workshop on Current Problems in Particle Theory, Johns Hopkins University, Baltimore, 1982;
in *Vertex Operators in Mathematics and Physics*, eds. J. Lepowsky, S. Mandelstam and I. Singer, (Springer Verlag, 1984)

[19] P. Breitenlohner, D. Maison, *Ann. Inst. H. Poincaré* **46** (1987) 215

[20] P. Breitenlohner, D. Maison, G.W. Gibbons, *Comm. Math. Phys.* **120** (1988) 295

[21] H. Nicolai, in *Recent Aspects of Quantum Fields*, eds. H. Mitter and H. Gausterer (Springer Verlag, 1991)

[22] B. Julia, H. Nicolai, *Nucl. Phys.* **B482** (1996) 431

[23] D. Bernard, B. Julia, hep-th/9712254

[24] H. Nicolai, *Phys. Lett.* **194B** (1987) 402

[25] S. Mandelstam, *Nucl. Phys.* **B64** (1973) 205;
S. Giddings, S. Wolpert, *Commun. Math. Phys.* **109** (1987) 177

[26] H. Nicolai, D. Korotkin, H. Samtleben, in *Quantum Fields and Quantum Space Time*, Series B: Physics Vol. 364, eds. G. 't Hooft, A. Jaffe, G. Mack, P. Mitter and R. Stora (Plenum Press, 1997), hep-th/9612065;
H. Samtleben, D. Korotkin, *Phys. Rev. Lett.* **80** (1998) 14

[27] A. Ashtekar, *Phys. Rev. Lett.* **57** (1986) 2244

[28] N. Marcus, J.H. Schwarz, *Nucl.Phys.* **B228** (1983) 145

[29] S. Helgason, *Differential Geometry and Symmetric Spaces* (Academic Press, 1962)

[30] B. de Wit, H. Nicolai, *Nucl. Phys.* **281** (1987) 211

[31] R. Geroch, *J. Math. Phys.* **13** (1972) 394;
W. Kinnersley, D.M. Chitre, *J. Math. Phys.* **18** (1977) 1538

[32] H. Nicolai, N.P. Warner, *Comm. Math. Phys.* **125** (1989) 384;
H. Nicolai, *Nucl. Phys.* **B414** (1994) 299

[33] D. Maison, *Phys. Rev. Lett.* **41** (1978) 521;
S. Belinskii, V. Zakharov, *Sov. Phys. JETP* **48** (1978) 985

[34] N. Bourbaki, *Groupes et Algèbres de Lie*, chapters 4–6 (Hermann, Paris, 1968);
V.G. Kac, *Infinite Dimensional Lie Algebras*, 3rd edition (Cambridge University Press, 1990)

[35] F. Levstein, *J. of Algebra* **114** (1988) 489;
J. Bausch, G. Rousseau, *Algèbres de Kac-Moody affines: automorphismes et formes reelles*, Rev. de l'Institut E. Cartan **11** (1989)

[36] B. Julia, in *Lectures in Applied Mathematics*, Vol. **21** (1985) 355

[37] W. Nahm, *Nucl. Phys.* **B135** (1978) 149

[38] H. Nicolai, *Phys. Lett.* **B276** (1992) 333

[39] A. Feingold, I.B. Frenkel, *Math. Ann.* **263** (1983) 87

[40] S. Mizoguchi, hep-th/9703160

[41] H. Verlinde, *Nucl. Phys.* **B337** (1990) 652

[42] D. Friedan, S. Shenker, *Nucl. Phys.* **B281** (1987) 509

[43] D. Korotkin, H. Nicolai, *Phys. Rev. Lett.* **74** (1995) 1272

[44] B. Julia, H. Nicolai, *Nucl. Phys.* **B439** (1995) 291

[45] E. Witten, *Int. J. Mod. Phys.* **A1** (1986) 39

[46] I. Bakas, hep-th/9411118, hep-th/9606030

[47] A. Sen, *Nucl. Phys.* **B447** (1995) 62

[48] R.W. Gebert and S. Mizoguchi, hep-th/9712078

[49] D. Korotkin, H. Samtleben, hep-th/9710210

Perfect and Quasi-Perfect Lattice Actions

W. Bietenholz [1]

[1] HLRZ c/o Forschungszentrum Jülich, D-52425 Jülich, Germany

Abstract: Perfect lattice actions are exiting with several respects: they provide new insight into conceptual questions of the lattice regularization, and quasi-perfect actions could enable a great leap forward in the non-perturbative solution of QCD. We try to transmit a flavor of them, also beyond the lattice community.

1 Introduction

For many field theoretic models, in particular in $d = 4$, the lattice is the only regularization so far, which provides non-perturbative results. One discretizes the Euclidean space and introduces matter fields only on the lattice sites, and gauge fields on the links connecting them. By means of some Monte Carlo procedure one generates a number of configurations, and for large statistics we can numerically perform a functional integral (to some accuracy). However, even if the statistical error is under control, we still have to worry about systematic errors. Simulations take place at a finite lattice spacing a and in a finite size L, so the final limits $\bar{\xi}/a,\ L/\bar{\xi} \to \infty$ require extrapolations of the simulation results ($\bar{\xi}$ is the correlation length). The finiteness of these ratios causes artifacts, and in practice those due to $a > 0$ are most troublesome.

It is very expensive in computer time to approximate the continuum limit by "brute force", i.e. by using finer and finer lattices. The computational effort grows at least like a^{-6}, in some cases this factor can rise up to a^{-10}. When he was about to quit lattice physics, K. Wilson estimated that a convincing solution of QCD requires a lattice of $256^3 \times 512$ sites. This appears hopeless for generations of supercomputers as well as human beings; what the up-coming supercomputers ("teraflops") can process is around $24^4 \dots 32^4$ for full QCD. (Larger lattices can be used in the so-called quenched approximation: there one sets the fermion determinant equal to a constant, which means physically that sea quark effect are neglected. Hence one brings in another systematic error, which is often difficult to control.)

However, Wilson referred to his standard lattice action, which handles derivatives by nearest lattice site differences (including a possible gauge variable living on the connecting link). The pure gauge part is described by a "plaquette action": from the paths around single plaquettes one constructs a (gauge invariant) lattice version of

$F_{\mu\nu}$ [1]. Nowadays, there is a consensus that *improved lattice actions* are a ray of hope for a much faster progress in the near future, than the gradual increase in computer power provides. Of course, many lattice actions have the correct continuum limit, and a better choice could suppress the artifacts significantly. Non-standard lattice actions involve additional terms, which make the simulations more complicated and slower. Still the ratio gain/cost can be huge, if they allow us to, say, double or triple the lattice spacing, which has to be chosen typically $\leq 0.1 fm$ for Wilson's QCD action.

At present, there are essentially two improvement programs under investigation. One of them has been formulated by K. Symanzik [2], and tries to cancel the cut-off artifacts order by order in a, by adding irrelevant operators to Wilson's action. This is very similar to the Runge and Kutta procedure for solving differential equations. In QCD (with Wilson fermions) it has been realized to $O(a)$, first on the classical level by adding a so-called "clover term" (a plaquette version of $\sigma_{\mu\nu}F_{\mu\nu}$) [3]. More recently, the renormalization of the clover coefficient has been estimated by a mean-field approach [4], and finally the non-perturbative $O(a)$ improvement has been completed by extensive simulations, taking the PCAC relation as a guide-line [5]. At present, many tests are being performed with the resulting action, but it is not clear yet to which extent it enables the use of coarse lattices [6]. As a preliminary impression, there is progress in the scaling on fine lattices, but it does not enable the use of really coarse lattices (it seems that this $O(a)$ improvement accidently amplifies the $O(a^2)$ artifacts).

These notes are devoted to the alternative improvement program, which goes under the name "perfect actions". It is based on renormalization group concepts, and it is non-perturbative with respect to a. It works beautifully in principle, as we know from a sequence of 2d toy models: the $O(3)$ model, the Gross-Neveu model, the $CP(3)$ model, and the Schwinger model. Moreover, it has theoretically fascinating properties. In particular, it allows us to reproduce symmetries of the continuum theory exactly on the lattice, even in cases where this is apparently impossible. Examples are the continuous Poincaré invariance, as well as chiral symmetry. It is even possible to introduce a perfect lattice topology. Hence the program is very attractive from the conceptual point of view, also apart from the practical aspect of reducing the computational effort in simulations.

However, the implementation in full QCD is very tedious. In practice, a crucial aspect is the struggle for an excellent locality of the action, so that the truncation of the couplings – which is required at some point – does not do too much harm to it. In this context, a good parameterization and truncation of the action are now major issues in this program, which have not yet been solved in a really satisfactory way. Still, the potential of this method is great; if it can really be applied, then it can do by far better than an $O(a)$ improvement, although it requires more non-standard terms than just a clover term. In this program, the improvement can be extended to better and better approximations to perfection. In contrast, it is hardly feasible to carry on Symanzik's program to $O(a^2)$, so if $O(a)$ should be insufficient with some respect, then that program is at a dead-lock.

2 Perfect and classically perfect actions

It has been known for a long time that there exist so-called *renormalized trajectories* in parameter space [7], [1] which represent *perfect lattice actions.* These are actions without any lattice artifacts; they display the continuum values of scaling quantities at any lattice spacing.

This can be understood from the consideration of block variable renormalization group transformations (RGTs). As a simple example, we start from a hypercubic lattice with spacing $1/n$, and divide it into disjoint blocks of n^d sites, where d is the Euclidean space-time dimension, and n is going to be the blocking factor. The block centers form a coarse lattice of unit spacing. There we define new variables, collectively denoted by ϕ', which we relate to the corresponding block averages of the fine lattice variables ϕ. The action on the coarse lattice is given by

$$e^{-S'[\phi']} = \int D\phi \; K[\phi', \phi] e^{-S[\phi]}. \tag{1}$$

The kernel K must be chosen such that the partition function, and all expectation values – hence the physical contents of the theory – remain invariant. This requires

$$\int D\phi' \; K[\phi', \phi] = 1, \tag{2}$$

which still leaves quite some freedom. The simplest choice,

$$K[\phi', \phi] = \prod_{x'} \delta(\phi'_{x'} - \frac{1}{n} \sum_{x \in x'} \phi_x), \tag{3}$$

determines the δ function RGT (the sum $x \in x'$ runs over the fine lattice sites x in the block with center $x' \in \mathbb{Z}^d$). An obvious generalization of the kernel (3) "smears" the δ function to a Gaussian, so that the transformation term appears in the exponent. The RGT reduces the correlation length *in lattice units*, ξ, by a factor n, $\xi' = \xi/n$.

Assume that we are on a "critical surface" in parameter space, where $\xi = \infty$. For suitable parameters we arrive – after an infinite number of RGT iterations (which include permanent re-scaling to the newest lattice units) – at a finite fixed point action (FPA) S^*, which is invariant under the considered RGT: $S^{*\prime} = S^*$.

Let us now perform a tiny step away from the FPA – and from the critical surface – in a relevant direction, and then keep on iterating RGTs. Thus we follow a trajectory in parameter space, which takes us to shorter and shorter correlation length. This is a renormalized trajectory, each point of which is related to the vicinity of the FPA solely by the renormalization group. There is no way irrelevant operators can contaminate the actions represented by the points on such a trajectory, hence these actions are perfect. Scaling quantities extracted from such actions are therefore completely free of lattice spacing artifacts. Needless to say that the identification of (quasi-)perfect actions at moderate or even short ξ – where the simulations take

[1] One may think of the parameter space as being spanned by all possible couplings between the lattice variables.

place – is a dream of humanity (or at least of the lattice community). However, it is very difficult to find good approximations to perfect actions, which are tractable in simulations.

As an alternative description, one may start on a suitable point close to the critical surface, and after many RGTs a renormalized trajectory is approximated asymptotically. This property suggests an explicit construction by Monte Carlo RGTs, which has been tried for some time, but which did not prove very useful: such RGT steps are tedious to perform, and the iteration is restricted to very few (often hardly one) reliable steps, which do not suppress the artifacts dramatically.

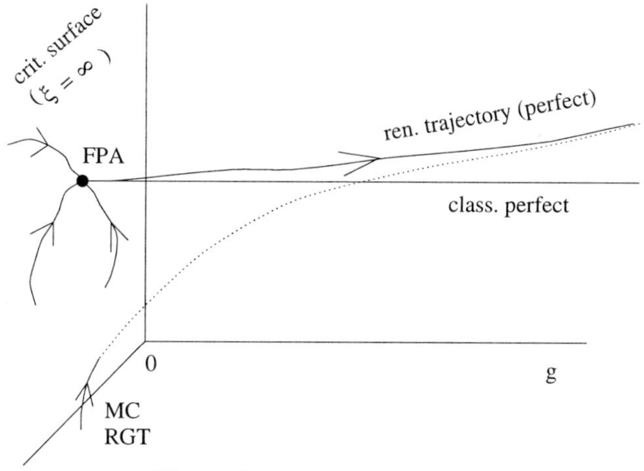

Figure 1: A caricature of the picture in parameter space.

A few years ago, P. Hasenfratz and F. Niedermayer suggested a new trick to construct approximately perfect actions, which is particularly designed for asymptotically free theories [8]. We write the RGT (in a slightly modified notation) as

$$e^{-\frac{1}{g'^2}S'[\phi']} = \int D\phi \; e^{-\frac{1}{g^2}\{S[\phi]+T[\phi',\phi]\}}, \tag{4}$$

where T is now the transformation term, which specifies the RGT. Here the critical surface – and hence the FPA – is situated at $g = g' = 0$, where the RGT is determined simply by minimization,

$$S'[\phi'] = \min_{\phi}\{S[\phi] + T[\phi',\phi]\}, \quad S^*[\phi'] = \min_{\phi}\{S^*[\phi] + T[\phi',\phi]\}. \tag{5}$$

Thus the search for a FPA simplifies enormously to a classical field theory problem; no (numeric) functional integral is needed. [2] [3] But we still need to proceed to moderate ξ in order to control the finite size effects. And here the authors of Ref. [8]

[2]In the literature, this is often called a "saddle point problem", although one just deals with minima.

[3]In practice one may proceed as follows: make a parameterization ansatz for the action S'; choose some configurations ϕ'; (numeric) minimization yields $S'[\phi']$. For a sufficient number of configurations, the parameters in the ansatz for S' are determined.

suggest to just switch on g and use the "*classically perfect action*" $(1/g^2)S^*[\phi]$. Thus we follow the weakly relevant (in leading order marginal) direction away from the FPA. If we multiply g^2 by $\hbar$, we understand the notion "classically perfect"; in the limit $\hbar \to 0$ the minimization trick persists at finite g. In the full quantum theory, however, this is not the case; the renormalized trajectory deviates from its classically perfect approximation. The hope – and the one uncontrolled assumption in this program – is that the latter is still an approximately perfect action at moderate ξ. (Intuitively we may support this hope by associating the quantum deviations of renormalized trajectories with the β functions, which tend to be smooth for asymptotically free theories.) In fact, toy model studies confirm the excellent quality of this approximation in a striking way: in a scaling test for the 2d $O(3)$ model (in a small volume $L \simeq \xi$) no lattice artifacts at all could be seen down to $\xi \simeq 5$. A study of the Gross-Neveu model revealed that in the large N limit the classically perfect action is also quantum perfect (N suppresses the quantum fluctuations) so that, for instance, the dynamically generated fermion mass divided by the chiral condensate is a constant, independent of ξ [9]. The classically perfect 2d $O(3)$ action, together with a classically perfect topological charge, confirmed accurately the absence of scaling of the topological susceptibility [10]. Decent topological scaling was found, however, using a classically perfect action for the 2d $CP(3)$ model [11]. In the 1d XY model even a quantum perfect topology was worked out: for a given lattice configuration, we integrate over all possible continuum interpolations, each with a well defined topological charge. Thus we attach to a lattice configuration an ensemble of charges with appropriate Boltzmann weights [12]. Furthermore, the scaling artifacts of the classically perfect approximation could be studied: they become negligible around $\xi \simeq 3$, whereas the standard lattice formulation still suffers from severe artifacts at large ξ, see Fig. 2.

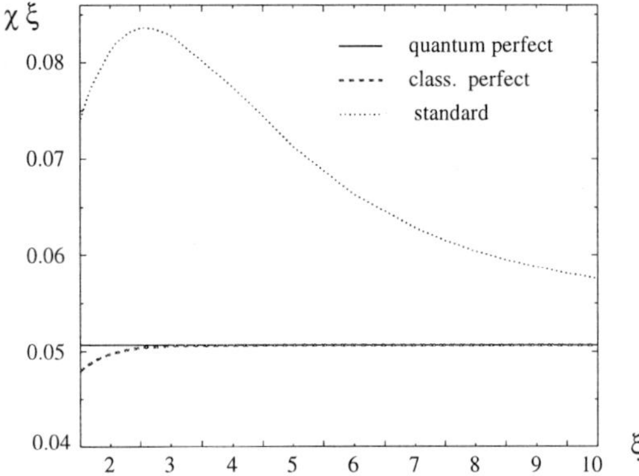

Figure 2: The scaling of the topological susceptibility χ (multiplied by the correlation length in lattice units, ξ) in the 1d XY model for the (quantum) perfect action, the classically perfect action and the standard lattice action.

In the Schwinger model with Wilson-type fermions, the classically perfect action was again very successful, in particular in view of the scaling of "meson" masses; also the dispersion relations and the rotational invariance of the correlation functions look good [13]. Approximations to FPAs were also constructed for non-Abelian gauge fields in $d = 4$. They have been applied to studies of topology of $SU(2)$ [14], and various types of strongly truncated FPAs have been suggested for $SU(3)$ [15] (but recently those authors reported certain problems with their actions).

Let us emphasize again that the improvement is designed to improve the *scaling*, i.e. scaling quantities (dimensionless ratios of observables) should converge to their continuum limit at smaller ξ than it is the case for the standard action. The impact on *asymptotic scaling* can not be predicted. However, it has been observed that also asymptotic scaling tends to set in much earlier for quasi-perfect actions [9, 15, 16]. In particular, Λ_{QCD} is much larger – hence much closer to its continuum value – for the classically perfect action (compared to Wilson's action) [17].

2.1 Classically perfect operators

For a given configuration Φ on the coarse lattice, the first eq. in (5) singles out one configuration ϕ_c on the fine lattice, which minimizes the right-hand side. This can be iterated to finer and finer interpolating lattices (in units of the original coarse lattice), until we arrive at a minimizing continuum field $\varphi_c[\Phi]$, which we call the "classically perfect field" [18]. It can be viewed as a particularly sophisticated interpolation of the initial lattice field. In contrast to ordinary interpolations, this inverse blocking process is based on the renormalization group, at least in the classical limit, and its artifacts tend to be suppressed exponentially. It can be used to define classically perfect topological objects on the lattice, by requiring their stability under inverse blocking. This implies a better foundation for long range stability than just smoothing the lattice field locally by hand.

Now consider an operator $\mathcal{O}[\varphi]$ given as a functional of the continuum fields φ. We build a classically perfect version of such operators simply by the substitution

$$\mathcal{O}[\varphi] \to \mathcal{O}[\varphi_c[\Phi]]. \tag{6}$$

Thus the operator is given in terms of lattice fields, but still defined in the continuum, again as a sophisticated interpolation.

As an application, this procedure has been applied to the free gauge field, referring to a lattice with unit spacing [18]. From the classically perfect field, we constructed Polyakov loops, the correlation function of which yields a static quark-antiquark potential $V(\vec{r})$. Indeed, the classically perfect potential converges to the continuum value with increasing $r = |\vec{r}|$ much faster than the potential arising from Wilson's plaquette action, it is continuously defined and – most importantly – it has an amazing degree of rotational invariance, even at very short distances, see Fig. 3.

3 Perturbatively perfect actions

As another approximation to perfection, perfect actions can be calculated analytically for free and perturbatively interacting fields. As potential applications, one can hope that such actions are immediately useful in simulations, or – more modestly

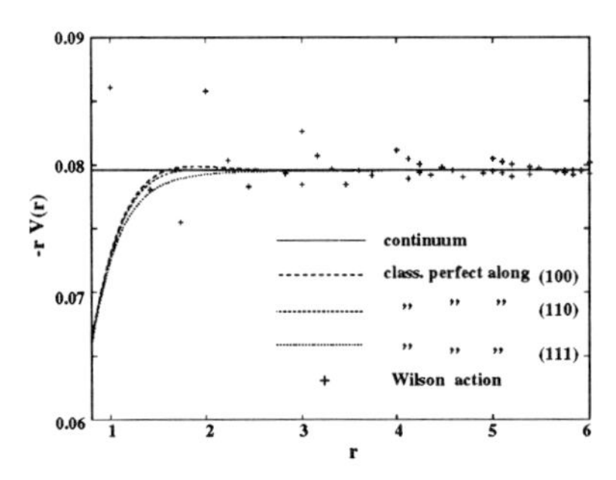

Figure 3: The static quark-antiquark potential (multiplied by the negative distance $-r$) obtained from the classically perfect action and from the plaquette action.

– that they single out the most important non-standard lattice terms, which gives a handle for a good parameterization. Moreover, they provide a promising starting point for the minimization program, leading to a classically perfect action. This is important, because even the minimization can not easily be iterated.

3.1 Free fermion and gauge fields

In cases where the blocking step can be performed analytically, it is more efficient to send the blocking factor $n \to \infty$ and to perform only one step, which yields the perfect action directly, instead of tedious iterations. Hence (in coarse lattice units) the initial action lives in the continuum, so we don't need to worry either which lattice action to use on the fine lattice.

Consider the free fermion. As described in Sec. 2, one usually relates $\psi'_x \sim (1/n^d) \sum_{x' \in x} \psi_x$, where ψ', ψ are fermionic fields on the coarse resp. fine lattice. In the limit $n \to \infty$ this relation turns into $\Psi_x \sim \int_{C_x} dy \; \psi(y)$, where Ψ_x lives on a unit lattice, C_x is the unit hypercube with center x, and ψ is a continuum field. In momentum space, this relation reads

$$\Psi(p) \sim \sum_{l \in Z\!\!\!Z^d} \psi(p + 2\pi l) \Pi(p + 2\pi l), \quad \Pi(p) = \prod_\mu \frac{\hat{p}_\mu}{p_\mu}, \quad \hat{p}_\mu = 2 \sin \frac{p_\mu}{2}, \tag{7}$$

where $p \in B =]-\pi, \pi]^d$. If we compute the Gaussian type RGT [4]

$$\begin{aligned} e^{-S[\bar{\Psi},\Psi]} &= \int D\bar{\psi} D\psi \exp \Big\{ -\int \frac{dp}{(2\pi)^d} \, \bar{\psi}(-p)[i\not{p} + m]\psi(p) \\ & -\frac{1}{\alpha} \int_B \frac{dp}{(2\pi)^d} \, [\bar{\Psi}(-p) - \sum_{l \in Z\!\!\!Z^d} \bar{\psi}(-p - 2\pi l)\Pi(p + 2\pi l)] \times \end{aligned}$$

[4] We ignore constant factors in the partition function. The RGT parameter α is arbitrary. The mass m is given in lattice units, even in the continuum action.

$$[\Psi(p) - \sum_{l\in \mathbb{Z}^d} \psi(p+2\pi l)\Pi(p+2\pi l)] \Big\} , \qquad (8)$$

we obtain the perfect lattice action

$$S[\bar\Psi,\Psi] = \int_B \frac{dp}{(2\pi)^d} \bar\Psi(-p)G(p)^{-1}\Psi(p), \quad G(p) = \sum_{l\in \mathbb{Z}^d} \frac{\Pi^2(p+2\pi l)}{i(p_\mu + 2\pi l_\mu)\gamma_\mu + m} + \alpha, \quad (9)$$

which we write in coordinate space as

$$S[\bar\Psi,\Psi] = \sum_{x,r\in \mathbb{Z}^d} \bar\Psi_x[\rho_\mu(r)\gamma_\mu + \lambda(r)]\Psi_{x+r}. \qquad (10)$$

The couplings $\rho_\mu(r)$, $\lambda(r)$ have been evaluated numerically [18].

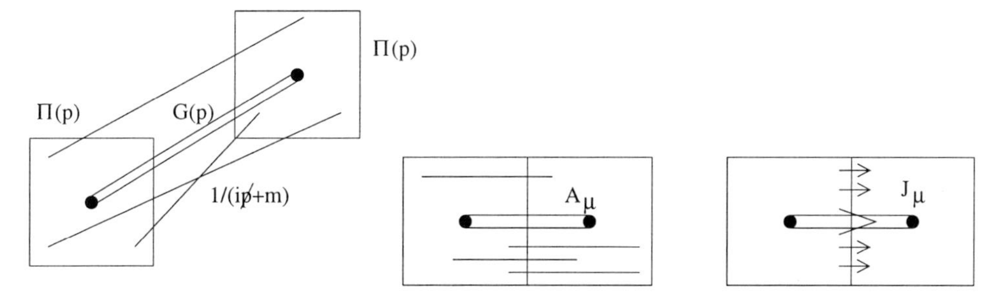

Figure 4: The blocking scheme for the free fermion propagator, for the Abelian gauge field, and for any sort of currents.

The spectrum of this perfect lattice fermion can be read off from the pole structure of the propagator $G(p)$,

$$E(\vec p)^2 = (\vec p + 2\pi \vec l\,)^2 + m^2, \qquad (11)$$

hence it is the *exact* continuum spectrum (plus 2π periodic copies, which are omnipresent on the lattice). This reveals the perfect character of this action: the continuous rotation and translation invariance is exactly present *in the observables*, though not in the form of the action itself. In the latter, the hypercubic structure of the lattice is visible – a perfect action on a triangular lattice, for instance, looks different [19] – but what matters are the observables.

The procedure described above, which we call "blocking from the continuum", is also applicable to the Abelian gauge field. Here we integrate over all straight connections between corresponding continuum points in two adjacent lattice cells, and relate this integral to the non-compact lattice link variable A_μ. (A RGT in terms of "compact" link variables $U_\mu \in U(1)$ in $d=2$ has been carried out in Ref. [20].)

3.2 Chiral symmetry

A major issue in fermionic actions is the doubling problem: according to the Nielsen Ninomiya "No Go theorem", species doubling (unphysical extra fermions) always occurs under some mild assumptions about a lattice action, like locality, hermiticity and chiral invariance [21]. For instance, in the naive fermion formulation with the inverse lattice propagator $G^{-1}_{naive}(p) = i \sin p_\mu \gamma_\mu$ (at $m = 0$), this is obvious from the occurrence of 2^d zeros in the first Brillouin zone B. It is a notorious problem to put a single chiral fermion on the lattice.

It turns out that the perfect action (9) is not plagued by doubling. Moreover, in the case $m = \alpha = 0$, [5] the action is chirally symmetric, but in this case it is non-local ($|\rho_\mu(r)|$ only decays like $|r|^{1-d}$ at large distances) so there is no contradiction with the No Go theorem [22]. As soon as $m > 0$ or $\alpha > 0$ (or both), the action becomes *local*, i.e. the couplings in ρ_μ and λ decay exponentially in $|r|$, and at the same time the chiral symmetry is explicitly broken in the action.

But this is *not* the end of the story. If we start from a chiral fermion in the continuum, then the chiral symmetry is supposed to be preserved under the RGT, due to its very nature, also for finite α. And this is in fact the case, as we see if we focus on the *observables* again: as an example, it has been shown explicitly in the Schwinger model that the axial anomaly is reproduced correctly, if we map the entire continuum theory in a consistently perfect way on the lattice [23]. This includes the blocking of the fermion and gauge field, of the fermion-gauge interaction term to the first order, and of the axial current. A perfect lattice current is identified by a procedure analogous to the treatment of the fields. The blocking scheme here integrates the continuum flux through the face between adjacent lattice cells, and a coupling to an external source incorporates the current in the RGT. [6] The axial charge is blocked from the continuum too, and all these perfect quantities match to reproduce the correct axial anomaly on the lattice.

By construction, the chiral symmetry comes out correctly to all orders in perturbation theory, where the blocking from the continuum can be carried out. So we can put the system on the lattice without doing any harm to it, i.e. the lattice regularization is – with this respect – as good as any regularization in the continuum. (This is in contrast to a wide-spread feeling that it has a particular weakness due to the fermionic doubling problem.)

The construction of *local* chiral fermions we sketched above – the fixed point action for $\alpha > 0$ – is a nice way around the Nielsen Ninomiya theorem, which refers to a chiral symmetry of the lattice action itself. We don't need it to be manifest in the action, [7] but we can still reproduce it in the observables, which really matter (like the continuous rotation and translation symmetry, which we recovered in the spectrum). However, since this construction by "blocking from the continuum" is perturbative, it does not directly imply any claim about a non-perturbative formulation of the lattice chiral fermion (which is not known in the continuum either).

While these notes are being written down, however, it has been shown that also

---

[5] The case $\alpha = 0$ corresponds to the δ function RGT.

[6] The perfect currents, which have been worked out in this context, gave also rise to a study of perfectly discretized hydrodynamics [25].

[7] The violation in the form of the action is due to the non chirally symmetric transformation term for $\alpha > 0$.

the Atyiah Singer index theorem holds for a FPA [24], confirming again that the perfect fermion is perfect.

3.3 Truncation

The perfect actions mentioned above all include couplings over infinite distances. Even if the long range couplings are exponentially suppressed (locality), they are still needed to reproduce the continuum symmetries exactly. Since we can't work with them in simulations, the nice properties described above may seem somewhat academic, rather far from application (is it comparable to the string theory talks at this symposium ?).

To proceed to practical applications, we need to truncate the couplings, and this does necessarily some harm to the perfect properties. In the 2d applications, it was permissible to truncate only couplings, which are suppressed by various orders of magnitude [8, 13]. However, in $d = 4$ – and with non-Abelian gauge fields, where many more lattice paths have to be distinguished – we can not be so generous. The number of additional terms, that the practitioner can work with, is strongly limited.

Hence it is very important to choose the RGT such that the perfect action is *as local as possible*, i.e. the couplings decay as fast as possible. Then there is hope that perfection survives the truncation in a good approximation. In this sense, we have tuned the RGT parameter α in the RGT (8) in order to optimize locality and we could achieve that in the special case $p = (p_1, 0 \dots 0)$ (mapping on $d = 1$) the couplings are restricted to nearest neighbors. This is a successful optimization criterion for the locality in the general 4d case, for fermions as well as scalars [18, 19]. For the gauge field we have chosen the RGT parameter, which is analogous to α, to be momentum dependent. Thus we obtained the standard plaquette action in $d = 2$, and again the same RGT parameters also provide excellent locality in $d = 4$ [18].

The truncation itself was performed by means of periodic boundary conditions: we construct a perfect action in a small volume of 3^4 sites, and then use the same couplings – which are restricted to a unit hypercube – in larger volumes too. This truncation keeps all normalizations exact. To check the quality of the truncated free actions, we considered the spectrum after truncation, which is not perfect any more, but still drastically improved compared to the standard action [26], see Fig. 5. The same holds for thermodynamic scaling quantities at finite temperature [26, 27] and at finite chemical potential [28], as illustrated in Fig. 6. A different truncation of the perfect free gauge field in terms of closed loops has been given in Ref. [26].

For an immediate application, the truncated perfect "hypercube fermion" can be "gauged by hand": we connect the coupled sites by the link variables on the shortest lattice paths, and average over these paths. This is of course a drastic short-cut and not the consistently perfect construction, but with some respect it already leads to a remarkable progress. In particular, the meson dispersion relation reaches an impressive quality, as shown in Fig. 7. To handle such complicated actions efficiently, work is in progress for an adequate optimization of the algorithm on a parallel machine. In particular, we are working on a quick evaluation of the fermion

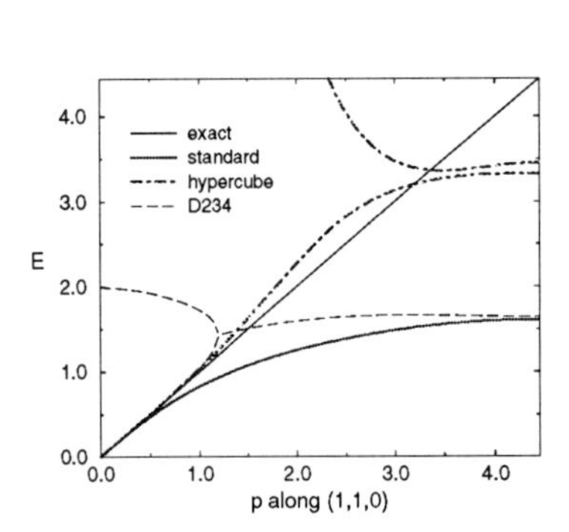

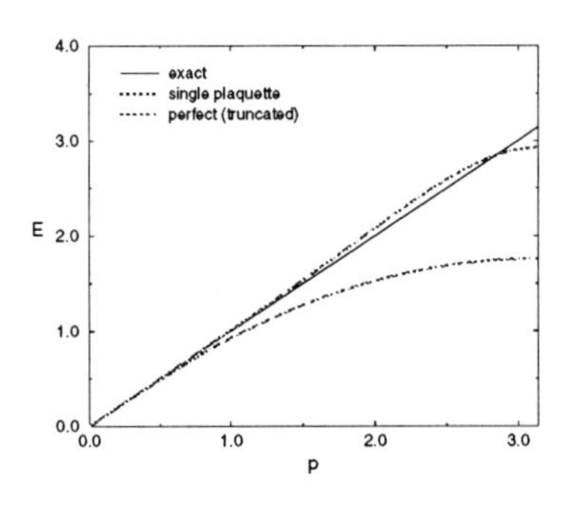

Figure 5: The dispersion relation for free massless fermions (left) and gauge fields (right), compared to the Wilson action and – for the fermion – to a Symanzik improved action called D234.

matrix [29], which involves many more non-zero off-diagonal elements than it is the case for Wilson's action.

3.4 Vertex function

As we proceed to perturbatively interacting fields, we can incorporate the interaction into the RGT and still block from the continuum. If we realize this to the first order in the gauge coupling g, we obtain a perfect quark-gluon and a 3-gluon vertex function. The $\bar{q}qg$ vertex has been evaluated in $d = 2$ [26] and its couplings have been reproduced numerically from a multigrid [13]. In spite of good locality, we obtain in the 4d case still inconveniently many couplings, which seem to contribute significantly. By means of a rather tough truncation, their number was reduced to 5 short ranged extra terms, in addition to those present in the action of the "hypercube fermion gauged by hand" [30].

The best candidate for an immediate use of this action is heavy quark physics. This optimism is supported from a consideration of the non-relativistic expansion

$$E = m_s + \frac{1}{2m_{kin}} \vec{p}^{\,2} + \frac{1}{2m_B} \vec{\Sigma}\vec{B} + \dots \ , \quad \Sigma_k = \epsilon_{ijk}\sigma_{ij}/2. \tag{12}$$

For the periodically truncated perfect fermion, the mass parameter m coincides with the static mass m_s. The lattice parameters tend to deviate from the continuum relation $m_s = m_{kin} = m_B$. Fig. 8 shows that $m_s = m_{kin}$ is approximated well for the hypercube fermion, and if we gauge it by hand then $m_B(m_s)$ is somewhat improved. We can achieve a drastic improvement also for m_B by including terms of the truncated perfect vertex function.

As a first experiment, we performed quenched simulations with the truncated perfect vertex function for the charmonium spectrum. The computational overhead

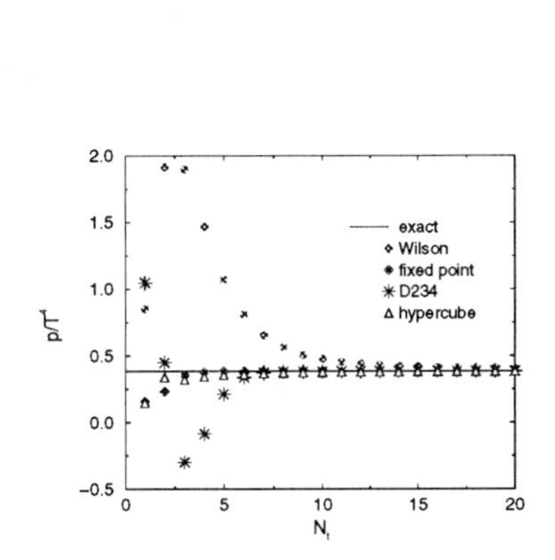

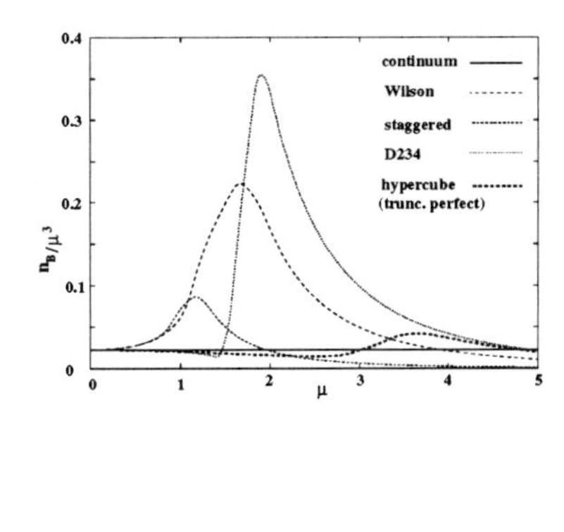

Figure 6: The thermodynamic scaling quantities *pressure/temperature*^{4} with N_t sites in Euclidean time (at $\mu = 0$), and (*baryon number density*)/(*chemical potential*)^3 at $T = 0$, for massless free fermions.

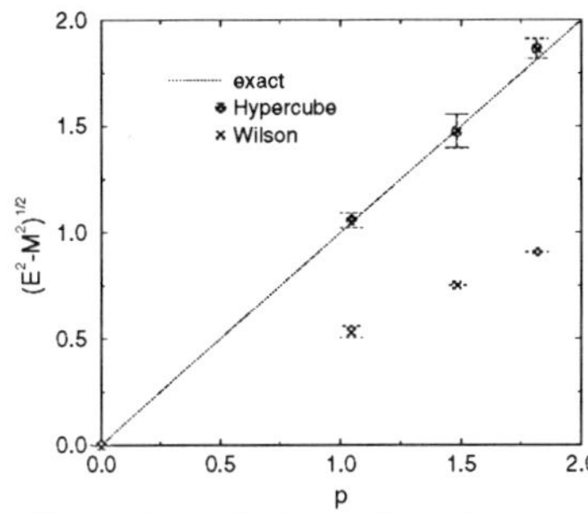

Figure 7: The meson dispersion relation using the truncated perfect "hypercube fermion", and the Wilson fermion.

(compared to Wilson's action) is around a factor 20, but we simulated on an coarse lattice of spacing $a = 0.24fm$ (about 3 times larger than usual), which could clearly out-do this factor. Here a lattice of $8^3 \times 16$ sites appears to be sufficient. The physical units were determined from the string tension, and the 1s η_c ground state was matched to the experimental value. The further states were predictions of the simulation. The 2s states of η_c and J/ψ were predicted successfully, but the gap to the 1s J/ψ state (hyperfine splitting) was clearly too small [30], see Fig. 9. A possible reason – except for quenching – is that the vertex couplings are still supposed to be renormalized (due to $O(g^2)$). As a rather ad hoc procedure, we have tested a "tadpole improvement" (mean field estimate of the renormalization à la Ref. [4]), which helps to some extent, but the hyperfine splitting – a quantity, which is extremely sensitive to lattice artifacts – is still too small. However, it is known from the clover action that the fermion part tends to be insufficiently renormalized by this method, so we are now incorporating the larger renormalization factor, translated from the result

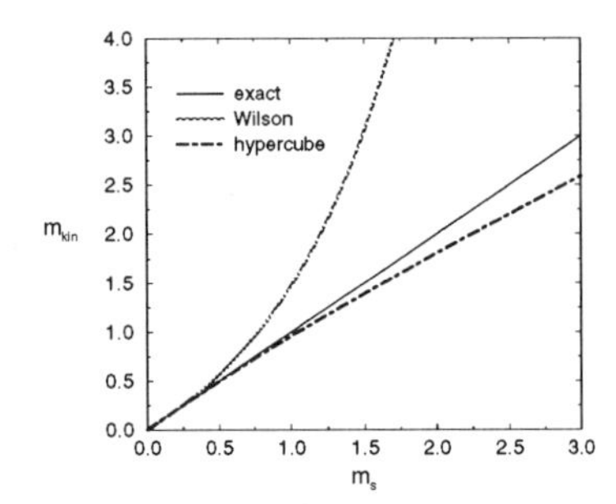

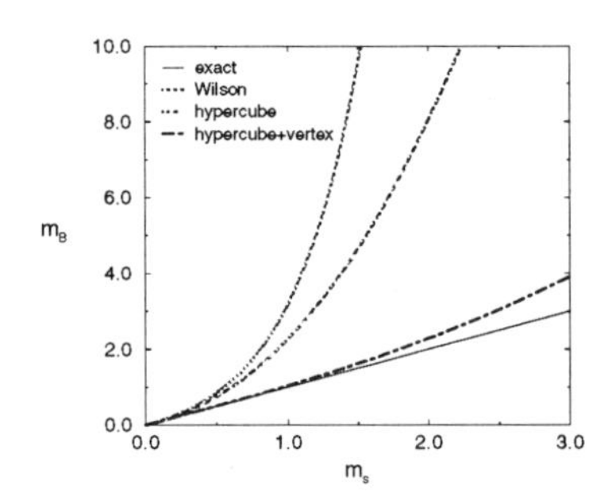

Figure 8: The kinetic and the magnetic mass as functions of the static mass for the Wilson fermion and the hypercube fermion. m_B improves strongly only after including elements of the perfect vertex function.

found by the ALPHA collaboration for the clover term [5].

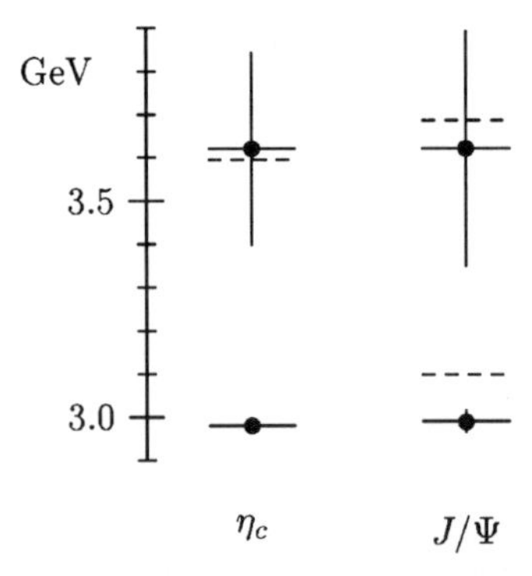

Figure 9: The charmonium spectrum. The experimental values are dashed, and only the ground state of η_c is fitted.

The form of the perfect ggg vertex function is very similar to the $\bar{q}qg$ vertex [31]. If we include it, then we obtain an action which is entirely perfect (before truncation) to $O(g)$; any artifacts of $O(a^n)$ and $O(ga^n)$ are eliminated, so that the leading artifacts occur in $O(g^2a)$. It is not obvious how this compares to a Symanzik improved action, which erases all artifacts in $O(ag^n)$, but which is plagued for instance by cut-off effects in $O(ga^2)$.

The climax of the improvement is of course a QCD action, which is improved non-perturbatively in both, a and g. The classically perfect action, to be constructed from a multigrid minimization, is an action of this type, but it could not be worked out in an "ultimate" form yet. In practice one would perform inverse blocking steps with a small blocking factor (see footnote 3), but iteration rapidly leads to large lattices, which can not be handled any more. Of course one could put a well identified action, say after a blocking factor 2 RGT, back to the fine lattice and block again, hence avoiding the use of large lattices. However, small lattices are only instructive if the action is extremely local. In any case, a good starting point for the iteration, as well as a good parameterization ansatz for the blocked action, are highly desirable, and the perfect vertex functions could help with this respect.

4 Status and outlook

Although we are still working on better charmonium results, it seems that the direct application of small field perfect actions is not as successful as we hoped. This is also confirmed by a study of the anharmonic oscillator [16]. Apparently, at some point the minimization trick must come into play. In simple cases one might think about extending it to a semi-classically perfect action à la WKB, i.e. taking the quadratic fluctuations around the classical solution into account. Also one full RGT step with a small blocking factor, starting from a classically perfect action at moderate ξ, could be very helpful.

One point that is still very important to work on, is the optimization of locality: in the case of the vertex functions, it seems that the interaction term in the RGT can be further optimized with this respect. In other cases, e.g. for so-called staggered fermions (see second and third Ref. in [1]), it turned out that one can do better even for the free particles by taking a blocking scheme different from the block average described in Sec. 2 [33]. The fine lattice (or continuum) variables can be blocked in many ways, not only by a piece-wise constant weight factor as in eq. (7). Such an optimization led to significant progress for the truncated perfect staggered fermion, which has been tested in simulations of the Schwinger model (with naive gauging plus "fat links"); in particular the "pion" mass gets strongly suppressed.

In thermodynamics, where lattice artifacts are especially bad, a remarkable improvement has been observed using approximately perfect actions [34]. As a further step, one could put those actions on anisotropic lattices, which hardly poses additional complications [19]. It has also been shown how to include a chemical potential in a perfect action, which is needed for simulations at finite baryon density [28] (a field, which had only little success so far).

In further experiments for QCD, pionic systems were investigated and some improvement in the scaling behavior was found [35], but that study also tends to involve more and more ad hoc elements, which are not related to the perfect program. From that side, there is no claim for an "ultimate" version of a quasi-perfect action either. A problem, which our attempts have in common, is a strong additive mass renormalization. This is a effect of truncation (and simplified gauging), which affects the reliability of the "perfect couplings".

According to our experience, it is worthwhile proceeding in very small steps to higher degrees of complications, in order to elaborate a really clean treatment. All attempts at short-cuts to QCD have led to half-cooked proposals, which have to be reconsidered afterwards. Thus the program is quite time-consuming, but if it really works one day, a quasi-perfect action (given by a set of couplings) can be used by the practitioner for any problem in QCD, without worrying about its lengthy derivation. Following these piecemeal tactics, we are now working on a FPA for 2d and 3d non-Abelian gauge theories, starting with $SU(2)$. As a scaling test, the 3d glueball spectrum can be measured.

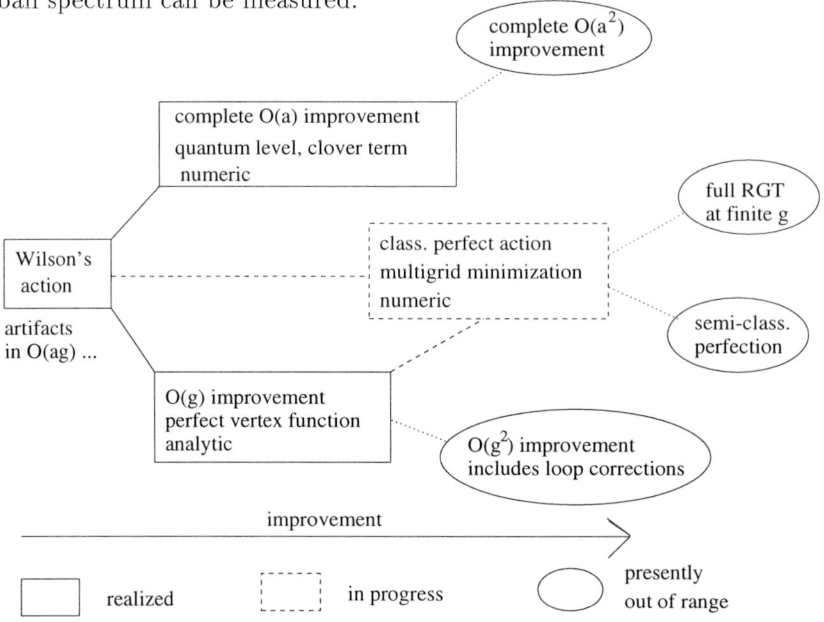

Figure 10: A schematic overview over the status of the improvement for QCD lattice actions with Wilson-type fermions.

On the conceptual side, it is straightforward to apply the procedure of blocking from the continuum for instance to supersymmetry. This results in a SUSY lattice action, which is completely different from the one presented by I. Montvay at this symposium. Again it is more complicated, but it could represent all symmetries exactly. Even applications to quantum gravity are conceivable: also here it would be desirable to integrate out the short-range details between the lattice sites in a way, which is based on the renormalization group. Again, such a method would differ from those discussed by J. Ambjørn and B. Pertersson in their talks.

At last, as another project in progress, it would be interesting to combine the perfect treatment of fermions with the domain wall concepts of lattice chiral fermions. So far, that construction has always been based on Wilson fermions [36], although this can be generalized. Inserting a truncated perfect fermion instead, one cumulates all sort of virtues: small lattice artifacts, an arbitrary number of flavors (unlike staggered fermions), and the absence of additive mass renormalization (unlike Wilson fermions). Note that the domain wall formalism for chiral fermions is related to the subject, which was most celebrated at this symposium, the theory of D-branes (I thank J. Schwarz for this remark).

Acknowledgment It is a pleasure to thank R. Brower, S. Chandrasekharan, H. Dilger, E. Focht, K. Orgions, T. Struckmann, U.-J. Wiese and the co-authors of Ref. [29] for their collaboration.

References

[1] M. Creutz, "Quarks, Gluons and Lattices", Cambridge University Press, 1983. H. Rothe, "Lattice Gauge Theories", World Scientific, 1992. I. Montvay and G. Münster, "Quantum Fields on the Lattice" Cambridge University Press, 1994.

[2] K. Symanzik, Nucl. Phys. B226 (1983) 187, 205.

[3] B. Sheikholeslami and R. Wohlert, Nucl. Phys. B259 (1985) 572.

[4] G.P. Lepage and P. Mackenzie, Phys. Rev. D48 (1993) 2250. G.P. Lepage, hep-lat/9607076 and references therein.

[5] M. Lüscher, S. Sint, R. Sommer, P. Weisz and U. Wolff, Nucl. Phys. B491 (1997) 323.

[6] R. Horsley, talk presented at this symposium (hep-lat/9801034).

[7] K. Wilson and J. Kogut,, Phys. Rep. C12 (1974) 75. K. Wilson, Rev. Mod. Phys. 47 (1975) 773.

[8] P. Hasenfratz and F. Niedermayer, Nucl. Phys. B414 (1994) 785.

[9] W. Bietenholz, E. Focht and U.-J. Wiese, Nucl. Phys. B436 (1995) 385.

[10] M. Blatter, R. Burkhalter, P. Hasenfratz and F. Niedermayer, Phys. Rev. D53 (1996) 923.

[11] R. Burkhalter, Phys. Rev. D54 (1996) 4121.

[12] W. Bietenholz, R. Brower, S. Chandrasekharan and U.-J. Wiese, Phys. Lett. B407 (1997) 283.

[13] C. Lang and T. Pany, hep-lat/9707024.

[14] T. DeGrand, A. Hasenfratz and T. Kovacs, hep-lat/9801037 (this summarizes a lengthy series of papers by the Boulder group, see also talk by T. Kovacs presented at this symposium). M. Feurstein, E. Ilgenfritz, M. Müller-Preussker and S. Thurner, hep-lat/9611024. E. Ilgenfritz, H. Markum, M. Müller-Preussker and S. Thurner, hep-lat/9801040.

[15] T. DeGrand, A. Hasenfratz, P. Hasenfratz and F. Niedermayer, Nucl. Phys. B454 (1995) 587; 615. M. Blatter and F. Niedermayer, Nucl. Phys. B482 (1996) 286.

[16] W. Bietenholz and T. Struckmann, hep-lat/9711054.

[17] T. DeGrand, A. Hasenfratz, P. Hasenfratz, F. Niedermayer and U.-J. Wiese, Nucl. Phys. B (Proc. Suppl.) 42 (1995) 67.

[18] W. Bietenholz and U.-J. Wiese, Nucl. Phys. B464 (1996) 319.

[19] W. Bietenholz, hep-lat/9709117.

[20] I. Bender and D. Gromes, preprint HD-THEP-94-18.

[21] H. Nielsen and M. Ninomiya, Nucl. Phys. B185 (1981) 20; Nucl. Phys. B193 (1981) 173. L. Karsten, Phys. Lett. B104 (1981) 315.

[22] P. Ginsparg and K. Wilson, Phys. Rev D25 (1982) 2649. S. Cronjäger, diploma thesis, Hamburg University (1985). U.-J. Wiese, Phys. Lett. B315 (1993) 417.

[23] W. Bietenholz and U.-J. Wiese, Phys. Lett. B378 (1996) 222; Nucl. Phys. B (Proc. Suppl.) 47 (1996) 575.

[24] P. Hasenfratz, V. Laliena and F. Niedermayer, hep-lat/9801021.

[25] E. Katz and U.-J. Wiese, comp-gas/9709001.

[26] W. Bietenholz, R. Brower, S. Chandrasekharan and U.-J. Wiese, Nucl. Phys. B (Proc. Suppl.) 53 (1997) 921.

[27] F. Niedermayer, Nucl. Phys. B (Proc. Suppl.) 53 (1997) 56.

[28] W. Bietenholz and U.-J. Wiese, hep-lat/9801022.

[29] N. Eicker et al., hep-lat/9709143, and in preparation.

[30] K. Orginos et al., hep-lat/9709100., and in preparation.

[31] K. Orginos, Ph.D. thesis, Brown University (1997).

[32] W. Bietenholz, R. Brower, S. Chandrasekharan and U.-J. Wiese, Nucl. Phys. B495 (1997) 285.

[33] W. Bietenholz and H. Dilger, in preparation.

[34] A. Papa, Nucl. Phys. B478 (1996) 335. S. Spiegel, Phys. Lett. B400 (1997) 352.

[35] T. DeGrand, hep-lat/9709052.

[36] D. Kaplan, Phys. Lett. B288 (1992) 342. Y. Shamir, Phys. Lett. B305 (1993) 357.

The Superalgebraic Approach to Supergravity

C.R. Preitschopf [1] , M.A. Vasiliev [2]

[1] Institut für Physik, Humboldt-Universität,
Invalidenstr. 110, D-10115 Berlin
[2] I.E. Tamm Theoretical Department, Lebedev Physical Institute
Leninsky Prospekt 53, 117924 Moscow, Russia

Abstract: We formulate classical actions for N=1 supergravity in D=(1,3) as a gauge theory of $OSp(1|4)$. One may choose the action such that it does not include a cosmological term.

Introduction

The theory of supergravity [1, 2] was discovered more than 20 years ago, but in spite of determined efforts we do not understand some very basic properties of this theory. There is, as of now, no manifestly supersymmetric action for the most interesting cases, namely in $D = 10$ and $D = 11$. It seems that ordinary superspace becomes exceedingly difficult to handle beyond $N = 1$ in $D = 4$. The introduction of harmonic variables [3] helps significantly, but they are introduced more as a (very clever) trick than from first principles. A further mystery arises from nonperturbative phenomena in string theory: the supersymmetry algebra includes tensor charges which point to $OSp(1|32)$, but the relation of this symmetry group to $D = 11$ supergravity [4], though suspected from the beginning, remains unclear.

It seems that we are missing some essential ingredients in the formulation of supergravity, which probably are not strictly necessary in the case $N = 1$, $D = 1$, but which become indipensable in higher dimensions or for higher N. In this paper we take a first step in our search for those ingredients by (partially) disentangling gauge symmetries and reparametrizations.

This was first done by McDowell and Mansouri for gravity and supergravity [5]. The gauged superalgebra approach to supergravities they used is developed and carefully explained in [9]. A formulation in terms of compensating fields was given by Chamseddine [7] for gravity coupled to Spins 3/2, 1 and 1/2. The gravity case was examined in detail by Stelle and West [8]. We present supergravity as a partially compensated gauge theory of $OSp(1|4)$.

Gravity as a gauge theory

In order to describe gravity as a gauge theory of $SO(2,3) = Sp(4, R)$ (in the anti-de Sitter case) or $SO(1,4)$ (in the de Sitter case), one introduces a tangent space metric

$\eta^{MN} = (-1, 1, 1, 1, \mp 1)$ and a connection 1-form ω^{MN}, with $N = 0, 1, 2, 3, 5$. The connection should decompose as the usual fourdimensional Lorentz connection ω^{mn} and the vierbein $e^m = \rho\omega^{m5}$. We may perform this split in a gauge-covariant way at the cost of introducing a compensator field U^M that satisfies $U^M U_M = \mp\rho^2$. Then the role of the vierbein is played by the frame field

$$E^M = DU^M , \tag{1}$$

and the covariant Lorentz-connection

$$\omega_{\mathcal{L}}^{MN} = \omega^{MN} \mp \frac{1}{\rho^2}\left(U^M E^N - E^M U^N\right) , \tag{2}$$

which is defined by $D_{\mathcal{L}}U^M = dU^M + \omega_{\mathcal{L}}^{MN}U_N = 0$. U^M describes locally a fourdimensional subspace of the fivedimensional tangent space we started with. The de Sitter curvature $\frac{1}{2}R^{MN} = d\omega^{MN} + \omega^M{}_K\omega^{KN}$ decomposes as $\frac{1}{2}R^{MN} = \frac{1}{2}R_{\mathcal{L}}^{MN} \pm \frac{1}{\rho^2}\left(U^M T^N - T^M U^N\right) \pm \frac{1}{\rho^2}E^M E^N$ with $T^M = DE^M = R^{MN}U_N$. The action

$$\begin{aligned} S &= -\tfrac{1}{16\pi\rho}\textstyle\int_{M_4}\epsilon_{N_1\ldots N_5}U^{N_1}R^{N_2N_3}R^{N_4N_5} \\ &= -\tfrac{1}{16\pi\rho}\textstyle\int_{M_4}\epsilon_{N_1\ldots N_5}U^{N_1}\left(R_{\mathcal{L}}^{N_2N_3}R_{\mathcal{L}}^{N_4N_5} \pm \tfrac{4}{\rho^2}E^{N_1}E^{N_2}R_{\mathcal{L}}^{N_4N_5} + \tfrac{4}{\rho^4}E^{N_1}E^{N_2}E^{N_3}E^{N_4}\right) \end{aligned} \tag{3}$$

is manifestly reparametrization invariant and de Sitter gauge invariant. Upon gauge fixing $U^M = \mp\rho\delta_5^M$ we obtain

$$S = 2\pi\chi(M_4) - \frac{1}{2\kappa^2}\int_{M_4}|e|R(e,\omega) \mp \frac{6\pi}{\kappa^4}\int_{M_4}|e| \tag{4}$$

with $\kappa^2 = 2\pi\rho^2$, i.e. the usual Einstein action with a cosmological constant and a topological term. The presence of such terms is due to the simple choice of S above, and has nothing to do with the fact that we formulated a gauge theory of the (anti-) de Sitter group. We may write a slightly more complex action that reproduces precisely Einstein gravity:

$$S_{\mathcal{E}} = \mp\frac{1}{8\kappa^2\rho}\int_{M_4}\epsilon_{N_1\ldots N_5}U^{N_1}E^{N_1}E^{N_2}R_{\mathcal{L}}^{N_4N_5} = -\frac{1}{2\kappa^2}\int_{M_4}|e|R(e,\omega) , \tag{5}$$

which is again reparametrization and gauge covariant. We learn that the vacuum algebra, i.e. the symmetry algebra of (anti-) de Sitter or Minkowski space has little to do with the gauge algebra: we may choose the Poincare or the (anti-) de Sitter algebra for either one, and independently. The main advantage of choosing (3) or (5) is that the vacuum solutions $R^{MN} = 0$ resp. $R_{\mathcal{L}mn}{}^{np} = 0$ can be read off the action immediately.

$Sp(4, R)$

In order to formulate supergravity as a gauge theory it will prove useful to convert the vector notation of the previous section to a spinor one. We define the fundamental representation of $Sp(4, R)$ in terms of an $Sl(2, C)$-spinor as

$$\Psi^a = \left(\Psi^\alpha, \overline{\Psi}^{\dot\alpha}\right) , \tag{6}$$

with $(\Psi^\alpha)^* = \overline{\Psi}^{\dot\alpha}$. The invariant tensor $C_{ab} = (\epsilon_{\alpha\beta}, \epsilon_{\dot\alpha\dot\beta})$ and its inverse $\tilde{C}^{ab} = (\tilde\epsilon^{\alpha\beta}, \tilde\epsilon^{\dot\alpha\dot\beta})$ may be used to raise and lower spinor indices. The $Sp(4,R)$-generators J_{ab} are symmetric bispinors. The covariant derivative $D\Psi^a = d\Psi^a + \omega^a{}_b\Psi^b$ leads to the curvatures $R^{ab} = d\omega^{ab} + \omega^{ac}\omega_c{}^b$, which decompose as $\mathrm{R}^\alpha{}_\beta = \frac{1}{4}\sigma_{mn}{}^\alpha{}_\beta\, R^{mn}$, $\overline{\mathrm{R}}^{\dot\alpha}{}_{\dot\beta} = \frac{1}{4}\bar\sigma_{mn}{}^{\dot\alpha}{}_{\dot\beta}\, R^{mn}$, $\mathrm{R}^\alpha{}_{\dot\beta} = \frac{1}{2}\sigma_m{}^\alpha{}_{\dot\beta}\, R^{m5}$ and $\mathrm{R}^{\dot\alpha}{}_\beta = \frac{1}{2}\bar\sigma_m{}^{\dot\alpha}{}_\beta\, R^{m5}$. The vector field U^M now appears as an antisymmetric traceless bispinor:

$$U^{[ab]} = i\begin{pmatrix} U^5\epsilon^{\alpha\beta} & U_m\sigma^{m\alpha\dot\beta} \\ -U_m\bar\sigma^{m\dot\alpha\beta} & -U^5\epsilon^{\dot\alpha\dot\beta}\end{pmatrix} . \tag{7}$$

After gauge fixing we obtain the action (3) in the form given by Mac Dowell and Mansouri [5]:

$$S = -\frac{i\rho^2}{4\kappa^2}\int_{M_4} R^\alpha{}_\beta R^\beta{}_\alpha - R^{\dot\alpha}{}_{\dot\beta} R^{\dot\beta}{}_{\dot\alpha} . \tag{8}$$

$OSp(1|4)$ and Supergravity

We now upgrade the gauge algebra to $OSp(1|4)$, with fundamental representation $\Psi^A = (\Psi^a, \Psi^j) = (\Psi^A)^*$, grading

$$(-)^A = \begin{cases} +1 & \text{for } A \in \{\alpha, \dot\alpha\} \; ; \quad \alpha, \dot\alpha \in \{1,2\} \\ -1 & \text{for } \quad A = j \; ; \qquad j \in \{1\} \end{cases} \tag{9}$$

and invariant tensors

$$\begin{aligned} C_{[AB)} = -(-)^A C_{BA} = (C_{ab}, i\delta_{ij}) \quad &; \quad \tilde{C}^{[AB)} = -(-)^A\tilde{C}^{BA} = \left(\tilde{C}^{ab}, -i\delta^{ij}\right) \\ C_{AB}\tilde{C}^{BC} = \tilde{C}^{CB}C_{BA} \; &= \; \delta_A^C = (\delta_a^c, \delta_i^k) , \end{aligned} \tag{10}$$

which raise and lower indices as follows:

$$\Psi_A = C_{AB}\Psi^B \quad ; \quad \Psi^A = \tilde{C}^{AB}\Psi_B . \tag{11}$$

The standard index contraction is

$$\Psi_A\Phi^A = -(-)^A\Psi^A\Phi_A \quad ; \quad \left(\Psi_A\Phi^A\right)^* = \Phi^A\Psi_A , \tag{12}$$

and upon introducing the graded symmetric gauge connection $\omega^{(AB]}$ we arrive at curvatures R^{AB}, which now contain fermionic (gravitino) curvatures R^{aj} in addition to those listed in the previous section. The compensator field $U^{[AB)}$ is graded antisymmetric and traceless, and we require it to satisfy the conditions [7]

$$U_A{}^A = 0 \quad ; \quad U_{AB}U^{BA} = 4\rho^2 \quad ; \quad U_{AB}U^B{}_C U^{CA} = 0 . \tag{13}$$

The action

$$S = \frac{\rho}{2\kappa^2}\int_{M_4} U^A{}_B R^B{}_C \left(\delta^C{}_E + \frac{1}{2\rho^2}U^C{}_D U^D{}_E\right) R^E{}_A (-)^A \tag{14}$$

is then manifestly reparametrization invariant (since it is a 4-form), and $OSp(1|4)$-gauge invariant. In addition, it possesses a supersymmetry which stays hidden because we are not working in superspace. We gauge fix $U^A{}_B = -i\rho(\delta^\alpha{}_\beta, \delta^{\dot\alpha}{}_{\dot\beta}, 0)$ and (14) takes the form

$$S = -\frac{i\rho^2}{4\kappa^2}\int_{M_4} R^\alpha{}_\beta R^\beta{}_\alpha + 2R^\alpha{}_j R^j{}_\alpha - R^{\dot\alpha}{}_{\dot\beta} R^{\dot\beta}{}_{\dot\alpha} - 2R^{\dot\alpha}{}_j R^j{}_{\dot\alpha} , \tag{15}$$

which may be rewritten as

$$\begin{aligned} S &= -\frac{\rho^2}{4\kappa^2}\int_{M_4} d\left[\epsilon_{mnpq}(\omega^{mn}d\omega^{pq}+\frac{2}{3}\omega^{mn}\omega^{pl}\omega_l{}^q)+\psi^\alpha D^{\mathcal{L}}\psi_\alpha+\overline{\psi}^{\dot\alpha}D^{\mathcal{L}}\overline{\psi}_{\dot\alpha}\right] \\ &-\frac{1}{2\kappa^2}\int_{M_4}|e|R(e,\omega)+\frac{1}{2}\int_{M_4}|e|\epsilon^{mnpq}\left(\overline{\psi}_m{}^{\dot\alpha}\bar\sigma_{n\dot\alpha}{}^\beta D^{\mathcal{L}}_p\psi_{q\beta}-\psi_m{}^\alpha\sigma_{n\alpha}{}^{\dot\beta}D^{\mathcal{L}}_p\overline{\psi}_{q\dot\beta}\right) \\ &+\int_{M_4}|e|\left(\frac{96}{\kappa^2\rho^2}-\frac{2i}{\rho}\left[\psi_{m\alpha}\sigma^{mn\alpha}{}_\beta\psi_n{}^\beta+\overline{\psi}_{m\dot\alpha}\bar\sigma^{mn\dot\alpha}{}_{\dot\beta}\overline{\psi}_n{}^{\dot\beta}\right]\right), \end{aligned} \tag{16}$$

where $D^{\mathcal{L}}_m$ contains only the Lorentz connection. It is well known that (16) is supersymmetric.

Supersymmetry

In order to make this symmetry transparent, we compute the variation of the above action under infinitesimal variations of the fermionic compensators:

$$\delta S \propto \int_{M_4} R^\alpha{}_{\dot\beta}\left(\delta U^{\dot\beta}{}_j R^j{}_\alpha + R^{\dot\beta}{}_j \delta U^j{}_\alpha\right) \tag{17}$$

and compare it with the effect of arbitrary infinitesimal variations of the Lorentz connection:

$$\delta S \propto \int_{M_4} R^\alpha{}_{\dot\beta}\omega^{\dot\beta}{}_\gamma\delta\omega^\gamma{}_\alpha - R^{\dot\alpha}{}_\beta\omega^\beta{}_{\dot\gamma}\delta\omega^{\dot\gamma}{}_{\dot\alpha}\,. \tag{18}$$

we may either determine $\delta\omega^\alpha{}_\beta, \delta\omega^{\dot\alpha}{}_{\dot\beta}$ such that the sum of both variations vanishes (1st order formalism) or we may solve the algebraic equations of motion for $\omega^\alpha{}_\beta, \omega^{\dot\alpha}{}_{\dot\beta}$, i.e. set $R^\alpha{}_{\dot\beta}=0$ (1.5 order formalism). In either case we have shown that there is a hidden fermionic symmetry, which turns out to be supersymmetry:

$$\delta\omega^\alpha{}_{\dot\beta} = \omega^\alpha{}_j\lambda^j{}_{\dot\beta} - \lambda^\alpha{}_j\omega^j{}_{\dot\beta} \quad , \quad \delta\omega^\alpha{}_j = D^{\mathcal{L}}\lambda^\alpha{}_j + \omega^\alpha{}_{\dot\beta}\lambda^{\dot\beta}{}_j\,. \tag{19}$$

In this picture supersymmetry looks like a gauge symmetry, because it is inherited from a bona fide gauge theory of a graded Lie algebra. The hidden aspect of this symmetry is, that in the sense expressed by (17) and (18), the action is in fact independent of the fermionic fields U^{aj}, and hence the fermionic gauge symmetry is true and uncompensated.

Projectors

The compensators U^{AB} allow us to distinguish an $Sl(2,C)$ subgroup within $OSp(1|4)$ in a gauge covariant way. This is most clearly seen by the construction of projection operators:

$$\begin{aligned} \Pi^{(a)}{}_{(b)} &= -\frac{1}{\rho^2}\,U^A{}_C U^C{}_B \quad , \quad \Pi^{(i)}{}_{(j)} = \delta^A{}_B + \frac{1}{\rho^2}U^A{}_C U^C{}_B \\ \Pi^{(\alpha)}{}_{(\beta)} &= \frac{1}{2}\left(\frac{i}{\rho}U^A{}_B - \frac{1}{\rho^2}U^A{}_C U^C{}_B\right) \quad , \quad \Pi^{(\dot\alpha)}{}_{(\dot\beta)} = \frac{1}{2}\left(-\frac{i}{\rho}U^A{}_B - \frac{1}{\rho^2}U^A{}_C U^C{}_B\right). \end{aligned} \tag{20}$$

Proving the projection property requires some computation, but relies only on the properties (13) of the compensator field, which imply among other things $U^a{}_a =$

$\frac{1}{\rho^2}U^a{}_bU^b{}_jU^j{}_a$ and $U^A{}_BU^B{}_CU^C{}_D = -\rho^2 U^A{}_D$. We use the notation $\Psi^{(\alpha)} = \Pi^{(\alpha)}{}_{(\beta)}\Psi^B$, and may then formulate (14) as

$$S = -\frac{i\rho^2}{4\kappa^2}\int_{M_4} R^{(\alpha)}{}_{(\beta)}R^{(\beta)}{}_{(\alpha)} + 2R^{(\alpha)}{}_{(j)}R^{(j)}{}_{(\alpha)} - R^{(\dot\alpha)}{}_{(\dot\beta)}R^{(\dot\beta)}{}_{(\dot\alpha)} - 2R^{(\dot\alpha)}{}_{(j)}R^{(j)}{}_{(\dot\alpha)}\ , \quad (21)$$

which is the $OSp(1|4)$-invariant form of the MacDowell-Mansouri action. Once one knows the projectiors it is also possible to perform the $Sl(2,C)$-decomposition covariantly. We start with

$$\begin{aligned}\omega^A{}_B &= \omega_{\mathcal{L}}{}^A{}_B + U^A{}_CE^C{}_D(1+U^2)^D{}_B - (1+U^2)^A{}_CE^C{}_DU^D{}_B \\ &\quad -\frac{1}{4}\left(U^A{}_CE^C{}_D(U^2)^D{}_B - (U^2)^A{}_CE^C{}_DU^D{}_B\right)\ , \end{aligned} \quad (22)$$

where the Lorentz connection $\omega_{\mathcal{L}}{}^A{}_B$ is defined by $D^{\mathcal{L}}U^{AB} = 0$, and $E^A{}_B = DU^A{}_B$. Then (14) is decomposed as follows:

$$\begin{aligned} S &= \frac{\rho}{4\kappa^2}\int_{M_4}(-)^AU^A{}_B\big[R_{\mathcal{L}}R_{\mathcal{L}}\big]^B{}_A + 8d\left((-)^AU^A{}_BE^B{}_CD^{\mathcal{L}}E^C{}_A\right) \\ &\quad -4(-)^AU^A{}_B\big[2EUE(1+U^2)R_{\mathcal{L}} + E(1+U^2)EU^2E(1+U^2)E\big]^B{}_A \\ &\quad -(-)^AU^A{}_B\big[EU^2ER_{\mathcal{L}} - 8EUED^{\mathcal{L}}E\big]^B{}_A \\ &\quad +(-)^AU^A{}_B\Big[\frac{1}{4}EUEUEUEU + 4EUEUE(1+U^2)E\Big]^B{}_A\ . \end{aligned} \quad (23)$$

The first line of (24) is topological, the second line vanishes upon gauge fixing, since the index j can take only one value, the third line yields the kinetic terms for gravitons and gravitinos, and the last line describes the supersymmetric cosmological constant.

The Cosmological Term

The decomposition of the action (14) in the form (16) shows that we have obtained a supersymmetric cosmological term. One may be tempted to regard this as an inherent feature of any $OSp(1|4)$ gauge theory of gravity. (24) implies that this is not so. We easily extract pure supergravity without cosmological constant:

$$\begin{aligned} S_{\mathcal{E}} &= \frac{1}{4\rho\kappa^2}\int_{M_4}(-)^AU^A{}_B\Big(4DU^B{}_{(c)}DU^{(c)}{}_DR^D{}_A - 3DU^B{}_{(c)}DU^{(c)}{}_{(d)}R^{(d)}{}_A \\ &\quad -\frac{2}{\rho^2}DU^B{}_CDU^C{}_DDU^D{}_EDU^E{}_A + \frac{3}{2\rho^2}DU^B{}_{(c)}DU^{(c)}{}_{(d)}DU^{(d)}{}_{(e)}DU^{(e)}{}_A\Big) \\ &= -\frac{\rho}{4\kappa^2}\int_{M_4}(-)^AU^A{}_B\big[EU^2ER_{\mathcal{L}} - 8EUED^{\mathcal{L}}E\big]^B{}_A \\ &= -\frac{1}{2\kappa^2}\int_{M_4}|e|R(e,\omega) + \frac{1}{2}\int_{M_4}|e|\epsilon^{mnpq}\left(\overline{\psi}_m{}^{\dot\alpha}\bar\sigma_{n\dot\alpha}{}^{\beta}D^{\mathcal{L}}_p\psi_{q\beta} - \psi_m{}^{\alpha}\sigma_{n\alpha}{}^{\dot\beta}D^{\mathcal{L}}_p\overline{\psi}_{q\dot\beta}\right)\ . \end{aligned} \quad (24)$$

Admittedly the first line of (25) does not look particularly elegant. We understand it only by rewriting it in the form of line two. Without mentioning compensators Chamseddine and West wrote this action quite early in the development of supergravity [6]. It proves our point that the gauge algebra is independent of the vacuum algebra, i.e. the algebra of isometries of the vacuum state.

Conclusions

We have presented both gravity and supergravity as partially compensated gauge theories. The results of section make it easy to formulate any Lorentz-covariant theory in terms of $OSp(1|4)$, without the need for a group contraction. The significance of this gauge group lies then in the simplicity of (14) and the hidden supersymmetry.

We believe that one can extend this anaysis to include additional fermionic as well as bosonic coordinates. This should lead to a natural and simple form of supergravity in superspace.

Acknowledgments

Part of this work was done at Utrecht. We thank B. de Wit and the theory group for hospitality and a productive work environment. This work was supported in part by the following grants: DFG 436 RUS 113, EC network ERBFMRXCT96-0045, INTAS 93-633 ext. and RU 96-01-01144.

References

[1] S. Deser and B. Zumino, Phys. Lett. **B62** (1976) 335.

[2] D. Freedman, S. Ferrara and P. van Nieuwenhuizen, Phys. Rev. **D13** (1976) 3214.

[3] A. Galperin, E. Ivanov, S. Kalitzin, V. Ogievetsky, E. Sokatchev, Class. Quant. Grav. **1** (1984) 469; A. Galperin, E. Ivanov, V. Ogievetsky, E. Sokatchev, Class. Quant. Grav. **2** (1985) 601,617.

[4] E. Cremmer, B. Julia and J. Scherk, Phys. Lett. **B76** (1978) 409.

[5] S. MacDowell and F. Mansouri, Phys. Rev. Lett. **38** (1977) 739.

[6] A. Chamseddine and P. West, Nucl. Phys. **B129** (1977) 39.

[7] A. Chamseddine, Nucl. Phys. **B131** (1977) 494.

[8] K. Stelle and P. West, Phys. Rev. **D21** (1980) 1466.

[9] P. van Nieuwenhuizen, Lectures in Supergravity Theory, in: Recent Developments in Gravitation - Cargese 1978, edited by M. Levy and S. Deser, Plenum Press, 1979.

LIST OF PARTICIPANTS

J. Ambjorn
Niels Bohr Institute
Blegdamsvej 17
DK-2100 Copenhagen

D. Antonov
Humboldt Universität
Institut für Physik
Theorie der Elementarteilchen
Invalidenstr. 110
D-10115 Berlin

Dr. V. Azcoiti
Departamento de Fisica Teórica
Universidad de Zaragoza,
Plaza San Francisco
E- 50009 Zaragoza

C. Bachas
Centre de Physique Theorique
Ecole Polytechnique
F-91128 Palaiseau

J. Balog
Research Institute for Particle and
Nuclear Physics
H-1525 Budapest Pf. 49

E. Bergshoeff
Institute for Theoretical Physics
University of Groningen, Nijenborgh 4
NL-9747 AG Groningen

K. Behrndt
Humboldt Universität
Institut für Physik
Theorie der Elementarteilchen
Invalidenstr. 110
D-10115 Berlin

C.J. Biebl
DESY Zeuthen
Platanenallee 6
D-15738 Zeuthen

W. Bietenholz
HLRZ, Forschungszentrum Jülich
D-52425 Jülich

W. Bock
Humboldt Universität
Institut für Physik
Computational Physics
Invalidenstr. 110
D-10115 Berlin

V. Bornyakov
Institute for High Energy Physics IHEP
RU-142284 Protvino
Russia

I. Brunner
Humboldt Universität
Institut für Physik
Theorie der Elementarteilchen
Invalidenstr. 110
D-10115 Berlin

J. Buchbinder
Department of Theoretical Physics
Tomsk State Pedagogical University
Tomsk 634041, Russia

G. Lopes Cardoso
Institute for Theoretical Physics
Utrecht University
NL-3508 TA Utrecht

B. Dörfel
Humboldt Universität
Institut für Physik
Theorie der Elementarteilchen
Invalidenstr. 110
D-10115 Berlin

H. Dorn
Humboldt Universität
Institut für Physik
Theorie der Elementarteilchen
Invalidenstr. 110
D-10115 Berlin

D. Ebert
Humboldt Universität
Institut für Physik
Theorie der Elementarteilchen
Invalidenstr. 110
D-10115 Berlin

A. Di Giacomo
Dipartimento di Fisica and INFN
2 Piazza Torricelli
I-56100 Pisa

M. Douglas
Department of Physics and Astronomy
Rutgers University
Piscataway, NJ 08855-0849
U.S.A.

N. Dragon
Institut für Theoretische Physik
Universität Hannover
Appelstraße 2
D-30167 Hannover

K. Förger
Centre de Physique Théorique
Ecole Polytechnique
F-91128 Palaiseau Cedex

S. Förste
Sektion Physik
Universität München
Theresientraße 37
D- 80333 München

Z. Fodor
KEK Theory Division
Tsukuba, 1-1 Oho, Ibaraki, 305
JAPAN

J. Fuchs
Max-Planck-Institut für Mathematik
Gottfried-Claren-Str. 26
D-53225 Bonn

B. Geyer
Naturwissenschaftlich-Theoretisches
Zentrum und Institut für
Theoretische Physik
Universität Leipzig
Augustusplatz 10-11
D-04109 Leipzig

M. Hasenbusch
Humboldt Universität
Institut für Physik
Computational Physics
Invalidenstr. 110
D-10115 Berlin

S. Hollands
Humboldt Universität
Institut für Physik
Theorie der Elementarteilchen
Invalidenstr. 110
D-10115 Berlin

R. Horsley
Humboldt Universität
Institut für Physik
Theorie der Elementarteilchen
Invalidenstr. 110
D-10115 Berlin

E.-M. Ilgenfritz
Institute for Theoretical Physics
Faculty of Science
Kanazawa University
Kakuma-machi
920-1192 Kanazawa
Japan

E. Ivanov
Bogoliubov Laboratory of
Theoretical Physics, JINR
141980 Dubna, Russia

F. Jegerlehner
DESY Zeuthen
Platanenallee 6
D-15738 Zeuthen

J. Jersak
Institute of Theoretical Physics E
RWTH Aachen
D-52056 Aachen

S. Kachru
Department of Physics
University of California, Berkeley and
Lawrence Berkeley National Laboratory
University of California
Berkeley, CA 94720, USA

H. Kaiser
DESY Zeuthen
Platanenallee 6
D-15738 Zeuthen

A. Karch
Humboldt Universität
Institut für Physik
Theorie der Elementarteilchen
Invalidenstr. 110
D-10115 Berlin

W. Kerler
Fachbereich Physik
Universität Marburg
D-35032 Marburg

S. Ketov
Institut für Theoretische Physik
Universität Hannover
Appelstraße 2
D-30167 Hannover

T. Kovacs
Department of Physics
University of Colorado
Boulder, CO 80309-390, USA

S. Kuzenko
Institut für Theoretische Physik
Universität Hannover
Appelstraße 2
D-30167 Hannover

O. Lechtenfeld
Institut für Theoretische Physik
Universität Hannover
Appelstraße 2
D-30167 Hannover

D. Lüst
Humboldt Universität
Institut für Physik
Theorie der Elementarteilchen
Invalidenstr. 110
D-10115 Berlin

H. Markum
Institut für Kernphysik
Technische Universität Wien
Wiedner Hauptstr. 8-10
A-1040 Wien

A. Miemiec
Humboldt Universität
Institut für Physik
Theorie der Elementarteilchen
Invalidenstr. 110
D-10115 Berlin

A. Mironov
P.N. Lebedev Institute of Physics
Russian Academy of Sciences
Leninsky Pr. 53
RU-117924 Moscow

V. Mitrjushkin
Bogoliubov Laboratory of
Theoretical Physics, JINR
141980 Dubna, Russia

T. Mohaupt
Fachbereich Physik
Fachgruppe Theoretische Physik
Martin-Luther Universität
Halle Wittenberg
D-06099 Halle

S. Mahapatra
Physics Department
Utkal University
Bhubaneswar-751004, India

I. Montvay
Deutsches Elektronen Synchrotron
DESY,
Notkestr. 85
D-22603 Hamburg

M. Müller-Preußker
Humboldt Universität
Institut für Physik
Theorie der Elementarteilchen
Invalidenstr. 110
D-10115 Berlin

G. Münster
Institut für Theoretische Physik I
Universität Münster
Wilhelm-Klemm-Str. 9
D-48149 Münster

H. Nicolai
Max-Planck-Institut
für Gravitationsphysik
Schlaatzweg 1
D-14473 Potsdam

H.P. Nilles
Physikalisches Institut
Universität Bonn
Nussallee 12
D-53115 Bonn

B. Nilsson
Institute of Theoretical Physics
Göteborg University and
Chalmers University of Technology
S-412 96 Göteborg

S. Olejnik
Institute of Physics
Slovak Academy of Sciences
SK-842 28 Bratislava
Slovakia

H.-J. Otto
Humboldt Universität
Institut für Physik
Theorie der Elementarteilchen
Invalidenstr. 110
D-10115 Berlin

B. Petersson
Institut für Theoret. Physik
Universität Bielefeld
Universitätsstr.
D-33615 Bielefeld

J. Plefka
NIKHEF
P.O. Box 41882
NL-1009 DB Amsterdam

M. Polikarpov
Institute for Theoretical
and Experimental Physics
B.Cheremushkinskaya 25
Moscow, 117259
Russia

C. Preitschopf
Humboldt Universität
Institut für Physik
Theorie der Elementarteilchen
Invalidenstr. 110
D-10115 Berlin

W. Rühl
Fachbereich Physik
Universität Kaiserslautern,
Postfach 3049
D-67653 Kaiserslautern

N. Sasakura
Department of Physics
Kyoto University
Kyoto 606-8502, Japan

K. Scharnhorst
Humboldt Universität
Institut für Physik
Computational Physics
Invalidenstr. 110
D-10115 Berlin

G. Schierholz
DESY Zeuthen
und HLRZ
D-15735 Zeuthen

K. Schilling
HLRZ, Forschungszentrum Jülich
D-52425 Jülich

J. Schulze
Institut für Theoretische Physik
Universität Hannover
Appelstraße 2
D-30167 Hannover

J.H. Schwarz
California Institute of Technology
Pasadena, CA 91125
U.S.A

C. Schweigert
CERN
Theory division
CH-1211 Genève 23

H. Simma
DESY Zeuthen
Platanenallee 6
D-15738 Zeuthen

Yu. Simonov
Institute for Theoretical
and Experimental Physics
B.Cheremushkinskaya 25
Moscow, 117259
Russia

I. Stamatescu
FESt
Schmeilweg 5
D-69118 Heidelberg

K. Stelle
The Blackett Laboratory
Imperial College,
Prince Consort Road
London SW7 2BZ, UK

A. Tseytlin
The Blackett Laboratory
Imperial College,
Prince Consort Road
London SW7 2BZ, UK

A. Van Proeyen
Instituut voor theoretische fysica
Katholieke Universiteit Leuven
B-3001 Leuven

M. Vasiliev
I.E.Tamm Department of
Theoretical Physics
Lebedev Physical Institute,
Leninsky Prospect 53
117924 Moscow

G. Weigt
DESY Zeuthen
Platanenallee 6
D-15738 Zeuthen

B. de Wit
Institute for Theoretical Physics
Utrecht University
NL-3508 TA Utrecht

U. Wolff
Humboldt Universität
Institut für Physik
Computational Physics
Invalidenstr. 110
D-10115 Berlin

V. Zhukovski
Physics Faculty
Moscow State University
119899 Moscow
Russia

G. Zoupanos
Physics Department
National Technical University,
Zografou
GR-15780 Athens

AUTHOR INDEX

WILEY-VCH